# Laser Annealing Processes in Semiconductor Technology

Woodhead Publishing Series in Electronic and Optical Materials

# Laser Annealing Processes in Semiconductor Technology

## Theory, Modeling, and Applications in Nanoelectronics

Edited by


Fuccio Cristiano

Antonino La Magna



ELSEVIER

WP
WOODHEAD PUBLISHING
An imprint of Elsevier

Woodhead Publishing is an imprint of Elsevier
The Officers' Mess Business Centre, Royston Road, Duxford, CB22 4QH, United Kingdom
50 Hampshire Street, 5th Floor, Cambridge, MA 02139, United States
The Boulevard, Langford Lane, Kidlington, OX5 1GB, United Kingdom



**Library of Congress Cataloging-in-Publication Data**
A catalog record for this book is available from the Library of Congress

**British Library Cataloguing-in-Publication Data**
A catalogue record for this book is available from the British Library

ISBN: 978-0-12-820255-5
ISBN: 978-0-12-820256-2

For information on all Woodhead publications
visit our website at https://www.elsevier.com/books-and-journals

*Publisher:* Matthew Deans
*Acquisitions Editor:* Kayla Dos Santos
*Editorial Project Manager:* Joshua Mearns
*Production Project Manager:* Vignesh Tamil
*Cover Designer:* Miles Hitchen

Typeset by SPi Global, India

# Contents

## 5. Materials science issues related to the fabrication of highly doped junctions by laser annealing of Group IV semiconductors

*Ray Duffy, Enrico Napolitani, and Fuccio Cristiano*



## 6. Continuum modeling and TCAD simulations of laser-related phenomena in CMOS applications

*Salvatore Francesco Lombardo, Ioannis Deretzis, Alberto Sciuto, and Antonino La Magna*

# Contributors

**María Aboy** (79), Department of Electricity and Electronics, Technical School of Telecommunications Engineering, University of Valladolid, Valladolid, Spain

**Frédéric Antoni** (293), ICube, D-ESSP, Strasbourg, France

**Francesca Chiodi**, Centre for Nanoscience and Nanotechnology, Université Paris Saclay, CNRS, Palaiseau, France

**Fuccio Cristiano** (357), LAAS-CNRS, University of Toulouse, Toulouse, France

**Richard Daubriac** (357), CEA-LETI, Université Grenoble-Alpes, Grenoble, France

**Ioannis Deretzis** (251), Consiglio Nazionale delle Ricerche (National Research Council), Istituto per la Microelettronica e Microsistemi CNR-IMM, Catania, Italy

**Ray Duffy** (175), Tyndall National Institute, University College Cork, Cork, Ireland

**Giuseppe Fisicaro** (79), CNR Institute for Microelectronics and Microsystems (CNR-IMM), Catania, Italy

**Guglielmo Fortunato** (1), IMM-CNR, Rome, Italy

**Karim Huet** (137), Laser Systems and Solutions of Europe, SCREEN Semiconductor Solutions, Co., Ltd, Gennevilliers, France

**Sébastien Kerdilès** (357), CEA-LETI, Université Grenoble-Alpes, Grenoble, France

**Shao Qi Lim** (323), The Australian National University, Canberra, ACT, Australia

**Salvatore Francesco Lombardo** (251), Consiglio Nazionale delle Ricerche (National Research Council), Istituto per la Microelettronica e Microsistemi CNR-IMM, Catania, Italy

**Pedro López** (79), Department of Electricity and Electronics, Technical School of Telecommunications Engineering, University of Valladolid, Valladolid, Spain

**Antonino La Magna** (251), Consiglio Nazionale delle Ricerche (National Research Council), Istituto per la Microelettronica e Microsistemi CNR-IMM, Catania, Italy

**Luigi Mariucci** (1), IMM-CNR, Rome, Italy

**Luis A. Marqués** (79), Department of Electricity and Electronics, Technical School of Telecommunications Engineering, University of Valladolid, Valladolid, Spain

**Enrico Napolitani** (175), Department of Physics and Astronomy, University of Padua, Padua, Italy

**Alessandro Pecora** (1), IMM-CNR, Rome, Italy

**Lourdes Pelaz** (79), Department of Electricity and Electronics, Technical School of Telecommunications Engineering, University of Valladolid, Valladolid, Spain

**Vittorio Privitera** (1), IMM-CNR, Catania, Italy

**Iván Santos** (79), Department of Electricity and Electronics, Technical School of Telecommunications Engineering, University of Valladolid, Valladolid, Spain

**Alberto Sciuto** (251), Consiglio Nazionale delle Ricerche (National Research Council), Istituto per la Microelettronica e Microsistemi CNR-IMM, Catania, Italy

**Frank Simon** (1), Coherent Laser System, Göttingen, Germany

**Spyros Stathopoulos** (49), Department of Applied Mathematical and Physical Sciences, National Technical University of Athens, Athens, Greece

**François Stock** (293), ICube, D-ESSP, Strasbourg, France

**Dimitris Tsoukalas** (49), Department of Applied Mathematical and Physical Sciences, National Technical University of Athens, Athens, Greece

**Jim S. Williams** (323), The Australian National University, Canberra, ACT, Australia

**Wenjie Yang** (323), The Australian National University, Canberra, ACT, Australia

# Preface

In the history of semiconductor device fabrication, annealing is one of the critical process steps used to achieve targeted device performances. For the majority of semiconductor manufacturing process steps, the requirements in terms of thermal budget, defined as the integrated temperature field in time (physical units [K × m] or [°C × m]), have always followed a descending trend through the reduction of the time spent at the highest process temperature. Today, the current developments move toward 3D integrated circuits to answer the growing demand for increased functionality and performance. Such new drivers impact the whole supply chain, and low thermal budget solutions are now key to manage local strain and stress, limit species diffusion and interdiffusion, and control layer deposition, etch, and quality. Moreover, the conceptual transition from the usual local evaluation of the annealing impact to a global one, i.e., from the single device level to 3D complex structures and eventually the whole processed wafer, is now mandatory for advanced circuit design. More generally, complexity is one of the keywords in the field of future nanosystems manufacturing, including nanoelectronics. Among other needs, the manufacturing of complex objects requires the implementation of complementary and flexible process options that selectively act in a subportion of the system under fabrication.

Within this context, laser annealing stands as an excellent solution for achieving ultra-low thermal budget heating with nanometer resolution by locally modifying the inner properties of extremely small materials' regions. Indeed, the beneficial (and timely) convergence between the current (and future) nanofabrication challenges and the impressive recent advances in the laser annealing field has made the use of laser annealing technology into a real and promising option to achieve highly controlled manufacturing processes in the nanoelectronics domain and beyond. What's more, when compared to conventional annealing, laser irradiation is somehow a revolutionary process, thanks to the numerous parameters that determine its features, including wavelength, light polarization, beam extension, pulse duration, and pulse repetition rate. This enlargement of the tunable parameters list meets the abovementioned flexibility requirement, whereas the process conditions can be adjusted to the materials and structures of the designed device.

Considering the relatively high level of maturity reached within this research field, an up-to-date reference book that synthesizes the knowledge of scientific and technological advancement occurred in the last few decades is therefore expected to provide a valuable help to the scientific community. This will include postgraduate students and academic researchers interested in laser physics and technology as well as manufacturing and applications engineers in the semiconductor industry seeking to fill the gap between a fundamental understanding of the laser annealing process and the shortcomings of the industrial applications. The semiconductor industry is surely the main driver for the scientific and technological advancement of the laser annealing processing. Yet, we believe that the community of potentially interested readers is not limited to the specialists in the field and that ideas and methods stimulating new technological solutions and research pathways can be found in any chapters of this book. Indeed, selected examples of such a successful approach are given in the last part of the book, which includes reviews of recent researches in which the consolidated expertise in semiconductor laser processing is exploited toward novel and exciting applications.

The book is organized in three parts dedicated respectively to the general aspects of the laser annealing technology, the review of its main applications in the "mainstream" nanoelectronics domain, and, finally, the survey of some novel applications of laser annealing to semiconductor materials.

In the first part, a historical perspective of the laser annealing technology and its evolution since its early days in the mid-1970s will be first presented (Chapter 1). A particular emphasis is given to the presentation of the equipment engineering issues as well as to the first successful industrial application of laser annealing for the fabrication of low-temperature polycrystalline silicon films in the late 1980s. This will be followed by a review of the basic aspects of laser-matter interaction from the point of view of group IV semiconductors (Chapter 2), including the radiation absorption mechanisms and its modification in the presence of heavy extrinsic doping (typical of semiconductor processing), the heating mechanisms, and heat diffusion. The final chapter of the first part (Chapter 3) will set the theoretical framework for the simulation of the evolution of a laser-irradiated system to its final state. In particular, atomistic simulation techniques, particularly accurate for studying detailed interactions at both electronic and atomic levels, will be discussed in this chapter. Recent results on atomistic modeling of laser-related phenomena in Si will finally be reported, including melting and regrowth processes, defect and dopant kinetics, dopant segregation, and the anomalous formation of extended defects.

In the second part of this book, Chapter 4 will first provide an up-to-date review of laser annealing applications in current CMOS technology, with a focus on realistic candidates for integration in a high-volume semiconductor manufacturing flow, based on electrically relevant results. These will include power devices, CMOS logic and 3D integration, and, finally, memory applications. In Chapter 5, the most relevant materials science issues related to the fabrication of highly doped junctions by laser annealing will be initially reviewed. These will include melt and recrystallization kinetics, solute trapping as well as residual damage of Si, Ge, and SiGe materials submitted to nanosecond laser anneals in the melt regime. In the second part, the concepts of above-equilibrium solubility and thermal stability will be discussed, followed by a detailed review of the most relevant experimental results demonstrating the ability of nanosecond laser annealing to achieve dopant activation well above solid solubility limit. Finally, Chapter 6 deals with coupled simulations of the evolving fields (e.g., electromagnetic field, temperature, phases, dopant, and alloy densities) based on the technology computer-aided design tool for the laser annealing process. Accuracy of these tools is based on dedicated developments of the numerical formalism and on the reliability of the material calibration. Their application is fundamental for expanding the flexibility of the laser process solutions over a wide combination of materials and 3D integrated systems.

As a recognition of the "cross-fertilization" character of advanced research, the third part of the book is devoted to "beyond-CMOS" applications of laser machines and technologies with similar features to those discussed in the previous parts, mostly dedicated to nanoelectronics. In Chapter 7, the use of laser annealing for the engineering of carbon-based metamaterials in view of novel optoelectronics applications is reviewed. In this case, the unique features of the laser annealing process are exploited in combination with the unique bonding character of the carbon element, capable of forming stable homoligands in different coordination states (namely the mixed sp./sp3 or polyynes, sp2 of graphene, and sp3 of diamond), for the synthesis of a new class of transparent electrodes. Similarly, Chapter 8 also looks at optoelectronic applications by reviewing recent research works aiming at transforming group IV semiconductor elements and alloys into active materials in the near-infrared frequency range (i.e., the spectral range of telecommunications). Hyperdoping (i.e., the inclusion of high density of activated impurities) is invoked for the scope, and again, laser process is a key technique. Finally, Chapter 9 reports the current state of laser annealing techniques dedicated to achieve an ultra-high density of activated dopant impurities, but unlike Chapter 8, the ultra (hyper)-doping is aimed in this case at transforming the most known semiconductor material, silicon, into a superconductor one. The results reported in this chapter provide realistic prospects for the integration of semiconductor- and superconductor-based devices on the same Si substrate, thanks to the implementation of laser annealing.

Our final thoughts go to Elsevier for having welcomed our project, the Elsevier publishing team who supported us throughout the process, and all chapters' authors who accepted to share their knowledge. We thank you all for your excellent human, technical, and scientific contribution to this book, especially considering the difficult times we have been facing since the beginning of the pandemic crisis.

**Fuccio Cristiano**
**Antonino La Magna**

Chapter 1

# Historical evolution of pulsed laser annealing for semiconductor processing

Guglielmo Fortunato[a], Luigi Mariucci[a], Alessandro Pecora[a], Vittorio Privitera[b], and Frank Simon[c]
[a]*IMM-CNR, Rome, Italy,* [b]*IMM-CNR, Catania, Italy,* [c]*Coherent Laser System, Göttingen, Germany*

## 1.1 Section 1: Introduction

Laser annealing, as a new method to activate dopants and remove damage in ion-implanted semiconductors, was originally proposed by Russian scientists in the middle 70s [1, 2]. By the end of the 70s, it became a hot topic and laser processing was mainly applied for annealing ion-implanted layers, at that time a major issue for the semiconductor industry. Early experimental studies were performed by using pulsed Q-switched ruby (694 nm, 1.79 eV) laser, with tens of nanoseconds pulse duration. Irradiation of ion-implanted Si produced, above a certain fluence, recrystallization of the amorphized silicon layer and activation of the implanted dopant impurities.

Originally, to explain these findings, two main models were proposed: the melting model and a possible mechanism based on a highly excited electronic system that might retain its excitation for tens or hundreds of nanoseconds in the form of a "hot dense plasma." However, a number of experimental evidences, including time-resolved reflectivity, time-dependent Raman scattering, time-dependent X-ray Bragg scattering, and time-dependent electrical resistivity, clearly indicated that the melting model correctly describes the phenomena. In fact, the energy provided by the laser field to the electronic system is transferred quickly to phonons in time less than about 1 picosecond and thus leads to melting the near-surface region.

Initial interest in laser semiconductor processing was devoted to the annealing of ion-implanted layers, to overcome some problems connected to the implants, i.e., the heating of the entire sample to recover the crystalline order. Laser irradiation allows to heat only the surface of the sample without bulk degradation of the substrate. Among the advantages of pulsed laser processing are a

Laser Annealing Processes in Semiconductor Technology. https://doi.org/10.1016/B978-0-12-820255-5.00006-4

"near-zero" thermal budget (since laser pulses last only for tens of nanoseconds); the extent of dopant diffusion is controlled by the melt depth (with negligible diffusion in the adjacent solid substrate); and active dopant concentrations can exceed the solid solubility limit [3, 4].

Between the late 70s and early 80s, laser annealing was a very popular topic, and thanks to many studies, it was possible to precisely determine the phase diagram of crystal, amorphous and liquid Si in a wide range of temperature, quenching rate, and solidification velocity. It was shown that a-Si melts at a temperature ~200 K below the crystalline melting point and that a phenomenon, called explosive crystallization (EC), is triggered by the latent heat released by the resolidifying surface producing a self-sustained melting front toward the underlying a-Si [5, 6]. The EC phenomenon will be discussed in more detail in Chapters 5 and 6. Measured quenching rates of pulsed laser processing were reported in the order of $10^{10}$ K/s, and at high interface velocities, above 20 m $s^{-1}$, undercooling can induce the formation of a-Si. In general, ion implantation produces damage or amorphization of the crystalline structure so that the structural changes induced by laser irradiation mainly concern the a-Si layer on crystalline silicon (c-Si). Depending upon the energy density, different scenarios can be observed: (1) As the energy density reaches the melting threshold of the a-Si (primary melt, i.e., molten layer produced directly by laser irradiation), the severely undercooled liquid layer near the surface suddenly resolidifies, triggering EC, which transforms the a-Si layer into fine-grained polycrystalline silicon (FG-polysilicon) [7, 8]; (2) as the energy density increases above the energy density required to melt the FG-polysilicon (~1500 K [9]) produced by EC, the surface melts a second time and primary melt propagates into the film; and (3) by further increasing the energy density, the maximum primary melt depth propagates into the FG-polysilicon layer. If partial melting of the FG-polysilicon layer is achieved, the resolidification will start at the solid-liquid interface with a random regrowth, due to intrinsic inhomogeneities of the FG-polysilicon seed layer [8]. If the energy density is high enough to produce the melting of the entire FG-polysilicon, then the resolidification will start from the c-Si, producing an epitaxial regrowth of the implanted layer [10, 11]. Melting a part of the c-Si substrate ensures the epitaxial growth of a nearly defect-free crystal with minimum "quenched-in" interstitials [12].

In spite of the advantages offered by the pulsed laser annealing of ion-implanted wafers, laser irradiation in the melt-regrowth regime was not adopted in microelectronic industries as rapid thermal annealing (RTA), developed in the solid-phase regime, was able to guarantee the technology requirements at that time. Indeed, by RTA, it is possible to heat wafers of any size, cassette to cassette, in the timescale of seconds to remove the damage and to activate the implanted species with limited diffusion and to form thin dielectric and/or metallic layers.

However, by the middle 80s, a novel application of pulsed laser annealing was proposed to form the active layer of polycrystalline silicon (polysilicon) thin-film transistors (TFTs). Polysilicon TFT technology has been deeply investigated over the last 40 years for applications in active-matrix liquid-crystal displays and, more importantly, in active-matrix organic light-emitting displays, a technology that today is widely used in mass production of mobile phone displays. The Sony group [13] was among the first ones to demonstrate the possibility to transform thin (of the order of 10–100 nm) a-Si layers, deposited on glass, into polysilicon by excimer laser irradiation. Introduction of such technology produced a substantial improvement in the polysilicon TFT performances, leading to electron field-effect mobility $>100\ cm^2\ (V\ s)^{-1}$ [14–19].

Excimer lasers proved to be a key tool for such tremendous improvement of the polysilicon TFT technology thanks to their main features: UV light emission (output wavelengths 193, 248, and 308 nm for ArF, KrF, and XeCI gas mixtures, respectively) with a short pulse duration (10–30 ns). The industrialization of the polysilicon TFT technology was largely due to the progress obtained in excimer lasers and optical systems to manipulate the laser beams. We will discuss in the next section some of the more relevant aspects of such technologies. The combination in Si of strong optical absorption of the UV light ($\alpha > 10^6\ cm^{-1}$) and small heat diffusion length during the laser pulse (~100 nm) implies that high temperatures can be developed in the Si surface region, causing melting, without appreciable heating (<400°C) of the substrate. This makes the excimer laser annealing (ELA) process compatible with glass substrates or even plastic substrates [20]. Another advantage is that the polysilicon obtained from the melt-regrowth process has a good crystallinity with very few in-grain defects, allowing excellent performances of the devices. A deeper discussion of the application of pulsed laser annealing to low-temperature polycrystalline silicon for large area electronic applications is presented in Section 1.3 of this chapter.

In the early 2000s, the increasingly stringent requirements of the c-Si metal-oxide-semiconductor technology, driven by the downscaling of the device geometries, posed the issue to find an alternative process to RTA to form ultra-shallow junctions. Indeed, for nodes below 90 nm, source-drain (S/D) abrupt junctions shallower than 50 nm [3, 21, 22], with sheet resistance lower than 500 Ω/sq, were needed to reduce short-channel effects (SCEs) and the ON resistance of the transistor. Conventional RTA or even high-ramp-rate (400°C/s) short-time (<1 s) high-temperature (>1000°C) "spike" annealing [23] failed in forming ultra-shallow junctions, due to nonequilibrium diffusion that leads to increased diffusivity during postimplantation annealing (transient-enhanced diffusion) [24]. This produced a renewed interest in pulsed laser annealing for MOS applications [3, 21, 22], not only for the main features of the process but also thanks to the tremendous progress of the excimer laser technology, pushed by the display industry, which occurred in the meantime. ELA in the melting regime also offers many advantages compared to other nonmelting-based methods, such as exact control over the junction depth, higher dopant activation, and profile abruptness. In fact, when irradiating Si by laser light with sufficient energy density, a well-defined melted zone, with sharp transition from liquid to crystal phase, is formed. The diffusivity of dopants is raised in the liquid state ($\sim 2 \times 10^{-4}$ $cm^2/s$, [25]), and the dopants are able to redistribute uniformly within the melted layer, giving rise to boxlike profiles after regrowth. Due to the steep thermal gradient between the liquid and solid phases, immediately after irradiation, the liquid-crystal interface advances toward the surface at a rate of $\sim 3$ m/s [26]. As a result of such rapid solidification (low-temperature solid-phase regrowth is typically $1.5 \times 10^{-10}$ m/s at 550°C), less dopant is segregated into the liquid phase at the liquid-crystal interface, and enhanced dopant trapping occurs. The fraction of the implant dose, which is retained within the semiconductor during ELA, is governed by ablation (during energy deposition), evaporation (during the melted phase), and segregation (during regrowth). The electrical activation of the retained dopant in the regrown layer following ELA is eventually limited by morphological instability at the liquid-crystal interface during regrowth, lattice strain, and thermodynamic limit [27]. Several groups have reported a number of studies for the formation of ultra-shallow junctions, addressing also process integration issues into MOS technology. In Section 1.4 of this chapter, some of the main results are reviewed.

## 1.2 Section 2: Excimer laser technology

This section will focus on some details of the tool that is required to enable the semiconductor annealing processes described in this book. The tool is a laser beam, and for its generation, a laser and suitable optics are required. Since the light-material interaction happens to be very effective if the wavelength is short, the laser is a UV laser, operating at 308 nm or below, 248 or 193 nm. The laser is an excimer laser, the rare species among the light-emitting devices that have been developed in recent decades. The description will outline the characteristics of the beam that excimer lasers emit and the parameters that are relevant for the material interaction. The second part will briefly describe the optics that can be used to shape the beam and prepare it for various industrial applications.

Excimer lasers have been developed 50 years ago by Nikolai Basov [28]. They consist of a sealed gas volume with overpressure, exchanged from pulse to pulse in the area between two electrodes by an integrated blower. The electrodes allow to excite the gas by quickly repeated high-voltage discharge. The gas contains a halogen and an inert, that under this condition temporarily can share an electron with the halogen to form an excited dimer. Falling back into separation—spontaneously or by stimulation—UV photons are emitted very efficiently. The high laser gain of this setup requires a resonator that just consists of a flat high-reflection mirror and a well transmittive flat mirror for outcoupling of the beam.

This pulsed laser beam has unique characteristics, one of which is its high pulse energy that is hard to reach by other laser concepts in this wavelength regime.

Emerging from a laser resonator as described above, the beam quality is limited and the spatial coherence is low: The difference in terms of the quality parameter $M^2$ is significant if compared to lasers with optimized beam quality and does not allow efficient generation of small beam waists by simple focusing. This means a significant difference between excimer-based optical systems and systems implementing other lasers. The low quality is regarded as benefit, though, because the low spatial coherence allows efficient homogenization.

A closer look into the beam parameters allows to better understand the typical concepts of optical beam shaping.

### 1.2.1 Characteristics of the beam emitted by excimer lasers

**(i)** Near-field distribution: In contrast to beams that show axial symmetry, an excimer laser beam's cross section is easier to describe by rectangular symmetry. It has an axis parallel to the current flow through the gas and an axis perpendicular to this. Usually, the first one is longer than the latter and (after averaging) shows super-Gaussian intensity distribution in the corresponding cross sections, while in the shorter axis, the intensity distribution comes close to a Gaussian shape.

The intensity distribution is shown by diagrams in Fig. 1.1 that correspond to camera images from the beam as it would look on a fluorescent screen, and it is called near-field distribution.

**(ii)** Far-field distribution: While the two-dimensional intensity distribution is identified with the near field, the far-field distribution is supposed to show the angular distribution of the beam. Again, it is represented by a camera image that shows the beam on a screen. The screen has to be placed "far" from the laser, which is uncomfortable, so instead some optical beam shaping is required: A spherical lens is used to focus the beam on the screen. The lens is supposed to introduce no aberrations, which requires a sufficiently long focal length. The position of the screen is crucial for the appearance of the image—only if placed well in the real focal distance, the image can be called "far field" and represents the beam's angular distribution at the position of the lens. This is important because the focal length usually does not coincide with the position of the beam waist, which again may be different for short axis and long axis of the beam.

**(iii)** Angular characteristic: The beam waist of the respective beam axis represents the inherent divergence, which cannot be further reduced by optical means, whereas the far field in addition to the inherent divergence contains a second type of divergence, which could be better called decollimation because it can be reduced to zero by collimating optics.

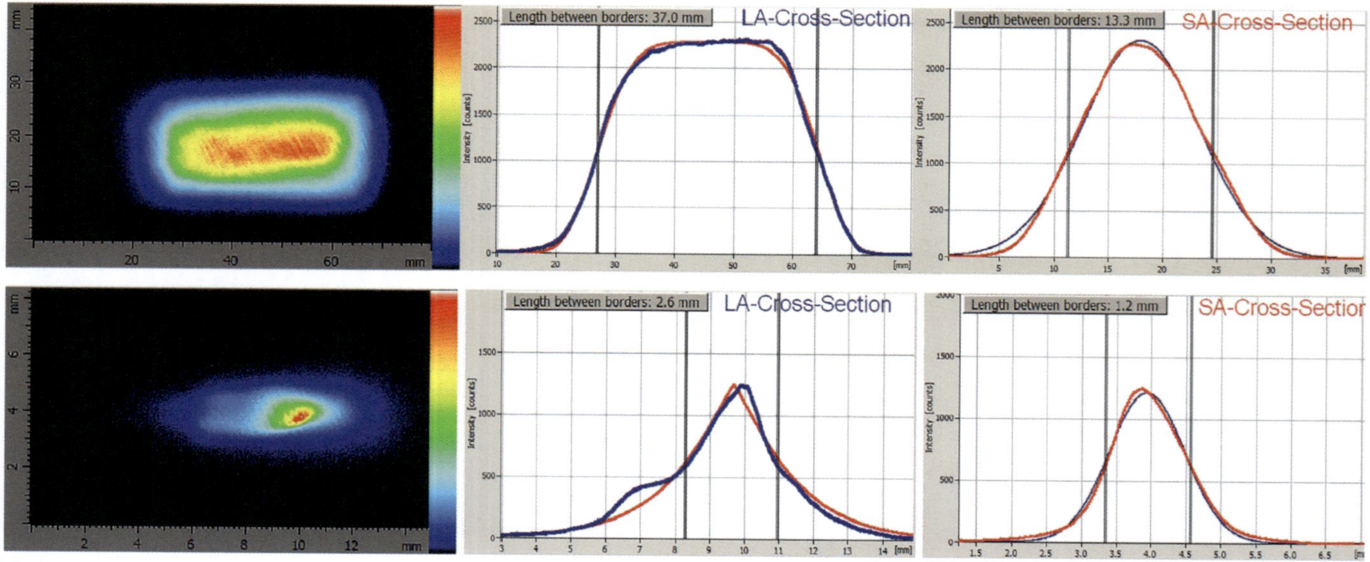


**FIG. 1.1** Near-field (top) and far-field (bottom) images of a high-power XeCl excimer laser beam. LA: long axis; SA: short axis.

For modeling the beam (e.g., for raytracing software), it is interesting to collect enough information about intensity and angular distribution at a given position to allow calculating intensity distributions at other positions. Near-field and far-field distributions are not enough for this because the angular distribution is not the same throughout the near field. Typically, the edge points further outward than the rest of the beam.

Due to the high gain of the laser gas and the high output coupler transmission, the first part of the pulse contains a high percentage of light that has not been reflected by the other resonator mirror. Later, in the pulse, the beam is dominated by the light that has passed the gas several times and can only leave the resonator under low angles. This makes the angular distribution typically decrease throughout each pulse.

**(iv)** $M^2$: Most lasers allow to describe their beam quality roughly by the parameter $M^2$, which expresses their beam waist diameter $\omega$ as multiple of the waist diameter $\omega_G$ a Gaussian beam with the same wavelength would have, if focused under angle $\theta$ by the same lens:

$$\omega = M^2 \bullet \omega_G = M^2 \bullet \lambda/\theta\pi$$

$M^2$ is correlated with the *beam parameter product* (BPP), which is introduced by the equation

$$\text{BPP} = \omega \bullet \theta = M^2 \cdot \lambda/\pi$$

which explains that BPP is an invariant quantity for a given beam—even if passing lenses and mirrors—just depending on the wavelength $\lambda$ and the beam quality parameter $M^2$. For Gaussian beams, $M^2$ is 1—for excimer laser beams, it has different values for both axes of the profile, but both are significantly bigger than 30 and for high-power excimer lasers easily larger than 100.

A high $M^2$ value corresponds to low spatial coherence: Many independent modes are excited at the same time and diffractive effects are limited, which is especially useful for homogenization of the beam. Nevertheless, if necessary, the quality of a beam with a high $M^2$ value can be increased by cutting the beam into segments on one axis and arranging the segments next to each other on the other axis. For the axis of a beam with sufficiently high $M^2$ (at least $M^2 \gg 5$), segmented in n parts, this results in an improved quality according to $M^2_{\text{new}} = M^2/n$. Thus, the design of the respective beam delivery system can benefit from a great flexibility.

**(v)** Pulse energy: Near-field and far-field images, received after averaging over several single pulses, should not feign that the distributions are constant from pulse to pulse. Instead, a spatial fluctuation of the energy throughout the cross section can be observed that resembles a burning and flickering flame. Although the integrated pulse energy can be very constant from pulse to pulse (e.g., 0.2 %$\sigma$), some energy is shifted back and forth throughout the cross section. This explains why excimer laser beams are usually homogenized before being applied for a material process.

### 1.2.2 Parameters that are relevant for the light-material interaction

While pulse energy and geometry of the beam should be considered for efficient beam shaping, the following parameters describe what makes excimer laser beams suitable for light-material interaction.

(i) Wavelength: Due to high difference in potential energy between inversion state and ground state of excited dimers, the photon energy reaches 4eV (XeCl) to 6.4eV (ArF) corresponding to a wavelength in the deep UV.

The spectral bandwidth of a free-running laser covers 400–1000 pm as shown in Fig. 1.2. For reduction of the bandwidth, additional internal optics

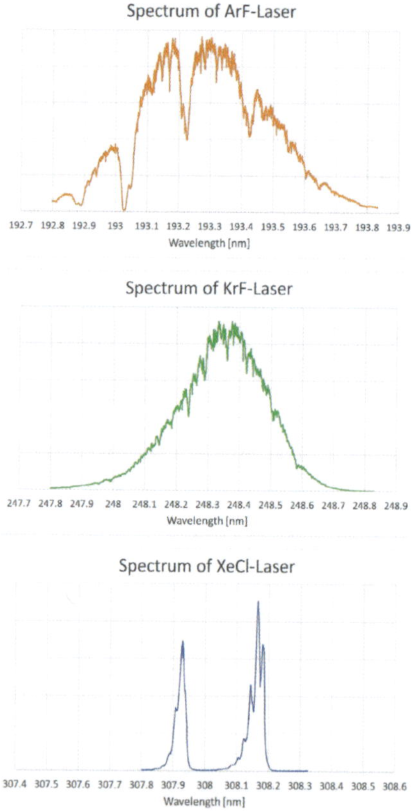


FIG. 1.2 Wavelength Spectra of ArF, KrF, and XeCl lasers (top to bottom).

inside the resonator are necessary to selectively amplify a single wavelength. This can be helpful to avoid chromatic aberrations in high-resolution imaging, e.g., as applied for mask projection during photolithography.

**(ii)** Pulse shape: Although a short pulse duration of approximately 30–40 ns is typical for excimer lasers, the exact shape depends on the design of the discharge circuit. After optimization of the laser efficiency, some typical pulse shapes are shown in Fig. 1.3.

The ns regime of the pulses is too short for the absorbed energy to dissipate into the substrate. As a consequence, most of the excimer laser processes are low-temperature processes, which is important for working on sensitive materials such as plastic.

**(iii)** Polarization: Excimer laser beams optimized for high power are unpolarized, although it was shown that with small losses only the beam can be polarized. Indeed, unpolarized light is required for maximum isotropy of laser processes.

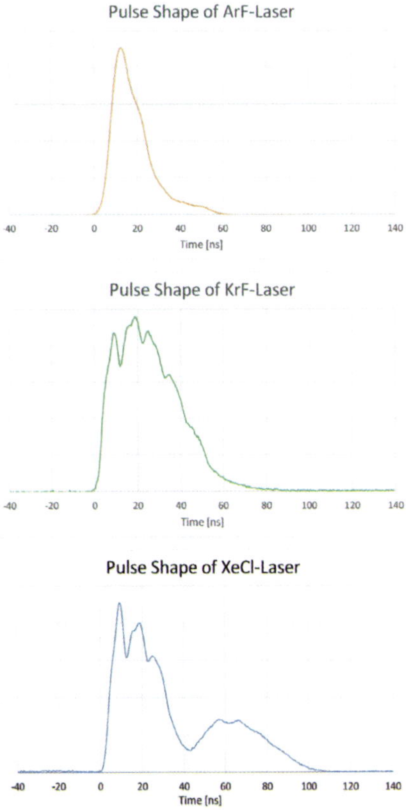


**FIG. 1.3** Pulse shapes of ArF, KrF, and XeCl lasers (top to bottom).

### 1.2.3 Optics that can be used to shape the beam and prepare it for industrial applications

Excimer lasers without optical beam shaping are not very helpful. In addition to conventional optics for bridging the distance from the laser beam exit to the substrate, specific optical subsystems are also required for

- shaping the substrate beam,
- homogenization,
- mask projection, and
- moving the beam.

As a consequence of the rectangular beam symmetry, in many cases, the optics have cylindrical instead of spherical shape and both beam axes are treated independently.

The challenges of building UV-optical systems comprise the choice of materials, the sensitivity of the UV beam to surface defects and bulk defects, and the competing requirements for precision and high power throughput. In particular, the lens material must be suitable for high transmission while withstanding long-time penetration by UV light. And, it must allow to make large components to handle the high pulse energy. In the past 25 years, numerous systems have been realized to adapt the excimer lasers to applications like ablation, marking, annealing, and activation—not only in display and semiconductor industry. This section will introduce three typical types of optical systems as representative examples:

- Rectangular beam for homogeneous substrate irradiation:

Homogenization typically makes use of fly's eyes or cylinder lens array homogenizers. Various manufacturing techniques are available to build such homogenizer; the preferred way is the assembly from precisely polished cylindrical elements. If the homogenizer is designed to be capable of processing a BPP bigger than that of the laser and smaller than what is required at the substrate, it fits well to the system. Homogeneities of 5% ($2\sigma$) within the rectangular flattop beam on the substrate can easily be achieved.

- Line beam for efficient substrate scanning:

For scanning of big substrates, a rectangular beam with high aspect ratio—such as a long line—is suitable, because it allows to deliver energy to the whole substrate, while this is moving slowly. Aspect ratios of up to $1500 \times 0.5\ mm^2$ are possible and have been realized already, while for the long axis of the line beam, the homogeneity was better than 1%($2\sigma$). If the pulse energy available from a single laser is not enough, the cylinder lens homogenizer concept allows several laser beams to be combined on the substrate, adding their pulse energies, e.g., to 6J, and reducing fluctuations by their incoherent overlap.

- High-resolution mask projection for precise substrate treatment:

If the beam on the substrate must be precisely structured or must have sharp transitions to nonirradiated areas, a mask can be projected after homogeneous illumination by the excimer laser beam. For this type of beam shaping, a special objective is required that has sufficient projection capabilities and can handle the laser power. More than 100 W UV power has been used this way for substrate treatment, while resolution, field size, and demagnification were well controlled by the layout of the projection lens, e.g., $2\mu$ resolution within ø60 mm image field, corresponding to ø150 mm on the mask. The depth of focus for this type of optical system is limited though, to, e.g., $50\mu$ depending on the process. Together with a precise substrate stage, an excimer laser-based mask projection system can be a very flexible tool: Just changing the mask can adapt it to different requirements.

In conclusion, it can be summarized that using excimer laser beams for semiconductor processes not only benefits from spectral characteristics and short pulse shape, but also flexible beam shaping and good homogenization can be achieved by suitable optics.

## 1.3 Section 3: Excimer laser annealing for low-temperature polycrystalline silicon technology

ELA has been widely used for the crystallization of amorphous silicon deposited on glass or plastic substrates. Indeed, the very short laser pulse (of the order of few tens of nanoseconds) allows melting and regrowth of the silicon in few hundreds of nanoseconds. In this short time, the substrate temperature remains well below 500°C, making possible the use of low melting temperature substrates, such as glass or polyimide. Furthermore, the melt and regrowth process produces polycrystalline silicon with low in-grain defects and grain size that can exceed 1 μm, enabling the manufacture of devices with characteristics that can approach the performance of crystalline devices.

The crystallization mechanisms and the resulting characteristics of the polycrystalline silicon depend on laser energy. Usually, two energy regimes and a subregime are identified: For laser energy densities higher than the melting threshold of the amorphous silicon, EC is triggered and the film is only partially melted (partial melting regime), with melting depth and grain size increasing with the laser energy; at higher energy, the amorphous silicon can be completely melted (complete melt regime) and the crystallization occurs via homogeneous nucleation in the supercooled liquid, resulting in small grain silicon. A subregime has been identified [29] corresponding to the near-complete melting when the silicon film is almost totally melted and only few crystalline clusters produced by EC, sparsely distributed, remain unmelted at the liquid-substrate interface. In this condition, the so-called superlateral growth (SLG) is triggered and

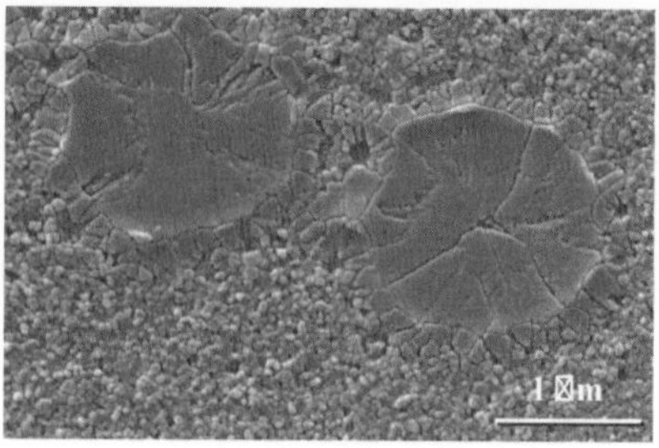


**FIG. 1.4** Polysilicon grains obtained in superlateral growth subregime.

grain size exceeding 1 μm can be obtained (see Fig. 1.4), resulting from the lateral growth of isolated unmelted solids at the bottom interface. This subregime appears to be particularly interesting in terms of the structural and electrical characteristics of the polycrystalline silicon.

However, due to its peculiar mechanism, the SLG regime corresponds to a narrow processing window and large nonuniformities are observed in the grain size distribution and, consequently, in the performance of the devices. The nonuniformity of the grain size, although particularly evident in the SLG regime, is characteristic of the whole partial melt regime. To reduce the impact of this variability, several techniques have been suggested, starting from multishot crystallization [30] (that also reduces the in-grain and grain boundary defects) and the regrowth control techniques that aimed to control the position of the grains and to increase their size. Some of these techniques [sequential lateral solidification (SLS), two-pass excimer laser crystallization (ELC) process, phase-modulated ELA, and micro-Czochralski] will be described in more detail in this section.

In order to study the ELC mechanisms, structural (TEM and SEM) and transient measurements were carried out. In particular, in situ time-resolved front and backside transient optical reflectance (FTR and BTR) measurements [6, 8, 9, 29, 31], sometimes in conjunction with transient electrical conductance [6, 9] and thermal emission measurements [9], allowed to analyze the dynamic of melting and resolidification of thin amorphous and poly-Si films. The reflectivity and the conductive properties of silicon show a noticeable variation when melting occurs, allowing to determine the onset of melting, the melt duration, the melt depth, and possible melt propagation into the silicon layer together with the liquid-solid interface velocity. In particular, FTR measurements show that, for laser energy density above the melting threshold, the signal rapidly increases (see Fig. 1.5) due to the high reflectivity of the molten silicon.

The reflectivity amplitude remains high until the solidification is completed. After few nanoseconds, an oscillation of the signal is sometimes observed [8, 29] and attributed to interference effects induced by the EC [6], caused by the different melting temperature of amorphous silicon (about 200 K below the melting temperature of c-Si). EC is triggered by the latent heat released

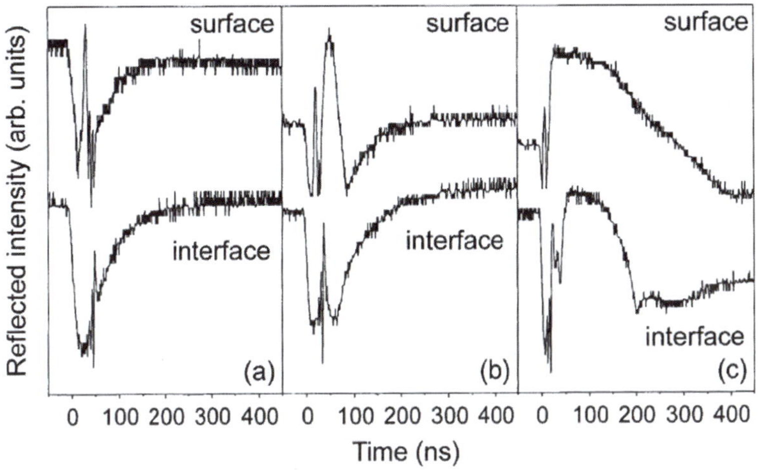


**FIG. 1.5** FTR and BTR profiles during ELA at energy densities of (A) 200 mJ/cm$^2$; (B) 250 mJ/cm$^2$; and (C) 700 mJ/cm$^2$. *(After F.C. Voogt, R. Ishihara, A combined TEM and time-resolved optical reflectivity investigation into the excimer-laser crystallization of a-Si films, Thin Solid Films 383 (2001) 45–47, 2001.)*

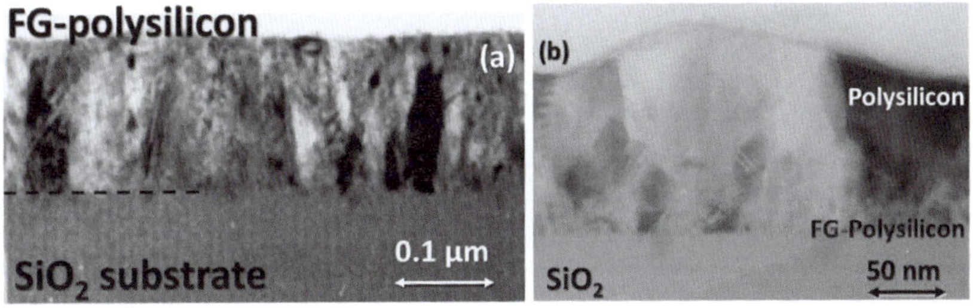


**FIG. 1.6** Cross-sectional TEM images of a-Si films after ELA at energy densities just above the threshold for a-Si melting (A) and under the partial melting regime (B). *(A: After F.C. Voogt, R. Ishihara, A combined TEM and time-resolved optical reflectivity investigation into the excimer-laser crystallization of a-Si films, Thin Solid Films 383 (2001) 45–47, 2001. B: After C.-C. Kuo, Evolution of microstructure in polycrystalline silicon thin films upon excimer laser crystallization, Laser Phys. 18 (2008) 464–471, 2008.)*

by the resolidifying surface toward the underlying a-Si [7, 8], with the self-propagating buried liquid layer moving into the a-Si at a speed of 11–13 m/s [32, 33], and it will be discussed in more detail in Chapters 5 and 6. Here, we can note that, as shown by TEM analysis (Fig. 1.6A), EC produces fine-grained (FG) (around 10 nm) polysilicon that extends into the whole silicon film even for partial melting regime.

According to FTR measurements, the fine-grained surface region can sequentially undergo remelting and the primary liquid melting front propagates

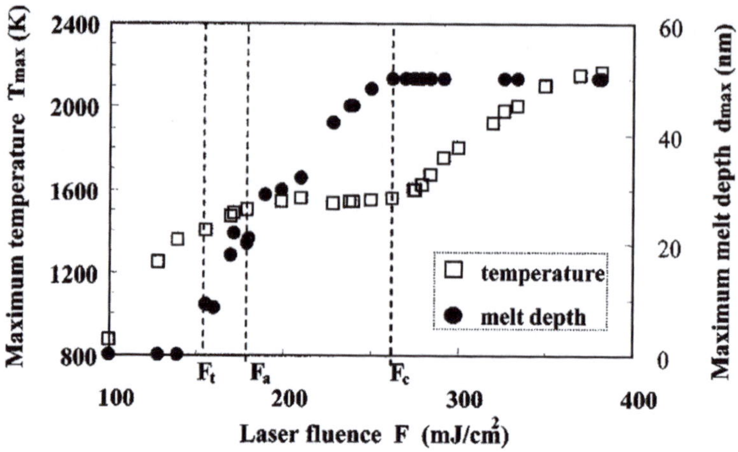


FIG. 1.7 Dependence of maximum temperature and melt depth on laser energy density. $F_a$ and $F_c$ correspond to the laser fluences for surface and complete melting, respectively. *(After M. Hatano, S. Moon, M. Lee, K. Suzuki, C P. Grigoropoulos, Excimer laser-induced temperature field in melting and resolidification of silicon thin films, J. Appl. Phys. 87 (2000) 36.)*

toward the substrate for a melt depth that increases with the laser energy density, as shown by the transient conductivity measurements (see Fig. 1.7) [9].

This results in the typical double-layer structure observed in the TEM cross section of polysilicon film crystallized in partial melting regime, consisting of a FG-polysilicon layer near the substrate and with larger grain polysilicon on top (see Fig. 1.6B).

Transient thermal emission measurements show that in the partial melting regime (laser fluence, F, between $F_a$ and $F_c$ in Fig. 1.7), the liquid temperature remains constant, since the absorbed excess of laser energy is transformed in the latent heat of phase change from solid to liquid Si. The temperature is estimated to be of 1535 K, well below the melting point, of crystalline-Si (1685 K).

Since the experimental evidence is that the a-Si is transformed into an intermediate FG-polysilicon by the EC, this implies that the melting temperature, $T_m$, of FG-polysilicon is still lower than that of c-Si. For higher laser energy density ($F > F_c$ in Fig. 1.7), the silicon film is completely melted and the liquid temperature increases with the laser fluence. After the end of laser pulse, the silicon remains melted for several tens of nanoseconds (Fig. 1.8), while the temperature rapidly decreases below the $T_m$ of c-Si (region II in Fig. 1.8). Indeed, due to the lack of seeds for heterogeneous nucleation, substantial supercooling is observed until homogeneous nucleation occurs. The beginning of solidification is clearly visible in the thermal emission curve since the latent heat released induces an increase in the temperature (Fig. 1.8) [9].

This complex melt-regrowth process induced by laser irradiation results in a polysilicon film with grain size that changes in a nonlinear way with the laser energy density (Fig. 1.9) [29]. In the partial melting regime, the average grain

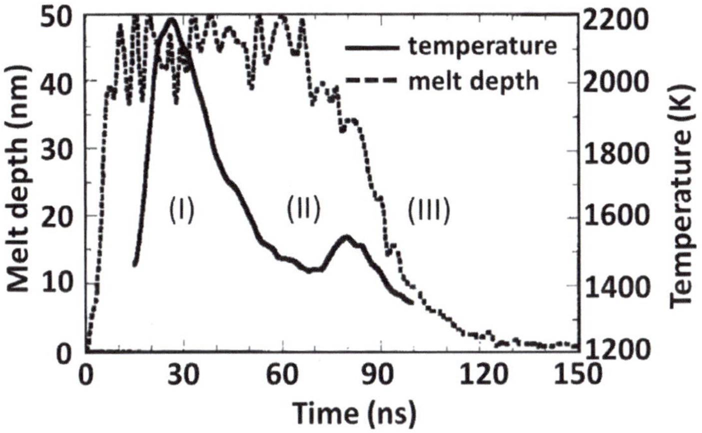


**FIG. 1.8** Temperature and melt depth variation during ELA carried out at energy density inducing complete melting. *(After M. Hatano, S. Moon, M. Lee, K. Suzuki, C.P. Grigoropoulos, Excimer laser-induced temperature field in melting and resolidification of silicon thin films, J. Appl. Phys. 87 (2000) 36.)*

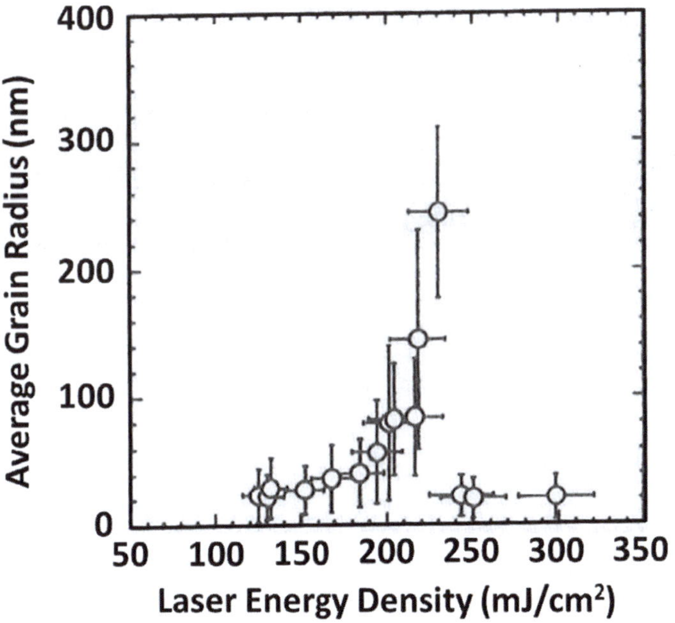


**FIG. 1.9** Variation of average grain radius of crystallized Si film on laser energy density. *(After J. S. Im, H.J. Kim, M.O. Thompson, Phase transformation mechanisms involved in excimer laser crystallization of amorphous silicon films, Appl. Phys. Lett. 63 (1993) 1969.)*

size gradually increases with the fluence until a sharp increase in grain size, with size in excess to 1–2 μm, can be observed for laser energy density corresponding to the SLG subregime. In the complete melting regime, the grain size abruptly decreases down to few tens of nanometers due to homogeneous crystallization.

It should be noted that the analyses of the above time-resolved measurements were carried out considering a uniform liquid layer. For instance, while the conductance increase is representative of the volume fraction of liquid phase, it is usually reported as an indicator of the melt depth. However, a more detailed analysis carried out by SEM and AFM on polycrystalline silicon film after one-shot laser irradiation suggests a more complicated scenario. Indeed, the laser crystallization induces three-dimensional processes that result in grains with disk-shaped structure not only in the SLG subregime but even in partial melting regime [34–38]. As shown in Fig. 1.10A [34], for laser fluence corresponding to the SLG subregime, SEM analysis shows that the polysilicon consists of radial disk-shaped structures with diameter >1 μm. Inside the disks, some defects are evident possibly related to low-angle grain boundaries [39]. AFM analysis (Fig. 1.11A) shows at the edges of the disks the presence of protrusions, where two or more expanding grains collapse, resulting from the mass transport during the lateral solidification [40]. The presence of such protrusions is clear evidence that the grain growth occurs mainly in the lateral direction

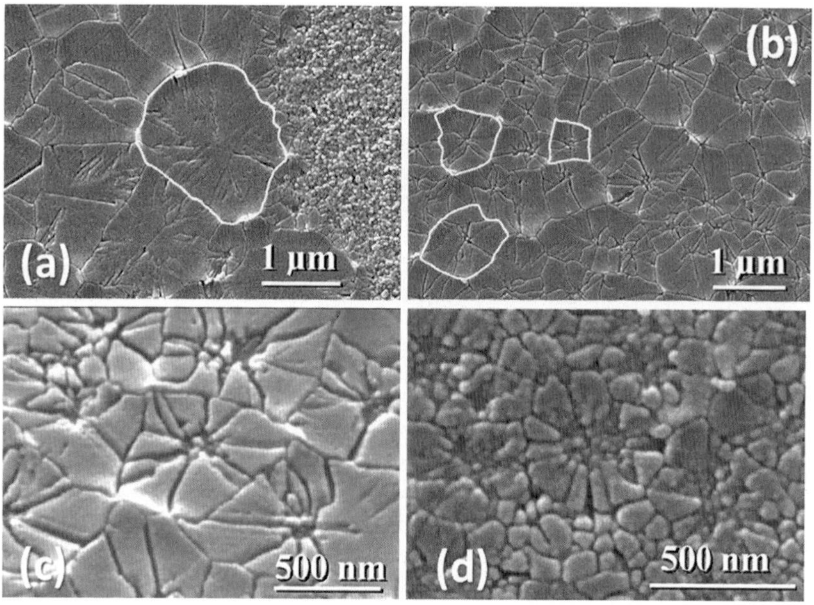


**FIG. 1.10** SEM micrograph of Secco-etched polysilicon samples crystallized at different laser energy densities: (A) 300 mJ/cm$^2$ (SLG regime); (B) 270 mJ/cm$^2$; (C) 250 mJ/cm$^2$; and (D) 220 mJ/cm$^2$. For eye guide, some disk structures are highlighted.

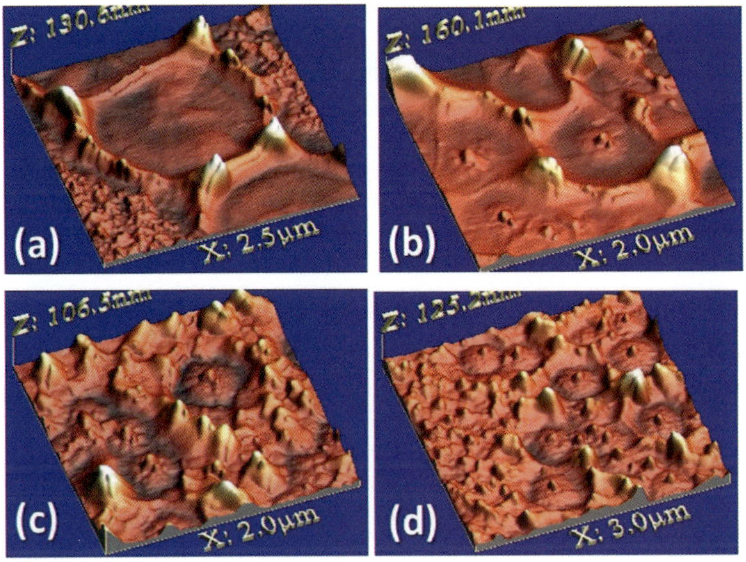


**FIG. 1.11** Surface topography, measured by AFM, of polysilicon samples crystallized at different laser energy densities: (A) 300 mJ/cm$^2$ (SLG regime); (B) 270 mJ/cm$^2$; (C) 250 mJ/cm$^2$; and (d) 220 mJ/cm$^2$.

[40]. The protrusions are as high as 100 nm where three grains meet (triple junction) and they are the main cause of the roughness of the film. In the partial melting regime (Fig. 1.10B), SEM analysis also shows disk-shaped structures consisting of an external ring of radially arranged relatively defect-free grains and an inner small grain region that is approximately circular in shape. The presence of protrusions at the edge of the external ring indicates that the grains grew laterally starting from small grain central regions (Fig. 1.11B). Furthermore, SEM analyses show that the grains of the external rings can extend, for sufficiently small thickness, over the entire thickness of the film (Fig. 1.12A).

For thicker sample (Fig. 1.12B), double-layer film can be observed with large grain layer on top of FG-polysilicon extending throughout the film down to the substrate. As discussed before, this double-layer structure of the polysilicon film is a strong evidence that EC occurred at the beginning of the laser crystallization process, with the FG layer originating from EC. For decreasing laser fluences, the disk structures show reduced diameters and increasingly larger inner small grain central regions (Fig. 1.10C and D). The reduction of the protrusions at the edges of the external rings (Fig. 1.11C and D) also suggests a decrease in mass transport during resolidification. The disk structures and some mass transport are still observed for laser energy densities corresponding to an average melt depth of about 50% of film thickness [34].

Starting from the experimental evidence that EC took place in these a-Si films, as confirmed by the data reported in Fig. 1.12B, it has been suggested

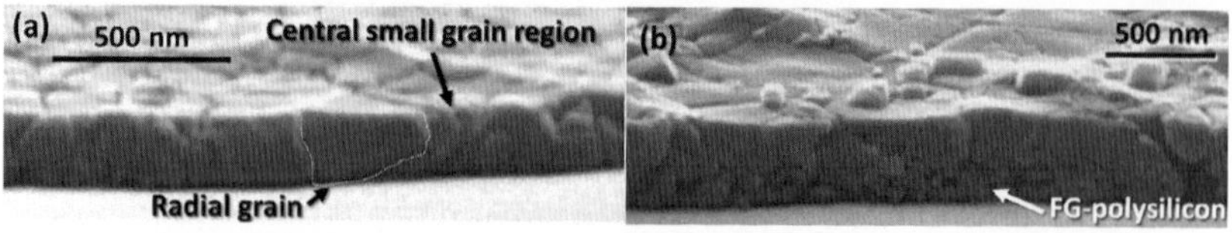


**FIG. 1.12** SEM image of the cross section of polysilicon films crystallized by a single-shot excimer laser in partial melting regime at fluence of 440 mJ/cm$^2$ (A) (SLG subregime was at 600 mJ/cm$^2$, film thickness of 200 nm) and of 700 mJ/cm$^2$ (B) (SLG threshold was at 950 mJ/cm$^2$, film thickness of 400 nm).

[34] that, after the primary melt has reached the maximum depth and resolidification starts, the small grains in the disk core regions originate by the regrowth exploiting as seed layer the FG-polysilicon produced by the EC at the beginning of laser irradiation. The regrowth process will start randomly due to the intrinsic inhomogeneities of the FG-polysilicon layer. During the growth of these small grains, latent heat is released reducing further nucleation in the surroundings. The observation that the external grains of the disks show lateral growth and extend throughout the film indicates that, for sufficient heating, the FG-polysilicon is melted (recalescence) and the small core grains act as seeds for the lateral epitaxial growth of the external grains of the disks.

More recently, Hu et al. [35] suggested a different crystallization scenario on the basis of the absence of the initial oscillation of FTR signal during the laser crystallization of their PECVD a-Si films. Under laser irradiation, solidification of a-Si film starts by heterogeneous nucleation at the liquid-solid interface and proceeds via a defective mode of crystal growth [41], producing fine-grained polysilicon observed in the core region. The latent heat released during the crystallization increases the temperature and eventually melts the surrounding region (recalescence, as proposed by Mariucci et al. [34] in the previous model) leading to less defective epitaxial growth, mainly in the lateral direction, that accounts for the grains of the external rings.

The above discussion about the physics of ELC showed that single-shot annealing results in polysilicon with nonuniform grain size, rough surface, and random crystallographic orientation. It has been shown that the increase in the number of shots per point (multishot process) can improve the properties of the final polysilicon [30]. Large number of shots per point increases the grain size of polysilicon film and induces a strong crystallographic orientation, reducing the grain boundary energy barrier and increasing the carrier mobility. Furthermore, the films show a very smooth surface and minimal in-grain defects, reducing the carrier traps inside the semiconductor and at the semiconductor-dielectric interface.

SEM and AFM analyses (Fig. 1.13) [34] clarify the effect of multishots on the polysilicon structure that was obtained after the first shot showing that the disk-like structures were substantially modified by the successive shots. Indeed, already after two shots (Fig. 1.13), the number of inner small grains is reduced

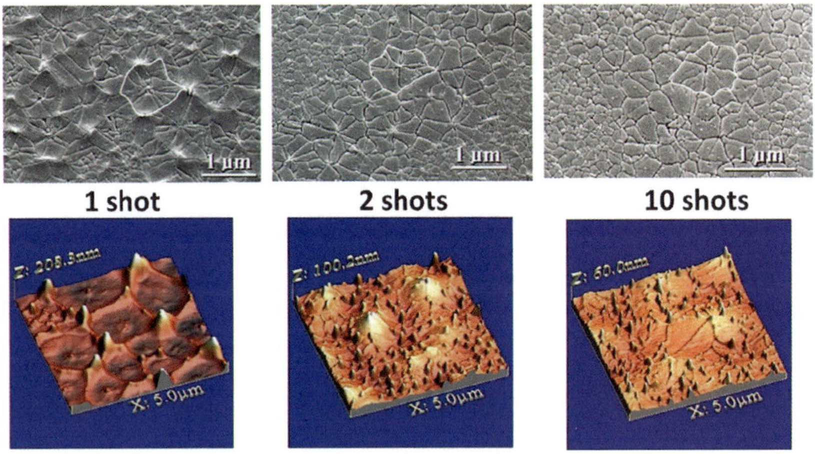


**FIG. 1.13** SEM micrograph and AFM measurements of polysilicon samples crystallized at 270 mJ/cm$^2$ by different number of laser shots.

**FIG. 1.14** Cross-sectional TEM micrograph showing the columnar grain structure of polysilicon film crystallized by multishot irradiation in partial melting regime.

as a consequence of the expansion of the external radial grains toward the central regions. This is evidenced by the increase in the peaks in the center of disks, observed in AFM images (Fig. 1.13).

At the same time, the second shot induces the reduction of peaks at the triple junctions and the growth of ridges at the radial grain boundaries of the external ring of grains. The increase in the shot number (Fig. 1.13) induces a general reduction of the height of the ridges and of peaks at the center of the rings and at the triple junction, improving the flatness of the surface. Multishot irradiation produces effects also in the vertical structure of the polysilicon film that consists, even in the partial melting regime (Fig. 1.14), of columnar grains extending from the surface down to the polysilicon-substrate interface.

### 1.3.1 Excimer Laser process engineering

As previously discussed, the ELA process is able to produce very large grain size but with very high nonuniformity in terms of size distribution at the same time, mainly related to pulse-to-pulse fluctuations and nonperfect beam homogenization.

To solve these problems, different techniques, to improve the uniformity of the crystallized films and to maximize the grain size, have been explored with the aim to obtain higher performance polysilicon TFTs, with improved field-effect mobility and homogeneous electrical characteristics. These advanced ELA techniques are usually exploiting multishot regime, the effects of specific laser beam energy density distribution, the partial overlapping of the sequential shots, with the concomitant movement of the sample to cover the entire area to crystallize.

Some of these techniques mainly aim at controlling the uniformity of the grain size rather than maximizing grain size, while the other techniques, operating in the SLG regime, aim to control and extend the lateral grain growth.

Among the advanced excimer laser procedures, the main ones include the trailing-/leading-edge irradiation mode [42], SLS [43], two-pass ELC process [44], phase-modulated excimer laser annealing (PMELA) [45], and micro-Czochralski process [46], for location-controlled single grain growth. In the following, these approaches are briefly described.

#### *1.3.1.1 Trailing-/leading-edge scanning mode*

A very effective and relatively simple technique to improve the uniformity of the crystallized films and maximize the size of the crystalline grains is to scan, on the amorphous silicon film, an appropriately profiled energy beam. In particular, the use of semi-Gaussian energy density laser beam profiles, swept along the direction of the Gaussian energy distribution, can produce very uniform grain size distribution of the film with good medium grain size [47].

By using this technique, two main crystallization regimes can occur, depending on whether the maximum energy density of the Gaussian distribution ($E_{TOP}$) is larger or not than the threshold energy density ($E_{th}$) needed to induce complete melting of the polysilicon film, formed by the advancing leading edge of the beam (see Fig. 1.15a1).

When scanning a beam with $E_{TOP} > E_{th}$, the final structure of the film is controlled by the crystallization induced by the trailing edge of the beam (*trailing-edge* mode). In fact, when the fine-grained polysilicon, produced by complete melting induced by the high-energy part of the beam, is further irradiated at lower energy densities, as the beam is finely stepped forward, the right conditions to trigger the SLG mechanism are achieved somewhere along the trailing edge of the beam. After SLG material is formed, further irradiations at lower energies, as the beam is further stepped, will not change the SLG structure, since only partial melting will occur.

In order to homogeneously crystallize the film in the SLG regime, the SLG energy density window has to be compared with the Gaussian energy distribution, to determine the step size. Homogeneous crystallization is normally obtained with a high shot density (~100 shots/point), limiting the throughput

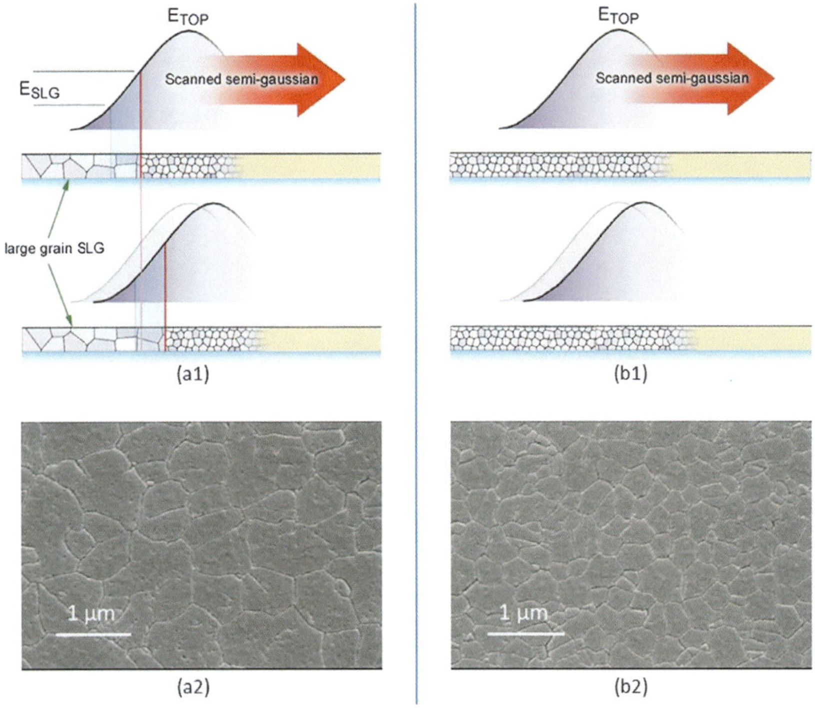


**FIG. 1.15** Scheme of laser beam scanning in (a1) "trailing-edge" mode and (b1) "leading-edge" mode. SEM micrographs of Secco-etched samples irradiated in trailing-edge ($E_{TOP} = 350$ mJ/cm$^2$) (a2) mode and in leading-edge ($E_{TOP} = 280$ mJ/cm$^2$) (b2) mode, both with 10 shots/point.

of this technique. In order to reduce the number of shots, a broad energy distribution beam profile can be adopted along the scanning direction.

On the other hand, if $E_{TOP} < E_{th}$, the final structure of the film is controlled by the leading edge of the beam (*leading-edge* mode) and the grain size will depend on $E_{TOP}$ (see Fig. 1.15b1).

In Fig. 1.15, the SEM micrographs of two samples, one irradiated in the trailing-edge mode (Fig. 1.15a2) and the other in the leading-edge mode (Fig. 1.15b2), both with 10 shots/point, are shown. As can be noted, a uniform grain size distribution is obtained for the two regimes, with a grain size ranging between 0.7 and 1 µm for the trailing-edge mode and 0.3 and 0.4 µm for the leading-edge mode.

### *1.3.1.2 Sequential lateral solidification*

SLS is a procedure that uses a spatially varying excimer laser beam intensity, which consists of a number of beamlets, few micrometers wide in one direction and several millimeters in the other [48]. These beamlets are produced by projecting the homogenized excimer laser beam through a mask having

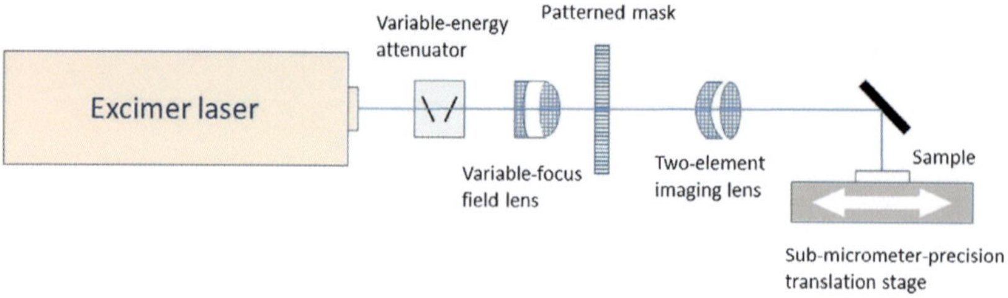


**FIG. 1.16** Schematic diagram of an ELC projection system configured for the SLS process. *(After R.S. Sposili, J.S. Im, Sequential lateral solidification of thin silicon films on $SiO_2$, Appl. Phys. Lett. 69 (19) (1996).)*

alternate clear and occluded stripes. Within the high-intensity beamlet areas, full film melting takes place and, at the edge of these regions, there will be an SLG region, which seeds the lateral growth into the molten region (see Fig. 1.16).

In the single-shot irradiation, two scenarios occur, depending upon the relative widths of the beam and of the seeded lateral growth distances. In particular, to avoid that the center of the molten region undergoes random nucleation before the lateral growth front reaches this region, resulting in fine-grained material in the center of each exposed stripe, the beamlet size needs to be ~5 μm, so that the two lateral growth fronts meet in the center of the previously melted region and produce a collision-front grain boundary. In the orthogonal direction, there are low-angle subgrain boundaries running parallel to the lateral growth direction. Subsequently, in this first shot, the sample is moved by a distance shorter than the beam half widths prior to the next irradiation, such that the original melt-front-collision grain boundary is melted by the second pulse, and the remaining polycrystalline material seeds further lateral growth, propagating the grain to the right. With an appropriate projection mask design, this process can be indefinitely repeated to produce grains tens or hundreds of micrometers long (see Fig. 1.17A).

If the sample is moved, after the first shot, by a distance greater than the beam half widths, the original collision-front grain boundary is retained, and a similar grain structure is generated by the second irradiation. With an appropriate mask design, this process can be implemented as a two-shot process, giving the highest throughput obtainable with SLS (Fig. 1.17B).

The grain structures resulting from these two SLS implementations are shown in Fig. 1.17A and B, respectively. The above discussed techniques produce the formation of "directional" material by scanning line beamlets. The technique can also be implemented to give localized single crystal regions by the use of chevron-shaped beamlets, or by implementing the two-shot process in two orthogonal directions.

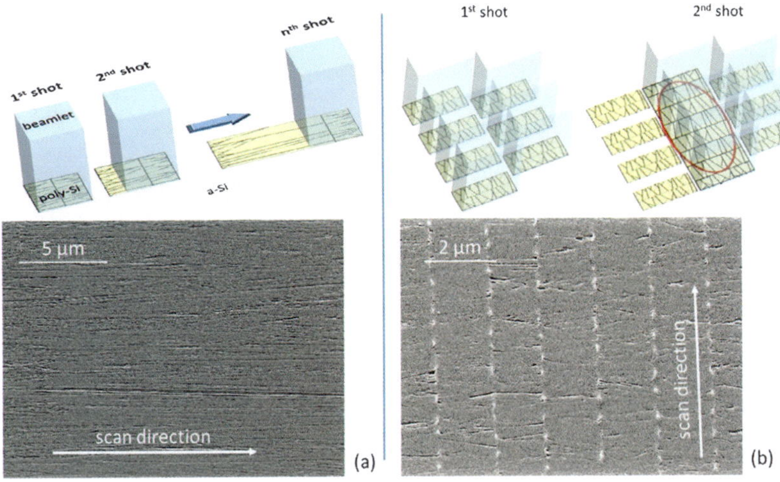


**FIG. 1.17** Schematic illustration of the SLS grain propagation with the corresponding SEM micrographs of poly-Si obtained in the condition of (A) extended grain material and (B) two-shot process.

### *1.3.1.3 Two-pass excimer Laser process*

This technique is characterized by a reduced number of laser shots and a rather wide energy density window [49]. In the first pass of the process (schematically shown in Fig. 1.18), control of the lateral-growth location is achieved by an opportune spatial modulation of the absorbed laser energy via a contact mask.

Lateral growth is triggered when the irradiated regions are fully melted and resolidification begins at the surrounding solid-liquid interface, which is located at the boundary between masked and unmasked areas. Seeding at the solid-liquid interface is strongly favored, because heterogeneous nucleation has a much lower activation energy compared to homogeneous nucleation. To crystallize the sample homogeneously, the film can be reirradiated (second pass) without the mask. Due to thermodynamic properties of amorphous and polycrystalline silicon, it can be demonstrated that a useful energy window exists to induce a complete melting of the remaining a-Si regions (masked areas during the first pass), while partially melting the polysilicon regions (unmasked areas during the first pass). In this way, the partially melted polysilicon grains act as seeds, allowing the regrowth and expansion of the grains formed during the first pass (see Fig. 1.18).

### *1.3.1.4 Phase-modulated excimer laser annealing*

The PMELA technique consists in exposing the substrate to a fluctuation in spatial intensity of the energy density [45]. This intensity fluctuation must vary from a threshold energy density lower than melting of the a-Si film, to a value

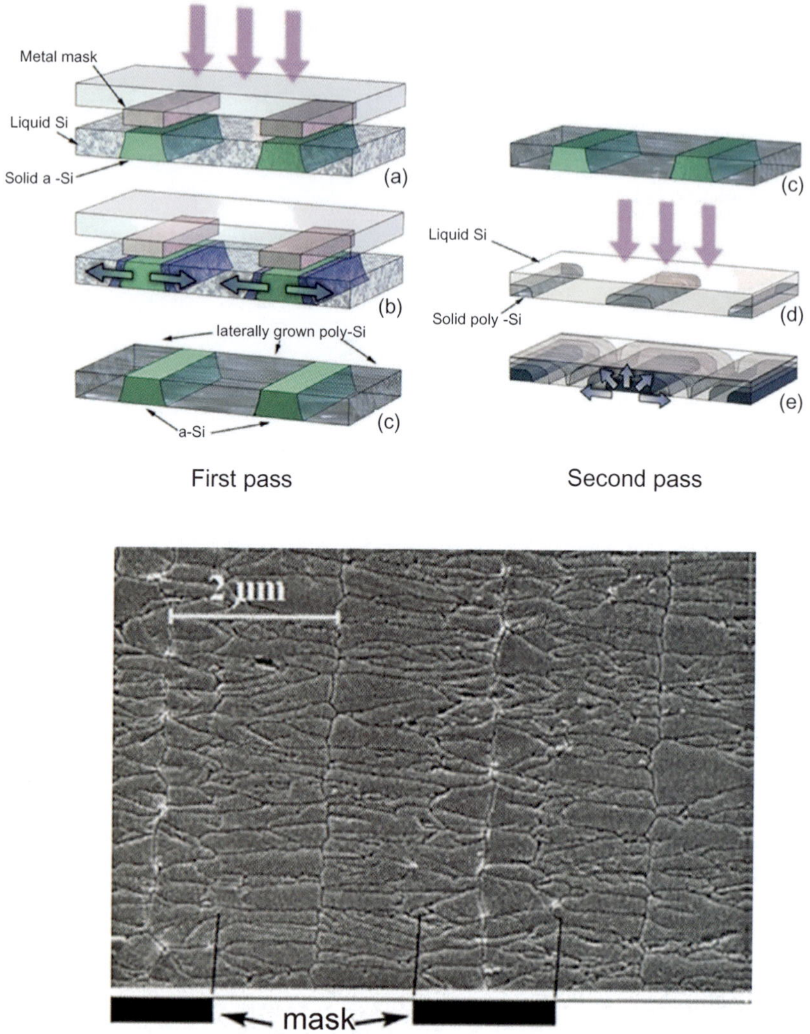


**FIG. 1.18** The top of the picture shows the schematic illustration of two-pass excimer laser crystallization process. First pass: laser irradiation through the contact mask (A) and lateral grain growth (B). Second pass: irradiation without the mask (D) and propagation of the grains formed during the first pass (E). At the bottom of the picture, it is shown the SEM micrograph of a Secco-etched sample after the second pass irradiation performed at $E_1 = 360$ mJ/cm$^2$, for liner/space geometry 1.5/2.5 μm. The position of the mask during the first pass irradiation, performed at $E_1 = 400$ mJ/cm$^2$, is also shown.

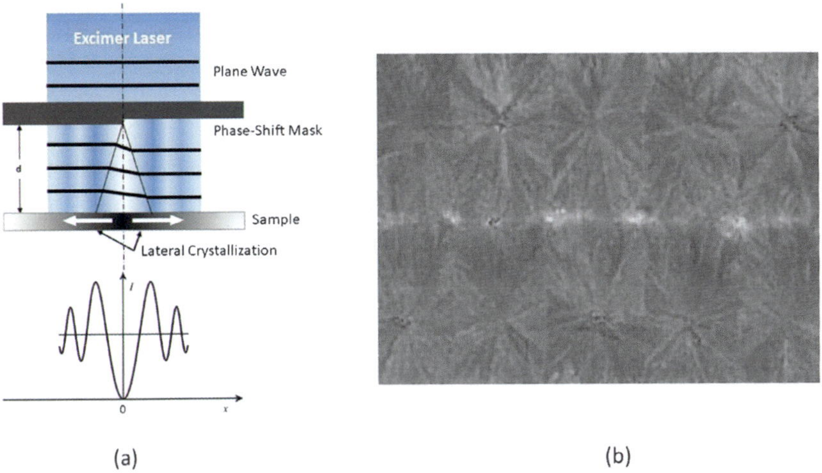


**FIG. 1.19** (A) Schematic of the PMELA method. (B) SEM image of location-controlled large grain arrays grown by PMELA. *(A: After O.H. Chang-Ho, M. Ozawa, M. Matsmura, A novel phase-modulated excimer-laser crystallization method of silicon thin films, Jpn. J. Appl. Phys. 37 (1998) L492–L495. B: After Y. Taniguchi, T. Katou, M. Hiramatsu, M. Matsumura, New phase modulators for next generation low-temperature crystallization method of Si films, Jpn. J. Appl. Phys. 45 (10A) (2008) 7753–7760.)*

high enough to completely melt the silicon film. Between these two regions, there will be a part of material that will experience near-melting conditions. This portion of material acts as seed, for the SLG process that will occur in the adjacent fully melted region, resulting in grain size of several microns.

In the PMELA technique, the laser beam is passed through a phase shift mask, so that the optical interference in the emerging beam produces the required intensity fluctuation (see Fig. 1.19A).

PMELA is essentially a single-shot technique capable of producing large areas consisting in a large grain matrix, typically around 5 μm square, as shown in Fig. 1.19B. The picture shows a combination of large grain areas with lateral growth areas plus the nucleation area. To achieve this well-defined location-controlled grain structure, the phase shift mask has been modified, with a robust fabrication process, from its original simpler form (composed of etched depressions in a quartz plate), to a two-dimensional bipolar phase modulator plate. This consists in a complex matrix with a base cell of variably sized pits and humps on the quartz plate [50].

### *1.3.1.5 Micro-Czochralski technique*

The micro-Czochralski technique requires a preliminary processing of the substrate in order to model the a-Si film. For this reason, it differs from all other techniques based on modulating the incident laser beam [51].

**FIG. 1.20** Cross-sectional TEM image of location-controlled gain formed by the micro-Czochralski process (A); SEM micrograph of an array of location-controlled grains grown by the micro-Czochralski technique (B). *(After V. Rana, R. Ishihara, Y. Hiroshima, D. Abe, S. Inoue, T. Shimoda, W. Metselaar, K. Beenakker, Dependence of single-crystalline Si TFT characteristics on the channel position inside a location-controlled grain, IEEE Trans. ED 52 (12) (2005) 2622–2628.)*

Basically, the technique consists in creating points, through the use of photolithographic techniques, in which the thickness of the a-Si is much greater than in the surrounding regions. These regions are obtained by realizing holes of very small diameter (<100 nm) in the $SiO_2$ substrate. The holes are subsequently filled by depositing the a-Si film of suitable thickness. Then, the amorphous film is exposed to a laser intensity sufficient to melt the surrounding film, while the film remains solid at the bottom of the hole. Hence, there will be a transition region in the hole, where the normal ELA process will form some crystallites that will generate lateral grain growth in the surrounding molten Si, as shown in Fig. 1.20A.

By controlling the diameter of the hole, it is possible to reduce the number of grain seeds, with the aim of producing individual crystalline areas. This procedure was implemented by appropriately spacing the holes in the substrate, in order to produce a series of adjacent square grains, reaching single grain size up to about 6 μm, with the seed regions located in the middle of the single grains [51]. In Fig. 1.20B, the grain structure obtained is shown.

## 1.3.2 Excimer laser-crystallized thin-film transistors

Excimer laser-crystallized polysilicon has been exploited to fabricate TFTs that show electrical characteristics largely improved compared to the amorphous silicon TFTs and that can even approach those of SOI devices. The melt-regrowth process induced during excimer laser irradiation produces polycrystalline silicon with low in-grain defects, while most of the defects are located at the grain boundaries. As a consequence, field-effect mobility of ELC TFTs depends on the laser energy density and increases with polysilicon grain size. For a standard laser process, the mobility values range from about 10 up to 200 $cm^2/Vs$, with the maximum value obtained for active layer crystallized in SLG regime [42], when the maximum grain size is achieved. However, due to nonuniformities of grain size in this regime, the mobility values show large fluctuations that make difficult the application of these high mobility materials for electronic circuits. ELC process engineering tried to improve both the performance and the uniformity of the electrical characteristics of ELC-TFTs.

Fig. 1.21 shows the electrical characteristics of TFTs that used polysilicon crystallized in trailing-edge scanning mode [52]. The large grain size and good uniformity of the polysilicon film result in very high field-effect mobility for both n- and p-channels, as well as in low threshold voltage and sharp subthreshold slope, making possible the fabrication of complex circuits.

Higher device performance was obtained by using the SLS crystallization process, which allowed to obtain n-channel TFTs with field-effect mobility, $\mu_n$, up to 600 $cm^2/Vs$ [53]. It must be pointed out that such high mobility values are obtained for TFTs with current flow parallel to the direction of grain

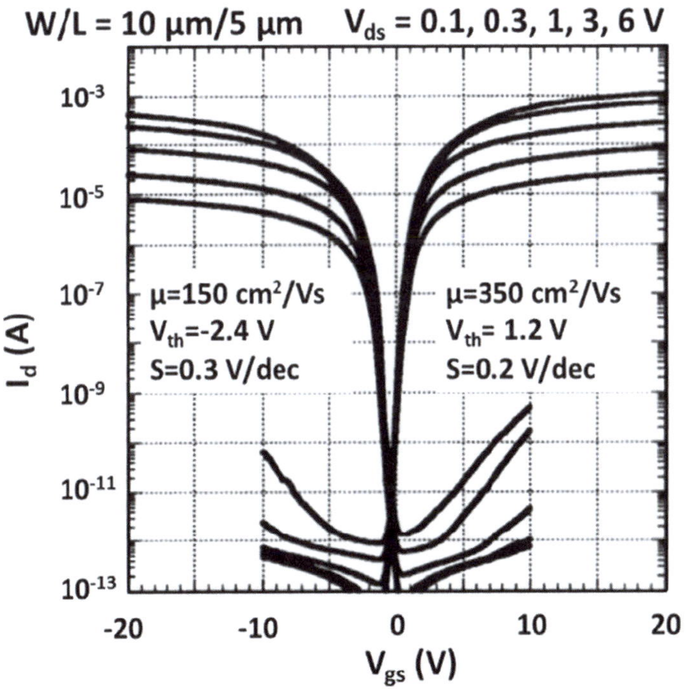


**FIG. 1.21** Electrical characteristics of TFTs that used polysilicon crystallized in trailing-edge scanning mode.

propagation, while for perpendicular current flow, the $\mu_n$ was reduced by even a factor four. Very high mobility values, up to $\mu_n = 900\ cm^2/Vs$ for n-channel and $\mu_p = 250\ cm^2/Vs$ for p-channel, have obtained using PMELA crystallization [54]. Such performance approaches the corresponding values of SOI devices, but these high mobility TFTs can be obtained in a small pseudo-single-crystal region of the polysilicon film with {100}-orientation, making these devices hardly suitable for circuit applications. Even higher field-effect mobilities have been demonstrated for single-grain thin-film transistors (SG-TFTs) fabricated inside location-controlled silicon grains using the $\mu$-Czochralski process. The high tensile strain of the silicon grains, obtained by the careful control of the melt-regrowth process, led to the carrier mobility enhancement of polysilicon TFTs [55]. Short-channel devices showed peak mobilities of 1400 $cm^2/Vs$ for n-channel and 500 $cm^2/Vs$ for p-channel, values that correspond to about 1.6 times over the SOI-TFTs mobilities. However, also in this case, the TFT performance was largely dependent on the position of the devices inside the grains and the mean field-effect mobilities were much lower ($\mu_n = 880\ cm^2/Vs$ and $\mu_p = 320\ cm^2/Vs$) [55]. Despite the large spread of the mobility values, operational amplifier and RF circuits, with cutoff frequency of 5.5 GHz, were demonstrated by using SG-TFTs with 1.5 μm channel length [56].

TFTs made by the different technologies available (amorphous silicon, polycrystalline silicon, oxide organic semiconductors) have their main application for the active-matrix backplane in AMLCD and AMOLED displays for the addressing of single pixel or for peripheral column and row driving circuits. Indeed, the high performance and the electrical stability of the polysilicon TFTs, together with the possibility of a CMOS process, enable the use of polysilicon TFTs for peripheral addressing circuits as an alternative to the external Si-ICs. The use of polysilicon driving circuits is particularly convenient for small displays, where the more expensive polysilicon process is compensated by the reduction of the cost of the external ICs (independent on the display dimensions). Single-pixel addressing of AMLCD is usually carried out by a-Si TFTs, while polysilicon TFTs are preferred for the pixel circuit of OLED displays, providing performances hardly achieved by the other TFT technologies. In fact, OLEDs are current-controlled devices and are very sensitive to variation and drift of TFT parameters and polysilicon TFTs can provide higher drive currents and higher electrical stability compared to a-Si TFTs. Furthermore, polysilicon process allows to develop the multiple TFT pixel circuits required to compensate for threshold voltage variations of TFTs and OLED degradation, so improving the quality and the reliability of the display.

As already mentioned, the ELC process has been also shown to be compatible with plastic substrates (for a review of this process see Ref. [20]), allowing the fabrication of polysilicon TFTs on polyimide with excellent characteristics [57] and opening the doors to flexible AMOLEDs [58, 59].

## 1.4 Section 4: Excimer laser annealing in MOS technology

### 1.4.1 Logic applications

By the end of the 90s–early 2000s, the semiconductor industry, producing at that time a worldwide revenue of around 250 billion euro and supporting a global market of more than 6 trillion euro in terms of electronic systems and services, started the transition from microelectronics, where critical device dimensions were measured in microns, to nanoelectronics. The new era of nanoelectronics elevated in that period the information and communications technology (ICT) society to unprecedented levels and enabled Europe to realize its vision of ambient intelligence, i.e., living environments aware of our presence and responsive to our needs.

Innovative solutions were hence needed for the major device, process, and material integration challenges, for the technologies planned in the coming years.

However, in order to achieve the technology nodes below 65 nm, many issues had to be tackled, such as SCE, gate leakage, and mobility degradation, and simple scaling down without proper optimization based on experiments and simulation would have been worthless. Moreover, several device parameters were approaching their physical or technical limits: gate oxide thickness, source-drain junction depth and parasitic resistance, including contact and wiring resistances. This caused the end of the conventional scaling scheme and the beginning of a new one. In order to further improve the VLSI performances, new concepts relative to materials and/or processing had to be presented.

Sub-90 nm MOSFET architecture needed ultra-shallow junctions and 2D dopant profiles design to control the SCE and the subthreshold performances. Indeed, the formation of ultra-shallow source-drain extensions with high doping level and an optimal dopant distribution in the channel zone allows to reduce the parasitic effect of threshold voltage roll-off versus channel length decrease. Alternative annealing techniques were definitely needed for the technology nodes in the early 2000s years (see Fig. 1.22).

On the one hand, flash lamp annealing and/or nonmelting LTP, with ultraviolet or infrared sources, were an extension of the by now surpassed RTA method, with the choice of lamp illumination (FLA) or laser radiation (nonmelt LTP) being regarded as a matter of hazard. On the other hand, the "innovative" use of melting ELA required a noteworthy effort of integration, but provided doping profiles with unique characteristics.

Conventional RTA faced the problems of thermal diffusion and low dopant activation, which are limited by thermal budget and solid solubility, respectively. That RTA was surpassed was also the general message launched at the 13th IEEE International Conference on Advanced Thermal Processing of Semiconductors (RTP 2005), held in Santa Barbara (USA) in 2005. While keeping pace with Moore's law had been accomplished by decreasing processing

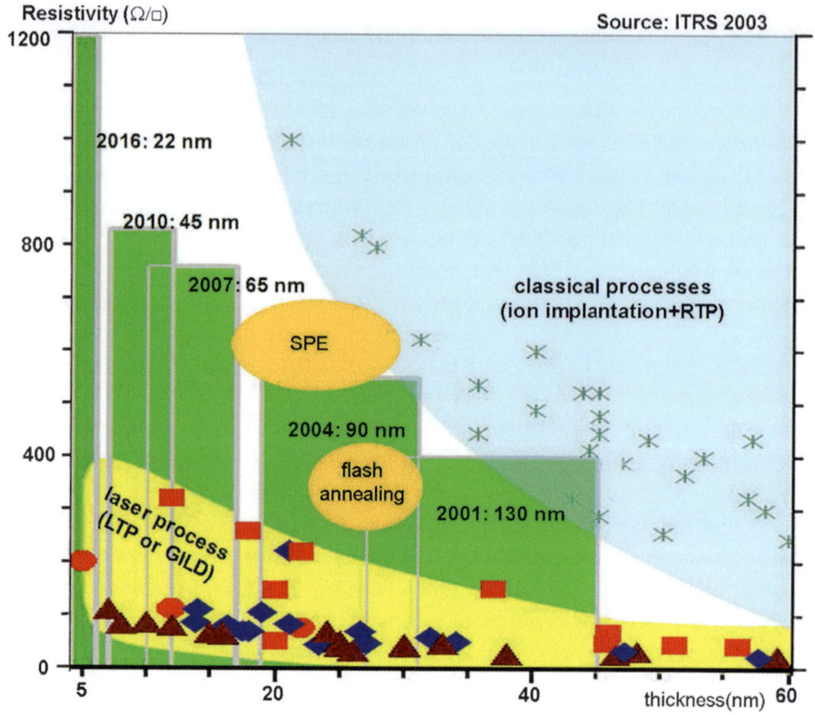


FIG. 1.22 Junction depth and sheet resistance requirements according to the ITRS 2003 for the CMOS technology nodes. The red, blue, and brown symbols refer to many literature data collected by the ITRS, relative to laser thermal processing or gas immersion laser doping, whereas the green stars correspond to sheet resistance/junction depth data obtained by conventional RTA. A region of values achieved by solid-state epitaxy (SPE), or by flash annealing, is also reported in the graph.

times till that period, fundamental limits were foreseen for the 45 nm node, requesting processing time of the order of 10 ms or even less and with nearly zero diffusion. This was beyond the processing regime currently provided by lamp-based RTP systems, limited to processing time on the order of 1 s. The fabrication of such ultra-shallow $p^+/n$ junctions was hence one of the main challenges for the production of sub-65 nm pMOSFET. Apart from the junction depth and the high doping level, also an extreme lateral abruptness of 2.5 nm/decade was required, in order to decrease the spreading resistance, and hence the total resistance, of the current path in the MOS circuit. The requirements for the CMOS technology in those years were met by the use of ion implantation with energies below 1 keV and the so-called spike annealing technique. The latter gives the possibility of extremely fast ramp rates, >250°C/s, and a soak annealing time close to zero. Nevertheless, the junction depth, the electrical activation and the profile abruptness achievable by this process were not suitable for the subsequent technology nodes.

The alternative proposed was represented by the use of ELA for the formation of ultra-shallow abrupt junctions with high electrical activation. This technique appeared quite promising, with ELA systems quite mature at that time, thanks to the strong developments boosted by the display applications (see the previous section), but exhibited severe technology integration issues.

Potentially, this method was very attractive to meet the requirements of the technology nodes. In the ITRS, many items were indicated in red for the technology nodes lower than 100 nm. Critical items included gate stack, junction depth, sheet resistance at contact and extension, power dissipation, and cost-effective integration. The ELA method could hence contribute to break the "red brick wall" present in the road map, but a major issue related to the ELA method remained its integration with the process steps and the materials used for the device fabrication. An intense activity was hence devoted to the study of the interaction of the laser beam both with doped silicon and with the structures forming the device, like dielectrics, metal contacts, and polycrystalline silicon. Therefore, the doped layers were studied in one and two dimensions to obtain data also on the lateral diffusion of the doping impurities beneath a mask. A process flow, aimed to the realization of a MOS device, compatible with the inclusion of the ELA process was finally elaborated and implemented.

Furthermore, reliable computational packages were produced, to be coupled and integrated with existing codes for process simulation, capable of simulating the interaction between the laser beam and the silicon device, including the evolution of the thermal field and dopant distribution during ELA and the subsequent quenching. The choice of Si melting regime, even though predicted extremely sharp, shallow, and highly active doping profiles, nevertheless implied hard challenges relative to device integration.

A signal of interest was represented by the rich production of patents by a leading company such as AMD. A significant activity on the use of laser processing for the semiconductor industry was performed in the USA, where both semiconductor industry and equipment suppliers were quite active in this field. The activity in Europe was rather intense in academic research as well as among equipment suppliers, whereas the involvement of the semiconductor industry seemed less heavy than in the USA.

By the end of the 90s, several reports [3, 21, 60, 61] were produced showing the feasibility of introducing ELA in the CMOS process, and MOSFETs, both n-channel [3] and p-channel [21], were demonstrated. Several issues appeared to limit the application of ELA including the melting of the polysilicon gate, which produced an undesired rounding of the profiles of the gate edges; melting under isolation, causing increase in leakage due to defect formation and undesired dopant diffusion; small process window to define junction depth and gate overlap if compared to ELA variability; deactivation and diffusion of dopants during thermal steps subsequent to ELA.

A possible way to enhance process margins consisted in selectively lowering the melt temperature of the source and drain regions, if compared to other device areas including the polysilicon gate. This was achieved by preamorphizing Si substrates prior to dopant ion implantation of the source-drain regions [22, 60–62]. Preamorphization implantation (PAI) of c-Si substrates can be performed by using either Si+ or Ge+ ion implantation: For instance, an a-Si layer of about 30 nm can be produced by Si ion implantation at 10 keV with a dose of $2 \times 10^{15}$ $cm^{-2}$. By using low laser energy densities, it is therefore possible to confine the melt depth within the a-Si layer formed by preamorphization, producing boxlike profiles, in contrast to the case of c-Si samples.

The origin of the energy density window to completely melt the preamorphized Si while keeping the underlying c-Si substrate solid is due to the nearly 200°C melt temperature difference between amorphous and crystalline Si. In this case, the junction depth is controlled by the thickness of the a-Si layer rather than by the laser energy density. This leads to an improved process window for the laser energy density (nearly 7% [61]) used for the dopant activation in a CMOS structure. For energy densities well above the threshold for melting the c-Si substrate, the preamorphization process no longer influences the junction depth, and consequently, similar results in terms of junction depth, junction abruptness, and dopant activation have been observed in samples with and without preamorphization. The effectiveness of PAI is clearly shown in Fig. 1.23, where it can be observed that by adopting a multishot process, the process window for the energy density is largely improved, as the junction depth and doping profile are basically equivalent for both 500 and 800 $mJ/cm^2$ irradiations, and with the junction depth (40 nm, as determined at the atomic concentration level of $3 \times 10^{18}$ $cm^{-3}$) coincident with the thickness of the amorphized region [22]. In addition, it can be also noted that junction depth and doping profile obtained after ELA are not influenced by sequential RTA annealing at 900°C for 20 s, making possible the reduction of dislocations in the recrystallized layer after ELA without any transient enhanced diffusion effect [22]. In order to improve the process window and reduce polysilicon gate rounding, it has been proposed to introduce a thermally conductive "absorber" layer [61], deposited over the entire device so that source-drain regions and gate are thermally shorted. This layer can also prevent melting under isolation and agglomeration of the polysilicon gates if an appropriate material is chosen: Under these circumstances, it has been shown that the device process window can be increased up to 14% [61]. Flowcharts for MOSFET fabrication were adapted by introducing the ELA process after high-temperature RTA annealing, as those needed to activate the S/D deep implant as well as the polysilicon gate. In this way, high-thermal budget annealing was performed prior to the ELA process avoiding possible dopant deactivation. Yu et al. from AMD [3] proposed a process where ELA was used to activate shallow S/D extension and demonstrated excellent device characteristics of N-MOSFETs with channel length of 70 nm. Alternatively, Goto et al. from Fujitsu [21] adopted a different device integration flow to reduce contact resistance, forming first S/D extension followed by RTA and subsequently, after sidewall formation, S/D deep junction was implanted and

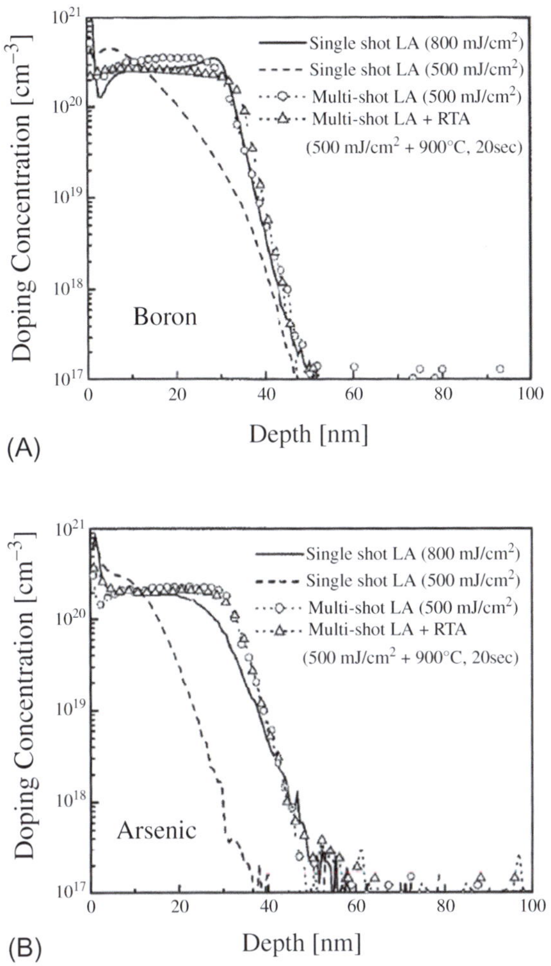


**FIG. 1.23** SIMS profiles of (A) boron and (B) arsenic after laser annealing with various conditions. *(After S.-D. Kim, C.-M. Park, J.C.S. Woo, Advanced source/drain engineering for box-shaped ultrashallow junction formation using laser annealing and pre-amorphization implantation in sub-100-nm SOI CMOS, IEEE Trans. Electron Dev. 49 (10) (2002) 1748-1754.)*

laser annealed. By adopting this process, P-MOSFETs with channel lengths down to 50 nm were demonstrated with well-suppressed SCE and extremely low contact resistance [21].

In order to reduce the laser energies required for activating arsenic and boron impurities, Kim et al. [22] developed a dual-PAI process to fabricate CMOS devices on SOI substrate. The energy densities were lowered to one-fourth of those required on bulk silicon, so that a large process window margin alleviating undesirable integration problems could be achieved. Improved short-channel characteristics were obtained in the fabricated devices with sub-100-nm gate lengths as reported in Fig. 1.24. $V_{th}$ roll-off <50 nm

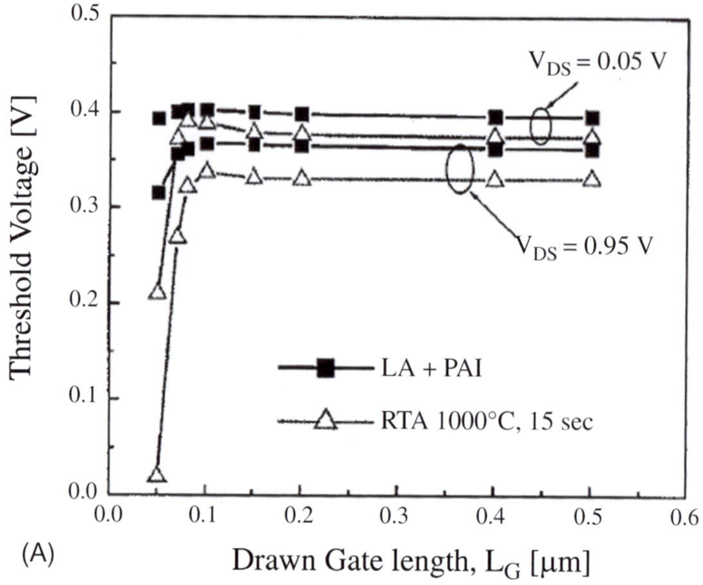


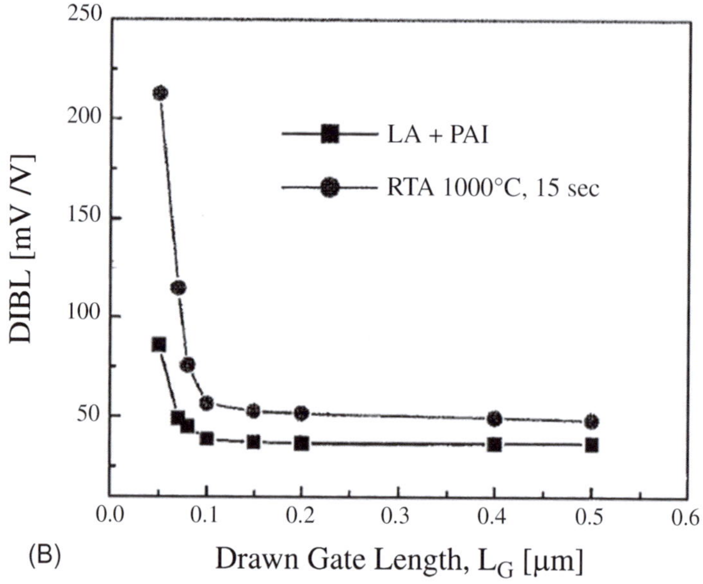


**FIG. 1.24** (A) $V_{th}$ roll-off and (B) DIBL characteristics of NMOS transistors as a function of gate length for LA-PAI and RTA devices. *(After S.-D. Kim, C.-M. Park, J.C.S. Woo, Advanced source/drain engineering for box-shaped ultrashallow junction formation using laser annealing and pre-amorphization implantation in sub-100-nm SOI CMOS, IEEE Trans. Electron Dev. 49 (10) (2002) 1748-1754.)*

(A) and substantially improved drain-induced barrier lowering (DIBL) (B) if compared to RTA were observed, due to the well-controlled S/D junction profiles.

To further improve the laser energy density process window, necessary to accommodate variations due to pattern density change and inhomogeneous thermal properties across the wafer, Shima et al. at Hitachi [63, 64] proposed a self-limiting laser process, which dramatically enhanced the process window. In particular, a phase-switch (PS) layer, consisting in a polysilicon layer deposited by CVD at 400°C, was introduced on top of the laser absorber layer: The PS layer has lower initial reflectance, but very high reflectance above the phase switching temperature. The switching temperature is optimized to be between amorphous and crystal Si melting temperature, ensuring that all preamorphized junctions would melt without any overmelt in c-Si. Fig. 1.25A shows simulated melt depth and surface reflectance as a function of laser fluence. Reflectance increases rapidly once the PAI junction completely melts and effectively cuts off further light absorption. Thus, ELA process becomes self-limiting, and as a result, the junction depths stay flat over a wide range of laser fluence (see Fig. 1.25B).

Both N- and P-channel devices were fabricated according to two different flowcharts including absorber and PS layers [63, 64]. In the process named "dual PAI," the ELA simply replaced the RTA and was performed after sequential PAI and ion implantation of S/D extension and deep junction and deposition of absorber and PS layers; in the process named "disposable spacer," S/D deep junctions were formed and RTA annealed and then the sidewalls were removed to form S/D extensions, which were activated using the self-limiting ELA process. Dual PAI process required a larger laser fluence compared to the "disposable spacer" process, due to the insufficient heating of the S/D extension caused by the presence of the sidewall spacers. Devices with improved performance, if compared to those fabricated using conventional RTA, were obtained (see Fig. 1.26).

However, the introduction of the materials used for the absorber layers (typically based on Ti, W, or TiN [64]) into mass production fabs has been a major obstacle for the industrialization of ELA into CMOS technology and the interest shifted to nonmelting laser-based processes. In particular, laser spike annealing (LSA), based on the scanning of a $CO_2$ laser, provided uniform heating, with temperatures up to 1300°C and dwell times in the submillisecond range, without the use of additional layers. LSA process can be easily integrated into conventional MOSFET flow as a direct replacement of RTA, and excellent device performance was demonstrated for sub-30 nm technology [65–67].

### 1.4.2 Power MOS and RF/microwave applications

The feasibility and reliability of the ELA process in the melting regime were also tested in low-voltage power MOSFET [68, 69]. Even though the

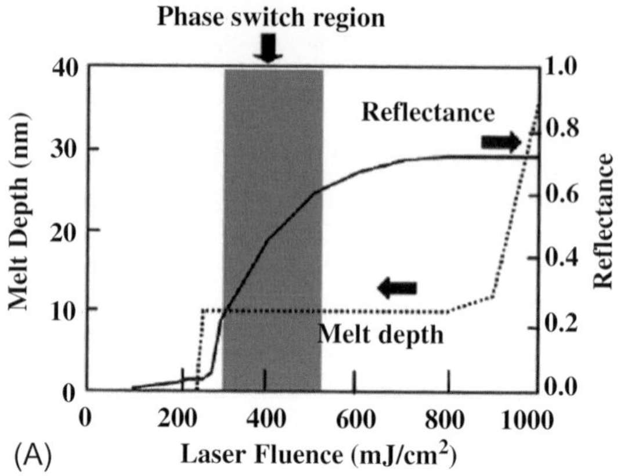


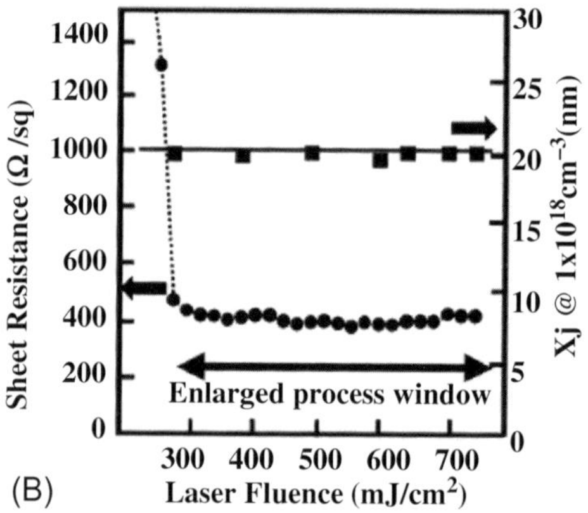


**FIG. 1.25** (A) Simulated melt depth and surface reflectance using SL-LTP. The reflectance increases rapidly once the PAI junction completely melts; (B) $X_j$ at which the profile has a concentration of $1 \times 10^{18}$ cm$^2$ and sheet resistance for B profiles as a function of laser power in the case of SL-LTP. *(After A. Shima, H. Ashihara, A. Hiraiwa, T. Mine, Y. Goto, Ultrashallow junction formation by self-limiting LTP and its application to sub-65-nm node MOSFETs, IEEE Trans. Electron Dev. 52 (6) (2005) 1165–1171.)*

semiconductor players were more concerned with the application of new methods to the CMOS technology, the use of such innovative annealing in the power field could open new perspectives for the VLSI power devices, which definitely needed suitable thermal processing as well. On the other hand, testing ELA on larger MOS structures did not, by any means, exclude its extension to ULSI, by exploiting opportune solutions, relatively to the structure and the process itself.

Fig. 1.27A shows a schematic of the power MOS structure. The source regions were implanted with As and laser annealing was applied to activate them. In order to improve the coupling of laser radiation with the source regions,

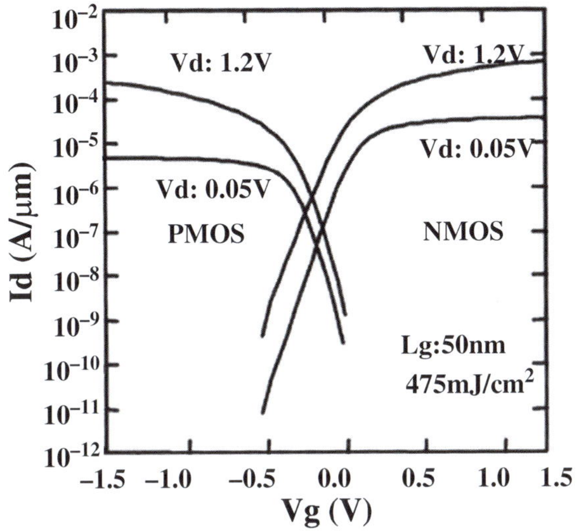


**FIG. 1.26** Subthreshold Id-Vg characteristics of nMOS and pMOS by SL-LTP: $V_d = 0.05$ and 1.2 V, $V_g = 1.2$ V, $L_g = 50$ nm. *(After A. Shima, H. Ashihara, A. Hiraiwa, T. Mine, Y. Goto, Ultra-shallow junction formation by self-limiting LTP and its application to sub-65-nm node MOSFETs, IEEE Trans. Electron Dev. 52 (6) (2005) 1165–1171.)*

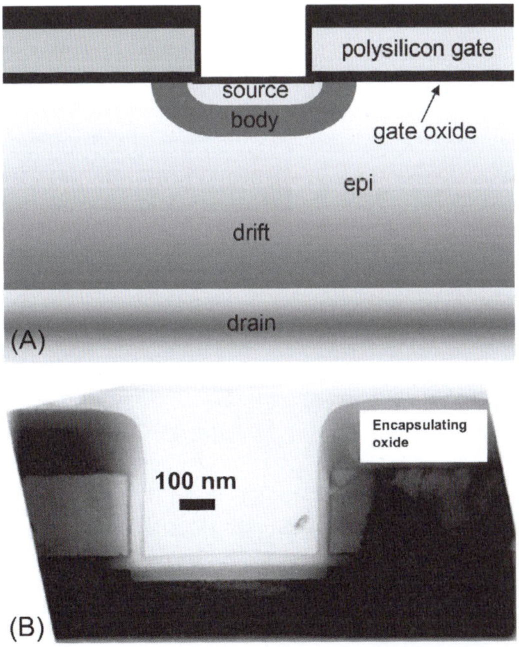


**FIG. 1.27** (A) Schematic of the MOS structure used for the experiments. (B) Cross-sectional TEM of the fabricated structure.

an oxide film was deposited, such that the reflectivity at 308 nm of the capped regions was reduced by the antireflective action of the oxide layer. The same oxide layer also covers the gate stack sidewalls, thus protecting the structure from mechanical deformations related to the melting of the polysilicon gate. A cross-sectional transmission electron microscopy (TEM) micrograph of the structure prior to irradiation is shown in Fig. 1.27B. For the As implant conditions (10 keV, $2 \times 10^{15}/cm^2$), an amorphous layer was generated, as clearly visible in the silicon region in between the polycrystalline silicon stripes. This made the PAI process step not necessary in the power MOS fabrication, as the source dopant implantation was simultaneously doping and amorphizing the Si substrate.

The laser apparatus used for irradiation was a Lambda Physik LPX 205 XeCl excimer laser (308 nm emitted wavelength, 30 ns pulse duration), delivering a homogeneous beam, with a variable spot size, that for our experiments was adapted to the chip size ($2.5 \times 3.5\ mm^2$), and with an energy density ($E_d$) that can be continuously changed, by using a beam attenuator, in the range 100–1200 $mJ/cm^2$.

An example of the carrier profiles as a function of depth, associated with the As doping after ELA, is reported in Fig. 1.28; they show a boxlike shape with a very high abruptness of 2.6 nm/dec. Due to the extremely high diffusivity of dopants in liquid Si, the As atoms are able to redistribute almost uniformly within the melted layer, up to the interface between the solid and liquid Si. The following rapid recrystallization from the liquid phase enhances dopant trapping and produces high electrical activation. The highest carrier

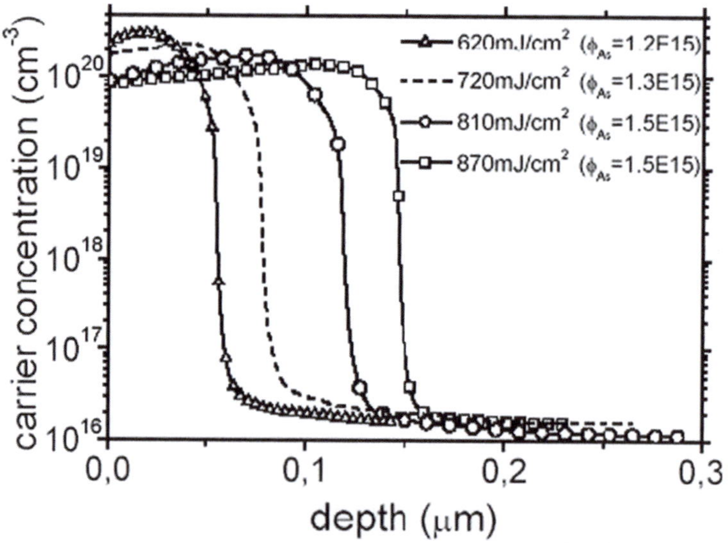


**FIG. 1.28** Carrier profiles, measured by SRP, after ELA at different energies of a 30 keV As.

concentration is achieved following shallow melting (620 mJ/cm$^2$) with a sheet resistance of 75 Ω/sq. Due to lateral heat dissipation occurring in the structured wafers, the junction depths obtained in such nonpatterned wafers can be rather different from the real device case. However, the one-dimensional analysis of doping obtained by laser annealing is useful to determine the shape of the dopant profiles and their evolution with the irradiation conditions.

While annealing the source layer, the remaining part of the doped layers composing the MOSFET, such as the boron-doped body, is not altered at all by the laser irradiation. This is due to the short duration of the process, of the order of tens of nanoseconds, as well as to the fast dissipation of the heat through the Si substrate, determining a thermal process in the solid phase of the underlying doped layers that lasts a few milliseconds, inducing a negligible diffusion not detectable by the analytical techniques.

In order to preserve the gate integrity from the morphologic modifications induced by the laser irradiation process, an oxide layer surrounding the polycrystalline silicon gate electrode was employed. The integrity of the complete device structure after laser irradiation is shown by the cross-sectional TEM analysis in Fig. 1.29, where the two-dimensional contour line of the doped region is identified by the white area in the implanted window; a shallower melting occurred at the edges of the implanted window, due to lateral heat loss and to the diffraction of the laser light through the narrow geometries of the patterned sample. The lateral diffusion underneath the gate oxide is negligible, so providing the possibility to shrink the channel length without incurring in SCEs.

The fabrication of a device implies the reliability of processing and patterning over a large scale, in order to ensure a high yield, necessary for a real industrialization of the method under investigation. A SEM image of a power MOSFET chip irradiated by laser is shown in Fig. 1.30, attesting the integrity of the structure subjected to ELA when operating under optimized conditions.

MOSFET wafers were fabricated according to the structure described in Fig. 1.27A, based on a process flow defined according to the indications coming from the experiments. The laser annealing process was performed through a

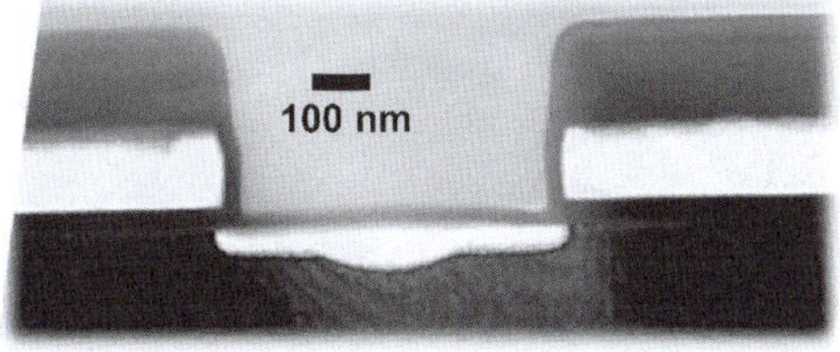


FIG. 1.29 MOSFET cross sections observed by TEM. Selective chemical etch exhibits the source (white region) region obtained by ELA of a 10 keV As implant.

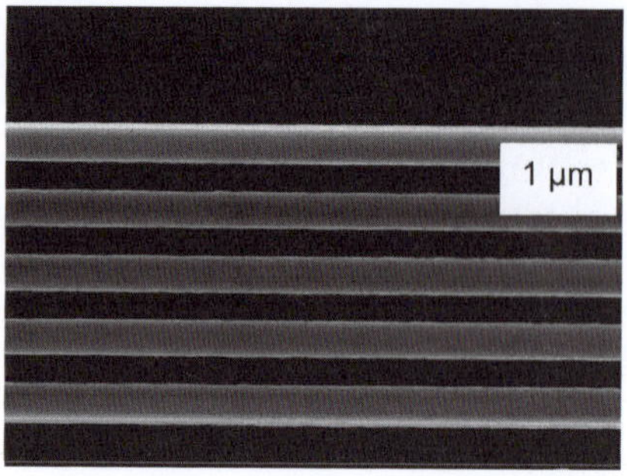


FIG. 1.30 SEM image of polycrystalline silicon gate stripes after ELA.

mask projection tool, in order to irradiate a single device with a uniform laser beam, avoiding beam tails.

Based on the evaluation of the best yield, as well as on process refinements, ELA-MOSFETs were fabricated, showing threshold voltage comparable with the conventionally processed device and, if driven at 10 V, the MOSFETs are completely switched on and the ON resistance is comparable with the standard one. Superior electrical performances could also be achieved by means of the ELA process [68]. In particular, the potential of the ELA process as a "zero lateral diffusion" thermal treatment was worth exploiting. MOSFETs with a shallower body layer and a shorter channel could be fabricated, aiming to a lower ON resistance. The "on" characteristics are shown in Fig. 1.31 for the 900 mJ/cm$^2$ case. Thanks to the reduction of the channel length and of the JFET component between adjacent body wells, made possible by the use of ELA, zero lateral diffusion was achieved, and a 10% decrease of the ON resistance of the

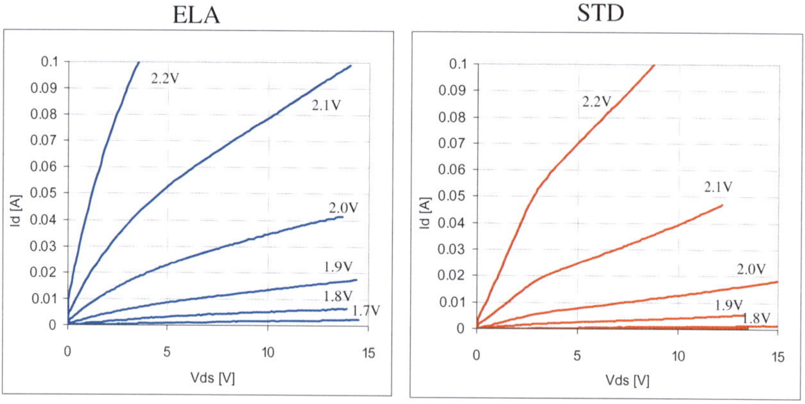


FIG. 1.31 Drain current increase for the ELA treated MOSFET, with respect to the device processed by conventional thermal processing. A significant increase at low current levels is clearly evidenced.

transistor was obtained at high current. This is a clear advantage of using ELA, revealing itself as a suitable process to scale the channel length.

Even though a satisfactory degree of reliability of the ELA-MOSFET was accomplished, further efforts were pursued in order to design a new gate stack structure, which could simplify the device fabrication and increase the ELA process window. A new concept was proposed, where the gate stack is protected by a multilayer mirror optimized at 308 nm, fabricated on top of the polygate, while the source regions were capped with 50 nm thick oxide layer, acting as antireflective coating [69].

The dielectric mirror (Bragg reflector) consisted of identical alternating layers of high and low refractive index, $n_h$ and $n_i$, with optical thickness equal to quarter wavelength: $n_h d_h = n_l d_l = 308$ nm/4.

It should be also pointed out that the dielectric mirror was designed using $SiO_2$ and $Si_3N_4$ layers, as these are compatible with IC processing.

In Fig. 1.32, a dielectric mirror formed of a sequence of six layers is shown, including an additional 100 nm $SiO_2$ layer deposited on top of the dielectric mirror structure itself, the latter not altering the dielectric mirror reflectivity (being the optical thickness of this layer, $n_l{}^*d$, equal to 308 nm/2).

In Fig. 1.33, the measured reflectivity is compared with simulations using two different values for the $Si_3N_4$ refractive index. In order to reproduce the measured data, a refractive index value of 1.9 is required, in agreement with that measured for PECVD-deposited $Si_3N_4$.

In Fig. 1.34, the TEM image of the structure with the dielectric mirror patterned on top of the gate stack, after ELA, is shown; the polygate was well confined and no outflows were present.

MOSFETs were fabricated by adopting the new structure for the gate stack and including a six-layer dielectric mirror.

The electrical results showed a great improvement in terms of gate-source isolation as well as yield with respect to the former structure [69], confirming the enhanced robustness of the process. This is due to the substantial increase in

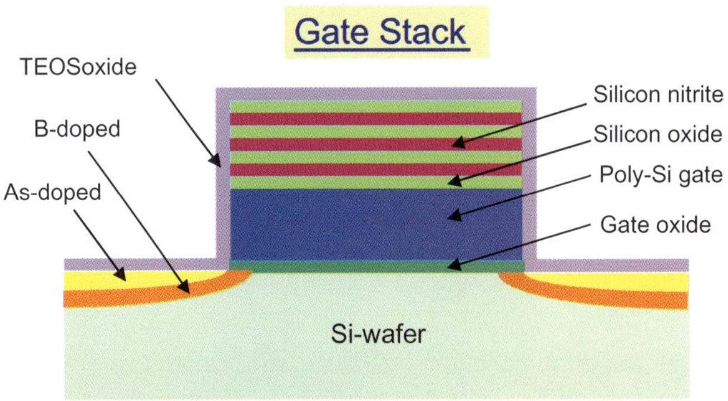


**FIG. 1.32** Gate stack structure with dielectric mirror.

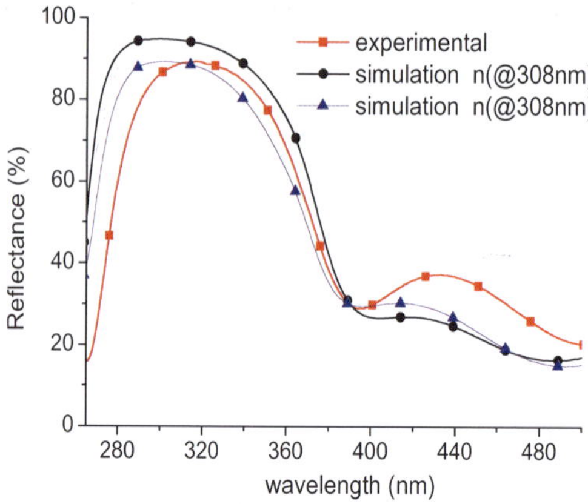


**FIG. 1.33** Experimental and simulated reflectivity of a dielectric mirror, including six alternating layers of $SiO_2$ and $Si_3N_4$ and a final 100 nm thick $SiO_2$ layer.

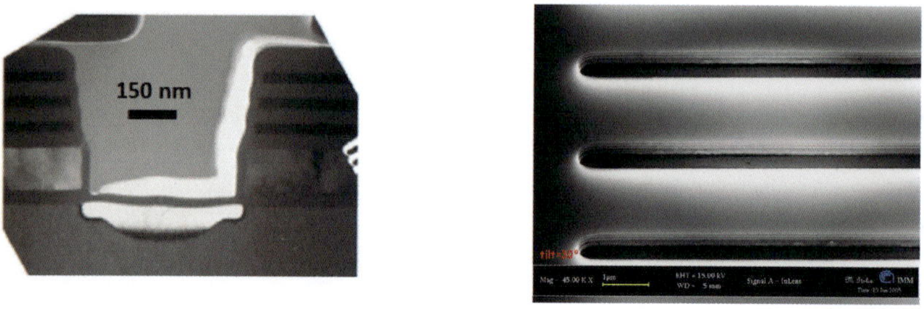


**FIG. 1.34** TEM and SEM images of irradiated source regions in the test structure with dielectric mirrors.

the ELA process window provided by the large difference between the reflectivity at 308 nm of the gate stack covered by the Bragg reflector (90%) and the source region covered by the AR coating (34%), so that less than one-sixth of the laser energy coupled into the source regions is absorbed by the gate stack. It should be pointed out that the Bragg reflector-based gate stack architecture can be applied to other device structures, such as scaled CMOS transistors, avoiding the use of absorber layers, which can represent an obstacle in process integration, and extending the possibilities of process integration of ELA, as discussed in Chapter 4.

Another interesting application of ELA to RF/microwave silicon devices was demonstrated by the group at the Delft Technical University. In particular, they showed ELA integration in a process that allows two-sided contacting of the devices: the back-wafer contacted silicon-on-glass (SOG) substrate-transfer technology (STT) [70]. In this technology, metal transmission lines can be placed on the low-loss glass substrate, while the resistive/capacitive parasitic of the silicon devices can be minimized by a direct two-[illegible] contacting. ELA was adopted to solve a major issue for the fabrication o[illegible]mic contacts on the back-wafer, i.e., the very limited thermal budge[illegible]ed in the back-wafer processing. The Delft TU group successfully c[illegible]ted sub-50 nm junction depth $p^+n$ and $n^+p$ diodes, formed by ELA in t[illegible]regime of $BF_2^+$ and $As^+$ implants, respectively, performed directly in[illegible]t windows. The latter are etched through a stack composed of a refle[illegible]sking layer deposited on a silicon oxide isolation layer. Diodes with [illegible] ideality factors in the range 1.06–1.16 were obtained for an isolation stack of 300 nm TEOS on 30 nm thermal oxide in combination with a 100 nm reflective Al layer [71]. The use of ELA allowed the formation of low-ohmic contact to the back of the diodes as well as contacts to the resistors in RF/microwave device fabricated using the SOG STT, leaving the underlying devices unaffected [70].

## 1.5 Conclusions

Since the introduction of pulsed laser annealing in the late 70s as a tool for semiconductor processing, the technique has been actively investigated and the historical evolution over more than 30 years has been reviewed in this chapter. Two major applications to the semiconductor industry have been considered in this chapter: the first dealing with the amorphous silicon transformation into polycrystalline silicon for the TFT technology and the second aiming to the dopant activation in CMOS technology for both logic and power MOS applications.

In the case of polysilicon technology, the advantages offered by ELA technology and described in Section 3 have guaranteed the establishment of industrial production. Today, several major companies, including Samsung, LG Display, CSOT, Tianma, Visionox and BOE, are producing AM-OLED displays, based on polysilicon TFT active matrix, mainly for mobile phone applications. This has boosted the technology of ELA systems, enabling unprecedented excimer laser performance (energy up to 6 Joule per pulse, max. power up to 3600 W at 600 Hz of repetition rate) and related optics (line beam up to 1500 mm).

In the case of CMOS applications, ELA in Si melting regime was shown to achieve extremely sharp doping profiles, in terms of depth and abruptness, with high electrical activation, and to suppress the lateral diffusion of the dopant, with consequent advantages for the reduction of the SCEs and the size scaling. A real advantage with respect to the other annealing methods could be established, with a significant impact on the requirements of the coming technology nodes. In spite of the ELA advantages, including the maturity of the ELA systems achieved through the polysilicon TFT technology, industrial production has not yet started. Indeed, several issues have prevented ELA integration, including melting of already formed structures, such as the polysilicon gate or metallization, melting under isolation, small process window, and possible deactivation, and diffusion of dopants during thermal steps subsequent to ELA. Several techniques, such as preamorphization of the ion-implanted layers, or the introduction of protective layers, phase-switch layers, or Bragg reflectors, have demonstrated to solve many of the issues above mentioned, at the expense of an increased complexity of the process. These technological achievements have paved the way to further developments and renewed interest in ELA, as thoroughly described in Chapter 4.

## References


[1] G.A. Kachurin, N.B. Pridaehin, L.S. Jmimov, Annealing of radiation-induced defects by pulsed laser irradiation, Fizika I Tekhnika Poluprovodnikov 9 (1975) 1429.

[2] I.B. Khaibullin, E.I. Shtyrkov, M.M. Zaripov, M.F. Galijautdinov, R.M. Bajazikov, Laser Annealing of Implanted Layers, VINITI N, 1974, pp. 2061–2074.

[3] B. Yu, Y. Wang, H. Wang, Q. Xiang, C. Riccobene, S. Talwar, M. Lin, 70 nm MOSFET with ultra-shallow, abrupt and super-doped S/D extension implemented by laser thermal process (LTP), in: Tech. Dig. Int. Electron Devices Meet, 1999, p. 509.

[4] Y.F. Chong, K.L. Pey, A.T.S. Wee, A. See, L. Chan, Y.F. Lu, W.D. Song, L.H. Chua, Annealing of ultrashallow p+/n junction by 248 nm excimer laser and rapid thermal processing with different preamorphization depths, Appl. Phys. Lett. 76 (2000) 3197.

[5] R.F. Wood, G.A. Geist, Modeling of nonequilibrium melting and solidification of laser-irradiated materials, Phys. Rev. B 34 (1986) 2606.

[6] M.O. Thompson, G.J. Galvin, J.W. Mayer, P.S. Peercy, J.M. Poate, D.C. Jacobson, A.G. Cullis, N.G. Chew, Melting temperature and explosive crystallization of amorphous silicon during pulsed laser irradiation, Phys. Rev. Lett. 52 (1984) 2360.

[7] J.S. Im, H.J. Kim, M.O. Thompson, Phase transformation mechanisms involved in excimer laser crystallization of amorphous silicon films, Appl. Phys. Lett. 63 (1993) 1969.

[8] F.C. Voogt, R. Ishihara, A combined TEM and time-resolved optical reflectivity investigation into the excimer-laser crystallization of a-Si films, Thin Solid Films 383 (2001) 45–47.

[9] M. Hatano, S. Moon, M. Lee, K. Suzuki, C.P. Grigoropoulos, Excimer laser-induced temperature field in melting and resolidification of silicon thin films, J. Appl. Phys. 87 (2000) 36.

[10] R.F. Wood, C.W. White, R.T. Young, in: R.K. Willardson, A.C. Beer (Eds.), Semiconductors and Semimetals, vol. 23, Academic Press, Inc, Cambridge, MA, 1984.

[11] Y.F. Chong, K.L. Pey, Y.F. Lu, A.T.S. Wee, T. Osipowicz, H.L. Seng, A. See, J.Y. Dai, Liquid-phase epitaxial growth of amorphous silicon during laser annealing of ultrashallow p+/n junctions, Appl. Phys. Lett. 77 (2000) 2994.

[12] Y.F. Chong, K.L. Pey, A.T.S. Wee, T. Osipowicz, A. See, L. Chan, Control of transient enhanced diffusion of boron after laser thermal processing of preamorphized silicon, J. Appl. Phys. 81 (2002) 4048.

[13] T. Sameshima, S. Usui, M. Sekiya, XeCl Excimer laser annealing used in the fabrication of poly-Si TFT's, IEEE Electron Device Lett. 7 (1986) 276.

[14] T. Sameshima, M.H.A.S. Usui, XeCl excimer laser annealing used to fabricate poly-Si TFT's, Jpn. J. Appl. Phys. 28 (1989) 1789.

[15] K. Sera, F. Okumura, H. Uchida, S. Itoh, S.K.A.K. Hotta, High-performance TFTs fabricated by XeCl excimer laser annealing of hydrogenated amorphous-silicon film, IEEE Trans. Electron. Dev. 36 (1989) 2868.
[16] R.Z. Bachrach, K. Winer, J.B. Boyce, S.E. Ready, R.I.J.A.B. Anderson, Low temperature crystallization of amorphous silicon using an excimer laser, J. Electron. Mater. 19 (1990) 241.
[17] H. Kuriyama, S. Kiyama, S. Noguchi, T. Kuwahara, S. Ishida, T. Nohda, K. Sane, H. Iwata, S. Tsuda, S. Nakano, High mobility poly-Si TFT by a new excimer laser annealing method for large area electronics, IEEE IEDM 91 (1991) 563.
[18] K. Shimizu, H. Horoya, O.S.A.M. Matsumura, High-mobility bottom-gate thin-film transistors with laser-crystallized and hydrogen-radical-annealed polysilicon films, Jpn. J. Appl. Phys. 30 (1991) 3704.
[19] S.D. Brotherton, D.J. McCulloch, J.B. Clegg, J.P. Gowers, Excimer-laser-annealed poly-Si thin-film transistors, IEEE Trans. Electron. Dev. 40 (1993) 40.
[20] G. Fortunato, A. Pecora, L. Maiolo, Polysilicon thin-film transistors on polymer substrates, Mater. Sci. Semicond. Process. 15 (2012) 627–641.
[21] K. Goto, T. Yamamoto, T. Kubo, M. Kase, Y. Wang, T. Lin, S. Talwar, T. Sugii, Ultra-low contact resistance for deca-nm MOSFETs by laser annealing, in: Tech. Dig. Int. Electron Devices Meet, 1999, p. 931.
[22] S.-D. Kim, C.-M. Park, J.C.S. Woo, Advanced source/drain engineering for box-shaped ultra-shallow junction formation using laser annealing and pre-amorphization implantation in sub-100-nm SOI CMOS, IEEE Trans. Electron Dev. 49 (10) (2002) 1748–1754.
[23] M. Current, D. Lopes, M. Foad, W. Boyd, Ultra-shallow junction technology for 100 nm CMOS: xR LEAP implanter and RTP-centura rapid thermal annealer, Mater. Chem. Phys. 54 (1–3) (1998) 33–36.
[24] A. Claverie, L.F. Giles, M. Omri, B.D. Manduit, G.B. Assayag, A.D. Mathiot, Nucleation, growth and dissolution of extended defects in implanted Si: impact on dopant diffusion, Nucl. Instrum. Methods Phys. Res. B; Beam Interact. Mater. At. 147 (1–4) (1999) 1–12.
[25] R.F. Wood, J.R. Kirkpatrick, G.E. Giles, Macroscopic theory of pulsed laser annealing, Phys. Rev. B; Condens. Matter 23 (6) (1981) 2923–2942.
[26] H. Tsukamoto, H. Yamamoto, T. Noguchi, H. Masuya, T. Suzuki, Improved characteristics of p(+)-n junctions formed by excimer laser annealing with low temperature pre-annealing, Jpn. J. Appl. Phys. 35 (7) (1996) 3810–3813.
[27] S. Earles, M. Law, R. Brindos, K. Jones, S. Talwar, A.S. Corcoran, Nonmelt laser annealing of 5-KeV and 1-KeV boron-implanted silicon, IEEE Trans. Electron Dev. 49 (7) (2002) 1118–1123.
[28] N.G. Basov, V.A. Danilychev, Y. Popov, D.D. Khodkevich, Laser for vacuum region of the spectrum with excitation of liquid xenon by an electron beam, Zh. Eksp. Fiz. 1970 (12) (1970) 473–474.
[29] J.S. Im, H.J. Kim, On the super lateral growth phenomenon observed in excimer laser-induced crystallization of thin Si films, Appl. Phys. Lett. 64 (1994) 2303–2305.
[30] H. Kuriyama, T. Nohda, S. Ishida, T. Kuwahara, S. Noguchi, S. Kiyama, S. Tsuda, S. Nakano, Lateral grain growth of poly-Si films with a specific orientation by excimer laser annealing method, Jpn. J. Appl. Phys. 32 (1993) 6190–6195.
[31] C.-C. Kuo, Characterization of polycrystalline silicon thin film fabricated by excimer laser crystallization, J. Russ. Laser Res. 28 (2007) 383–392.
[32] J.J.P. Bruines, R.P.M.V. Hal, H.M.J. Boots, Time-resolved reflectivity measurements during explosive crystallization of amorphous silicon, Appl. Phys. Lett. 49 (1986) 1160.

[33] A. Polman, S. Roorda, P.A. Stolk, W.C. Sinke, Triggering explosive crystallization of amorphous silicon, J. Cryst. Growth 108 (1–2) (1991) 114–120.
[34] L. Mariucci, A. Pecora, G. Fortunato, C. Spinella, C. Bongiorno, Crystallization mechanisms in laser irradiated thin amorphous silicon films, Thin Solid Films 427 (2003) 91–95.
[35] Q. Hu, C.S. Lee, T. Li, Y. Deng, U. Chung, A.B. Limanov, A.M. Chitu, M. Thompson, A.J.S. Im, Excimer-laser-induced melting and solidification of PECVD a-Si films under partial-melting conditions, Mater. Res. Soc. Symp. Proc. Vol. 1321 (2011) 197–202.
[36] S. Hazair, P.C.V.D. Wilt, Y. Deng, U.-J. Chung, A.B. Limanov, A.J.S. Im, Nucleation-initiated solidification of thin Si films, Mater. Res. Soc. Symp. Proc. Vol 979 (2007).
[37] C.-C. Kuo, A reliable approach to a rapid calculation of the grain size of polycrystalline thin films after excimer laser crystallization, Mater. Technol. 49 (5) (2015) 687–691.
[38] D.H. Lowndes, S.J. Pennycook, G.E. Jellison, S.P. Withrow, A.D.N. Mashburn, Solidification of highly undercooled liquid silicon produced by pulsed laser melting of ion-implanted amorphous silicon: time resolved and microstructural studies, J. Mater. Res. 2 (1987) 648–680.
[39] R. Ishihara, A. Burtsev, P.F.A. Alkemade, Location-control of large Si grains by dual-beam excimer-laser and thick oxide portion, Jpn. J. Appl. Phys. 39 (2000) 3872–3878.
[40] D.K. Fork, G.B. Anderson, J.B. Boyce, R.I. Johnson, P. Mei, Capillary waves in pulsed excimer laser crystallized amorphous silicon, Appl. Phys. Lett. 68 (1996) 2138–2140.
[41] H.S. Cho, D. Kim, A.B. Limanov, M.A. Crowder, A.J.S. Im, Sequential lateral solidification of ultra-thin a-Si films, Mater. Res. Soc. Symp. Proc. 621 (2001) Q9.9.1.
[42] S.D. Brotherton, D.J. McCulloch, J.P. Gowers, J.R. Ayres, A.M.J. Trainor, Influence of melt depth in laser crystallized poly-Si thin film transistor, J. Appl. Phys. 82 (8) (1997) 4086–4094.
[43] J.S. Im, M.A. Crowder, R.S. Sposili, J.P. Leonard, H.J. Kim, J.H. Yoon, V.V. Gupta, H. Jin Song, H.S. Cho, Controlled super-lateral growth of Si films for microstructural manipulation and optimisation, Phys. Stat. Sol. 166 (2) (1998) 603–617.
[44] G. Fortunato, L. Mariucci, R. Carluccio, A. Pecora, V. Foglietti, Excimer laser crystallization techniques for polysilicon TFTs, Appl. Surf. Sci. 154–155 (2000) 95–104.
[45] O.H. Chang-Ho, M. Ozawa, M. Matsmura, A novel phase-modulated excimer-laser crystallization method of silicon thin films, Jpn. J. Appl. Phys. 37 (1998) L492–L495.
[46] P.C. Van der Wilt, B.D. van Dijk, G.J. Bertens, R.M. Ishihara, C.I. Beenakker, Formation of location-controlled crystalline islands using substrate-embedded seeds in excimer-laser crystallization of silicon films, Appl. Phys. Lett. 79 (12) (2001) 1819–1821.
[47] M. Cuscunà, A. Bonfiglietti, R. Carluccio, L. Mariucci, F. Mecarini, A. Pecora, M. Stanizzi, A. Valletta, G. Fortunato, A novel fabrication process for polysilicon thin film transistors with source/drain contacts formed by deposition and lift-off of highly doped layers, Solid-State Electron. 46 (2002) 1351–1358.
[48] R.S. Sposili, J.S. Im, Sequential lateral solidification of thin silicon films on $SiO_2$, Appl. Phys. Lett. 69 (19) (1996) 2864–2866.
[49] L. Mariucci, A. Pecora, R. Carluccio, G. Fortunato, Advanced excimer laser crystallization techniques, Thin Solid Films 383 (2001) 39–44.
[50] Y. Taniguchi, T. Katou, M. Hiramatsu, M. Matsumura, New phase modulators for next generation low-temperature crystallization method of Si films, Jpn. J. Appl. Phys. 45 (10A) (2008) 7753–7760.
[51] V. Rana, R. Ishihara, Y. Hiroshima, D. Abe, S. Inoue, T. Shimoda, W. Metselaar, K. Beenakker, Dependence of single-crystalline Si TFT characteristics on the channel position inside a location-controlled grain, IEEE Trans. ED 52 (12) (2005) 2622–2628.
[52] P. Legagneux, F. Petinot, O. Huet, F. Plais, C. Reita, D. Pribat, R. Carluccio, A. Pecora, L. Mariucci, G. Fortunato, High quality polycrystalline thin films transistors made by excimer

laser crystallisation, in: ESSDERC '96: Proceedings of the 26th European Solid State Device Research Conference, Bologna, Italy, 1996, 1996, pp. 1071–1074.

[53] A. Voutsas, Assessment of the performance of laser-based lateral-crystallization technology via analysis and modeling of polysilicon thin-film-transistor mobility, IEEE Trans. Electron Dev. 50 (2003) 1494.

[54] M. Mitani, T. Endo, Y. Taniguchi, T. Katou, S. Shimoto, T. Ohno, S. Tsuboi, T. Okada, K. Azuma, G. Kawachi, M. Matsumura, Ultrahigh-performance polycrystalline silicon thin-film transistors on excimer-laser-processed pseudo-single-crystal films, Jpn. J. Appl. Phys. 47 (2008) 8707–8713.

[55] A. Baiano, R. Ishihara, J.V.D. Cingel, K. Beenakker, Strained single-grain silicon n- and p-channel thin-film transistors by excimer laser, IEEE Trans. Electron Dev. 31 (2010) 308.

[56] N. Saputra, M. Danesh, A. Baiano, R. Ishihara, J.R. Long, N. Karaki, S. Inoue, An assessment of μ-Czochralski, single-grain silicon thin-film transistor technology for large-area, sensor and 3-D electronic integration, IEEE J. Solid-State Circ. 43 (2008) 1563.

[57] A. Pecora, L. Maiolo, M. Cuscuna, D. Simeone, L. Minotti, L. Mariucci, G. Fortunato, Low-temperature polysilicon thin film transistors on polyimide substrates for electronics on plastic, Solid State Electron. 52 (2008) 348–352.

[58] S. An, J. Lee, Y. Kim, T. Kim, D. Jin, H. Min, H. Chung, S.S. Kim, 2.8-Inch WQVGA Flexible AMOLED Using High Performance Low Temperature Polysilicon TFT on Plastic Substrates, SID 10 Digest, 2010, pp. 706–709.

[59] M. Kim, H. Chung, J. Lee, Y. Park, S. An, T. Kim, D. Jin, H. Min, C. Yu, S. Kim, J. Jang, World-Best Performance LTPS TFTs With Robust Bending Properties on AMOLED Displays, SID 11 Digest, 2011, pp. 194–197.

[60] S. Talwar, S. Felch, D. Downey, Y. Wang, Study of laser thermal processing (LTP) to meet sub 130 nm node shallow junction requirements, in: 2000 International Conference on Ion Implantation Technology Proceedings. Ion Implantation Technology, 2000, pp. 175–177. vol. Alpbach, Cat. No. 00EX432.

[61] S. Talwar, Y. Wang, C. Gelatos, Laser thermal processing (LTP) for fabrication of ultra-shallow, Proc. Electrochem. Soc. 9 (2000) 95.

[62] K. Pey, P. Lee, Pulsed laser annealing technology for nano-scale fabrication of silicon-based devices in semiconductors, in: J.R. Lawrence (Ed.), Advances in Laser Materials Processing: Technology, Research and Applications, Woodhead Publishing, 2017 (Chapter 12).

[63] A. Shima, H. Ashihara, T. Mine, Y. Goto, M. Horiuchi, Y. Wang, S. Talwar, A. Hiraiwa, Self-limiting laser thermal process for ultra-shallow junction formation of 50-nm gate CMOS, in: IEEE International Electron Devices Meeting 2003, Washington, DC, USA, 2003, pp. 20.4.1–20.4.4.

[64] A. Shima, H. Ashihara, A. Hiraiwa, T. Mine, Y. Goto, Ultrashallow junction formation by self-limiting LTP and its application to sub-65-nm node MOSFETs, IEEE Trans. Electron Dev. 52 (6) (2005) 1165–1171.

[65] A. Shima, T. Mine, K. Torii, A. Hiraiwa, Enhancement of drain current in planar MOSFETs by dopant profile engineering using nonmelt laser spike annealing, IEEE Trans. Electron Dev. 54 (11) (2007) 2953–2959.

[66] T. Yamamoto, T. Kubo, T. Sukegawa, K. Hashimoto, M. Kasee, Advanced junction profile engineering featuring laser spike annealing and co-implantation for sub-30-nm strained CMOS devices, in: Symposium on VLSI Technology, 2006, Digest of Technical Papers., Honolulu, HI, 2006, pp. 186–187.

[67] T. Yamamoto, T. Kubo, T. Sukegawa, E. Takii, Y. Shimamune, N. Tamura, T. Sakoda, M. Nakamura, H. Ohta, T. Miyashita, H. Kurata, S. Satoh, M. Kase, T. Sugii, Junction profile

engineering with a novel multiple laser spike annealing scheme for 45-nm node high performance and low leakage CMOS technology, in: IEEE International Electron Devices Meeting, Washington, DC, 2007, pp. 143–146.
[68] V. Privitera, et al., Integration of melting excimer laser annealing in power MOS technology, IEEE Trans. Electron Dev. 54 (4) (2007) 852–860.
[69] G. Fortunato, et al., Bragg reflector based gate stack architecture for process integration, Appl. Phys. Lett. 89 (2006) 253502.
[70] L.K. Nanver, et al., Improved RF devices for future adaptive wireless systems using two-sided contacting and AlN cooling, IEEE J. Solid-State Circ. 44 (9) (2009) 2322–2338.
[71] V. Gonda, A. Burtsev, T.L.M. Scholtes, L.K. Nanver, Nearideal implanted shallow-junction diode formation by excimer laser annealing, Proc. IEEE RTP 13 (2005) 93–100.

Chapter 2

# Laser-matter interactions

Spyros Stathopoulos and Dimitris Tsoukalas
*Department of Applied Mathematical and Physical Sciences, National Technical University of Athens, Athens, Greece*

## 2.1 Introduction

The introduction of laser radiation since its original inception has transformed multiple areas of manufacturing and processing. The directionality and monochromaticity of the laser beam combined with exceptionally high-power density and multiple available wavelengths have positioned lasers as an extremely versatile tool for localized processing. These very particular properties allow for fine-tuning the thermal interaction depth as well as the absorption ratio from a given material. On the other hand, CMOS processing has evolved to a point that very precise control of the dopant diffusion and activation is required to accommodate sub-10 nm FET channels. It is, therefore, natural to combine the localized heating effect of laser radiation to thermally activate dopants in semiconductors in a diffusionless manner. From that point of view, this chapter aims to present the laser-semiconductor interaction and emphasize the case of an infrared laser, where heat distribution is studied in detail and an application example for silicon technology is given.

In this chapter, we deal first with the basics of laser-matter interaction, which are categorized into physical and chemical. Physical interactions are a direct result of heating (e.g., melting, diffusion, etc.), whereas chemical interactions result in a direct change in the chemical properties of the material. In semiconductors (and by extension to insulators), laser radiation can lead to a multitude of effects ranging from extraband carrier transition to two-phonon absorption [1]. This is in stark contrast with the case of metals, where laser-matter interaction is mostly a surface-localized phenomenon [2] and the most prevalent result is intraband transitions of carriers. Following that we concern ourselves with the particulars of laser absorption from semiconductors as well as the dependency of absorption coefficient on the incident wavelength. Depending on the energy of the photons, absorption from the semiconductor can be either direct (a photon with $eV > E_g$ is absorbed directly) or indirect (through free carrier absorption) as we move from the deep ultraviolet to the

Laser Annealing Processes in Semiconductor Technology. https://doi.org/10.1016/B978-0-12-820255-5.00004-0

far infrared. As a result, heating of the material can occur through different processes leading to radically different absorption lengths as well as thermalization procedures. In the case of doped semiconductors, absorption coefficient is also strongly dependent on the extrinsic doping level [3]. The particular characteristics of laser-semiconductor interaction as well as the overall extremely steep temperature gradients that can be achieved with laser heating can have direct implications to semiconductor processing. More specifically, lasers are well suited for the fabrication of shallow junctions in group IV semiconductors. Due to the constraints of modern CMOS processing, the realization of diffusionless junctions has evolved to a primary concern in order to avoid issues of subthreshold leakage and drain-induced barrier lowering. As such laser radiation can be beneficial for this kind of processing as it can achieve high-level activation and reconstruction of amorphization with minimal diffusion while remaining in the submelt regime both in silicon [4, 5] and germanium [6]; a combination of effects that is otherwise not directly achievable with other means of annealing.

Finally, it is important to mention that throughout the course of this chapter it is assumed that the size of the irradiated surfaces is much larger than the spot size of the beams involved so that it is possible to ignore fringe effects that may arise due to geometric irregularities of the target.

## 2.2 Absorption of electromagnetic radiation

soLight interacts with matter with elementary excitations, which include electronic excitations, such as interband and intraband transitions, excitons, plasmons, etc., as well as excitation of phonons, polaritons, etc. High-power density laser beams result in the creation of extremely large numbers of electronic excitations, either due to thermal or nonthermal effects. The number of such transitions may exceed $10^{22}$ cm$^{-3}$. The combination of many elemental excitations can lead to simultaneous emergence of new phenomena, such as thermal runaway in metals and semiconductors (i.e., positive feedback leading to exponential increase in temperature). By increasing power density nonlinear optical phenomena arise while further increase leads to plasma formation.

Depending on the energy of the photons impinging the material the energy absorption and redistribution can be mainly realized by the following ways: (a) collisions between free carriers; (b) plasmon excitation; (c) electron-hole recombination (Auger phenomenon); (d) electron-hole generation due to ionization; and (e) photon emission (electron-phonon interaction). Fig. 2.1 summarizes some of the key electronic excitations generated by laser radiation on a solid material. Interband transitions from the valence band to the conduction band can occur if the energy of the incident photon, $h\nu$, is greater than the energy gap, $E_g$. In this process, electron-hole pairs are created. Interband transitions are usually in the near-infrared or visible spectrum for the semiconductors and in the ultraviolet for the insulating materials. The presence of defects in the

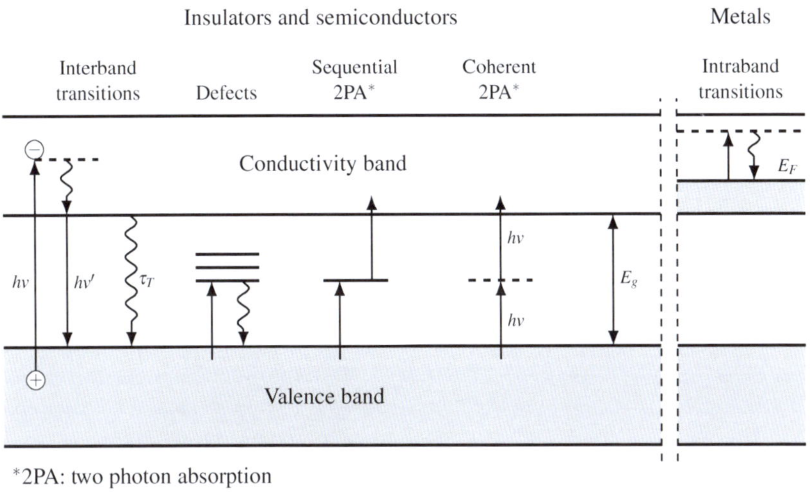


**FIG. 2.1** Typical electronic photon interactions with matter. *Continuous lines* indicate absorption or emission of photons, *hv*, and *wavy lines* indicate nonradiative transitions. *(From D.W. Bäuerle, Laser Processing and Chemistry, fourth ed., Springer Science & Business Media, 2013.)*

material creates energy states within the energy gap, which can lead to transitions in the conduction band with energies $hv < E_g$, mainly in high-power beams through two-photon absorption. Intraband transitions occur primarily in metals because the Fermi, $E_F$, level in metals is within the conduction band. In the case of metals, the radiation is absorbed by the conduction band electrons within a thickness of about 10 nm. The time between two subsequent collisions of a conduction band electron in the metal, e, is in the order of $10^{-14}$ to $10^{-12}$ s. The interaction between electrons and phonons, e p, usually takes longer than $10^{-12}$ to $10^{-10}$ s. Apart from the phonons, there is no other interaction of the excited electrons with the lattice. Since the absorption of metals is mainly a surface phenomenon, the optical properties of metals change dramatically when the surface characteristics are changed, which may be different from the bulk material. Such extreme cases are particularly evident in very thin metal films or metal nanoparticle films [2].

In semiconductors, interband transitions require significantly longer times of $10^{-12}$ to $10^{-6}$ s, especially in the case of transitions associated with defects. Depending on the wavelength of the incident radiation, the interband transition can be made either directly if the energy offered is greater than the energy gap, or by absorption of the free carriers.

The first case involves laser irradiation whose wavelength is within the optical and ultraviolet spectrum or lower (e.g., excimer). The very large energy of the photons ensures the absorption of optical energy through interband transitions creating direct electron-hole pairs. Because the quantum efficiency for interband absorption is close to unity, it is possible to generate very high carrier densities. For relatively high doping levels $N_c > 10^{18}$ cm$^{-3}$, the carrier recombination mechanisms are dominated by the Auger process. The energy transfer time to the lattice is of the order of a few picoseconds. This also has a direct impact on the absorption coefficient of silicon, which shows small dependence on both temperature and doping levels. The very high value of the absorption coefficient guarantees that the radiation energy will be converted to heat within the first nanometers near the surface. This amount of heat will not penetrate in large depth, since the average length of thermal diffusion is around 900 nm. Therefore, the energy distribution with the depth of the material will be exponentially decreasing, due to the form relating the decay of the beam within the volume of the material (Beer-Lambert's law), while the temperature distribution at each time slot will have a similar form (but of course more expanded). Short pulse durations will subsequently lead to a relatively adiabatic material heating. As such, the length, $l_\alpha$, required to attenuate the radiation to $1/e$ of its value is *much* smaller than the thermal diffusion length $l_T$. Assuming the size of the beam is much larger than the sum of these two values we can consider the radiation to be limited to the surface and can, therefore, be represented by a heat source located at the surface of the target. For silicon and germanium, this approach is valid for laser with wavelengths of less than 500 nm and pulse durations in the sub-μs regime.

In the second case, which is of most interest to us in this chapter, radiation from near-infrared laser (Nd:YAG) or deep infrared radiation such as the $CO_2$ laser is used as an example. We will then focus on how free carriers contribute to the optical absorption process at infrared wavelengths. Thus, in the context of this view, the absorption coefficient for a temperature, $T$, can generally be assumed to be [4]

$$\alpha(T) = \alpha_0 + N_{eh}(T)\sigma(T) \tag{2.1}$$

where $\alpha_0$ is the absorption coefficient of the lattice, $N_{eh}(T)$ is the density of the electron-hole pairs, and $\sigma$ is the effective cross-section absorption of such a pair. The lattice absorption coefficient is a fixed number for a given temperature and in the case of silicon is about 2 cm$^{-1}$ [5], whereas for the case of germanium varies from $10^{-4}$ to $10^{-1}$ cm$^{-1}$ depending on the temperature [6]. In any case, the absorption coefficient of the lattice contributes significantly less than the absorption from the carriers, especially in the case of heavily doped semiconductors and at the high temperatures, which are relevant during annealing. The effective cross-section, $\sigma(T)$, can be defined based on the Drude model as

$$\sigma(T) = \frac{q^2}{\epsilon_0 n c \omega^2}\left(\frac{1}{m_e^* \tau_e(T)} + \frac{1}{m_h^* \tau_h(T)}\right) \tag{2.2}$$

where $q$ is the elemental charge, $\epsilon_0$ is the dielectric constant of the vacuum, $n$ is the refractive index of the material, $c$ is the speed of light, and $\omega$ is the circular frequency, $2\pi\nu$. Parameters $m_{e,h}^*$ refer to the effective masses of electrons and holes, while $\tau_{e,h}$ are the collision times for electrons and holes [7]. Both the collision times, $\tau_{e,h}$, and carrier density, $N_{eh}$, are temperature dependent. For the density of electron-hole pairs, the general expression is given by the relation

$$N_{eh}(T) = 2\left(m_e^* m_h^*\right)^{3/4}\left(\frac{kT}{2\pi\hbar}\right)^{3/2}\exp\left(-\frac{E_g}{2kT}\right) \tag{2.3}$$

In the case of impurity-doped semiconductors, the term $N_{eh}$ can be broken down into two subterms, namely the *concentration of extrinsic carriers* due to implanted impurities, $N$, and the *intrinsic carrier density* $n_i(T)$, which is temperature dependent [8].

$$N_{eh}(T) = N + n_i(T) \tag{2.4}$$

For silicon, an estimate of the intrinsic carrier density is given by expression [8, 9]

$$n_i(T) = 3.87\times 10^{16} T^{3/2}\exp\left(-\frac{7020}{T}\right)\mathrm{cm}^{-3} \tag{2.5}$$

where $T$ is the temperature in Kelvin. In the case of germanium, we can also get the corresponding expression as per [9]

$$n_i(T) = 1.76\times 10^{16} T^{3/2}\exp\left(-\frac{4560}{T}\right)\mathrm{cm}^{-3} \tag{2.6}$$

A graphical representation of the concentration of free carriers for the two semiconductors can be seen in Fig. 2.2. For room temperature, the equations give the characteristic values of the intrinsic concentration, that is, $9\times 10^9$ cm$^{-3}$ and $2\times 10^{13}$ cm$^{-3}$ for silicon and germanium, respectively.

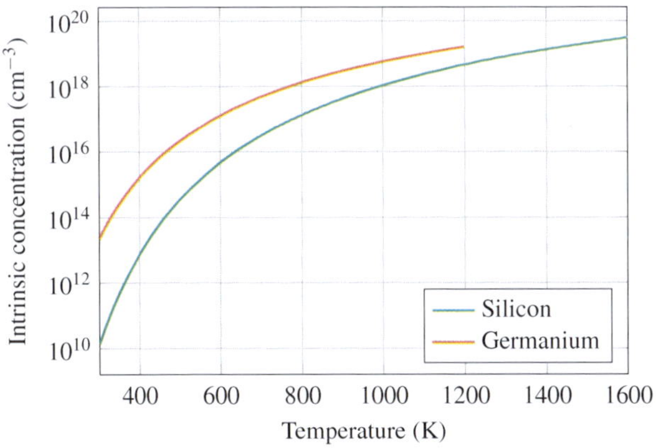


**FIG. 2.2** Intrinsic carrier concentration dependence on temperature for silicon and germanium based on Eqs. (2.5), (2.6).

Regarding the effective absorption cross-section of these free carriers, it may be initially estimated by the expression for a given wave number of the incident radiation, $\tilde{\nu}$, as [5]

$$\sigma(\tilde{\nu},T)=C(\tilde{\nu})T^{3/2} \quad (2.7)$$

Coefficient $C(\tilde{\nu})$ can be derived from theoretical calculations. For instance, for a wavelength of 10.6 μm ($\tilde{\nu}=943.4\ \mathrm{cm}^{-1}$) for a $CO_2$ laser, we have for the case of silicon [8, 10]:

$$\sigma_{\mathrm{Si}}(943.4\mathrm{cm}^{-1},T)=1.9\times10^{-19}T^{3/2}\mathrm{cm}^2 \quad (2.8)$$

while similar is the case for germanium [11]

$$\sigma_{\mathrm{Ge}}(943.4\mathrm{cm}^{-1},T)=1.15\times10^{-19}T^{3/2}\mathrm{cm}^2 \quad (2.9)$$

So based on the relations (2.1), (2.5), and (2.8), we can derive a relationship for the absorption coefficient of silicon for that specific wavelength as

$$\alpha_{Si}(T)=2+\left[1.9\times10^{-19}T^{3/2}\right]\times\left[N+3.87\times10^{16}T^{3/2}\exp\left(-\frac{7020}{T}\right)\right] \quad (2.10)$$

and, respectively, for germanium

$$\alpha_{Ge}(T)=1\times10^{-3}+\left[1.15\times10^{-19}T^{3/2}\right]\times\left[N+1.76\times10^{16}T^{3/2}\exp\left(-\frac{4560}{T}\right)\right] \quad (2.11)$$

with $T$ being the temperature in Kelvin and $N$ is the extrinsic doping concentration in $\mathrm{cm}^{-3}$.

Fig. 2.3 shows a representation of Eqs. (2.10), (2.11) up to the melting temperature of each semiconductor. An apparent hyperexponential dependence of the absorption coefficient on the temperature can be observed as well as the strong influence of extrinsic impurities. At very low doping levels, lattice absorption is prevalent in both cases, while the absorption coefficient rises sharply from around 700 K in silicon and from about 600 K in germanium due to the strong generation of electron-hole pairs (Eqs. 2.5, 2.6). When the extrinsic doping highly exceeds by far the intrinsic electron-hole pair concentration it mainly also determines the absorption of the material. The rate of increase of the absorption coefficient with temperature at such high doping levels is lower, though the absolute value of the coefficient itself is significantly higher. That plays an important role in the absorption of laser radiation in cases, where doping is pronounced and abrupt in the first few nm of the semiconductor as it is the case of low-energy and plasma implants.

As mentioned in the case of high doping, generally over $10^{17}\ \mathrm{cm}^{-3}$, the collisions between the carriers dominate over the ones with the crystal. The carriers in this case behave collectively in the form of plasma and the absorption of

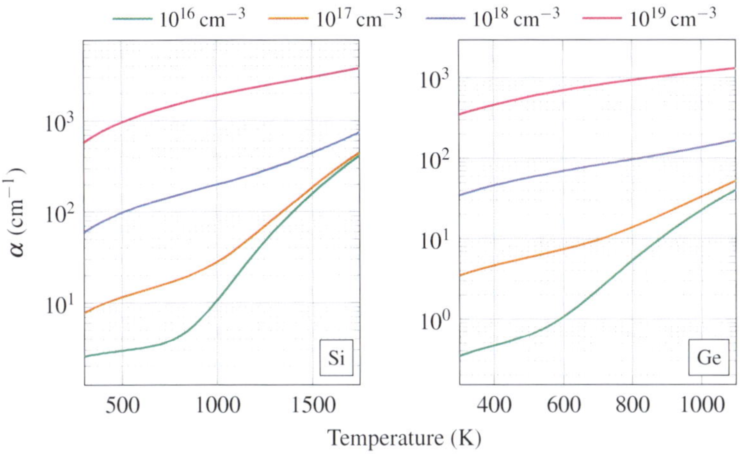


**FIG. 2.3** Absorption coefficient, $\alpha$, of doped semiconductors at deep infrared (10.6 μm) relative to temperature for various levels of doping according to Eqs. (2.10), (2.11).

radiation drives electrons high in the conduction band and the holes deep in the valence band. While at low doping, the carriers reach thermal equilibrium with the lattice, in the case of high doping we have the formation of a temperature distribution of the carriers before they reach a thermal equilibrium with the lattice. The exchange of energy between electrons and holes is done through recombination as well as through the Auger process. The typical times of these processes are quite long and constitute the main mechanism through which the temperature of the carriers eventually rises [12–14].

The free carrier plasma generally behaves as an independent system in the crystal that absorbs energy from radiation and exchanges energy with the lattice. The plasma itself receives energy through interband transitions as well as from the absorption through free carriers we analyzed earlier. In the case of Auger processes, additional energy ($E_g$) is added to the plasma energy after a recombination event. On the other hand, plasma loses energy due to loss of excess energy, $E_c$, of its carriers as well as through phonon absorption, which results in heating of the lattice and it is the main mechanism of heating the crystal due to absorption. The excess energy follows the expression [15]

$$
\begin{aligned}
N_{eh}\frac{\partial\langle E_c\rangle}{\partial t} = {} & I\left[\alpha_0\left(1-\frac{1}{h\nu}\right)+\sigma N_{eh}\right]+\frac{E_g}{\tau_r}\left(N_{eh}-\overline{N}_{eh}(T)\right)\\
& +\nabla(\langle E_c\rangle D_a \nabla N)-\left(\langle E_c\rangle-\frac{3kT}{2}\right)\frac{N}{\tau_E}
\end{aligned}
\tag{2.12}
$$

The first term refers to the energy that the lattice receives through interband transitions and absorption of the free carriers, while the second is the energy released through Auger recombination. The last two terms refer to the energy lost by the carriers and transferred to the lattice. The term $3kT/2$ is the excess energy in equilibrium and $\tau_E$ is the effective relaxation time for collisions of carriers with the lattice. The parameter $D_a$ is the bipolar diffusion coefficient and is characteristic of the attraction between electrons and holes, $\alpha_0$ is the absorption coefficient for the interband transition, and lastly, $\tau_r$ is the recombination life time. At the same time, the time evolution of electron-hole pairs follows the relationship [16]

$$\frac{\partial N_{eh}}{\partial t} = \alpha_0 \frac{I}{h\nu} + \nabla(D_a \nabla N_{eh}) - \frac{1}{\tau_r}\left[N_{eh} - \overline{N}_{eh}(T)\right] \tag{2.13}$$

For irradiation with a laser beam of power $I$, Eqs. (2.12), (2.13) determine the time evolution of the carrier plasma energy. It is important to note that the earlier derivation assumes that the density of the carriers is low relatively to the total density of states. Otherwise, the plasma degenerates, which limits the exchange energy through carriers due to lack of unoccupied energy states; as well as through Auger recombination. Also, beyond the thermal effects, the carrier plasma can store energy in the form of plasmonic oscillations. In the case of doped silicon with $10^{20}$ cm$^{-3}$, the percentage of energy corresponding to such plasmonic oscillations may reach up to 10% [13, 14]. Plasmons can be considered as pseudo-particles that lose energy and they can additionally interact with the lattice by means of collision. The interaction is stronger when their energy is comparable to the relative energies of phonons, usually at small densities of impurities. For higher concentrations, where the plasma is denser, the plasmons are independent of the lattice unless the plasmons energy crosses the semiconductor energy gap. In this case, the plasmons break down into pairs of electrons and holes [17].

Concluding this section, we must mention that in extremely short radiations plasma has a significant effect on the mechanical properties of the lattice. Removal of electrons from the valence band tends to decrease the phonon frequencies, which leads to an increased number of excited phonons at a given temperature. The extremely high density of excited carriers and phonons can lead the material to a fluid state well below the melting point of the material. On a nanosecond time scale, it is possible, for example, to recrystallize silicon at temperatures below the melting point [18].

## 2.3 Thermal effects of laser radiation

Although the effects of laser radiation on a material are quite complex, depending on the duration and power of the pulse, the predominant effect is to heat the material. In this case, both the time duration and the power of the laser are, respectively, neither extremely short nor extremely high. Since we have analyzed in the previous paragraph how the laser radiation is absorbed from semiconductors, we will now see a macroscopic view of distribution and heat dissipation in the material. We first consider a beam $w$ of a laser operating at the fundamental transverse electromagnetic ($TEM_{00}$) mode. In this case, the beam profile is circular and the radial distribution of intensity, $I$, on the wave front will be Gaussian following the form

$$I(r) = I_0 \exp\left(-\frac{r^2}{w^2}\right) \tag{2.14}$$

where $I_0$ the radiation intensity for $r = 0$. The total power, $P$, of the incident radiation will then be

$$P = 2\pi \int_0^\infty rI(r)dr = \pi w^2 I_o \tag{2.15}$$

For a uniform material, we can define the attenuation function, $f(z)$, as

$$f(z) = \alpha(T(z))\left[1 - \exp\left(-\int_0^z \alpha(z')dz'\right)\right] \tag{2.16}$$

with $\alpha(T(z))$ being the coefficient of absorption at current material temperature and depth. The integral in the exponential is also called the *optical thickness* of the material. In the simple case where the absorption coefficient does not depend on temperature, the attenuation function can be simplified to

$$f(z) = \alpha \exp(-\alpha z) \tag{2.17}$$

and in the case of lasers that have typical pulse times in the ns range, and the absorption in the material is predominantly surface-localized (as in the case of metals) the attenuation can degenerate into the simple function

$$f(z) = \delta(z) \tag{2.18}$$

which means that the effect of radiation is a predominantly surface effect.

Relation (2.17) is suitable for cases of semiconductor heating with deep ultraviolet laser (such as excimer), where the process is strongly adiabatic and surface constrained, but *not* in the case of deep infrared (such as $CO_2$ or similar lasers) where the thermal diffusion length is extremely long and the absorption coefficient varies strongly with temperature, and thus depth. In that case, the analytical relation (2.16) must then be used. After all, regardless of the form of the attenuation function in each case, we can write about the power density, $J$, which is finally acquired from the material, if $R$ is the reflectivity of the material (typically 0.30 for silicon and 0.38 for germanium; at 10.6 μm [1]):

$$J(z) = I(1 - R)f(z) \tag{2.19}$$

Finally, another parameter that we are interested in is the heat flow, $\Phi$. Once the energy is absorbed as heat does not cease to be strongly localized and tends to diffuse. The heat flow in this case at a point $z_0$ of the material will be

$$\Phi(z_0) = -K(T)\left.\frac{dT}{dz}\right|_{z=z_0} \tag{2.20}$$

with $K(T)$ being the *thermal conductivity* of the material.

Let us assume the heating of a thin uniform slab of material with density $\rho$ and heat capacity $c_p$, extending from depth $z$ to $z + \Delta z$. The energy balance across this slab will be

$$\Delta t[\Phi(z) - \Phi(z + \Delta z)] = \Delta T \rho c_p \Delta z \tag{2.21}$$

where $\Delta T$ is a differential temperature change, which results from the heat of a flow in the boundaries of the slab. For $\Delta z \to 0$, we will have

$$\Delta t \frac{\partial \Phi}{\partial z} \Delta z = \Delta T \rho c_p \Delta z \tag{2.22}$$

via Eq. (2.20) we can derive

$$\frac{\partial}{\partial z}\left(K \frac{\partial T}{\partial z}\right) = \rho c_p \frac{\partial T}{\partial t} \tag{2.23}$$

which is essentially the heat diffusion equation in one dimension for a material of thermal conductivity $K$, with specific heat capacity $c_p$ and density $\rho$. These three parameters determine the material in terms of its thermal behavior and in case it is uniform, they are stable at a given temperature. All three parameters are interrelated by means of the *thermal diffusivity* defined as

$$D(T) \equiv \frac{K(T)}{\rho c_p(T)} \tag{2.24}$$

In the case of the materials involved here, we can find the relevant numerical values of these parameters in the literature [1]. Concerning the densities, we have that

$$\rho_{\mathrm{Si}} = 2.33\ \mathrm{g\,cm^{-3}} \tag{2.25}$$

$$\rho_{\mathrm{Ge}} = 5.32\ \mathrm{g\,cm^{-3}} \tag{2.26}$$

for the thermal conductivities for silicon [19, 20]

$$K_{\mathrm{Si}}(T) = 0.235 + 4.45 \exp\left(-\frac{T}{247}\right)\ \mathrm{W\,cm^{-1}K^{-1}} \tag{2.27}$$

and germanium [21, 22]

$$K_{\mathrm{Ge}}(T) = 0.144 + 1.34 \exp\left(-\frac{T}{323}\right)\ \mathrm{W\,cm^{-1}K^{-1}} \tag{2.28}$$

Finally, for the heat capacity of silicon [2] and germanium [23, 24], we have

$$c_{p-\mathrm{Si}} = 0.81 + 1.3 \times 10^{-4} T - 1.26 \times 10^{4} T^{-2}\ \mathrm{Jg^{-1}K^{-1}} \tag{2.29}$$

$$c_{p-\mathrm{Ge}} = 0.36 - 11 T^{-1} - 107 T^{-2}\ \mathrm{Jg^{-1}K^{-1}} \tag{2.30}$$

A graphical representation of thermal conductivity, heat capacity, and of the thermal diffusivity for silicon and germanium based on the previous equations can be seen in Fig. 2.4.

Based on all the earlier, we can write the equation of thermal diffusion in its familiar one-dimensional (1D) form (2.31):

$$\frac{\partial T}{\partial t} = D(T)\frac{\partial^2 T}{\partial z^2} \tag{2.31}$$

Eq. (2.31) assumes no heat generation, that is, the system relaxes to equilibrium. If we assume heat generation due to absorption of external radiation, $I_0$, then we need to incorporate a *source term*, $Q(z, t)$.

$$Q(z,t) = (1 - R)I_0(z,t)f(z)q(t) \tag{2.32}$$

where $q(t)$ is a function that expresses the *time modulation* of the beam and $R$ is the reflectivity of target material. As such, Eq. (2.31) can be adjusted to accommodate the additional term, Eq. (2.32), as

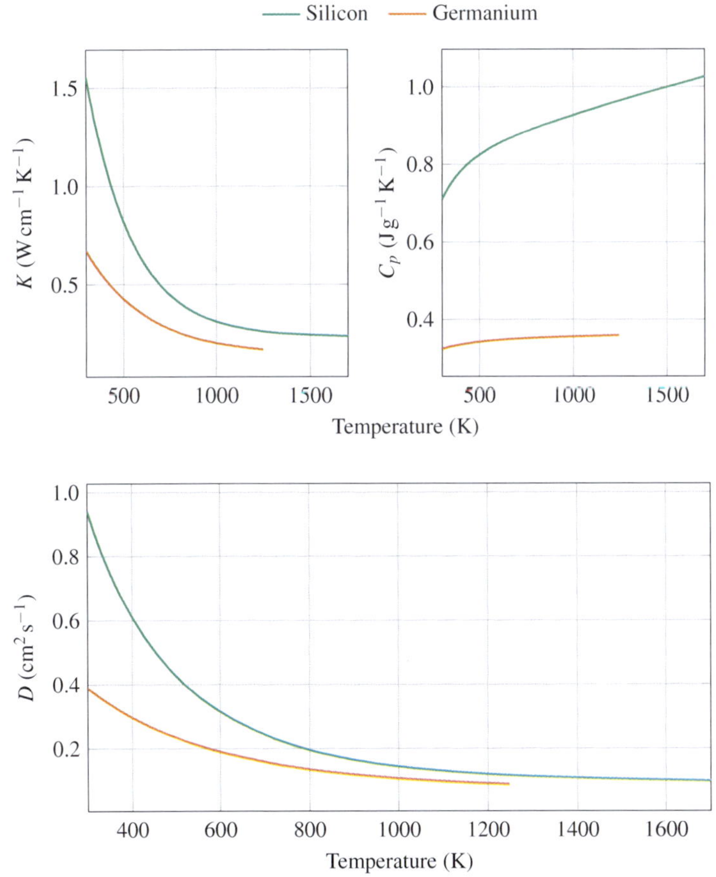


**FIG. 2.4** Dependence of thermal conductivity (*left*), heat capacity (*right*), and thermal diffusivity (*bottom*) on temperature for silicon and germanium up to their respective melting temperatures.

$$\rho c_p(T)\frac{\partial T(z,t)}{\partial t} - K(T)\frac{\partial^2 T(z,t)}{\partial z^2} = (1-R)I_0(z,t)f(z)q(t) \tag{2.33}$$

In the general case where all the parameters are temperature dependent, the solution of Eq. (2.33) can only be performed numerically. For parameters independent of temperature, we can use the method of Green's functions and we can find solutions for the case of a semiinfinite solid, where the thickness, $L$, is very much smaller than the other two dimensions. Initially for the case of a Gaussian shape beam with FWHM equal to $w$ at a depth $z$ from the surface of the material and at distance $r$ from the center of the Gaussian beam, the temperature field is given by

$$T(r,z,t) = \frac{P_a}{2K\pi^{3/2}}\int_0^{\delta} \exp\left(-\frac{r^2}{\beta^2+w^2}\right) Y_1 \frac{d\beta}{\beta^2+w^2} \tag{2.34}$$

If the beam is spatially extensible with respect to depth, which essentially means that the thermal diffusion length, $\delta = \sqrt{D\tau_l}$, is comparable with the thickness of the slab ($\tau_l$ being the irradiation time) the temperature field is then given by

$$T(r,z,t) = \frac{\alpha P_a}{4\pi K}\int_0^{\delta} \exp\left(-\frac{r^2}{\beta^2+w^2} + \frac{\alpha^2\beta^2}{4}\right) Y_2 \frac{\beta d\beta}{\beta^2+w^2} \tag{2.35}$$

The terms used in the earlier equations are

$$\begin{aligned}
P_a(t) &= (1-R)\int\int I(x',y',t)dx'dy' \\
\beta &= 2\sqrt{D|t-t'|} \\
Y_1 &= 2\sum_{n=-\infty}^{\infty} \exp\left(\frac{2nL-z}{\beta}\right)^2 \\
Y_2 &= \sum_{n=-\infty}^{\infty} \exp[\alpha(2nL-z)] \\
&\times \left[\operatorname{erfc}\left(\frac{(2n+1)L-z}{\beta} + \frac{\alpha\beta}{2}\right) - \operatorname{erfc}\left(\frac{2nL-z}{\beta}\right) + \frac{\alpha\beta}{2}\right]
\end{aligned}$$

The complete formalism around Eqs. (2.34), (2.35) can be found in M. Lax's extensive publications [25, 26].

## 2.4 Differences across the electromagnetic spectrum

It is worthwhile to refer to differences across the electromagnetic spectrum in terms of absorption and temperature distribution. We will summarize how a laser beam interacts with matter at infrared spectrum edge such as the $CO_2$ and we will compare it with a laser at the UV limit [3]. In the case of deep UV, for example, in the region of lasers of KrF (248 nm, 5.0 eV), ArF

(193 nm, 6.4 eV), or XeCl (308 nm, 4.0 eV), which are widely used in semiconductor heat treatment, the energy of the photons is large enough to cover the energy spectrum of the semiconductor. As a result, the absorption is carried out directly via interband transitions and formation of electron-hole pairs. Silicon as well as germanium have extremely high ultraviolet absorption coefficients. Indicatively at 250 nm the absorption coefficients are $1.67 \times 10^6$ and $1.43 \times 10^6\ cm^{-1}$ for silicon and germanium, respectively. As a result, radiation is converted very rapidly into heat at the first few nanometers of material. Due to high absorption coefficient, the pulse duration is in the range of 10–100 ns. During this time, the length of thermal diffusion is less than 1 μm, so the temperature distribution will decrease exponentially from the surface. Due to the small thermal diffusion, the distribution problem of temperature reduces to the redistribution of temperature reached at the surface following irradiation. As long as the size of the irradiated sample is much larger than the spot size itself, the treatment is practically 1D and the process is simplified. In the case of infrared radiation, 117 meV that corresponds to a laser beam with wavelength 10.6 μm is not enough to allow direct band transitions; consequently, all absorption is performed through the free carriers by the process described previously. The collisions of the carriers with the lattice are essentially the mechanism that leads to the heating of the material. The greatest difficulty in describing absorption and thermal diffusion in this case is mainly due to the extensive thermal effect the beam eventually creates. In contrast with a beam of nanosecond duration in the case of infrared, the small absorption requires longer pulses of $10^{-3}$ to $10^{-2}$ s. In that case, the thermal diffusion length can reach values even 800 μm as in silicon that is practically throughout the depth of a typical 6-in wafer. Therefore, an approach in which the temperature is simply distributed according to the heat equation from the surface to the bulk of the material does not apply. Any attempt to model the process must take into account both the extensibility of the temperature distribution and the strong dependence of the absorption from the temperature. Fig. 2.5 illustrates a summary of the thermal interactions in the two different irradiation regimes we have described.

## 2.5 Diffusion model extension to millisecond regime

The microelectronics industry has undoubtedly grown fast over the last 20 years and a very important share of this success is due the development of increasingly sophisticated technology computer-aided design (TCAD) tools. TCAD programs cover the entire spectrum of fabrication of a microelectronic device from the design of lithography masks, the simulation of front-end processes, back-end processes, modeling, and evaluation of devices, both in terms of device simulation and modeling of circuits (SPICE-type simulators).

Initial support for laser annealing has recently appeared for the major simulation frameworks; however, this is limited to nanosecond laser annealing for the deep UV irradiation regime. In this section, we describe a process which

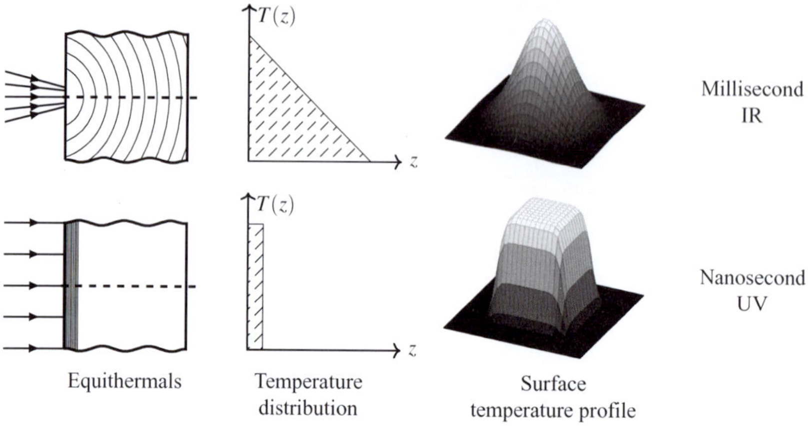


**FIG. 2.5** Comparison of two different irradiation regimes regarding temperature distributions. The *graphs* in the second column indicate the temperature distribution along the *dotted line*.

additionally takes into account (a) extrinsic doping dependency and (b) non-pulsed irradiation regimes. Again, $CO_2$ will serve as a typical CW deep infrared case. From a simulation point of view, irradiation in the CW regime is fundamentally different from a nanosecond-pulsed one. In these wavelengths and irradiation regimes, absorption depends heavily on the local temperature gradient. In addition, the spatial and temporal configuration is entirely different as the power flow is practically constant with time (the so-called *top-hut* distribution) and Gaussian in respect with the localized coordinate system. For comparison in a pulsed nanosecond configuration, the spatial distribution is practically constant but the temporal one is relatively constant. These aspects are summarized in Fig. 2.6.

A typical experimental setup representative of this simulation scenario is depicted in Fig. 2.7, based on a $CO_2$ laser at 10.6 μm. Samples rest atop a stage for preheating of samples to enhance beam-lattice coupling and aid absorption. This in turn leads to peak surface temperature. Given that such irradiations are in the millisecond regime a pyrometer allows for real-time monitoring of the irradiated area. $CO_2$ is a common industrial laser typically operated in CW mode. It was developed by Patel in the 1960s [27] and belongs to the overarching family of rotational-vibrational lasers. Despite the name, its active medium is actually a mixture of not only $CO_2$ but also $N_2$ and He. Applying a discharging voltage in that particular mixture of low-pressure gases leads to exciting nitrogen molecules, which in turn transfer their energy through molecular scattering to those of $CO_2$. This forms the basic *population inversion* of this particular laser. Deexcitation of the $CO_2$ molecules leads to the emission of light in two particular wavelengths: 9.6 and 10.6 μm [28].

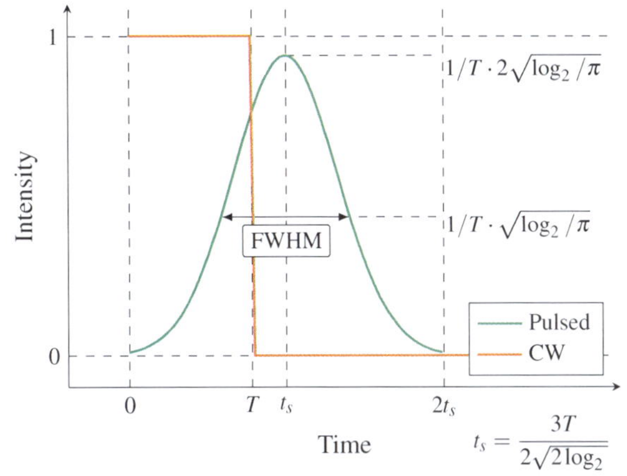


**FIG. 2.6** Comparison of the time evolution of a continuous laser pulse and from a pulsed laser for the same nominal maximum power (normalized to unit) and with pulse duration, *T*.

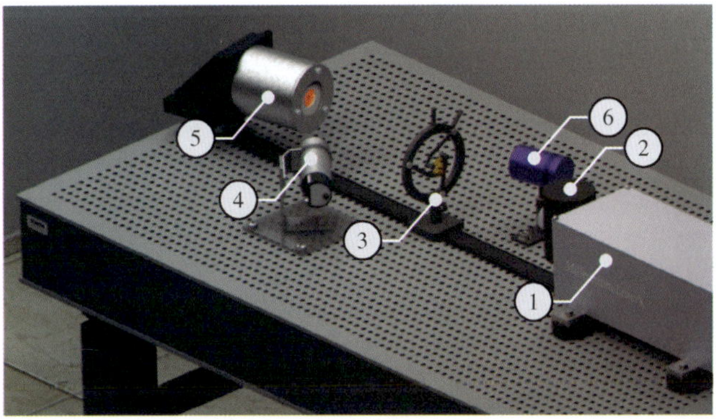


**FIG. 2.7** A typical $CO_2$ laser annealing setup: (1) $CO_2$ laser; (2) beam injector; (3) focusing ZnSe lens; (4) pyrometer; (5) beam injector; and (6) beam dump.

### 2.5.1 Thermal interaction

In order to solve the problem of dopant diffusion with laser heating, the introduction of the impurity profile after implantation is first required. Then, the problem of heat distribution needs to be solved. The heat equation that is typically solved in this case is

$$\rho c_P \frac{\partial T}{\partial t} = \nabla \cdot (K \nabla T) + G \tag{2.36}$$

where $\rho$, $c_P$, and $K$ are the *density*, *specific heat capacity*, and *thermal conductivity*, respectively. $G$ refers to source term and is calculated on the basis of the relation

$$G = I\alpha \exp\left(-\int_0^d \alpha d\mathbf{l}\right) = If(\mathbf{l}) \tag{2.37}$$

where $\alpha$ is the absorption coefficient and $d$ is the thickness of the target. In practice, the two previous equations are the three-dimensional (3D) expression of the general heat equation we have already introduced. At each time step of the simulation, the simulator would calculate the source term (*heat generation term*) and extract the depth-dependent temperature distribution. This thermal profile is used at each node of the grid to solve the diffusion equations for impurities.

During the simulation of a typical diffusion process, the temperature argument would refer to the temperature at which the annealing takes place. In the case of laser annealing, the temperature reflects only the initial temperature of the sample. Also by time, we denote the *total duration* of the simulation and not the duration of the pulse. This allows us to observe the time evolution of the process even after the laser pulse has stopped as the sample returns to its original temperature. In the case of high temperatures, this is important as diffusion effects would continue while the material is in cool down.

The challenges involved in this simulation are in two directions. The first concerns the correct description of the spatial configuration of the beam. For a typical pulsed-based laser irradiation power density would follow the equation

$$I(r,t) = (1-R)\frac{I_0}{\sqrt{2\pi}t_s}\exp\left(-\frac{(t-t_0)^2}{2t_s^2}\right) \tag{2.38}$$

with

$$t_s \equiv \frac{T}{2\sqrt{2\log_2}} \tag{2.39}$$

With $t_0$ we denote the point at which the irradiation has started; $R$ is the target reflectivity; $I_0$ is the incident power density; and $T$ is the total irradiation time. For a CW laser, the equivalent equation would be

$$I(r,t) = (1-R)I_0 \exp\left[-\frac{(r-r_0)^2}{2w_0^2}\right] \tag{2.40}$$

which means that the intensity profile modulation depends on the *radial distance*, $r$, from the center, $r_0$ of the incident beam. $w_0$ is called the *beam waist*, which for a beam with radius $r$ is defined as

$$w_0 \equiv \frac{r}{\sqrt{2}} \tag{2.41}$$

Beam waist is a typical characteristic of Gaussian beam and is essentially the size of the incident beam but as measured *at the focal point*.

The previous formulation of the incident power profile along with the heat equation and generation term as well as the dependence of the absorption coefficient described in the previous section *fully describes* the problem of heat absorption and diffusion in a doped semiconductor.

A typical output of such an irradiation regime in silicon can be seen in Figs. 2.8 and 2.9. The domain used is 100 μm in width; spot size is 1.1 mm; and preheating of the substrate is set at 600°C. Fig. 2.8 shows the isothermal distributions that resulted from these parameters. Maximum temperatures recorded are about 1300°C for 6 ms irradiation and 1400°C for 10 ms, just below the melting threshold. The Gaussian form of the temperature distribution is obvious resulting from nonuniformities of the incident radiation. It is also apparent the large thermal absorption length of the radiation extending beyond 80 μm in both cases. At the same time, in Fig. 2.9 we can see the temperature distribution across the computational domain, which reflects the Gaussian distribution of power.

The observation about the extent of the thermal effect is made more apparent through Fig. 2.10, which shows the temperature distribution in depth at the point of maximum temperature (i.e., at the center of the computational domain) just at the end of the pulse duration. We can see the uniform temperature gradient, which is consistent with the observations made before. The temperature develops smoothly practically throughout the depth of the material.

For comparison, a similar simulation procedure using an excimer KrF laser source [29] was performed. The conditions used were energy density of 580 mJ $cm^{-2}$ and pulse duration of 20 ns. In addition to the uneven temperature rise due to the Gaussian time-lapse behavior of the laser (the maximum temperature is reached at 32 ns, that is, 12 ns *after* the pulse expires), we can observe the rapid recovery of the sample to its original temperature. In the deep IR case, the thermal effect is even more extensive when compared to deep UV interaction, where the whole process is adiabatic and more localized. The thermal effect is limited to the top 3 μm unlike in the IR case, where the thermal effect

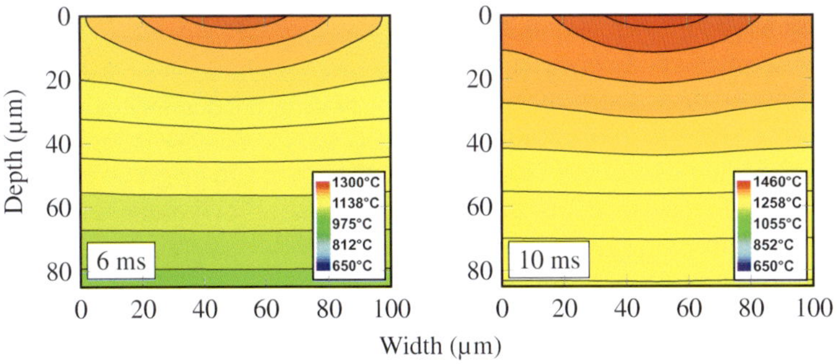


**FIG. 2.8** Isothermal temperature distribution curves for 9870 W $cm^{-2}$ pulse irradiation lasting 6 and 10 ms in silicon.

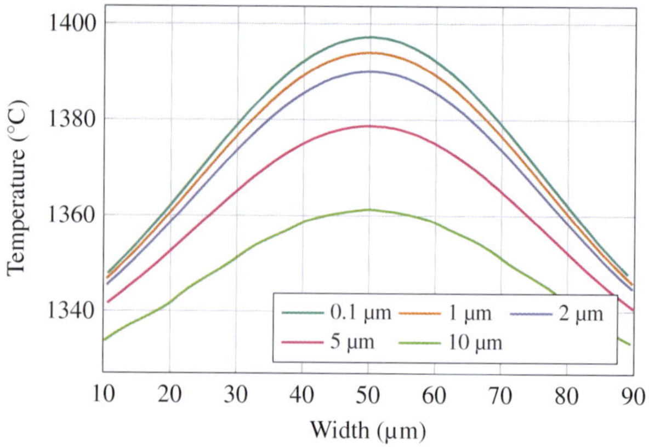


**FIG. 2.9** Temperature distribution across the entire computational domain at different depths for 10 ms irradiation.

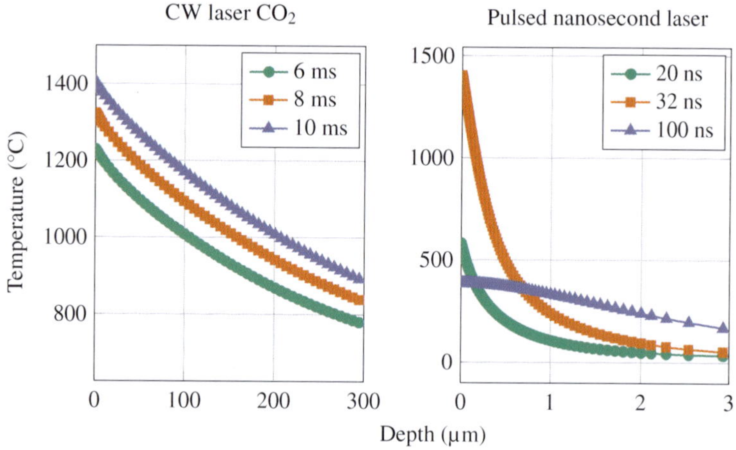


**FIG. 2.10** Depth-dependent temperature distribution at the center of the simulation domain (*left*) and compared with the case of UV irradiation in a 20 ns pulsed laser (*right*).

extends to the bulk of the material. Also interesting is the temporal evolution of the surface temperature. In Fig. 2.11, we can see the time evolution of the surface temperature at the center of the computational domain. The shape of the curve in the case of the IR laser is typical for the case of constant power pulses and quadratic time shapes as can be seen in Doubenskaia et al. [30]. Apart from the obvious difference in time scale, the temperature evolution is also radically different in shape than in the case of nanosecond-pulsed lasers. Still in Fig. 2.12 we can see the time evolution of the absorption coefficient in depth. It is evident that the very high impurity doping level at the first nanometers of the material results in an extremely large increase in the absorption coefficient. Essentially

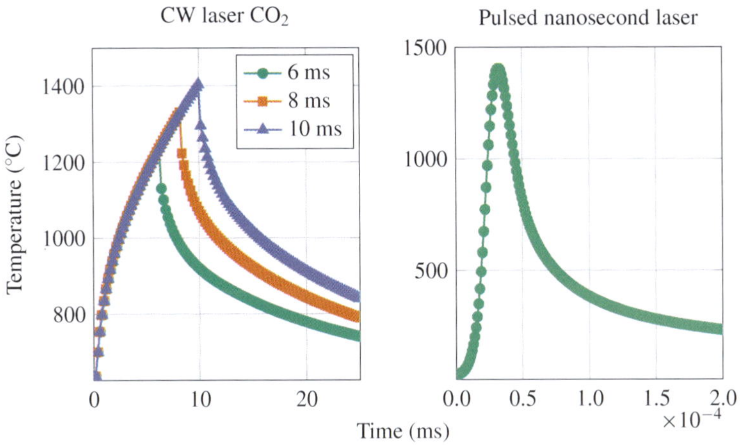


**FIG. 2.11** Time evolution of the surface temperature developed at the center of the simulated domain (*left*) and compared with irradiation in nanosecond UV regime by a pulsed laser of 20 ns (*right*).

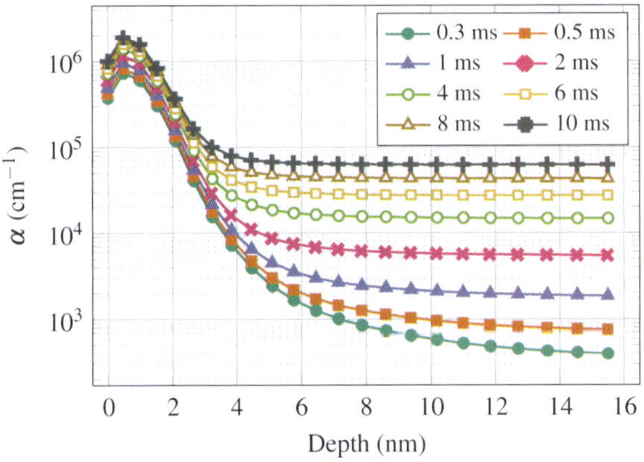


**FIG. 2.12** Silicon absorption coefficient vs. depth at various time steps during the first 10 ms of radiation. Absorptivity is appreciably more dependent on doping level rather than heating in the first few nanometers of the material.

most of the radiation is absorbed by very highly doped areas and as a result the form of the coefficient distribution somehow follows the distribution of boron impurities. As time passes on and the temperature in the sample increases, the absorption coefficient begins to increase. From equilibrium temperature up to 10 ms, the absorption coefficient has finally increased to two orders of magnitude.

Similar is the case for germanium with the main difference being the lower melting point of the material. In this case, in order to keep the process in the submelt regime and not trigger excessive diffusion, a lower-preheating temperature would be required. Fig. 2.13 shows the thermal distribution for germanium preheated at 270°C for the same irradiation characteristics as those illustrated for silicon.

Simulated results are compared against irradiated silicon samples using the setup described earlier. As shown in Fig. 2.14, which depicts real-time pyrometry data, there is an excellent agreement between simulated and experimentally measured temperatures. Since temperature is a major factor that ultimately guides the diffusion of impurities, the close matching between simulated and experimental data serves as an important initial validation that the process is indeed accurate.

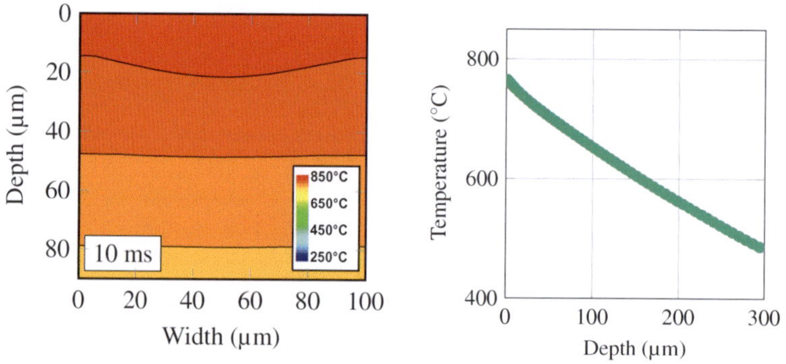


**FIG. 2.13** Thermal simulation results for germanium irradiation with a $CO_2$ laser beam with density of 5500 W $cm^{-2}$ and duration of 10 ms (*left*: isothermals and *right*: temperature distribution).

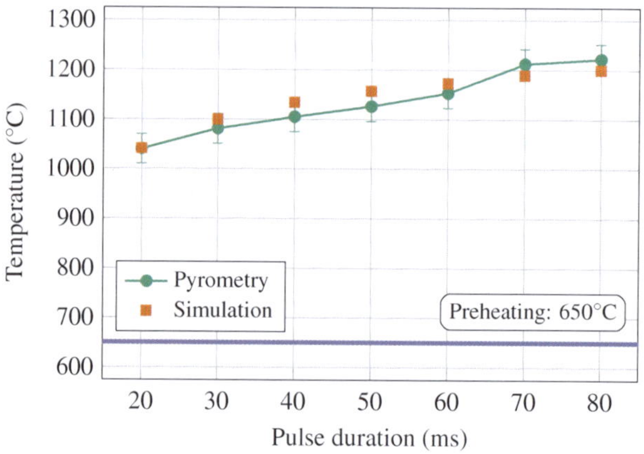


**FIG. 2.14** Maximum recorded surface temperature values and comparison with simulation. The spot size of the beam is 1.8 mm and the preheating temperature is 650°C.

### 2.5.2 Diffusion simulation and activation kinetics

Although temperature diffusion is a relatively straightforward process even when adjusted for the intricacies of the absorption coefficient modification due to doping, diffusion of dopants is a much more complicated topic. There are a multitude of dopant diffusion models developed over the years in accordance to demands imposed from the semiconductor industry. As the overall dimensions of devices are being scaled down transient phenomena that were of minimum importance to higher manufacturing nodes are now more relevant.

For instance, a particular case where laser annealing poses as an advantageous alternative to other forms of annealing is *shallow junction engineering*. Given the overall shrinkage of active transistors in modern circuits and microprocessors as well as the recent introduction of alternative device geometries, such as FinFETs, controlling the diffusion of dopants in the source/drain region is considered of prominent importance according to the most recent International Technology Roadmap for Semiconductors (ITRS) [31]. Several works illustrating the use of laser for such purpose have been presented before, for instance, using solid state [32–35] or more recently excimer [36–39] lasers.

One of the more comprehensive diffusion models is the *charged reaction diffusion model* (ChargedReact for sort), which is implemented by most commercial TCAD suites. ChargedReact (or *five-stream* model) [40] is regarded as the most generalized diffusion model tailored to take into account transient phenomena [41–46], which are extremely important for procedures that are based on steep temperature gradients, as is the case for laser annealing. ChargedReact serves as the basis for all examples used throughout this section.

Given the encouraging agreement between experimental and simulated data in the temperature distribution, we can proceed to the simulation of the diffusion of dopants under IR irradiation. Boron in silicon will serve as our example. Keeping the beam width at 1.1 mm and using power densities corresponding to irradiation power above 60 W, we perform a series of diffusion studies. Power densities lower than that threshold correspond to surface temperatures of lower than 900°C that are far too low to elicit any dopant activation within the limited time of the sub-100 ms irradiation. As diffusion dopant is a far more computationally expensive when compared to heat diffusion, we can take into account the temporal "rectangularity" of the laser pulse. Since dopant diffusion is temperature dependent, this means that the overall dopant movement and activation level will be the same whether we stop the simulation of a 20 ms pulse at 10 ms or simulate a 10 ms pulse. This allows us to have temporal snapshots of the diffusion process without recomputing everything from scratch. This is already evident, after all, from Fig. 2.11 presented before.

The gradual shift of the profile in relation to the incident power density can be seen in Fig. 2.15. In all cases, the shift of the boron profile is evident with increasing pulse duration, which is reasonable as the temperature increases. The same goes for increasing the pulse power density. It should be noted that in the

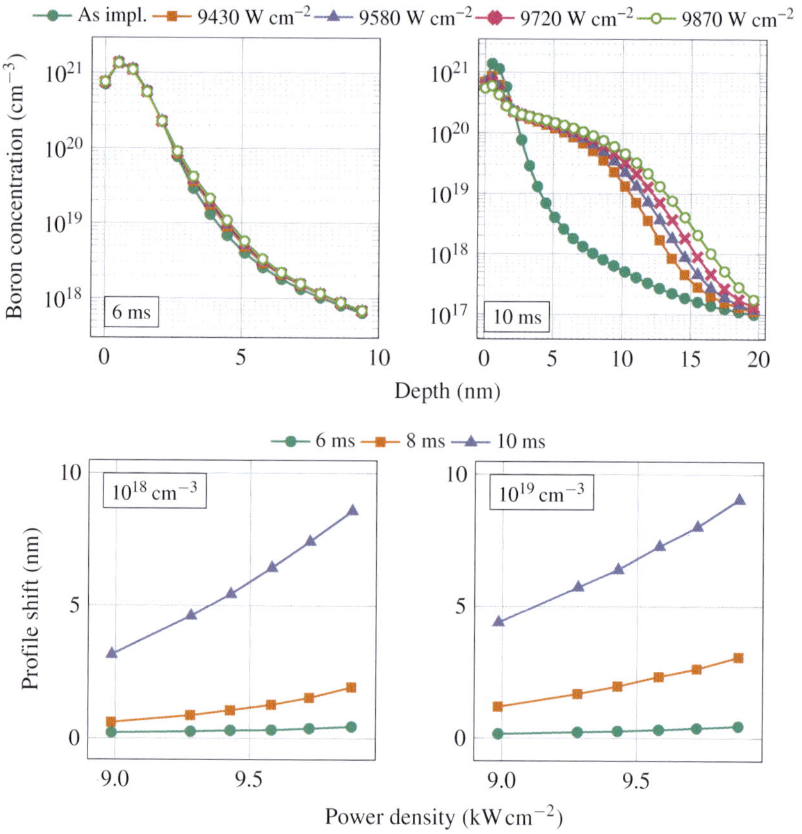


**FIG. 2.15** (*Top row*) Comparison of boron profiles for four different power densities at 6 and 10 ms. (*Bottom row*) Comparison of the boron dopant profile movement as a function of incident power density for two different doping levels. The as-implanted junction depth at $10^{18}$ cm$^{-3}$ is 7 nm, while at $10^{19}$ cm$^{-3}$ only 3.9 nm.

case of 6 ms the boron movement is extremely small giving junction depths very close to 7 nm of the as-implanted distribution. In contrast to 10 ms, we have a gradual shift of the profile inward as the power increases. It is apparent that the time span between 6 and 12 ms is the critical threshold beyond which the diffusion mechanism for this power range is clearly "activated." Obviously, corresponding results can be achieved with lower power and higher pulses but our goal is to maximize heat transfer with short pulses so as to reduce transient phenomena.

It is important to note a characteristic of all boron distributions is the existence of a relatively steady section at about $2 \times 10^{20}$ cm$^{-3}$. This indicates the existence of extensive boron-interstitial defects called *boron-interstitial clusters* (BICs) [43]. The form of the distribution is similar to other simulations found in the literature [47].

Fig. 2.15 (bottom row) also shows the movement of the profile at two different doping levels. Since the as-implanted profile has a junction depth at $1 \times 10^{18}$ cm$^{-3}$, we can deduce that the junction depth of the annealed silicon depending on the incident power varies from about 7.7 to 7.9 nm in the case of 6 ms, from 8.2 to 10.1 nm in the case of 8 ms, and from 10 to 18 nm in the case of 10 ms. It is clear from the simulation results that the degree of control provided by laser annealing even in small movements of the boron distribution profile is extremely high.

To compare with an experimental set of results, boron ions have been introduced into n-type silicon wafers by means of $BF_3$ plasma doping method with a dose $1 \times 10^{15}$ cm$^{-2}$ and energy 600 eV. In this particular case, the implantations were performed using the plasma ion implanter PULSION HP [48] but the methodology is still applicable to any shallow implantation technique.

First, the impact of the beam on the implanted silicon is presented. In the TEM cross-section image of Fig. 2.16A, we can see the level of amorphization introduced by the implantation. The amorphous depth is found at about 2.5–3 nm and indicates the extremely low level of damage caused by this implantation, which is comparable to corresponding energy boron implantation methods of less than 1 keV. The amorphous range is in very good agreement with the simulation, which estimated the depth of the amorphous pockets at approximately 2.7 nm. After irradiation with a power density of 9430 W cm$^{-2}$ (64 W) lasting 10 ms, we have the situation shown in Fig. 2.16B. As can be seen, the previously amorphous region has been recrystallized, which indicates that the $CO_2$ laser beam efficiently repairs the damage introduced by implantation. Nevertheless, the presence of darker areas in Fig. 2.16B probably indicates some residual stresses in the silicon crystal.

We can now proceed to the results of the boron distribution for the annealing conditions used. In Fig. 2.17, we can see the boron SIMS measurements. It is evident that with increasing pulse duration we have an increasing movement of the impurity profile as a result of the maximum temperature eventually developing. The diagrams show that the junction depths ($x_j$) achieved at $1 \times 18$ cm$^{-3}$

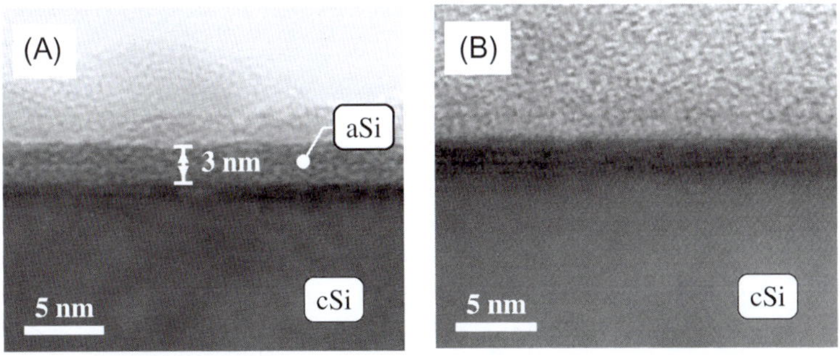


**FIG. 2.16** TEM images (A) before and (B) after 10 ms irradiation at 9430 W cm$^{-2}$ (64 W).

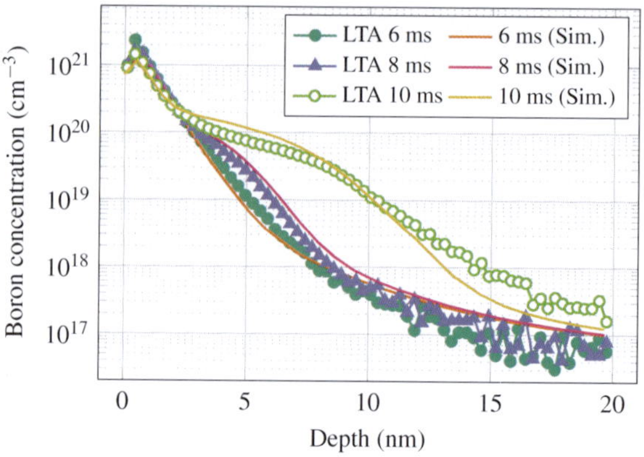


**FIG. 2.17** Comparison of the experimental results for the three different annealing times with the corresponding simulation results (*continuous lines*).

with annealing are 7.5 nm for 6 ms, that is, there is virtually no diffusion, 8.2 nm for 8 ms and 14.2 nm for 10 ms. It would seem that diffusion is actually initiated between 6 and 8 ms. Through the experiment, therefore, we scan all those critical temperatures by which we move from a diffusion-free process to one with a noticeable shift of the boron profile. The temperatures measured by the pyrometer for these times are 1250 ± 30°C for 6 ms, 1320 ± 30°C for 8 ms, and 1390 ± 30°C for 10 ms in very good agreement with the temperature simulation shown before.

In Fig. 2.17, one can also see a comparison of the experimental SIMS curves with the boron distributions obtained by the simulation. In the case of 6 and 8 ms irradiation, we have very good agreement of the two distributions, whereas in the case of 10 ms there is an underestimation of about 1.5–2 nm with respect to the contact depth; however, the shape of the curve is very good. In general, we can say that there is a good match between experiments and simulations, which indicate that the methodology we have adopted for modeling the phenomenon is correct. A summary of the experimental results can be seen in Table 2.1. It should be noted that there was also a check on the possible occurrence of parasitic diffusion due to preheating at 650°C for about 1–2 min but there appeared to be no diffusion in boron distribution between preheated and nonpreheated samples. Therefore, we do not consider that there is a possibility that the final distribution of boron would be affected by the preannealing conditions. A characteristic common to all extracted curves, and especially in the case of 10 ms, is a steady fraction of boron at concentrations above $2 \times 10^{20}$ cm$^{-3}$, which is very close to the solubility limit of boron in silicon [49]. The presence of this characteristic is attributed to the formation of extensive boron defects with interstitial silicon atoms called boron-interstitial clusters (BICs). The presence of BICs is analyzed by Cristiano et al. [43] who observed the formation of small defects by TEM microscopy at the boundary of the amorphous region after low-energy

**TABLE 2.1** **Summary of the experimental results (E) and comparison with the simulated ones (S).**

| Time [ms] | $x_j$ (E) [nm] | $x_j$ (S) [nm] | Temperature (E) [°C] | Temperature (S) [°C] |
|---|---|---|---|---|
| 6 | 7.5 | 7.3 | 1250 ± 30 | 1215 |
| 8 | 8.2 | 8.9 | 1320 ± 30 | 1340 |
| 10 | 12.9 | 14.2 | 1390 ± 30 | 1400 |

implants. Cristiano et al. also observed that although there is a tendency for the defects to dissolve at high temperatures, which reduces their extent to levels lower than the resolution of the TEM, the immobile profile does not appear to be fully dissolved which is shown by both the SIMS measurements and the simulation results.

Regarding electrical characterization, the tendency of surface resistance to decrease as the duration of irradiation increases was clearly shown. In the 6-ms pulse, the minimum measured sheet resistance was $2610 \pm 200\Omega/\square$, in the case of 8 ms, it was measured at $1730 \pm 30\Omega/\square$, and finally, in the case of 10-ms pulse, it was $620 \pm 50\Omega/\square$. The decrease in surface resistance implies that gradually increasing levels of activation are attained, which depend on the level of solid boron solubility at the attained temperature. In addition to the chemical distribution of boron, the distribution of active boron is also extracted. The comparison of the two distributions is shown in Fig. 2.18 in which it is apparent that the distribution of the active boron is very close to the solid solubility limit for the corresponding temperatures. This is an indication that in both cases of annealing we are very close to the maximum activation value.

To see, however, whether simulated active boron distributions are reflecting reality, we must extract the calculated surface resistance through them. To do this, we use the following mobility model for $p$-type impurities in silicon [50, 51]:

$$\mu(x) = \mu_{\min} + \frac{\mu_{\max} - \mu_{\min}}{1 + \left(\frac{N(x)}{C}\right)^r} \tag{2.42}$$

where $N$ is the total boron concentration and $\mu_{\min}$, $\mu_{\max}$, $C$, and $r$ are the material-dependent semiempirical parameters. For boron, these coefficients are $\mu_{\min} = 44.9\ \mathrm{cm^2\,V^{-1}\,s^{-1}}$, $\mu_{\max} = 470.5\ \mathrm{cm^2\,V^{-1}\,s^{-1}}$, $C = 2.23 \times 10^{17}\ \mathrm{cm^{-3}}$, and $r = 0.719$. Corresponding values have also been calculated for the other basic silicon impurities. Once the boron mobility is calculated relative to the depth, the surface resistance can also be calculated by integrating the active impurity distribution with the mobility down to the junction depth. If $N_A$

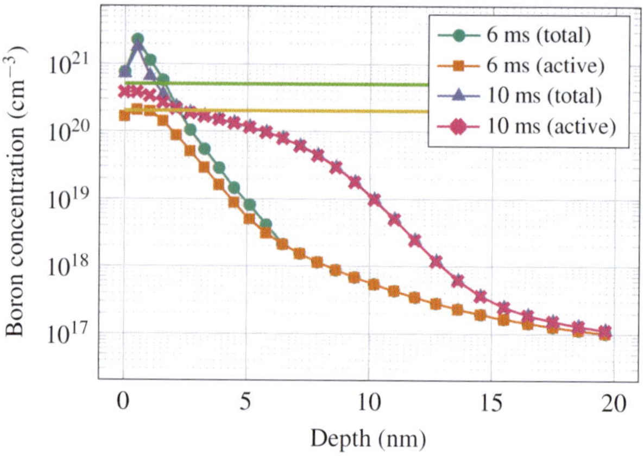


**FIG. 2.18** Comparison of chemical (total) and active boron distributions. The *parallel lines* indicate the corresponding solubility limits [49] at 1250 and 1400°C, which are the temperatures corresponding to these annealing conditions.

is the active boron distribution, $q$ is the elemental charge, and $x_j$ is the junction depth, then the surface resistance of the material is given by the relation

$$\frac{1}{R_S} = q \int_0^{x_j} N_A(x)\mu(N(x))dx \tag{2.43}$$

Based on the previous integral, we can derive the evolution of the surface resistance as the irradiation time increases. The result of this analysis is shown in Fig. 2.19. A good match between experiment and simulation is evident which indicates that the activation levels of Fig. 2.18 do indeed reflect reality. In general, both surface resistances and activation levels are similar to the literature [52–54] for laser annealing especially in the case of 10 ms.

## 2.6 Concluding remarks

Following a presentation of laser-matter interaction, we have focused this review on the interactions of infrared laser with group IV semiconductors, which are of particular interest for use in annealing studies of these materials. One critical issue that has to be considered in this case is the dependence of the absorption coefficient with doping level. The solution of heat equation then gives the temperature distribution throughout the sample, which is validated through a comparison with experimental data obtained using a $CO_2$ laser. The application on the influence of laser heating on the diffusion of dopants and the electrical characteristics of the samples has been then presented in the case of silicon substrate. This example has clearly illustrated the importance of $CO_2$ laser annealing in semiconductor technology for achieving a dopant diffusionless process with high levels of activation. The simulations which are in

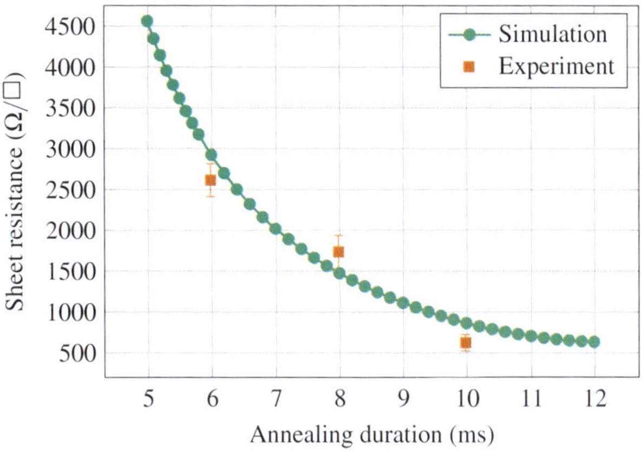


**FIG. 2.19** Evolution of the sheet resistance as calculated by the simulator compared to the experimental values for the three irradiation conditions.

very good agreement with experimental data underlie their predictive capability confirming our good understanding of the physical processes involved in laser-matter interactions and dopant diffusion phenomena.

## References

[1] D.W. Bäuerle, Laser Processing and Chemistry, fourth ed., Springer Science & Business Media, 2013.

[2] C.F. Bohren, D.R. Huffman, Absorption and Scattering of Light by Small Particles, Wiley-Blackwell, 1998.

[3] A. Florakis, E. Verrelli, D. Giubertoni, G. Tzortzis, D. Tsoukalas, Non-melting annealing of silicon by $CO_2$ laser, Thin Solid Films 518 (9) (2010) 2551–2554, https://doi.org/10.1016/j.tsf.2009.09.140.

[4] M.V. Allmen, A. Blatter, Laser-Beam Interactions With Materials, Springer-Verlag New York Inc., New York, NY, 1987.

[5] R.A. Smith, Semiconductors, second ed., Cambridge University Press, 1978.

[6] S.J. Fray, F.A. Johnson, J.E. Quarrington, N. Williams, Lattice bands in germanium, Proc. Phys. Soc. 85 (1) (1965) 153–158, https://doi.org/10.1088/0370-1328/85/1/321.

[7] H.Y. Fan, Effects of free carriers on the optical properties, in: R.K. Willardson, A.C. Beer (Eds.), Optical Properties of III–V Compounds, Semiconductors and Semimetals, vol. 3, Academic Press, New York, 1967.

[8] M. Blomberg, K. Naukkarinen, T. Tuomi, V.-M. Airaksinen, M. Luomajärvi, E. Rauhala, Substrate heating effects in CO2 laser annealing of ion-implanted silicon, J. Appl. Phys. 54 (5) (1983) 2327–2328, https://doi.org/10.1063/1.332343.

[9] J.R. Meyer, M.R. Kruer, F.J. Bartoli, Optical heating in semiconductors: laser damage in Ge, Si, InSb, and GaAs, J. Appl. Phys. 51 (10) (1980) 5513, https://doi.org/10.1063/1.327469.

[10] W. Spitzer, H.Y. Fan, Infrared absorption in n-type silicon, Phys. Rev. 108 (2) (1957) 268–271, https://doi.org/10.1103/physrev.108.268.

[11] H.Y. Fan, W. Spitzer, R.J. Collins, Infrared absorption in n-type Germanium, Phys. Rev. 101 (2) (1956) 566–572, https://doi.org/10.1103/physrev.101.566.
[12] R.G. Ulbrich, Low density photoexcitation phenomena in semiconductors: aspects of theory and experiment, Solid-State Electron. 21 (1) (1978) 51–59, https://doi.org/10.1016/0038-1101(78)90114-4.
[13] E.J. Yoffa, Role of carrier diffusion in lattice heating during pulsed laser annealing, Appl. Phys. Lett. 36 (1) (1980) 37, https://doi.org/10.1063/1.91306.
[14] E.J. Yoffa, Dynamics of dense laser-induced plasmas, Phys. Rev. B 21 (6) (1980) 2415–2425, https://doi.org/10.1103/physrevb.21.2415.
[15] A. Lietoila, Computer modeling of the temperature rise and carrier concentration induced in silicon by nanosecond laser pulses, J. Appl. Phys. 53 (4) (1982) 3207, https://doi.org/10.1063/1.331020.
[16] W.B. Gauster, Laser-induced infrared absorption in silicon, J. Appl. Phys. 41 (9) (1970) 3850, https://doi.org/10.1063/1.1659517.
[17] M. Rasolt, H. Kurz, Lifetime of an electron-hole plasma of silicon in the picosecond range, Phys. Rev. Lett. 54 (7) (1985) 722–724, https://doi.org/10.1103/physrevlett.54.722.
[18] J.A. van Vechten, R. Tsu, F.W. Saris, Nonthermal pulsed laser annealing of Si; plasma annealing, Phys. Lett. A 74 (6) (1979) 422–426, https://doi.org/10.1016/0375-9601(79)90242-1.
[19] S de Unamuno, E. Fogarassy, A thermal description of the melting of c- and a-silicon under pulsed excimer lasers, Appl. Surface Sci. 36 (1–4) (1989) 1–11, https://doi.org/10.1016/0169-4332(89)90894-5.
[20] V. Gonda, J. Slabbekoorn, L.K. Nanver, Silicon laser annealing by a two-pulse laser system with variable pulse offsets, in: 15th International Conference on Advanced Thermal Processing of Semiconductors, IEEE, 2007, https://doi.org/10.1109/rtp.2007.4383851.
[21] C.J. Glassbrenner, G.A. Slack, Thermal conductivity of silicon and germanium from 3 K to the melting point, Phys. Rev. 134 (4A) (1964) A1058–A1069, https://doi.org/10.1103/physrev.134.a1058.
[22] Y.S. Touloukian, E.H. Buyco, in: Y.S. Touloukian (Ed.), Specific Heat: Nonmetallic Solids, Thermophysical Properties of Matter, vol. 5, IFI/Plenum, New York, 1975.
[23] P. Sommelet, R.L. Orr, High-temperature thermal properties of germanium, J. Chem. Eng. Data 11 (1) (1966) 64–65, https://doi.org/10.1021/je60028a018.
[24] A.S. Okhotin, A.S. Pushkarskii, V.V. Gorbachev, Thermophysical properties of semiconductors, "Atom" Publishing House, Moscow, 1972.
[25] M. Lax, Temperature rise induced by a laser beam, J. Appl. Phys. 48 (9) (1977) 3919, https://doi.org/10.1063/1.324265.
[26] M. Lax, Temperature rise induced by a laser beam II. The nonlinear case, Appl. Phys. Lett. 33 (8) (1978) 786, https://doi.org/10.1063/1.90505.
[27] C.K.N. Patel, Interpretation of CO2 optical maser experiments, Phys. Rev. Lett. 12 (21) (1964) 588–590, https://doi.org/10.1103/physrevlett.12.588.
[28] O. Svelto, Principles of Lasers, third ed., Springer, USA, 1989.
[29] A. Florakis, N. Misra, C. Grigoropoulos, K. Giannakopoulos, A. Halimaoui, D. Tsoukalas, Non-melt laser annealing of plasma implanted boron for ultra shallow junctions in Silicon, Mater. Sci. Eng. B 154–155 (2008) 39–42, https://doi.org/10.1016/j.mseb.2008.09.035.
[30] M. Doubenskaia, P. Bertrand, I. Smurov, Pyrometry in laser surface treatment, Surf. Coat. Technol. 201 (5) (2006) 1955–1961, https://doi.org/10.1016/j.surfcoat.2006.04.060.
[31] Semiconductor Industry Association, International Technology Roadmap for Semiconductors, Sematech, 2015. http://www.itrs2.net.

[32] A. Gat, J.F. Gibbons, A laser-scanning apparatus for annealing of ion-implantation damage in semiconductors, Appl. Phys. Lett. 32 (3) (1978) 142, https://doi.org/10.1063/1.89960.
[33] A. Gat, J.F. Gibbons, T.J. Magee, J. Peng, V.R. Deline, P. Williams, C.A. Evans, Physical and electrical properties of laser-annealed ion-implanted silicon, Appl. Phys. Lett. 32 (5) (1978) 276, https://doi.org/10.1063/1.90046.
[34] J. Krynicki, J. Suski, S. Ugniewski, R. Grötzschel, R. Klabes, U. Kreissig, J. Rüdiger, Laser annealing of arsenic implanted silicon, Phys. Lett. A 61 (3) (1977) 181–182, https://doi.org/10.1016/0375-9601(77)90286-9.
[35] B. Mizuno, Y. Sasaki, Aiming for the best matching between ultra-shallow doping and milli-to femto-second activation, in: 15th International Conference on Advanced Thermal Processing of Semiconductors, IEEE, 2007, https://doi.org/10.1109/rtp.2007.4383811.
[36] V. Privitera, C. Spinella, G. Fortunato, L. Mariucci, Two-dimensional delineation of ultrashallow junctions obtained by ion implantation and excimer laser annealing, Appl. Phys. Lett. 77 (4) (2000) 552, https://doi.org/10.1063/1.127041.
[37] S. Whelan, A.L. Magna, V. Privitera, G. Mannino, M. Italia, C. Bongiorno, G. Fortunato, L. Mariucci, Dopant redistribution and electrical activation in silicon following ultra-low energy boron implantation and excimer laser annealing, Phys. Rev. B 67 (7) (2003), https://doi.org/10.1103/physrevb.67.075201.
[38] K.K. Ong, K.L. Pey, P.S. Lee, A.T.S. Wee, X.C. Wang, Y.F. Chong, Dopant activation in subamorphized silicon upon laser annealing, Appl. Phys. Lett. 89 (8) (2006), https://doi.org/10.1063/1.2335950.
[39] N. Misra, L. Xu, Y. Pan, N. Cheung, C.P. Grigoropoulos, Excimer laser annealing of silicon nanowires, Appl. Phys. Lett. 90 (11) (2007), https://doi.org/10.1063/1.2713774.
[40] S.T. Dunham, A quantitative model for the coupled diffusion of phosphorus and point defects in silicon, J. Electrochem. Soc. 139 (9) (1992) 2628, https://doi.org/10.1149/1.2221276.
[41] M.-J. Caturla, T. Diaz de la Rubia, J. Zhu, M. Johnson, Monte Carlo simulation of boron diffusion during low energy implantation and high temperature annealing, MRS Proc. 469 (1997), https://doi.org/10.1557/proc-469-335.
[42] P. Pichler, Current understanding and modeling of boron-interstitial clusters, MRS Proc. 717 (2002), https://doi.org/10.1557/proc-717-c3.1.
[43] F. Cristiano, X. Hebras, N. Cherkashin, A. Claverie, W. Lerch, S. Paul, Clusters formation in ultralow-energy high-dose boron-implanted silicon, Appl. Phys. Lett. 83 (26) (2003) 5407, https://doi.org/10.1063/1.1637440.
[44] I. Avci, Modeling extended defect ({311} and dislocation) nucleation and evolution in silicon, J. Appl. Phys. 95 (5) (2004) 2452, https://doi.org/10.1063/1.1645644.
[45] C.J. Ortiz, F. Cristiano, B. Colombeau, A. Claverie, N.E.B. Cowern, Modeling of extrinsic extended defect evolution in ion-implanted silicon upon thermal annealing, Mater. Sci. Eng. B 114–115 (2004) 184–192, https://doi.org/10.1016/j.mseb.2004.07.054.
[46] E. Lampin, F. Cristiano, Y. Lamrani, A. Claverie, B. Colombeau, N.E.B. Cowern, Prediction of boron transient enhanced diffusion through the atom-by-atom modeling of extended defects, J. Appl. Phys. 94 (12) (2003) 7520, https://doi.org/10.1063/1.1627461.
[47] M. Aboy, L. Pelaz, E. Bruno, S. Mirabella, S. Boninelli, Kinetics of large B clusters in crystalline and preamorphized silicon, J. Appl. Phys. 110 (7) (2011), https://doi.org/10.1063/1.3639280.
[48] S.B. Felch, F. Torregrosa, H. Etienne, Y. Spiegel, L. Roux, D. Turnbaugh, J. Matsuo, M. Kase, T. Aoki, T. Seki, PULSION® HP: Tunable, High Productivity Plasma Doping, AIP Publishing, 2011, https://doi.org/10.1063/1.3548413.

[49] P. Pichler, Intrinsic Point Defects, Impurities, and Their Diffusion in Silicon, Springer, Vienna, 2004.

[50] D.A. Antoniadis, Boron in near-intrinsic and silicon under inert and oxidizing ambients—diffusion and segregation, J. Electrochem. Soc. 125 (5) (1978) 813, https://doi.org/10.1149/1.2131554.

[51] G. Masetti, M. Severi, S. Solmi, Modeling of carrier mobility against carrier concentration in arsenic-, phosphorus-, and boron-doped silicon, IEEE Trans. Electron. Dev. 30 (7) (1983) 764–769, https://doi.org/10.1109/t-ed.1983.21207.

[52] J.A. Sharp, N.E.B. Cowern, R.P. Webb, K.J. Kirkby, D. Giubertoni, S. Gennaro, M. Bersani, M.A. Foad, F. Cristiano, P.F. Fazzini, Deactivation of ultrashallow boron implants in preamorphized silicon after nonmelt laser annealing with multiple scans, Appl. Phys. Lett. 89 (19) (2006) 192105, https://doi.org/10.1063/1.2385215.

[53] A. Florakis, A. Papadimitriou, N. Chatzipanagiotis, N. Misra, C. Grigoropoulos, D. Tsoukalas, Formation of silicon ultra shallow junction by non-melt excimer laser treatment, Solid-State Electron. 54 (9) (2010) 903–908, https://doi.org/10.1016/j.sse.2010.04.025.

[54] S.R. Aid, S. Matsumoto, G. Fuse, S. Sakuragi, Comparison of boron diffusion in silicon during shallow p+/n junction formation by non-melt excimer and green laser annealing, Phys. Status Solidi A 208 (12) (2011) 2772–2777, https://doi.org/10.1002/pssa.201127198.

Chapter 3

# Atomistic modeling of laser-related phenomena

Luis A. Marqués[a], María Aboy[a], Pedro López[a], Iván Santos[a], Lourdes Pelaz[a], and Giuseppe Fisicaro[b]

[a]*Department of Electricity and Electronics, Technical School of Telecommunications Engineering, University of Valladolid, Valladolid, Spain,* [b]*CNR Institute for Microelectronics and Microsystems (CNR-IMM), Catania, Italy*

## 3.1 Introduction

Although experimental tests are still essential for the design and optimization of new technologies in the semiconductor industry, modeling and simulation are becoming fundamental to reduce development times and costs. Nowadays, atomistic modeling, along with more traditional continuum-based tools, is being used for processing simulation [1]. Modeling is particularly important in order to study the effects of laser irradiation on semiconductor properties in both melted and solid phases [2], as laser processing is gaining relevance within the electronic industry. Many issues are still poorly understood, even for the widely investigated case of laser thermal annealing (LTA) of ion-implanted Si. For instance, the evolution of common dopant atoms in the melting phase, or the amount of dopant activation promoted by laser irradiation in solid regions [3–5]. Due to the intrinsic features of laser annealing treatments, i.e., localized irradiation with a space- and time-dependent (on the nanosecond scale) thermal field that leads to far-from-equilibrium conditions, experimental analysis can reveal only specimen postirradiation characteristics. In order to understand how the laser-irradiated system evolves and reaches its final state, a theoretical work based on full process simulation is required.

Traditional models for laser processing simulation were based on continuum techniques (see Chapter 6 of this book). However, they prove to be insufficient at the nanoscale. Atomistic simulation methods, in turn, are handy for studying detailed interactions, both at the electronic and atomic levels, including out-of-equilibrium situations as those present in laser processing. However, the extension of atomistic simulation approaches that are valid for conventional annealing is not straightforward for laser processing conditions. On the one

**Laser Annealing Processes in Semiconductor Technology https://doi.org/10.1016/B978-0-12-820255-5.00003-9**

hand, energy transfer to the lattice comes from the excitation of the electronic system, which now has to be carefully modeled (especially in ultrafast laser annealing) while it was completely ignored in conventional processing simulations. On the other hand, the characteristics of the thermal profiles induced by the laser are very demanding for numerical simulations, as they evolve very rapidly in time with substrate zones close to or above the melting point next to regions at room temperature. Additionally, atomistic parameters (diffusivities, bonding energies, prefactors, etc.), which have been calibrated in conventional annealing conditions, have to be revised in order to be applied to laser annealing simulation as entropy effects may become relevant.

In this chapter, we will review the main techniques for atomistic simulation to be applied for laser modeling, from the more fundamental ab initio and tight binding, to the less detailed Monte Carlo methods that allow reaching macroscopic size and timescales. In Section 3.2, we will introduce the general features of these techniques; to then consider more specifically in Section 3.3, the particular models used for laser annealing simulation in each case. In Section 3.4, we will present some recent results on atomistic modeling of laser-related phenomena in Si, in particular melting and regrowth processes, defect and dopant kinetics, dopant segregation, and the anomalous formation of extended defects. Finally, in Section 3.5, we will address some conclusions and future perspectives about the atomistic modeling of laser processing.

## 3.2 Atomistic simulation techniques

In this section, we will review the main techniques for atomistic simulation, and the need of using such techniques in a multiscale scheme in order to reach time and space scales typical of real materials and processes.

### 3.2.1 Ab initio

Ab initio methods allow determining the fundamental properties of materials based on the laws of quantum mechanics. The underlying core is the resolution of the many-body Schrödinger equation for systems containing from tens to thousands of atoms and their corresponding electrons. Given the position of a set of atom nuclei $R$ and their electrons $r$, the many-body Schrödinger equation of the system is:

$$H\Psi(R,r) = E\Psi(R,r) \tag{3.1}$$

where $\Psi(R,r)$ is the wave function of the system (atom nuclei and electrons), $H$ is its Hamiltonian, and $E$ the total energy of the system. The computational complexity grows exponentially with the number of atoms, and so several approaches have arisen to be able to solve it. Nonrelativistic Born-Oppenheimer approximation assumes that electrons move much faster than nuclei, as they are thousands of times lighter. Consequently, the electronic and nuclei motions are decoupled. By means of this ansatz, and together with the adiabatic approximation (nuclei move on the potential-energy surface of the electronic ground state), the electronic system can be studied separately through this time-independent Schrödinger equation:

$$-\frac{\hbar^2}{2m_e}\nabla^2\Psi_e(r;R)+V(r;R)\Psi_e(r;R)=E_e\Psi_e(r;R) \tag{3.2}$$

being $\hbar = h/2\pi$ where $h$ is Planck's constant, $m_e$ the mass of the electron, $\Psi_e(r;R)$ the wave function of the whole electronic system (with atom nuclei at positions $R$) once the electron and nuclei movement has been decoupled, and $V(r;R)$ is the potential affecting the electron movement (that includes electron-nuclei and electron-electron interactions).

A well-defined theoretical framework is the wave function-based approaches. The simplest one is the Hartree-Fock method, which produces reasonable results for many properties, by using the variational theorem to obtain an approximate solution. However, it neglects electron correlation in electronic structure calculations, which implies each electron accounts for the rest of them as a mean-field. Another drawback is the high computational effort: calculations for a system composed of $N$ atoms scale as $N^4$ [6]. A complementary approach to Hartree-Fock method is the density functional theory (DFT), based on the Hohenberg-Kohn theorem [7], which states that the ground-state electronic energy is uniquely determined by the electron density. Despite several approaches of the correct energy functional have to be made, DFT calculations are feasible for realistic models to accurately predict material properties.

Ab initio can be considered as the more fundamental atomistic technique since it gives an excellent insight into the underlying physics of systems with no free parameters. Nevertheless, the high computational cost of ab initio limits the accessible system sizes and simulation times. Most ab initio studies use simulation cells of about 200–300 atoms. Larger simulation cells are not commonly considered due to the computational overhead, although high-performance computing facilities can be used for treating larger systems. The timescale accessible to ab initio molecular dynamics is typically of the order of several picoseconds. Although diffusion paths can be studied with ab initio molecular dynamics, this timescale is, in general, not enough for observing complete diffusion events, and consequently, static methods such as the nudged elastic band method [8] are commonly employed.

The ab initio technique is best suited to provide essential information on defects and dopants, to be subsequently used in front-end process models. For example, it has been employed to study the energetics of Si self-interstitials (Is) and vacancies (Vs) [9–12] in order to find their most stable configuration and to obtain an atomistic interpretation of self-diffusion in Si [13], as well as to analyze the formation of Si self-I [14–16] and V [17] clusters. The diffusion of dopants [18–22] and the formation and dissolution of dopant-defect complexes [23, 24] have also been analyzed using ab initio techniques, providing information about which are the mobile species, and their formation and migration energies. Furthermore, ab initio simulations also enable studying the influence of charge states of dopant-defect complexes on the diffusion process [21].

### 3.2.2 Tight binding

As in the case of ab initio methods, the starting point of the tight binding (TB) method is the Schrödinger equation describing the system under study. After applying the Born-Oppenheimer approximation, the many-particle problem is reduced to the problem of just one particle (electron) moving in an effective field due to the other electrons and the ions. In the case of an extended system such a crystal, Bloch [25] provided a formal procedure for developing the crystalline one-electron wave function on the basis of the atomic orbitals, which constitutes the so-called linear combination of atomic orbitals, Bloch, or just TB method. Unlike the parameter-free ab initio technique, an adequate parameterization of the Hamiltonian matrix elements is necessary to describe the system under study. This fact makes TB less computationally expensive than ab initio, but also less transferable. An extensive review of the TB method can be found in Colombo [26].

The lower computational cost of TB enables accessing longer simulation times and larger systems with respect to ab initio. Typical cells contain about several thousand atoms, and it is possible to simulate their dynamics during times of the order of hundreds of picoseconds. Nevertheless, TB can be linked to temperature-accelerated dynamics schemes for easily accessing the nanosecond timescale [27]. Consequently, diffusion paths can be obtained directly from the simulation without making any initial guess or resorting to static techniques. The formation and migration mechanisms of native point defects in Si [28–31], the recombination of Si self-Is and Vs [31, 32], the formation energies of V and Si self-I clusters [27, 33–36], and the levels they introduce in the gap [35] are the examples of different studies performed using TB simulation techniques. Extended defects, such as planar {113} defects, have also been studied using TB simulations [37–39]. Furthermore, it has also been possible to calculate the binding and formation energies of boron-interstitial clusters in Si [40, 41].

### 3.2.3 Classical molecular dynamics

The classical molecular dynamics (CMD) simulation technique is based on the resolution of Newton equations of motion for all particles in the system. The forces among atoms ($F$) are derived from an empirical interatomic potential $U(r_1, r_2, \ldots, r_N)$, as described in Eq. (3.3) for a system with $N$ particles:

$$m_i \frac{d^2 r_i}{dt^2} = F_i = -\nabla_i U(r_1, r_2, \ldots, r_N), \quad i = 1, 2, \ldots, N \tag{3.3}$$

where $m_i$ is the mass of particle $i$, $r_i$ its position vector, and $F_i$ is the force acting on it. The only inputs for a CMD simulation are the interatomic potential and the initial positions and velocities of all particles, but no other parameters or previous assumptions are required. Once this information is known, Eq. (3.3) can be solved for each particle using a numerical integration scheme, providing as direct outputs the trajectories of the particles during a small time interval, or time step. The time step must be large enough to allow a long simulation time, but small enough to capture the dynamics of the system and conserve the energy between integration steps. Typical values for the time step are of the order of femtoseconds.

In the CMD technique, the electronic system is not considered, and atom dynamics is modeled by empirical interatomic potentials. Consequently, the accuracy of the CMD simulations relies on the ability of the potential to properly describe the material and the process under study. Interatomic potentials are analytical functions developed and fitted to reproduce thermodynamic properties of the equilibrium phases of a material [42–44]. Several potential parameterizations have been proposed for semiconductors, the most widely used being the Stillinger-Weber [45] and Tersoff [46] potentials. The choice of an appropriate potential is crucial in any CMD simulation, as each potential may be particularly suited for the description of a specific process in a given material. Recently developed machine-learning potentials aim at reducing the uncertainties of empirical potentials, as they can provide the accuracy of quantum mechanics without the computational cost of ab initio or TB techniques [47–49].

CMD allows the calculation of macroscopic properties that can be directly compared to experiments. As the trajectories of particles are simulated, not only structural but also dynamical quantities can be obtained, which is one of the main advantages of CMD with respect to other molecular simulation techniques and more fundamental methods. CMD is a suitable technique in analyzing the physical mechanisms underlying a process, even in nonequilibrium situations, as it can simulate the dynamics of a system without using additional input parameters or predefined equations. Besides, the exponential increase on the computational capabilities in the last decades has extended the range of application of CMD, expanding its temporal and length scales. With the use of large supercomputers or distributed computing schemes, the CMD technique has been able to simulate micron-size samples [50] and reach simulated times of hundreds of microseconds [51]. Nowadays, one of the most widespread codes for CMD simulations is LAMMPS, developed at Los Alamos [52, 53], which has somehow replaced the large number of preexisting codes developed by different research groups. A more detailed description of the CMD method and its areas of applicability can be found in several publications and books devoted to this technique [6, 54–57]. In particular, CMD simulations have been used to investigate a wide range of laser-related phenomena, such as thermoelastic deformation, melting and crystallization, spallation, or laser ablation. For further information on the use of the CMD technique to study laser applications, the reader is referred to recent reviews and books available in the literature [58–61].

### 3.2.4 Kinetic Monte Carlo

The dynamics of a system can also be simulated from an atomistic point of view by the use of kinetic Monte Carlo (KMC) techniques. KMC attempts to overcome the limitations of CMD by taking into account that the long-time dynamics of a system typically consists of a succession of events. These events are determined by several parameters (capture volumes, activation, migration, binding energies, etc.) whose values must be specified a priori.

All events simulated in KMC are thermally activated processes that require to overcome an energy barrier, $E_a$, to occur. The probability $P$ of occurrence follows a Boltzmann distribution:

$$P \propto \exp\left(-\frac{E_a}{k_B T}\right) \tag{3.4}$$

being $T$ the temperature and $k_B$ the Boltzmann constant. Once a given event has been selected from a distribution that depends on the physics of the system under study, a random number, $R$, is generated. If $R \leq P$, the event takes place. The frequency (or probability per unit time) at which a particular event $i$ occurs can be expressed as

$$\nu^i = \nu_0^i \exp\left(-\frac{E_a}{k_B T}\right) \tag{3.5}$$

where $\nu_0^i$ is the prefactor. For example, an atomic diffusion jump has a frequency of $\sim 10^9\ \text{s}^{-1}$, while it is of $\sim 10^3\ \text{s}^{-1}$ for the emission of a point defect by an extended defect. Then, rather than following the trajectory through every vibrational period as it is done in CMD, these events (diffusion jumps, emission-capture of point defects at extended defects, etc.) are treated directly. The result is that KMC can reach vastly longer timescales, typically seconds and often well beyond.

Some phenomena studied with KMC can take advantage of additional simplifications. For example, when studying the diffusion of dopants and defects, atoms placed at crystalline lattice positions do not contribute to the overall process. Therefore, these crystalline atoms can be removed from the simulation scheme, and only those which actively take part can be considered, i.e., only intrinsic (Si self-Is and Vs) and extrinsic (dopants and impurities) defects. This constitutes the underlying idea of the nonlattice KMC approach [62]. Other phenomena, like the liquid- or solid-phase epitaxial regrowth, for example, would need the presence of the lattice in the simulation if orientation effects have to be accounted for. In these cases, a lattice KMC approach, where all atoms in the lattice are included in the simulation, is required [63].

Once all the relevant events of a process have been identified and characterized with a frequency of occurrence (or probability per unit of time) given by their energetic features, the KMC simulation proceeds as follows:

- Depending on the particles present in the system and the different events they can perform (diffusion jumps, recombination of a Si self-I and a V, formation-dissolution of dopant complexes, emission of a particle from a cluster, etc.), a total rate for event $i$ is defined as $R_i = \sum_{j=1}^{i} \upsilon^j$, where $i = 1, \ldots, N$ ($N$ is the total number of possible events in the system).
- Two random numbers are generated, $r_1$ and $r_2$.
- The event that will take place is selected by finding an $i$ verifying $R_{i-1}$ - $< r_1 R_N < R_i$. If an event has a very high frequency associated with it, the event will be assigned a large fraction of the probability bar $R_i$ and it will be selected very often.
- The simulation time is increased by an amount $\Delta t = {}^{\ln(1/r_2)}\big/_{R_N}$.
- Event rates must be updated after each event takes place and the process begins again.

This loop is repeated millions of times in each simulation.

The KMC method has been extensively used for the simulation of processing in the semiconductor technology [64–67]. A recent review of the technique and its applications can be found in Martín-Bragado et al. [68].

### 3.2.5 Multiscale modeling

Most process simulators used in practical applications are based on continuum methods since they are fast and allow the consideration of large systems. However, as electronic device dimensions are reduced, the atomistic nature of the material arises and complex physical interactions show up. To include such new effects, a relatively high number of coupled partial differential equations have to be considered in continuum methods and the computational cost increases. Then, atomistic simulation techniques become a good alternative even for industrial applications. However, extracting meaningful information from mechanisms taking place at atomic level, and extending such information into the macroscopic world in a consistent manner, has inherent issues. To bridge the wide range of time and length scales, it is necessary to deal with a hierarchy of physical models.

Among all atomistic techniques, KMC can access to times and system sizes comparable with actual processing scales. This is because only the relevant events for the system under study are considered. However, the reliability of the obtained results depends on the models implemented in the KMC code. These models require first the correct identification of all the relevant events and interactions, and second their precise parametric characterization. Experiments can be very useful for these purposes. Nevertheless, sometimes it is not easy to design the appropriate experiment to measure the magnitude of interest since, in general, it is difficult to relate macroscopic observations to atomic properties. The required information for the models implemented in KMC can be in turn easily obtained from more fundamental atomistic simulation methods. Ab initio and TB can be employed to study defect configurations and energetics, material and electronic properties, and to optimize empirical potentials along with experimental data. CMD simulations using empirical potentials in turn can be used to determine interaction and diffusion mechanisms involving complex defects or to study the damage morphology generated after doping as well as its annealing behavior. All these data, along with experimental measurements if available, will define the relevant events and parameters to be considered in the KMC simulator.

Laser irradiation presents a clear multiscale nature, as it triggers several physical mechanisms affecting both the electronic subsystem and the nuclei of the material, with associated timescales that may range from femtoseconds to macroscopic times. Consequently, the simulation of laser-related phenomena requires the combination of different techniques as they give information on a diverse-scale level (see Fig. 3.1). On the one hand, ab initio and TB methods can describe the laser-induced electron excitation and the ultrafast heat transfer from electrons to lattice, but they are limited to systems with relatively few atoms and very short times. On the other hand, continuum methods can access macroscopic time and length scales, but frequently, their description is based on the equilibrium properties of the material, while laser-related phenomena are highly nonequilibrium in nature. Nevertheless, temperature and electromagnetic fields necessarily need continuum description for the simulation of laser processing in real devices (see Chapter 6). Large-scale CMD and KMC

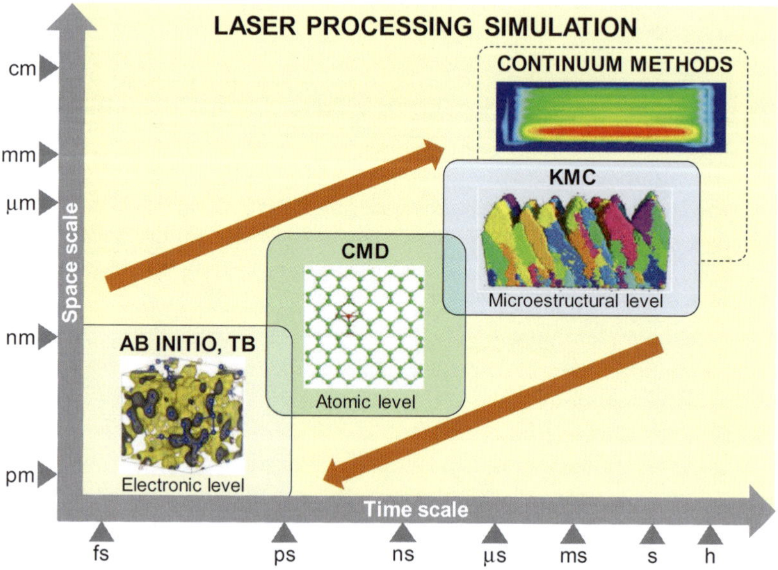


FIG. 3.1 Multiscale scheme for laser processing simulation.

simulations can contribute to bridge the gap between both extremes, as they can provide detailed insight into the atomic-level dynamics involved in laser processing, information that can be then fed into continuum simulators.

## 3.3 Laser annealing modeling from an atomistic perspective

In this section, we review the models used in the different atomistic simulation methods to address the study of laser-related phenomena.

### 3.3.1 Ab initio models

Ab initio modeling of Si laser heating has evolved with the improvement of computing capabilities as more sophisticated approaches could be applied, from the initial simulations based on finite temperature-density functional theory (FT-DFT) in the latter 90s of last century, to the recent coupling of Maxwell equations to time-dependent density functional theory (TD-DFT). In the following subsections, it will be given a historical review of the different approaches used and the main results obtained.

#### *3.3.1.1 Finite temperature-density functional theory*

Laser heating of Si was first studied using ab initio simulations by Parrinello and coworkers [69, 70]. In their work, they used the ab initio molecular dynamics simulation technique introduced by Alavi et al. [71] to study excited electrons.

In DFT, the energy of a system is defined as a functional of the electronic density. This energy functional is minimized at the ground state, and the ground-state electronic density determines the properties of the system [7, 72, 73]. As a ground-state theory, it cannot directly be applied to study the laser heating of a material where a fraction of electrons is excited from occupied to unoccupied electronic levels (in the case of semiconductors, from the valence band to the conduction band). In addition, right after laser illumination, the distribution of excited electrons into electronic levels is not in thermal equilibrium and therefore it cannot be described using a Fermi-Dirac distribution.

To circumvent these issues, Parrinello and coworkers considered that, since the relaxation time for electrons (from their initial excited distribution to an equilibrium distribution) is much shorter than the electron-ion relaxation time (after which energy is transferred from the electron subsystem to the ions), electrons rapidly thermalize at the initial laser-induced temperature, and their distribution can be described using a Fermi-Dirac distribution. Thus, the temperature of the electronic subsystem ($T_e$) is different and much higher from the temperature of the ions ($T_e \sim 10^4$ K $\gg T_{ion}$). The use of a Fermi-Dirac distribution for electrons with $T_e \neq 0$ K (i.e., different from the ground state electronic distribution) is called the finite temperature (FT) approach. It is commonly used to describe laser heating of materials in DFT studies, as it will be presented in this section, and in TB studies, as it will be shown in

Section 3.3.2. In addition, different temperatures for the electronic and ionic subsystems are the basis of the two-temperature model (TTM) commonly employed to describe the laser heating of materials [59, 74].

In their study, Parrinello and coworkers analyzed the stability of the Si lattice, initially thermalized at 300 K, when the electronic temperature was set to 15,000 and 25,000 K. For the lower electronic temperature, they observed large oscillations of atoms around their equilibrium positions, but melting was not observed within the simulated time (~0.4 ps). On the contrary, they observed a disordering process for the higher electronic temperature: the temperature of ions increased, and it was equilibrated at ~1700 K (temperature close to the experimental melting temperature of Si, 1688 K). The final structure was similar to liquid Si in terms of disorder and diffusivity, but with a higher atomic coordination (11–13 rather than the 6–7 of liquid Si), and the first peak of the pair correlation function was shifted to ~2.7 Å (this peak is at 2.4–2.5 Å in liquid Si). All the features of liquid Si were recovered in this case by setting $T_e = 1700$ K (i.e., electrons and ions in thermal equilibrium at the same temperature). The observed process of structural disordering at high $T_e$ was attributed to laser-induced changes in the interatomic potential due to a covalent bond weakening by the presence of a large concentration of excited electrons. Nevertheless, the temperature of the ions increased to the melting temperature, contrary to the cold melting suggested in the plasma annealing model for laser illumination [75, 76].

FT-DFT was also used by Garcia and coworkers to study laser heating of crystal Si [77–84]. They developed an ab initio code for highly excited valence electron systems, called CHIVES [77, 81, 82, 84], which uses a localized Gaussian basis set for atomic orbitals rather than plane waves as other DFT codes do. Their studies were focused on analyzing the effects of different electronic temperatures on the Si lattice. They found that the electronic temperature threshold for melting is 54 mHa, which corresponds to 1.47 eV and $T_e = 17{,}052$ K [78, 84]. For $T_e \leq 16{,}736$ K (53 mHa, 1.44 eV), there is no melting of the Si lattice although atoms oscillate around their equilibrium lattice positions. In this regime of $T_e$, there is a laser-induced modification of the potential energy surface and thus of the interatomic force constants. These modifications soften the covalent bonds of the Si lattice and cause phonon to be thermally squeezed [81]. For $T_e \geq 17{,}052$ K (54 mHa, 1.47 eV), they observed an ultrafast nonthermal melting process of the Si lattice due to the induced laser instability of the lattice. During this process, atoms show different diffusive behaviors with time [83, 84]. First, atoms behave superdiffusively in the initial stages after femtosecond laser illumination (i.e., after the electronic temperature is increased in the simulation) as bonds break due to laser illumination, which causes a lattice instability. This superdiffusive behavior is significantly relevant for $T_e$ above ~18,315 K [83]. Then, the behavior turns to fractional diffusive as the atomic scattering rate increases. Finally, atoms diffuse normally once the material has completely melted and atomic diffusion corresponds to that of a liquid.

This succession of diffusive behaviors is a consequence of the phonon squeezing for electronic temperatures above the melting threshold. In fact, Garcia and coworkers found that the observation of phonon squeezing in a material at electronic temperatures below the melting threshold indicates that the material can undergo nonthermal melting for high-fluence laser illumination. Nevertheless, they did not observe in their simulations any increment in the lattice temperature [78, 81], as it was reported by Parrinello and coworkers.

### *3.3.1.2 Finite temperature-density functional perturbation theory*

Laser illumination in Si causes bond softening, which results in the modification of the phonon spectrum of Si. For this reason, the study of the laser-induced changes in the phonon spectrum of Si was used in preliminary theoretical works to understand the lattice stability under laser illumination [85]. Phonon spectra and phonon dispersion relations can be obtained from the dynamical matrix of the system, which is calculated from the interatomic force constants (i.e., from the second derivative of the total energy with respect to atomic displacements). The harmonic approximation can be used in ab initio calculations to obtain the interatomic force constants from the changes in the forces acting on atoms when they are displaced from their equilibrium positions [86]. Another alternative and more versatile method is the linear response method through density functional perturbation theory (DFPT), i.e., the application of perturbation theory to DFT [87, 88]. This method is adequate when the external perturbation is small.

Different authors evaluated the Si lattice stability upon laser irradiation on the phonon spectra [89, 90]. In these studies, the effect of electronic excitation was incorporated in DFPT through the FT approach, i.e., through the electronic temperature in the Fermi-Dirac distribution. It was found that the transverse acoustic phonon frequencies along the Γ-X line become imaginary for $T_e \geq 1.5$ eV, which causes lattice instability. This electronic temperature range agrees with the results from Garcia and coworkers described in the previous section, who found the melting of the Si lattice for $T_e \geq 54$ mHa = 1.47 eV.

### *3.3.1.3 Time-dependent density functional theory*

TD-DFT constitutes a more appropriate method to study the interaction between a laser beam and matter as it is a time-dependent phenomenon. TD-DFT extends the basic ideas of ground-state DFT to solve the time-dependent Schrödinger equation, and it allows the treatment of excitations or more general time-dependent phenomena [91–93]. This method has been used to study the laser illumination of Si. The full modeling of the process is too complex and computationally demanding, and different approaches based on TD-DFT have been used.

Since laser-excited electrons rapidly relax into a thermal distribution after laser illumination, a FT approach can be used to describe the initial electronic distribution. This approach was used by Sato and coworkers in combination with TD-DFT to study the dielectric response of Si under laser [94]. They evaluated electronic temperatures up to 1.4 eV, and no melting of Si was reported.

Nevertheless, TD-DFT allows treating the laser interaction directly rather than using the FT approach. Lian et al. [95] considered the initial state of photoexcitation by changing the population of Kohn-Sham orbitals from a ground state to specific configurations. The population variation they considered was proportional to the optical transition probability associated with photons with an energy of 2.6 eV. Once the photoexcited electronic distribution was built, the system was studied using TD-DFT molecular dynamics simulations. Lian and coworkers found that when 10.16% (or 0.104 electrons per atom) of valence electrons are pumped to the conduction band, there is an electronic transfer from the bonding state to the antibonding state and the system undergoes an ultrafast disordering process at a temperature far below the thermal melting temperature of Si. They concluded that this ultrafast cold disordering of Si is due to a bond weakening associated with the laser-induced modifications in the electronic system, and it is consistent with the plasma annealing model for laser illumination [75, 76].

Results from Lian and coworkers show some common features from those obtained in theoretical works based on FT-DFT. The laser-induced increment of the electronic population of antibonding states in the Si lattice is a common result. Nevertheless, the percentage of valence electrons pumped to the conduction band for lattice disordering is slightly different: 10.6% (0.104 electrons per atom) [95], 17% (0.1 electrons per atom) [69, 70], 8.4% [79]. Garcia and coworkers also observed that the lattice temperature stays well below the melting temperature of Si 600 fs after laser illumination for electronic temperatures above the melting threshold of 54 mHa ($T_e = 17{,}052$ K) (see Fig. 5 of [78]). However, Parrinello and coworkers reported the increase of the lattice temperature up to the Si melting temperature after laser illumination [69, 70]. These discrepancies may arise from different theoretical methods used: in FT-DFT, the electronic temperature is not allowed to change, while the electronic distribution evolves with time in TD-DFT [96,97].

In TD-DFT formalism, the laser beam can also be described as an electromagnetic perturbation that interacts with the electrons of the material [98–101]. This approach was used to study laser-induced coherent phonon generation below the melting threshold [100] or the dielectric response of crystal Si following irradiation by high-intensity laser pulses [98]. It was also used to understand the electron transfer from the valence band to conduction band through laser illumination band gap reduction of Si [99]. This study also observed a laser-induced depletion of electronic density along the interatomic bonds, which was relocated in the interstitial regions associated with antibonding conduction band states. This study also reports the initial excitation of the electronic subsystem after laser illumination and the subsequent excitation of the lattice by the hot electrons that validate the phenomenological TTM. More recently, it was used to study the thermal effects on laser-assisted field evaporation from a Si surface [101]. This study revealed that dangling bonds present in the clean Si surface assist the laser excitation of electrons.

Recent developments for describing the laser illumination of materials are focused on coupling Maxwell equations (that describe the laser electromagnetic field) with TD-DFT [102]. In particular, Yabana and coworkers developed a multiscale simulation code called SALMON (Scalable Ab-initio Light-Matter simulator for Optics and Nanoscience) that combines Maxwell equations (for describing the laser electromagnetic field) with TD-DFT (for describing electron dynamics) and molecular dynamics (for describing lattice dynamics) [103–108]. This multiscale simulation has allowed observing how energy is transferred from the laser to the electrons and from the electrons to the lattice [108].

## 3.3.2 Tight binding models

TB methods have also been used for studying laser illumination of materials as they allow simulating larger systems for longer simulation times. As for ab initio, FT approach was used in early studies, while multiscale simulation tools are employed nowadays.

### *3.3.2.1 Finite temperature tight binding simulations*

As for ab initio methods, FT approach is used within TB methods to reproduce the excited electronic distribution after laser illumination. This approach assumes that the relaxation process from the initial excited nonequilibrium electronic distribution to a thermalized distribution is sufficiently fast. Stampfli and Bennenmann employed this approach to study the instability of the Si lattice induced by dense electron-hole plasma generated by laser illumination [109, 110]. Nevertheless, in their simulations, they considered the electronic entropy $S_e$ constant rather than $T_e$, and they determined $T_e$ through the condition of constant $S_e$. They found that the Si lattice becomes unstable if more than ~8% of the valence electrons are excited to the conduction band, as transverse acoustic phonons become instable. The percentage of excited valence electrons and the laser-induced instability of phonon modes are consistent with results obtained from ab initio methods.

Colombo and coworkers evaluated the effect of $T_e$ on the stability of the Si lattice [111]. They found that for $T_e \geq 15{,}000$ K, the system undergoes a melting transition within few hundreds of femtoseconds. The characteristics of the obtained liquid agree with the results of Parrinello and coworkers [69, 70]. They also found that the percentage threshold of excited valence electrons for ultra-fast melting is about 10%.

### *3.3.2.2 Time-dependent tight binding simulations*

Garcia and coworkers incorporated time evolution into the TB method through the time-dependent electronic occupation numbers [112–114]. This formalism allows the evolution of the electronic distribution (and therefore of $T_e$) with time. In addition, the initial excited electron distribution is not fixed as in the FT approach, but depends on the intensity of the laser pulse that is modeled with a function. Garcia and coworkers applied this method to study laser-illuminated Ge and C-based materials (graphite, diamond, and C nanotubes).

Medvedev and coworkers incorporated this time-dependent TB method in a multiscale scheme to model the effects of irradiation from extreme ultraviolet (photon energy above a few tens of eV) to hard X-rays (~40 keV photon energy). They developed a code called XTANT (X-ray-induced thermal and nonthermal transition), which combines TB molecular dynamics for atoms, with Monte Carlo treatment for high-energy electrons and core holes, and Boltzmann collision integrals for nonadiabatic (electron-phonon) coupling [115–121]. They applied their code to study the thermal and nonthermal melting of Si under femtosecond X-ray irradiation. They found thermal melting of Si for deposited energies of ~0.65 eV/atom that turns the crystalline structure into a low-density liquid by heating the lattice via an electron-ion-coupling mechanism. For doses above ~0.9 eV/atom, Si melts through a nonthermal mechanism into a high-density liquid phase within 300–500 fs via an interplay of lattice heating with the nonthermal modification of the interatomic potential due to the high density of excited valence electrons into nonbonding states.

### 3.3.3 Classical molecular dynamics models

In CMD, the electronic excitation effects are completely neglected. This implies that CMD cannot directly simulate the photon-electron interactions and the subsequent energy transfer to the nuclei that characterize the laser irradiation process. A first approach to overcome this limitation is to consider that the lattice has already thermalized as a consequence of the strong electronic excitation. Lattice temperature is then set following a pre-assumed heating profile, which specifies the spatial and temporal distributions of nuclei temperature. A more precise description of laser-excited semiconductors can be achieved by combining CMD with continuum models, in a so-called *hybrid approach*. Among these methods, the TTM considers that the electronic subsystem and the nuclei have different temperatures during the laser irradiation process, since a high electronic temperature is initially reached as a consequence of the laser-induced strong electronic excitation. Thus, the TTM simulates the heat transfer between the electrons and the lattice based on the time evolution of the electronic temperature, $T_e$, and the lattice temperature, $T_l$. Recently, electronic-temperature-dependent interatomic potentials have been developed for Si. These new parameterizations, based on quantum calculations, can simulate the local changes of material properties as a consequence of photon-electron interactions, which are responsible for phenomena such as nonthermal melting. In the following subsections, we will summarize different approaches for the modeling of laser irradiation in semiconductors by the CMD technique.

#### *3.3.3.1 Preassumed heating profiles*

Laser irradiation in semiconductors creates a nonequilibrium electron distribution with an excess of electron-hole pairs. The electrons excited to the conduction band will eventually relax and combine with holes, and the energy stored in the excited electronic subsystem will be transferred to the lattice. This process is relatively fast, within 100 femtoseconds to tens of picoseconds, compared to the characteristic timescale of the collective movement of atoms responsible for the structural and phase transformations triggered by the laser excitation. The electron relaxation effect can then be neglected, especially when laser duration (nanosecond pulse) is much larger than the electron thermal relaxation time. This serves as justification for a simplistic but effective CMD approach that assumes that the energy has already been transferred to the nuclei. Heating profiles resulting after laser irradiation are simulated by modifying the kinetic energy of the atoms within the irradiated area, according to the fast temperature increase induced by the laser.

Melting and rapid crystallization processes caused by laser irradiation in Si have been studied following this approach. The simulation box is divided into several layers whose temperature is separately set above and below the melting temperature, according to the power and duration of the laser pulse. System dynamics is determined by the high temperature gradients created, which result in phase transition and the subsequent crystallization during the cooling process. Several works have been devoted to analyze the melting, nucleation, and crystallization processes in crystal [122, 123] and amorphous Si [124, 125], in poly-Si on a $SiO_2$ substrate [126], and the melting and subsequent explosive crystallization of SiGe alloys [127].

A simplified heating profile has also been used to study the generation and growth of extended defects after laser annealing of ion-implanted Si [128, 129]. In these works, a nanosecond laser annealing of a highly damaged region resulting after ion implantation is simulated by considering temperatures from 0.5 to 0.9 times the Si melting point. The analysis of defect dynamics in this high-temperature regime has allowed to unveil the mechanisms responsible for the formation of some anomalous extended defects observed in experiments. Detailed information on this study is provided in Section 3.4.4.

In the *laser energy absorption model*, the amount of energy deposited in the lattice, and therefore the increase in the kinetic energy of the atoms, is obtained assuming that the laser beam is exponentially absorbed in the target. According to the Beer-Lambert law, the intensity of an electromagnetic wave $I(z)$ inside a material falls off exponentially from the surface as [130]:

$$I(z) = I_s \exp\left(-\frac{\Delta z}{\delta}\right) = I_{\max} \exp\left[-(t-t_0)^2/t_g^2\right] \exp\left(-\frac{\Delta z}{\delta}\right) \tag{3.6}$$

where $I_s$ is the incident laser beam intensity at the surface, $\delta$ is the optical penetration depth, and $z$ represents depth. As included in Eq. (3.6), the laser is usually assumed to have a Gaussian shape in time, with a pulse duration of $t_g$ centered at time $t_0$, and a maximal intensity of $I_{\max}$. A Gaussian distribution is also considered for the $x$ and $y$ directions perpendicular to the laser beam. This model allows a more detailed description of the temporal and spatial evolution of the heat deposited in the lattice under laser irradiation, and it has been used to study melting and resolidification processes [130] and thermomechanical and shock effects of laser irradiation in Si [131].

A different strategy to model the laser-induced heating of the lattice is the so-called *nonequilibrium molecular dynamics* (NEMD). This model developed by Chokappa and coworkers relies on the use of entities called "energy carriers" to supply a given energy fluence over a given pulse duration to the lattice [132]. The only information required is the power, the pulse duration, and the intensity-time profile of the laser. The energy carriers are fictitious particles with a kinetic energy that is determined by the integration of the intensity-time profile of the laser beam during a time step. These carriers transfer their energy by elastic collisions with substrate atoms, increasing their kinetic energy. A fixed number of atoms closest to the top face of the cell are selected for the energy transfer. However, this number is dynamically changed as the substrate expands and melts, and some atoms are located above a predefined threshold. This allows to take into account the reflectivity changes resulting from the phase change of the material. The NEMD technique has been employed to study rapid solidification in systems of Lennard-Jones particles [132], laser thermal processing of B-doped Si systems [133, 134], and crystal growth in SiGe/Si heterostructures [135].

### 3.3.3.2 Two-temperature model molecular dynamics

Ultrafast (pico- and femtosecond) laser pulses have a duration shorter than or comparable to the relaxation time of electrons to phonons. This drives the material into a nonequilibrium state, in which the highly excited electronic subsystem is at a much higher temperature than the nuclei during the first stages of laser irradiation. TTM simulates the heat transfer between the electrons and the lattice, by describing the temporal and spatial evolution of the electronic and lattice temperatures [136]. This model assumes that electronic thermalization is quickly reached (within the femtosecond timescale) and that electrons adopt a well-defined $T_e$. The initial heating of the electrons will cause the heat transfer to the lattice on a picosecond timescale, by means of electron-phonon coupling. The process will finish when electrons and atoms converge to the same temperature.

The TTM is a continuum model that uses two coupled thermal diffusion equations to describe the evolution of the electronic and lattice temperatures:

$$C_e \frac{\partial T_e}{\partial t} = \nabla k_e \nabla T_e - G(T_e - T_l) + A(r, t) \tag{3.7}$$

$$C_l \frac{\partial T_l}{\partial t} = \nabla k_l \nabla T_l + G(T_e - T_l) \tag{3.8}$$

where $C_e$ and $C_l$ are the electronic and lattice heat capacities, and $k_e$ and $k_l$ the electronic and lattice thermal conductivities, respectively. $G$ is the electron-phonon coupling constant and determines the rate at which energy is exchanged between the two subsystems. The term $A(r,t)$ represents the energy released in the electronic subsystem due to laser radiation absorption.

This model was developed by Kaganov et al. [136] and first applied to laser irradiation by Anisimov et al. [137]. It is a widespread method to analyze ultrafast laser irradiation experiments, mainly in metallic targets [138–140]. For the modeling of laser irradiation in semiconductors, the TTM must be modified to take into account that the density of free carriers $n$ (excited electrons in the conduction band and holes) is not spatially constant. The density-dependent TTM (nTTM) developed by Van Driel [74] includes Fourier's law for modeling heat conduction due to a temperature gradient, Ohm's law for electrical currents induced by a gradient in charge density, as well as Seebeck's and Peltier's effects. The nTTM was reformulated by Chen, Tzou, and Beraun [141], and it has been extended to simulate swift heavy ion implantation [142] and to account for the highly transient optical parameters induced during femtosecond laser pulses [143].

The TTM and CMD can be combined to obtain a realistic description of the nonequilibrium processes induced by short-pulse laser irradiation. The TTM molecular dynamics (TTM-MD) methodology is a continuum-atomistic approach in which the electron thermal transport and energy dissipation via electron-phonon coupling is described by the two-temperature model, while the movements of nuclei are simulated by CMD. This augmented CMD technique was proposed by Häkkinen and Landman [144] and further developed by other authors [145–147]. A detailed description of the TTM-MD technique can be found in Ivanov and Zhigilei [145, 146], Zhigilei et al. [148] and in a recent review by Darkins and Duffy [58], as well as examples of its application to laser interaction with metals in Zhigilei et al. [61].

In the TTM-MD method, the TTM Eq. (3.8) for lattice temperature $T_l$ is substituted by the Newton equations of motion within the CMD technique. A coupling term must be added to the classical equations of the movement of atoms to model the energy exchange between the electrons and the lattice. Several implementations for the energy exchange have been proposed [145, 147, 149, 150], but they all use a thermostat that depends on the local $T_e$. According to the model proposed by Ivanov and Zhigilei [145], an additional damping force is included, $F_{i,therm}^{e-l}$, that transfers the energy received by the lattice to the motion of atoms:

$$m_i \frac{d^2 r_i}{dt^2} = F_i + F_{i,therm}^{e-l} = -\nabla_i U(r_1, r_2, \ldots, r_N) + \xi m_i v_i^T, \quad i = 1,2,\ldots,N \qquad (3.9)$$

This new force is proportional to the momentum of atoms, obtained by considering the thermal velocity of each atom $v_i^T$, i.e., the resulting velocity after subtracting the velocity of the center of mass of the cell to which the atom belongs, as shown in Eq. (3.10). The coefficient $\xi$ is the coupling term and depends on $T_e$ [145].

$$v_i^T = m_i \frac{dr_i}{dt} - v^c, \quad i = 1, 2, \ldots, N \tag{3.10}$$

The diffusion equation for $T_e$ (equation 3.7) is usually solved by a finite-difference (FD) method, simultaneously with CMD integration of the equations of motion of atoms. The atomic subsystem in the CMD part and the electronic subsystem in the continuum TTM part are divided into volume cells, as illustrated in Fig. 3.2. Each CMD cell corresponds to the same space geometry in the TTM part. The electronic temperature, $T_e$, and other thermodynamic parameters as the density of carriers, $n$, are calculated by a FD scheme following a grid mesh. The time step used for the FD integration is usually smaller than that required for the integration of equations of motion in CMD. Time steps are chosen so that the one used in CMD corresponds to an integer number of FD time steps. For each cell in the TTM mesh, $T_e$ is calculated at each FD time step, and the energy transferred between the electrons and the lattice is accumulated for an integer number of integrations. Then, this energy is transferred to (or removed from) the kinetic energy of atoms belonging to the corresponding MD cell, following Eq. (3.9).

The TTM-MD technique has been implemented in the widely used CMD code LAMMPS [52, 53]. They use a Langevin thermostat that includes a stochastic force to model the energy gained by the nuclei from the electrons due to the electron-phonon coupling [151, 152]. This Langevin thermostat is inhomogeneous, since it is based on $T_e$ that has spatial and temporal variations.

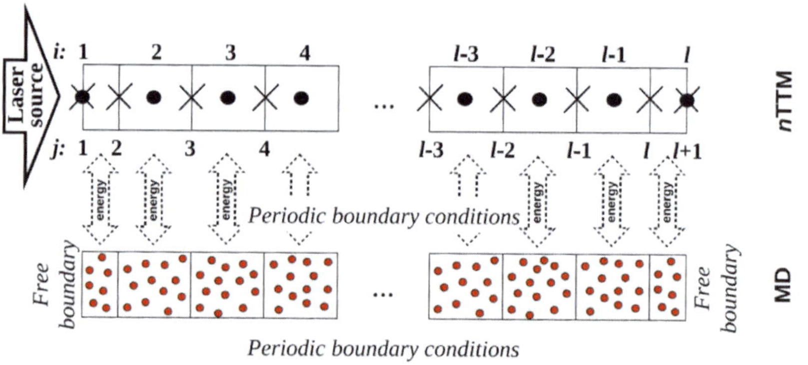


**FIG. 3.2** Schematic of the finite-difference grid mesh of the nTTM and the energy exchange between nTTM and CMD models. Black dots and crosses indicate the mesh points where $n$, $T_e$, and other thermodynamic parameters are calculated within the nTTM, while red dots represent atoms. *(Reprinted figure with permission from V.P. Lipp, B. Rethfeld, M.E. Garcia, D. S. Ivanov, Atomistic-continuum modeling of short laser pulse melting of Si targets, Phys. Rev. B 90 (2014) 245306, copyright (2014) by the American Physical Society.)*

A friction term $\gamma_i$ is also included that accounts for the energy loss by electronic stopping, especially for the modeling of swift heavy ion implantation.

In semiconductors, the combination of nTTM and CMD (nTTM-MD) has been used by Lipp and coworkers to study the melting of Si by a short laser pulse [153]. This work shows that the melting kinetics can be appropriately simulated within the nTTM-MD framework, providing a microscopic description of the melting phenomena induced by laser excitation. The nTTM-MD method was also employed by Gan and Chen [154–156] to analyze the thermal transport and the thermal stress wave in the ultrafast laser processing of semiconductors.

#### *3.3.3.3 Electronic-temperature-dependent interatomic potentials*

Femtosecond laser irradiation at high fluences can heat electrons up to temperatures well above 10,000 K, triggering several ultrafast nonthermal phenomena. In semiconductors, electrons are excited from the valence band to the conduction band, changing its bonding state into an antibonding state. This process causes bond weakening and may lead to nonthermal melting [157], which is a melting process that may happen at room temperature or at temperatures far below the melting point.

These nonthermal phenomena have been widely studied using $T_e$-dependent ab initio simulations (see Section 3.3.1). However, due to the computational cost of the ab initio technique, CMD is more suited for the modeling of femtosecond laser material processing. In the CMD methods described earlier, the heat transfer between the electrons and the lattice is implemented by electron-phonon coupling. None of them takes into account the effect of electron excitation on interatomic forces, which is crucial for modeling ultrafast laser irradiation at high fluences, when high $T_e$ are expected.

To overcome this limitation, dynamical interatomic potentials parameterized by $T_e$ have been developed. The conditions that this kind of potentials must fulfill are analyzed in Lipp et al. [158]. The number of available $T_e$-dependent interatomic potentials is reduced, and the simulations that use them are still limited. Potentials parameterized for $T_e$ were developed for W [159, 160], Mo [161], and Au [162]. The first $T_e$-dependent classical interatomic potential for Si was developed in 2010 by Shokeen and Schelling [163]. They used the so-called modified Tersoff potential (MOD) of Kumagai et al. [164] and made eight of the potential parameters to depend on $T_e$. These parameters were fitted to reproduce the cohesive energy at a given $T_e$ derived from FT-DFT calculations. Initially, coefficients were fitted to four selected values of $T_e$ [163], but then a polynomial of degree 6 was used to ensure a smooth fit [165]. Heterogeneous and homogeneous melting and ablation induced by femtosecond laser pulses in Si have been studied with this potential [166]. Its main drawback is that energy is not conserved [158, 167]. Ultrafast laser pulses may produce a rapid change in $T_e$, which results in an abrupt modification of the potential energy. This leads to a discontinuity in the energy of the atomic subsystem, which should be compensated by a change with the opposite sign of the energy of electrons.

The potential proposed by Darkins and coworkers is also based on the MOD potential, but follows a different approach [167]. They modified one of the analytical functions of the MOD potential by including an additional parameter. Only three parameters depend on $T_e$ and were fitted to the cohesive energy obtained by DFT calculations and some experimental values. Energy conservation is guaranteed by expressing the heat capacity in terms of the interatomic potential, and by introducing a potential energy correction to maintain a physically realistic heat capacity.

A self-learning $T_e$-dependent interatomic potential designed to reproduce ab initio molecular dynamics simulations was developed by Bauerhenne et al. [168]. The potential is a polynomial with 23 $T_e$-dependent parameters, whose overall shape is automatically adjusted by a self-learning procedure. In another work, Bauerhenne and Garcia [169] compared the performance of state-of-the-art $T_e$-dependent potentials for Si versus DFT molecular dynamics simulations on Si at high $T_e$. The best description was provided by the potential proposed by Bauerhenne et al. [168]. They also performed a fit of the parameters of several interatomic potentials that were developed to describe Si at the ground state, i.e., that are independent of $T_e$. The potentials analyzed were the first [170], second [46], and third [46] Tersoff potentials, the Stillinger-Weber potential [45], the MOD potential [164], and the Dodson potential [171]. These fittings provided a better description of Si at high $T_e$ when compared to the DFT molecular dynamics results. Among the potentials that were studied, the third parameterization of Tersoff exhibited the best behavior.

## 3.4 Laser-related phenomena in semiconductors

Ion implantation is routinely used for junction formation in Si technology. To minimize sheet resistance and achieve high dopant concentration, high implant doses are required leading to the amorphization of the Si substrate. In other cases, preamorphizing implants are carried out to insure the formation of an amorphous layer of controlled depth independent of the dopant implant. During subsequent solid-phase epitaxial regrowth, high dopant activation is achieved. Amorphous and crystal Si have different atomic densities and melting points. The temperature window between the amorphous and crystal melting point provides a precise way to control the melting depth combining the appropriate amorphizing implant and laser power. Melting followed by rapid liquid-phase epitaxy also provides a high dopant activation. Nevertheless, both solid-phase epitaxial regrowth and liquid-phase epitaxial regrowth often leave residual defects, which degrade device performance.

In the next subsection, we will focus on the atomistic models to reproduce recrystallization along the amorphous-crystal or liquid-crystal interfaces. Following subsections will deal with the simulation of dopant and defect kinetics, dopant segregation issues in the regrown interfaces, and the generation of anomalous extended defects.

### 3.4.1 Melting and regrowth

Given the typical space and timescales involved, most atomistic simulation studies on melting and regrowth processes in semiconductors, and particularly in Si, are based on CMD and KMC techniques. With respect to CMD, several strategies have been proposed to study microstructural evolution due to a phase change at the atomic level. They are based on the design of a theoretical system initially at equilibrium with solid-melt coexistence, i.e., at the melting temperature [172–176]. Then, the system is cooled down to promote recrystallization, which allows the study of the effect of the nonequilibrium liquid-solid interface kinetics in terms of the formation of point and extended defects. In those studies, it was assumed that the absorbed laser is converted into the kinetic energy of atoms, and as a result, the effect of laser irradiation is simulated by imposing a specific temperature on atoms to melt the solid phase.

The group of Landman and Luedtke successfully used CMD simulations for the study of Si liquid-phase epitaxial growth [173, 174]. Starting from a system at equilibrium with solid and liquid phases in contact, the system was then driven out of equilibrium by allowing the conduction of heat to the underlying substrate (by cooling two crystal layers at the bottom of the sample at a desired rate). Under these conditions, the system initially undercools, and subsequently, crystallization and growth occur. Their studies demonstrated that the crystal growth velocity from the melt as well as the quality of the grown crystal and the type of defects that develop depend on the crystallographic orientation and on the rate of heat transfer to the substrate. Indeed, the orientation dependence of high-speed Si crystal growth from the melt has been observed experimentally [177].

Figs. 3.3 and 3.4 show the real trajectories of particles in the Si(001) and Si (111) systems, both exhibiting solid and melt characteristics, from the CMD simulations of Landman et al. [173] and Luedtke et al. [174]. The equilibrium crystal-melt interface in the (001) orientation shows a broad and pronounced structure, with a tendency for the formation of microfacets on {111} crystal planes (Fig. 3.3). In contrast, the Si(111) crystal-melt equilibrium interface is characterized by a sharp, abrupt, and flat transition from the crystal to the melt region (Fig. 3.4). Simulations showed for both systems that crystallization did not start until the temperature at the interface dropped from the initial melting-point temperature by about 150 K. Both the broad microfaceted structure of the solid-melt interface in the (100) system and the flat, abrupt, solid-melt interface in the (111) system maintain during growth. Nevertheless, once crystallization starts, the crystal-melt interface propagates in different ways in both systems. The crystal front in the (001) system propagates at an approximately monotonous growth rate, whereas the crystallization in the (111) system is

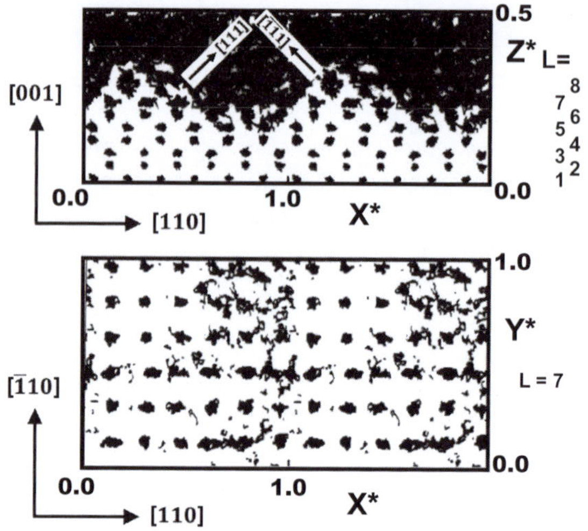


**FIG. 3.3** Real-space particle trajectories recorded over 2000 CMD simulation steps for the Si(001) system. Note the breakup of the interface into {111} microfacets. *(Extracted from U. Landman, W. D. Luedtke, M.W. Ribarsky, R.N. Barnett, C.L. Cleveland, Molecular-dynamics simulations of epitaxial crystal growth from the melt. I. Si(100), Phys. Rev. B 37 (9) (1998) 4637–4646.)*

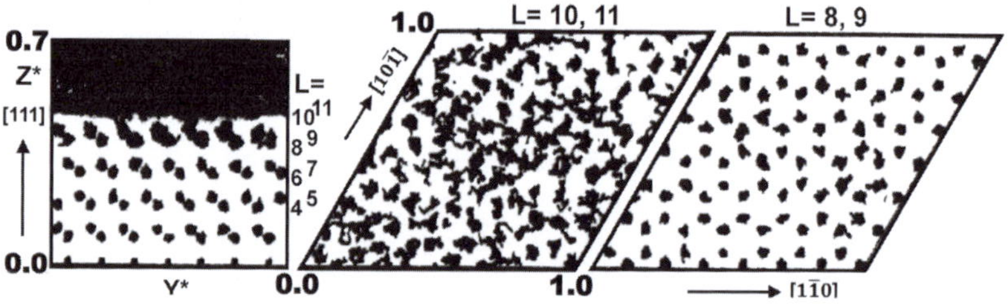


**FIG. 3.4** Real-space particle trajectories recorded over 2000 CMD simulation steps for the Si(111) system. The crystal-melt interface is flat and abrupt. *(Extracted from W.D. Luedtke, U. Landman, M.W. Ribarsky, R.N. Barnett, C.L. Cleveland, Molecular-dynamics simulations of epitaxial crystal growth from the melt. II. Si(111), Phys. Rev. B 37 (9) (1988) 4647–4655.)*

discontinuous (nonmonotonous) with periods of rapid crystallization interrupted by the stages of cessation of the propagation of the interface. This indicates a difference in the growth modes for these two crystalline orientations. Since in the (111) system the spatial variation of the crystalline potential extends over a rather narrow region, it affects only the melt particles in its immediate vicinity and growth occurs via a layer-by-layer mechanism. Furthermore, the growth is accompanied by dynamical self-annealing processes, which provide a mechanism for structural perfection. Occasionally, the propagation of the crystallization front is halted and then is resumed after a certain period.

These events are related to the occurrence of defects in the newly crystallized layer, which then induce the generation of defects in the subsequent crystallizing layers above it. The proliferation of these defects stops further motion of the interface. During the halting period, annealing processes take place. Once the structural perfection has been restored, vigorous crystallization resumes. These self-annealing processes are characteristic of the (111) system and are related to the ease with which defects can occur in this system as compared to that of the (001) orientation (associated with the smoothness of the interface and the fact that Si(111) requires the formation of triple and single bonds) [174, 178]. Such self-annealing processes become more predominant at high growth velocities.

By following the advance of the crystallization front, an averaged crystallization rate can be estimated. For the same cooling rate, growth on the (001) system proceeded at a higher rate (~18 m/s) than that on the (111) system (~14 m/s) in agreement with experimental observations [177–179]. Moreover, unlike the case for the (001) system, the crystal grown on the (111) surface at high velocity is imperfect, containing stacking faults, defective layers, and a wide region of disordered material. The mistakes in registry (stacking faults) involve perfectly ordered layers that occur close to the start of the crystallization stage. The generation of defective layers and disordered regions start to occur later, after the temperature of the system drops significantly and activation barriers for defect self-annealing cannot be surmounted. The perfection of the grown crystal is markedly improved by reducing the cooling rate, resulting in a lower growth velocity (~9 m/s) and yielding a grown crystal containing stacking faults and only a narrow region of disorder, limited to the immediate vicinity of the solid-vacuum interface. These findings correlate well with experimental observations [177–179] and provide an atomistic understanding of the nature of epitaxial growth.

Motooka and coworkers analyzed in more detail defect-formation processes during crystal growth in the Si(001) system [172, 175, 176]. The starting point for CMD simulations was a system where Si atoms were immersed in a heat bath with a fixed thermal gradient $G$ across the melting point. Their simulations indicated that hexagonal structures were formed near the solid-liquid interfaces and that these regions gave rise to microfacets composed of primarily {111} planes [172] that were previously reported by Landman and coworkers. Most of these hexagonal configurations annihilated during further crystal growth, but a part of them were left, which resulted in defect formation with five- and seven-member rings [172].

On the contrary, experiments have evidenced a macroscopic rule stating that the type of grown-in defect is determined by the ratio of $V_{SL}$ (solid-liquid interface velocity or growth rate) to $G$, $V/G$ [180, 181]. Indeed, there exists a critical value of $(V/G)_{\text{critical}} \approx 0.1\ \text{mm}^2/\text{minK}$ such as if $V/G > (V/G)_{\text{critical}}$, V-type grown-in defects or voids are formed in the crystal, whereas I-type defects are formed when $V/G < (V/G)_{\text{critical}}$. Based on CMD simulations, Motooka and coworkers showed that there exists a transition layer with ~10 Å thickness at the solid-liquid interface, also corroborated by high-resolution transmission electron microscopy (TEM). They suggested that the observed $G$-dependence of the defect type can be attributed to nonequilibrium atomic diffusion in the transition layer [175]. Thus, it was proposed that the defect type was controlled by the structural difference in the five-membered rings at the solid-liquid interface. Fig. 3.5 shows the typical five-membered rings that give rise to single I and V defects, respectively. In the case of the five-membered ring that initiates I defect formation, there exist two atoms, indicated by A, bonded to the B atoms in the five-membered ring, whereas one of these atoms is lost in the case of V formation. Fig. 3.6 shows an example of V formation process during melt growth in the [001] direction. The five-membered ring structures that lead to I defects were always seen for larger $G$, whereas structures conducing to V defects were only seen for smaller $G$, consistently with the macroscopic $V/G$ rule. This can be microscopically attributed to the nature of atomic motions in a transition layer at the solid-liquid interface. The nonequilibrium atomic diffusion is found to be faster for larger $G$ and thus the lower density region can be quickly filled by Si atoms from the liquid side, which results in I formation. On the contrary, the atomic diffusion constant becomes smaller for smaller $G$ and space filling is not effective, which increases the opportunity to create Vs. V generation in LTA of pure Si has been confirmed by positron annihilation measurements [182].

With respect to KMC models of amorphization and regrowth, the simplest ones assume that Si turns amorphous when a threshold defect (self-Is or Vs) concentration is reached during ion implantation [183]. Upon thermal

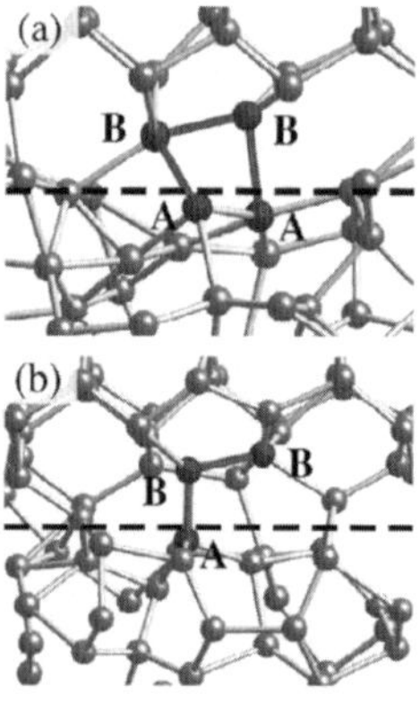


**FIG. 3.5** Typical five-membered rings resulting from CMD simulations that initiate the formation of (A) a single I defect and (B) a single V defect. Dashed lines indicate the solid-liquid interface. *(Extracted from T. Motooka, K. Nishihira, R. Oshima, H. Nishizawa, F. Hori, Atomic diffusion at solid/liquid interface of silicon: transition layer and defect formation, Phys. Rev. B 65 (2002) 081304.)*

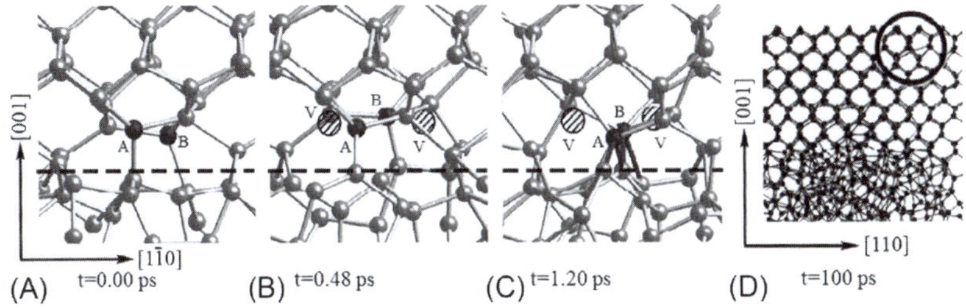


**FIG. 3.6** V formation process during a CMD simulation of melt growth in the [001] direction when a thermal gradient $G = 0.001$ K/Å is considered. (A) A pair of five-membered rings are observed at the solid-liquid interface; (B) and (C) motion of atoms labeled as A and B, belonging to each five-membered ring, initiate the formation of Vs (labeled as V); and (D) after further crystal growth, a single V defect is obtained (see encircled region). *(Extracted from T. Motooka, K. Nishihira, R. Oshima, H. Nishizawa, F. Hori, Atomic diffusion at solid/liquid interface of silicon: transition layer and defect formation, Phys. Rev. B 65 (2002) 081304.)*

treatment, regrowth is considered to be "instantaneous" and the material turns into perfect crystalline Si in the region that was fully amorphized, while the evolution of self-Is and Vs is followed in damaged but not completely amorphous regions left behind the initial amorphous-crystal interface [184].

In more detailed KMC models, the bond-defect or interstitial-vacancy (IV) pair was considered the building block of the amorphous phase since it leads to the five- and seven-membered rings characteristic of the amorphous phase, as shown in Fig. 3.7. This defect is the result of an incomplete self-I/V recombination, and it can also be formed directly by a ballistic impact during ion implantation. Si turns amorphous when the concentration of this defect exceeds 25% [185].

Successful detailed amorphization nonlattice KMC models that describe most features of amorphization and solid-phase epitaxial regrowth are based on this defect. The model developed by Pelaz and coworkers characterizes

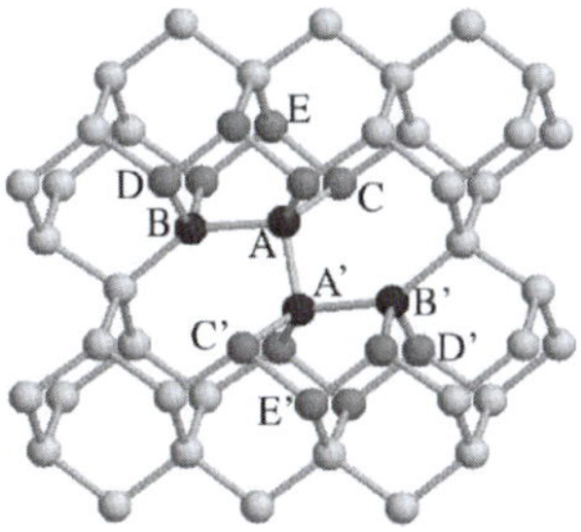


**FIG. 3.7** Atomic configuration of the IV-pair showing the five- and seven-membered rings. *(Extracted from L.A. Marqués, L. Pelaz, J. Hernández, J. Barbolla, G.H. Gilmer, Stability of defects in crystalline silicon and their role in amorphization,. Phys. Rev. B 64 (4) (2001) 045114.)*

IV-pairs by the number of local neighbors [186]. Regrowth occurs as IV-pairs are recombined, being the recombination rate lower as the local number of IV-pairs increases. The rates are set so that the experimental planar amorphous-crystal interface regrowth velocity is reproduced. This model naturally captures the fact that rough interfaces or small amorphous pockets regrowth faster than planar amorphous interfaces [187]. The model proposed by Mok and coworkers define the so-called *amorphous pockets* as agglomerates of self-Is and Vs and their recombination rate depends on their size (number of individual Is and Vs irrespectively of their geometry; [188]). A region is considered to turn amorphous when the density of self-Is and Vs within the amorphous pockets exceeds a critical concentration. In these nonlattice KMC models, the excess atoms (Is) or deficit atoms (Vs) within the amorphous layer are swept along the regrowth interface. They are released when the amorphous layer is buried and both fronts meet, or they are annihilated at the surface when the regrowth interface reaches it [189]. As a result, it is considered that the regrown amorphous layer is free of defects, and only end-of-range defects remain beyond the initial amorphous-crystal interface as shown in Fig. 3.8.

Since nonlattice KMC models do not take into account the Si lattice atoms, there is no information about crystal orientation in the recrystallization front. Lattice KMC models exploit their potential in solid-phase-epitaxial regrowth problems in which crystal orientation plays a significant role. Martin-Bragado and Moroz [67] have developed a lattice KMC model that assumes that each atom in the amorphous phase needs to form two undistorted bonds with the crystal. Depending on the crystal orientation, one adjacent atom (in (100) oriented surfaces), two adjacent atoms (in (110) surfaces), or three adjacent atoms (in (111) surfaces) are necessary for each atom to complete two undistorted bonds. This is modeled using different prefactors to simulate the different frequencies at which atoms in the amorphous phase join the crystalline one. Defective atoms are produced when Si atoms do not bond properly to the lattice, they have a different orientation and can propagate different fronts for recrystallization leading to the formation of pits and defective Si [190]. This model reproduces the different recrystallization speeds of different planar orientations by proper fitting of the prefactors, the arrow tip formation during solid-phase epitaxial regrowth of thin Si fins, and the formation of trenches in rectangular-shaped amorphized regions in (001) and (011) Si.

Crystal growth from the liquid phase is often addressed in KMC models similarly to solid-phase epitaxial regrowth, since enhanced atomic mobility in the liquid is not specifically considered. The KMC events are the transitions of lattice sites at the solid-liquid interface that is promoted from the liquid phase (L-site) to the solid one (S-site) and vice versa. The transition, occurring at a given time of the simulated evolution, is stochastically selected in the list of possible events according to its own probability assigned on the basis of the particle energetics. The transition probability of a solid-to-liquid event $P_{SL}$ is as follows [191]:

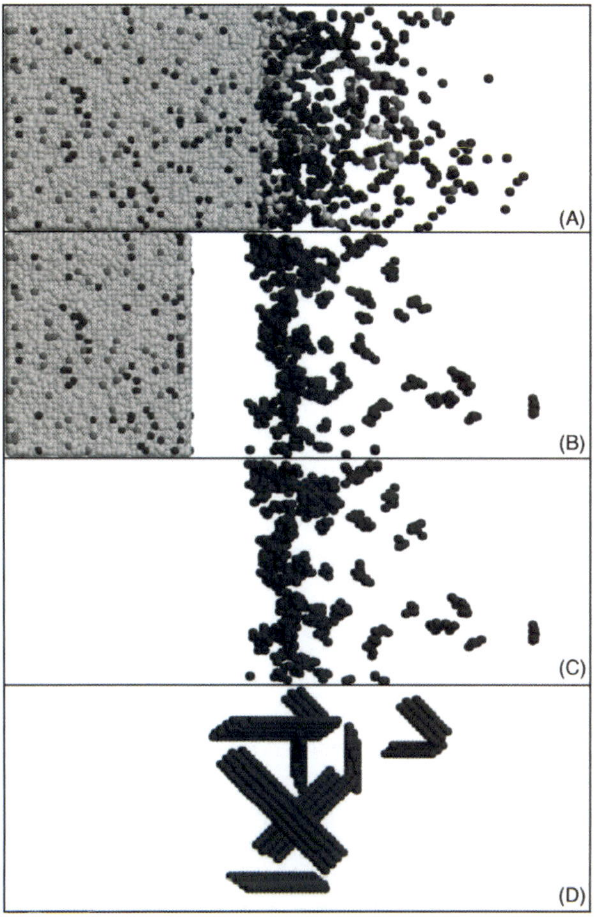


FIG. 3.8 Damage evolution of a 5-keV Si implant into Si showing the residual defects after regrowth beyond the initial amorphous-crystal interface. *(Extracted from P. López, L. Pelaz, L.A. Marqués, I. Santos, M. Aboy, J. Barbolla, Atomistic modeling of defect evolution in Si for amorphizing and subamorphizing implants, Mater. Sci. Eng. B 114–115 (2004) 82–87.)*

$$P_{SL} = P_0 \times \exp\left[-n_S(\Phi_S - \Phi_L)/(k_B T)\right] \tag{3.11}$$

where $n_S$ is the number of bonds which the interface site forms with atoms in the solid phase, $\Phi_S$ is the bond energy when the bond is formed by two atoms in the solid phase, and $\Phi_L$ is the bond energy when one of the two atoms forming the bond is in the liquid phase. Similarly, the transition probability of a liquid to solid event is given as follows:

$$P_{SL} = P_0 \times \exp\left[-\Delta S_B/k_B\right] \tag{3.12}$$

where $\Delta S_B/k_B = 2(\Phi_S - \Phi_L)/(k_B T_M)$ is the entropy of fusion and $T_M = 1688$ K is the melting point of Si. The value of bond difference for Si is $\Phi_S - \Phi_L = 0.96$ eV

[192], while the probability prefactor $P_0$ is calibrated by correlating the simulated and experimental interface speed $V_{SL}(T)$ as a function of temperature (i.e., the undercooling, see [193]). Additionally, the lattice KMC model proposed by La Magna et al. [182] accounts for the possibility of forming Vs during the liquid-crystal regrowth (as it is observed in LTA experiments on pure Si) by defining a V-site in the solid when a site marked in the liquid has all its next neighbors labeled as solid sites as the solid-liquid interface evolves.

### 3.4.2 Dopant and defect kinetics

Pulsed laser irradiation of damaged solids promotes ultrafast nonequilibrium kinetics, on the sub-microsecond scale, leading to microscopic modifications of the material state. The KMC technique has been applied to study dopant and defect kinetics under LTA from an atomistic point of view. However, due to the intrinsic features of a LTA treatment, i.e., localized irradiation with a space- and time-dependent thermal field $T(\mathbf{r},t)$ in the nanometer and nanosecond scale, KMC requires a particular formulation with respect to other standard constant-temperature annealing processes [194]. From the theoretical point of view, some relevant questions have to be addressed: the formalization of a reliable kinetic formalism for the evolving defect-impurity system in a nonuniform fast-varying thermal field [195–197]; the generalization of the kinetic parameters (e.g., diffusivity, clustering rate constants, etc.) in this regime; the inclusion in the formalism of eventual phase transitions [198]; and the interaction between the defects present in the as-prepared state (e.g., after an implantation process) and the defects generated by the process [182].

A key feature of the method here discussed is the coupling of the KMC code with a phase field (PF) model that properly simulates the thermal field evolution in the irradiated sample also when a liquid-solid phase transition takes place. The PF-KMC method allows simulating the material modification in the device structures at a macroscopic scale with a microscopic resolution, computing the phenomenon evolution in a nearly natural way.

The simulations here presented belong to the class of nonlattice KMC models. Dopant atoms can reside in a substitutional position $B_S$ or form a complex with I defects, i.e., boron-interstitial clusters (BICs, [64]). For a dopant-defect system, it is assumed that Is and Vs and the BI complex (one B dopant atom bound to one self-I point defect) are the only mobile species and that defect clusters $X_m$ ($X = I$ or V) as well as the dopant-defect cluster $B_nI_m$ (a cluster formed by n dopant atoms and m Is) can only absorb or emit mobile defects [1]. A two-dimensional scheme of the KMC environment has been reported in Fig. 3.9A with I and V defects in mobile or clustered configurations. The reactions accounted for in the PF-KMC code are as follows:

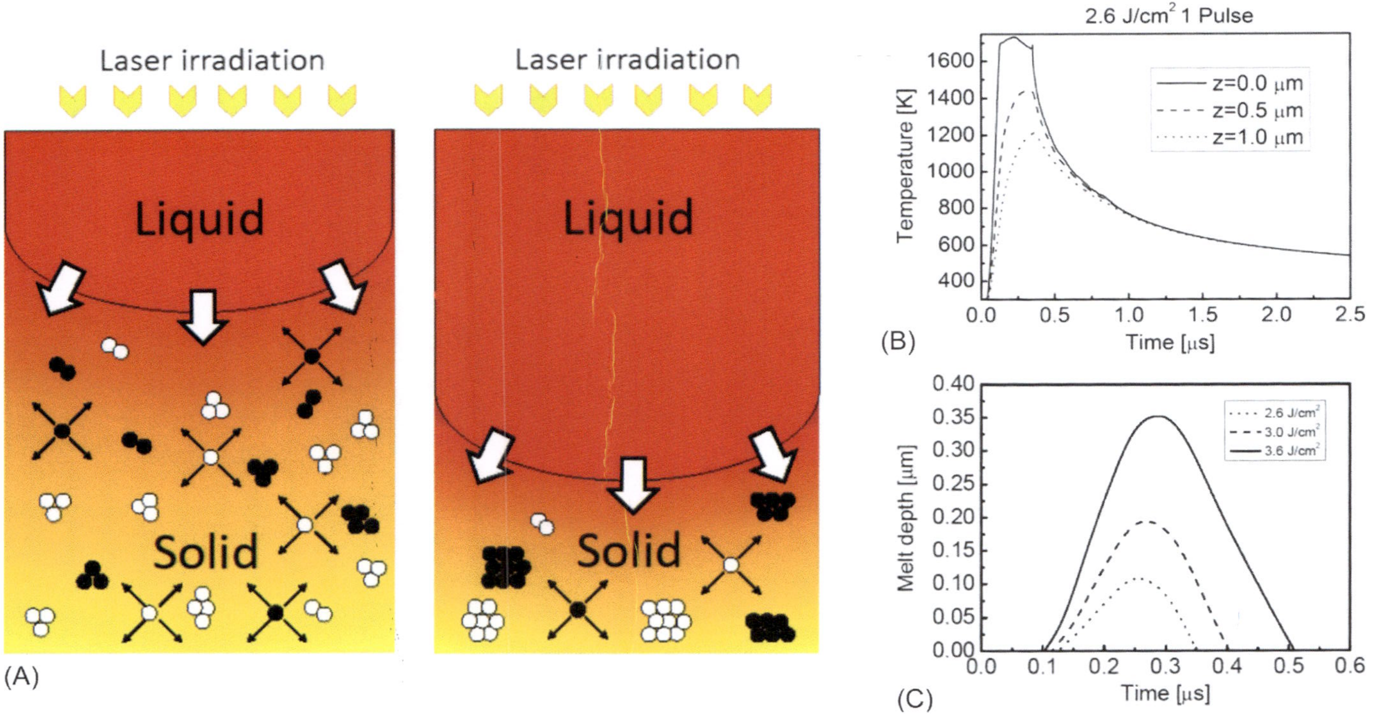


**FIG. 3.9** (A) Schema of the nonlattice KMC environment that considers I (black balls) and V (white balls) defects in mobile (surrounded by arrows) or clustered configurations. The two panels refer to the melting front evolution induced by a laser irradiation process. Particle annihilation in the molten region (red) and defect ripening process in solid phase (yellow) are also represented; (B) time evolution of the temperature at three different positions for the 2.6 J/cm$^2$ laser process: 0.0 (solid), 0.5 (dashed), and 1.0 (dotted) μm below the irradiated surface (at 0.0 μm); and (C) time evolution of the melt depth during the partial melting laser processes with 2.6 (dotted), 3.0 (dashed), and 3.6 (solid) J/cm$^2$ laser fluences. *(Reprinted from G. Fisicaro, L. Pelaz, P. Lopez, A. La Magna, Kinetic Monte Carlo simulations for transient thermal fields: computational methodology and application to the submicrosecond laser processes in implanted silicon, Phys. Rev. E 86 (3) (2012) 036705, copyright (2012) by the American Physical Society.)*

$$\mathrm{I} + \mathrm{V} \Longleftrightarrow 0 \tag{3.13}$$

$$\mathrm{X} + \mathrm{X_{m-1}} \Longleftrightarrow \mathrm{X_m} \tag{3.14}$$

$$\mathrm{I_m} + \mathrm{V} \Longleftrightarrow \mathrm{I_{m-1}} \tag{3.15}$$

$$\mathrm{V_m} + \mathrm{I} \Longleftrightarrow \mathrm{V_{m-1}} \tag{3.16}$$

$$\mathrm{I} + \mathrm{B_S} \Longleftrightarrow \mathrm{BI} \tag{3.17}$$

$$\mathrm{I} + \mathrm{B_nI_m} \Longleftrightarrow \mathrm{B_nI_{m+1}} \tag{3.18}$$

$$\mathrm{V} + \mathrm{B_nI_m} \Longleftrightarrow \mathrm{B_nI_{m-1}} \tag{3.19}$$

$$\mathrm{BI} + \mathrm{B_nI_m} \Longleftrightarrow \mathrm{B_{n+1}I_{m+1}} \tag{3.20}$$

where I, V, or BI annihilation has been considered when a mobile defect falls inside the capture radius of another point or clustered defect.

If temperature $T$ is uniform in the whole simulation box, there is a one-to-one correspondence between the event type and its associated rate, independently of the position of the particle in the simulation box. As a consequence, events can be classified into discrete classes and efficient sampling method for the selection of the occurring event from the set of possible ones can be straightly implemented. However, these algorithms cannot be applied for a KMC simulation of the laser annealing process since the temperature is a non-uniform variable field $T(\mathbf{r},t)$ (represented by means of the color scale in the schematic of Fig. 3.9A).

An evolution and a change of the thermal field require an update of all the frequencies associated with all the possible events. This particularity requires the introduction of a different sampling method. The event could be simply selected by directly using the probability distribution $\nu_i[T(\mathbf{r},t)]/R$, where the index $i$ runs over all the $N$ events and $R$ being the sum of all event probabilities. However, this trivial selection procedure is not efficient for large systems since it requires an $o(N)$ CPU time per KMC step. The use of a *binary tree storage* of $\nu_i[T(\mathbf{r},t)]$, where the root node is $R$ and the leaves are the single $\nu_i[T(\mathbf{r},t)]$ frequencies, can overcome these difficulties and deliver an $o(\log N)$ algorithm [194].

In a KMC method applied to LTA, the $T(\mathbf{r},t)$ field must be calculated by an external code in order to simulate the transient conditions of a real process [197–199]. Pulsed laser irradiation can induce a strongly varying $T(\mathbf{r},t)$ (on the sub-microsecond and sub-micrometer scale); moreover, if the energy density exceeds a proper threshold value, a localized melting can occur. Therefore, the evolution of the molten region (phase field) should also be simulated and reliably mapped in the KMC code. The typical size of a laser spot hitting the surface is in the range of centimeters. If compared with the adsorption length of the laser light of the irradiated material, i.e., hundreds of nanometers, the thermal and phase fields can be considered constant in the directions orthogonal to the laser beam. As a consequence, a one-dimensional problem in the direction of the laser beam can be set. The thermal and phase fields induced by a laser annealing process are obtained within the PF methodology described in Fisicaro et al. [194]. According to the PF approach, a continuous field $\varphi$ ($-1 \leq \varphi \leq 1$, with $\varphi = -1$ and $\varphi = 1$ in the pure liquid and solid phases, respectively) is introduced to describe smooth phase transitions. Then, the phase field and heat partial differential equations are solved concurrently.

In the case of low defect densities, the damage evolution does not influence the $T(\mathbf{r},t)$ and $\varphi(\mathbf{r},t)$ solutions. This holds, for example, when implant conditions are below the amorphization threshold [200]. In such conditions, we can assume that parameters for the irradiated substrate used in the heat transport equations (heat capacity, heat conductivity, density, adsorption coefficient, etc.) do not depend on the damage state. Therefore, a sequential coupling between PF and KMC models can be adopted. Once a particular laser process has been simulated with the PF model, the temperature field as a function of time and melt depth evolution in the processed region are extracted and stored at constant space intervals (in the direction orthogonal to the irradiated surface plane) with a given time resolution $\delta t$. Fig. 3.9B reports the time evolution of the temperature at three different positions for the 2.6 J/cm$^2$ laser process, while Fig. 3.9C shows the time evolution of the melt depth for several laser fluences. We note that the minimal value of $\delta t_{\rm min}$ is given by the inner discretization of the time variable in the partial differential equation solvers of the PF equations and that choice $\delta t = \delta t_{\rm min}$ is obligatory in the eventual case of interdependence between the optical and thermal parameters of the PF model and the damage state. The choice of $\delta t$ is related to the time variation of the event frequencies by the local temperature rate $\delta T$ [194]. The space dependence of the temperature fields at different times is stored in a grid representing equally spaced points of the simulation box. A linear interpolation allows calculating the temperature for the particles occupying a given position and the related local frequency can be properly updated.

Concurrently, by means of the storage of the evolving melt depth curve, the time-dependent molten region can be tracked, and all the particles that during their evolution fall inside this zone are fully annihilated. The melting process evolution and the correlated effects in the nonlattice KMC environment are represented in the schema of Fig. 3.9A. The two panels refer to the melting front evolution induced by a laser irradiation process. In particular, particle annihilation in the molten region (red) and the defect ripening process in the solid phase (yellow) are represented. Considering the laser pulse shape and analyzing the simulation output, for all the laser processes considered in the following, a value of $\delta t = 0.5$ ns ensures temperature variations lower than 10 K with a consequent acceptable variation of the total $R$.

The free defect migration energies were taken from the work of Pelaz et al. [62], whereas the frequency prefactors were fixed to achieve the proper diffusivity values near the Si melting point ($D_{\rm melt,I} = 3.6 \times 10^{-4}$ and

$D_{melt,V} = 4.0 \times 10^{-5}$ cm$^2$/s, [201]). This strategy allows a reasonable physical setting of the free defect parameters in the wide temperature range spanned during the simulated processes. This particular issue makes the LTA simulations more ambitious than those of conventional thermal processes [202, 203], where the physical properties have to be calibrated in a limited temperature range. The I and V cluster parameters (binding energies and capture radii) are taken from Pelaz et al. [1], Cowern et al. [204], Cristiano et al. [205], and Bongiorno et al. [206]. The $I_m$ ($V_m$) clusters can emit only an I-type (a V-type) point defect, whereas a $B_nI_m$ aggregate can emit a mobile BI or an I point defect. The binding energies for the release of a mobile defect from an immobile agglomerate $B_nI_m$ were taken from Aboy et al. [64]. In the liquid state, free and clustered defects are assumed to be entirely dissolved, whereas dopant atoms are positioned in a substitutional state $B_S$ during solid regrowth after melting.

The initial state (postimplant) for the dopant-defect system (i.e., $I_m$, $V_m$, and $B_nI_m$) was derived by the binary collision approximation and atomistic KMC simulations of their evolution at room temperature. The simulation results suggest a reduction of 60% in the estimated total implantation damage stored in small defect aggregates [207]. We note that the initialization of the dopant-defect system for the LTA simulation is a crucial issue. In conventional simulations of dopant-defect systems, it is a common practice to set all the implanted impurity atoms in an interstitial position and, concurrently, set distribution of self-I atoms equal to the implanted dopant distribution and stored as $I_2$ clusters [1, 202]. It is clear that this choice, which neglects the first stage of the annealing treatment, cannot be implemented for LTA simulations, where the postimplant evolution occurs only for hundreds of nanoseconds.

The PF-KMC code was applied to the study of the B activation dynamics under LTA treatment in ion-implanted Si substrates. KMC results are compared to experimental chemical and electrical measurements [208]. Laser irradiation was applied to a B-implanted Si substrate (30 keV B energy, $1 \times 10^{15}$ cm$^{-2}$ dose, projected range $R_p \sim 120$ nm). After implantation and/or LTA, all samples remained monocrystalline, as verified by TEM. The samples were annealed using the EXCICO LTA system (laser wavelength $\lambda = 308$ nm) at room temperature, with a pulse duration of 160 ns. Various laser energy densities were considered to achieve different thermal budgets and maximum melt depths $R_{melt}$. Conditions of nonmelt $R_{melt} = 0$ (1.5 J/cm$^2$), partial melt with $R_{melt} = 75$ nm $< R_p$ (2.3 J/cm$^2$), and $R_{melt} = 140$ nm $> R_p$ (2.6 J/cm$^2$) were reproduced. The chemical B profiles were analyzed using secondary ion mass spectroscopy (SIMS), and the electrical activation was evaluated experimentally on both qualified SSM150 spreading resistance profiling (SRP) system and electrochemical capacitance-voltage (ECV) equipment. The melt depths were extracted from the chemical SIMS analyses considering the relevant profile modification in the molten region [195].

During the evolution promoted by the laser-induced thermal budget, B atoms can reach substitutional lattice sites as well as form immobile aggregates

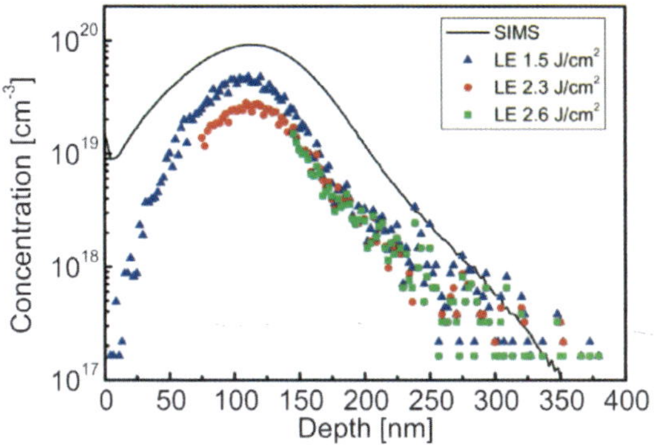


**FIG. 3.10** Simulated profiles of B atoms trapped in BICs for different laser fluences. The as-implanted SIMS profile (dark line) is also shown. *(From G. Fisicaro, L. Pelaz, M. Aboy, P. Lopez, M. Italia, K. Huet, F. Cristiano, Z. Essa, Q. Yang, E. Bedel-Pereira, M. Quillec, A. La Magna, Kinetic Monte Carlo simulations of boron activation in implanted Si under laser thermal annealing, Appl. Phys. Express 7 (2) (2014) 021301, copyright (2014) The Japan Society of Applied Physics.)*

with I defects (i.e., BICs) [208]. The first component will increase the carrier density, whereas the second will need an additional thermal budget to dissolve and then become active. Fig. 3.10 shows the B component stored in the $B_nI_m$ clusters for all the cited laser irradiation conditions. Comparing the regions (depth > 75 nm) where the samples remain solid for both 1.5 and 2.3 J/cm$^2$ cases, a pronounced decrease in the local density of BICs with increasing laser fluence can be observed. On the contrary, a comparable inactive impurity concentration is predicted by KMC simulations when the laser fluence increases from 2.3 to 2.6 J/cm$^2$. This comparable dopant trapping is connected with the similar local activation efficiency of the two processes, denoting the limit of the laser-induced thermal field to release further B atoms from the BICs [208]. Hence, the formed BICs play a central role in the solid-phase activation efficiency during the laser-induced kinetics in the highly damaged crystalline matrix.

Comparisons of the chemical and active (SRP or ECV) profiles demonstrate that significant activation levels are reached under all the annealing conditions [208]. In Fig. 3.11, the SRP and ECV electrically active distributions after LTA are presented and compared to the associated SIMS profiles for 1.5 J/cm$^2$ non-melting process. In this case, an activation level of 78% (ECV) and 12% (SRP) with respect to the total implanted B is recovered [208]. The discrepancy between the activation levels extracted using the ECV and SRP techniques can be rationalized in view of their measurement peculiarities. ECV gives direct access to the total charges by integrating with the capacitance-voltage scan the density of states associated with the impurity levels in a wide energy range of the Si gap; the SRP technique involves only the mobile charges in the measurement of the active concentration. The presence of the implant damage in the

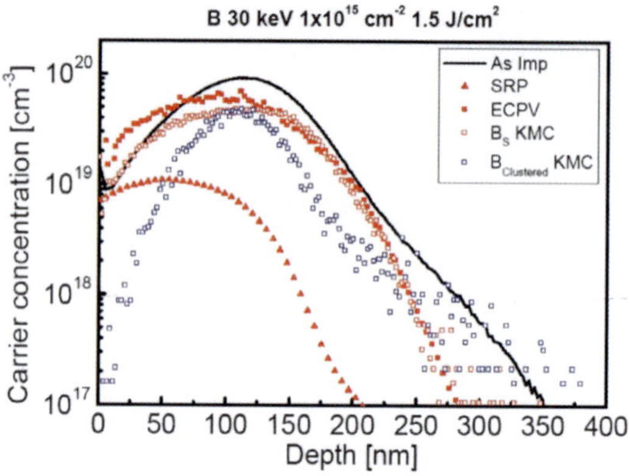


**FIG. 3.11** Experimental results from SRP (red filled triangles) and ECV (red filled squares) measurements for the 1.5 J/cm$^2$ laser process. Red and blue empty squares refer, respectively, to the simulated substitutional B atoms $B_S$ and the B atoms trapped in BICs. The SIMS profile after LTA (dark line) is also shown. *(From G. Fisicaro, L. Pelaz, M. Aboy, P. Lopez, M. Italia, K. Huet, F. Cristiano, Z. Essa, Q. Yang, E. Bedel-Pereira, M. Quillec, A. La Magna, Kinetic Monte Carlo simulations of boron activation in implanted Si under laser thermal annealing, Appl. Phys. Express 7 (2) (2014) 021301, copyright (2014) The Japan Society of Applied Physics.)*

nonmelting process can explain the difference between the ECV and SRP results in terms of both reduced mobility and incomplete activation of the $B_S$ component. Fig. 3.11 also shows the simulated substitutional B atoms $B_S$, and the B component stored in the $B_nI_m$ clusters. As a reference, the SIMS profile after LTA is also represented in the figure. No variation of the annealed B profile was observed with respect to the as-implanted one, indicating that a submicrosecond LTA process does not induce any macroscopic diffusion of the implanted impurity distribution in the nonmolten regions. The PF-KMC simulation reproduces the measured activation efficiency in the solid phase [208].

### 3.4.3 Dopant segregation

Anomalous impurity redistribution has been observed in Si and Ge that undergo a melting laser process during the melting-regrowth evolution [209, 210]. Atomistic simulations are particularly important to study the effect of the nonequilibrium interface kinetics on segregation mechanisms in alloyed systems.

Lattice KMC techniques have been widely applied to investigate the segregation mechanism during the crystallization of diluted Si-based alloy systems. The segregation coefficient $k$ relates the concentration of a component in the solid phase to the concentration of the same component in the liquid phase. It is used to provide the boundary conditions at the interface for the diffusion fields in solid and liquid. In this case, the bond counting in Eq. (3.11) has to be modified in order to consider the Si—Si bond $\Phi_S^{\text{Si-Si}}$, the Si-X impurity bond $\Phi_S^{\text{Si-X}}$ and the impurity-impurity bond $\Phi_S^{\text{X-X}}$ [211, 212]. The impurity diffusion in the liquid region can be considered by means of a random jump sequence where the jumping frequency is derived from the known diffusivity values of X in the liquid phase. After a proper calibration, lattice KMC simulations recover reliably the nonequilibrium impurity trapping effect in a good agreement with the know experimental data (see the work of M.T. Aziz and coworkers in [213, 214], and [215], for a quantitative experimental analysis using aperiodic step-flow models).

Jackson and coworkers applied Monte Carlo simulations for the study of alloy solidification, and the consequent nonequilibrium segregation effects such as solute trapping [191, 216, 217]. Simulations have been carried out for various growth temperatures, growth rates, surface roughness, liquid diffusivity, equilibrium segregation coefficient, entropy of fusion, and composition of the liquid. Typical Monte Carlo input parameters are the crystal structure, the bond energies usually taken from the equilibrium properties of the alloy, the temperature, and the entropy of fusion.

When the diffusivity of impurity atoms in the liquid is negligible with respect to the dynamics of the solid-liquid interface, the solid that forms has the same concentration as the liquid, so that the *k*-value is one. For Si doped with Bi, the dependence of the nonequilibrium segregation coefficient on growth rate, orientation, and liquid diffusivity in Monte Carlo simulations of rapid solidification are in substantial agreement with the experiment [191, 211, 216]. The Monte Carlo computer simulation results suggest that the nonequilibrium *k*-value can be expressed as a function of a parameter $\beta$, which in turn depends on the growth rate $v$, on the free energy $\Delta G$ driving the transformation, and on the diffusion coefficient $D$:

$$k \approx k_e{}^{1/(1+A\beta)} \tag{3.21}$$

$$\beta = ((av/D) \times (1 - \exp(\Delta G/(k_B T))))^{1/2} \tag{3.22}$$

being $k_e$ the equilibrium *k*-value, $a$ the interatomic spacing, and A a constant.

Simulation results for a solid-liquid interface lying at the (100) face of a diamond cubic lattice demonstrate that all *k*-value (for a variety of growth temperatures, diffusion rates, and entropies of fusion) superimposes when plotted against $\beta$ [212]. Fig. 3.12 shows the simulation results compared with experimental data. The KMC approach is also able to reproduce the experimental dopant segregation for three different dopants in Si (Ge, Sn, and Bi).

The estimated effective segregation coefficient $k[V_{SL}(T)]$, which determines the ratio of the impurity density at the solid and liquid sides of the interface, monotonically increases with the interface speed (i.e., with the undercooling $T_M - T$) starting from the equilibrium value $k[V_{SL}(T \approx T_M) \approx 0] = k_0$ and asymptotically reaching the limit value $k[V_{SL}(T) \gg 1 \text{ m/s}] = 1$. We note the $k_0$, which can be derived from the alloy equilibrium phase diagram, can be used to estimate the alloy bond energy $\Phi_S^{\text{Si-X}}$.

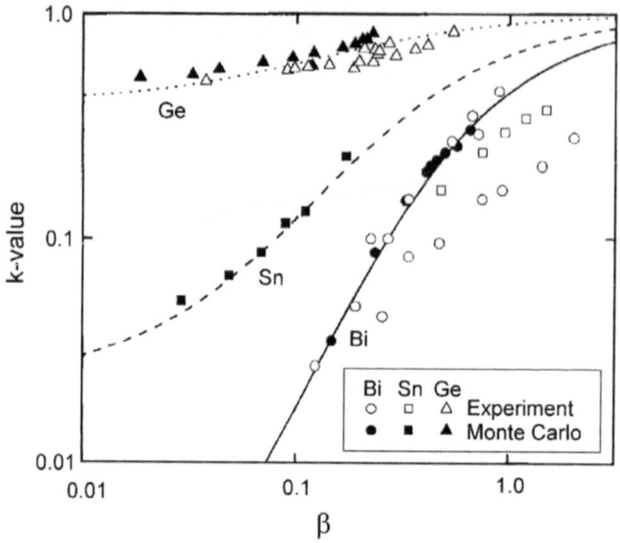


**FIG. 3.12** Segregation coefficient $k$ in Si plotted as a function of β for three different values of $k_e$: 0.4, 0.023, and $7 \times 10^{-4}$, which correspond to Ge, Sn, and Bi, respectively. Filled symbols represent simulation results, and open symbols, experimental data. The lines correspond to Eq. (3.21) with $A = 8$. Growth was normal to the (100) face of the diamond cubic structure in both simulations and experiments. *(Extracted from K.M. Beatty, K.A. Jackson, Monte Carlo modeling of dopant segregation, J. Cryst. Growth 271 (3–4) (2004) 495–512.)*

The dependence of the nonequilibrium trapping effect on the interface crystal plane is also qualitatively predicted by lattice KMC. In particular, for the same undercooling level, the nonequilibrium segregation coefficient increases when the solid-liquid interface lies to crystal planes with high-coordination number $\eta$, (e.g., (111) planes in Si; [191]). This enhanced trapping efficiency is related to the interface kinetics. Indeed, in the low-$\eta$ surface (e.g., (100) planes in Si), the solid-liquid interface is rough at the atomic level and continuous regrowth occurs. In the case of high-$\eta$ surfaces, the solid-liquid interface is flat and crystal regrowth proceeds by means of a random nucleation of a new crystal island (slow event) that extends to the whole plane with very high lateral speed. This particular regrow mechanism allows more efficient incorporation of impurities.

### 3.4.4 Anomalous generation of extended defects

Defects in electronic devices may be formed during its manufacturing and, in some cases, also during its operation [218, 219]. Defect generation occurs, for example, in dopant implantation for junction formation, or when devices are exposed to irradiation, such as it is the case of radiation detectors for particle accelerators or devices in space applications. When an energetic particle enters the semiconductor substrate, it collides with its atoms producing self-Is and Vs. At the processing level, these lattice defects can cause the deactivation of dopants, the increase of their diffusion, and also the reduction of carrier mobility [220–222]. From an electronic point of view, defects modify the distribution of semiconductor bands. Some defects can add deep levels in the gap, which has very important consequences: they increase leakage currents (critical in devices for mobile applications), they decrease the capacity of charge collection in radiation detectors, and they cause efficiency losses in photovoltaic and optical devices [223–228].

I defects are especially relevant in semiconductor technology. In the manufacture of junctions, a large amount of defects is generated during the dopant implantation step (the substrate can even become amorphous for high doses). A subsequent thermal annealing is needed to eliminate defects and to electrically activate dopants. Since dopants end up occupying perfect lattice positions, after annealing there is a net amount of self-Is in the substrate that diffuse, interact, and give rise to extended defects, very stable and difficult to eliminate. Given its technological relevance, a great research effort, both experimental and theoretical, has been dedicated for the study and characterization of these I defects, particularly in Si [204, 220, 229–239].

Experiments show that the morphology of I defects in Si depends on the particular processing conditions. In short anneals at relatively low temperatures, Is form small agglomerates whose structure is not resoluble by microscopy techniques. For longer treatments at higher temperatures, Is give rise to elongated structures called "$\{113\}$ defects," where they occupy regular positions within the lattice in planes perpendicular to the $\langle 113 \rangle$ direction. For longer treatments and even higher temperatures, Is form very large defects, which structurally are dislocation loops where excess atoms are grouped into disks perpendicular to the $\langle 111 \rangle$ direction. These three types of defects were identified during the 90s decade, and since then, they have been extensively studied and characterized [220, 231, 232, 237, 238].

Formation energy of I defects depends on their type and size, as shown in Fig. 3.13. For $\{113\}$ defects and $\{111\}$ dislocations, formation energies decrease monotonically with size [205, 240]. For sizes smaller than about 400 Is, $\{113\}$ defects are more stable, while for larger sizes the most stable are $\{111\}$ dislocations. Interestingly, for small agglomerates, there is not a clear monotonic reduction of formation energy with size, but rather there are evidences of the existence of "magic numbers," i.e., very stable cluster configurations of specific sizes, particularly for 4 and 8 Is [204, 236, 241].

It is well known that I defects grow through a mechanism known as "Ostwald ripening" [229, 230, 239]. This mechanism considers that extended defects are immobile, but they can exchange free self-Is that are a highly mobile species. The frequency $\nu$ of emission of free self-Is from a defect at temperature $T$ is given by

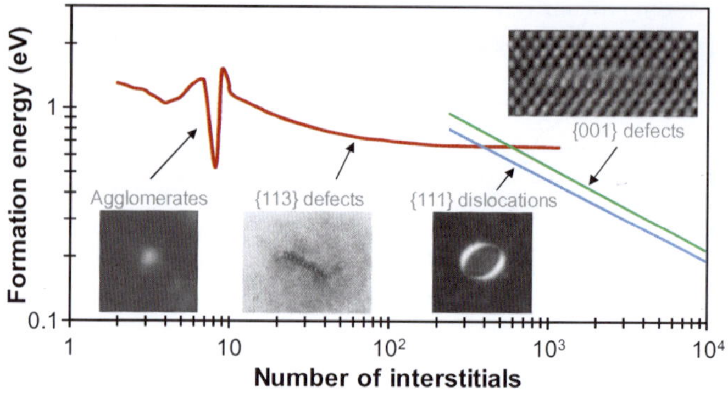


FIG. 3.13 Formation energies of I defects in Si as a function of size. *(Data and images extracted from D.J. Eaglesham, P.A. Stolk, H.-J. Gossmann, J.M. Poate, Implantation and transient B diffusion in Si: the source of the interstitials, Appl. Phys. Lett. 65 (1994) 2305–2307; N.E.B. Cowern, G. Mannino, P.A. Stolk, F. Roozeboom, H.G.A. Huizing, J.G.M. van Berkum, F. Cristiano, A. Claverie, M. Jaraíz, Energetics of self-interstitial clusters in Si, Phys. Rev. Lett. 82 (1999) 4460–4463; F. Cristiano, J. Grisolia, B. Colombeau, M. Omri, B. de Mauduit, A. Claverie, L.F. Giles, N.E.B. Cowern, Formation energies and relative stability of perfect and faulted dislocation loops in silicon, J. Appl. Phys. 87 (2000) 8420–8428; B. Colombeau, F. Cristiano, A. Altibelli, C. Bonafos, G. Ben Assayag, A. Claverie, Atomistic simulations of extrinsic defects evolution and transient enhanced diffusion in silicon, Appl. Phys. Lett. 78 (2001) 940–942; Y. Qiu, F. Cristiano, K. Huet, F. Mazzamuto, G. Fisicaro, A. La Magna, M. Quillec, N. Cherkashin, H. Wang, S. Duguay, D.C. Blavette, Extended defects formation in nanosecond laser-annealed ion implanted silicon, Nano Lett. 14 (2014) 1769–1775.)*

$$\nu = \nu_0 \exp\left(\frac{E_f - E_{f,I} - E_{m,I}}{k_B T}\right) \tag{3.23}$$

where $\nu_0$ is the vibration frequency of Si atoms in the lattice; $E_f$ the formation energy of the defect; $E_{f,\,I}$ and $E_{m,I}$ the formation and migration energies of the free self-I, respectively; and $k_B$ the Boltzmann constant. The emission frequency is, therefore, higher for defects of higher formation energy, which means that low-energy defects will grow at the expense of higher energy defects. Thus, for example, a large {113} defect will grow at the expense of smaller {113} defects and I agglomerates, and large {111} dislocations at the expense of all of them. The Ostwald ripening mechanism can be easily programmed in KMC simulators [207]. In fact, KMC models based on the Ostwald ripening process have worked very well over the past three decades to predict the formation and evolution of self-I defects in Si under technologically relevant conditions. For example, Fig. 3.14 shows the good agreement between experimental and KMC simulation results on the concentration of Is in extended defects and their size as a function of time and annealing temperature [68].

However, recent ultrafast laser annealing experiments on ion-implanted Si have yielded unexpected results regarding self-I defect formation [242]. The use of laser anneals for the monolithic integration of electronic devices in three

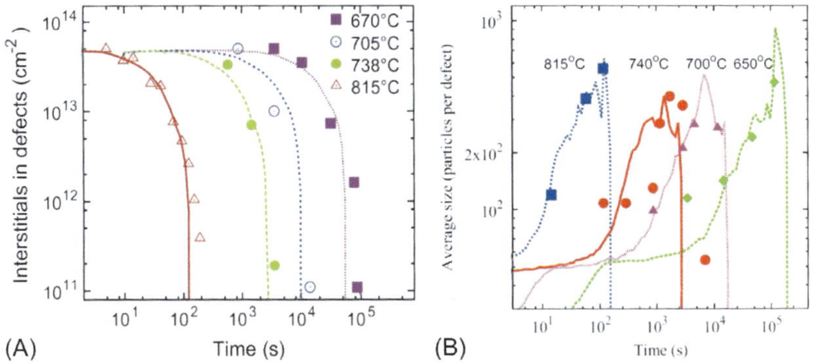


**FIG. 3.14** (A) I concentration in extended defects as a function of annealing time and temperature, after a 40 keV, $5 \times 10^{13}$ cm$^{-2}$ Si implant into Si. (B) I cluster average size evolution with annealing time at different temperatures, after a 40 keV, $6 \times 10^{13}$ cm$^{-2}$ Si implant into Si. Symbols are experimental data and lines are simulation results. *(Extracted from I. Martín-Bragado, R. Borges, J.P. Balbuena, M. Jaraiz, Kinetic Monte Carlo simulation for semiconductor processing: a review, Prog. Mater. Sci. 92 (2018) 1–32.)*

dimensions is being explored as a technological alternative to continue with the increase in integration density [243–245]. It has been shown that with ultrafast laser annealing, dopants implanted for the fabrication of junctions in the top-layer transistors can be electrically activated without negatively affecting lower layer transistors. These anneals by lasers allow to reach temperatures close to the Si melting for very short times, of the order of tens of nanoseconds, as opposed to conventional anneals, where lower temperatures are reached, at most 200°C below melting, with typical annealing times of the order of seconds. In the cited experiments of Qiu et al. [242], a crystal Si substrate was first self-implanted to generate a large concentration of self-Is in the substrate (but below the amorphization threshold) and then was annealed with a single laser pulse of about 200-ns duration. Fig. 3.15 shows a TEM image of the sample after annealing. In the unmelted regions of the substrate, large I-rich {001} loops were observed (with diameters around 10 nm, containing about $10^3$ Si self-Is). This kind of {001} I extended defects, while common in C and Ge [246], had not been observed in Si before. They were not expected from a thermodynamic point of view since their formation energy is higher than the corresponding to {111} dislocations (see Fig. 3.13). The generation of {001} loops was initially justified by the increase of the formation energy of {111} dislocations with respect to {001} defects due to the presence of stress in the sample as a consequence of the temperature gradient produced by the laser anneal [242]. However, later experiments on silicon-on-insulator (SOI) substrates, where thermally induced stress was negligible, discarded that justification as large {001} loops were again observed after laser annealing [128]. Generation of {001} loops was also unexpected from a kinetic point of view, since to form extended defects containing thousands of self-Is under the conventional

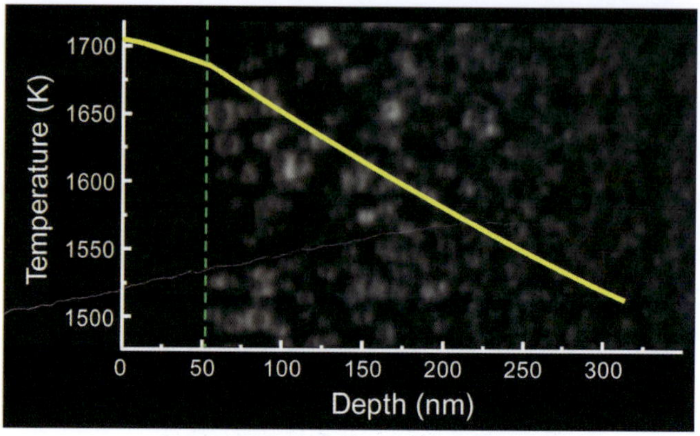


**FIG. 3.15** TEM image showing large {001} loops formed after laser annealing of a self-implanted crystal Si sample. Superimposed yellow line shows the local maximum temperature gradient generated in the substrate by the laser. Green dashed line indicates the melt depth. *(Extracted from L.A. Marqués, M. Aboy, I. Santos, P. López, F. Cristiano, A. La Magna, K. Huet, T. Tabata, L. Pelaz, Ultrafast generation of unconventional {001} loops in Si, Phys. Rev. Lett. 119 (2017) 205503.)*

Ostwald ripening mechanism would require times of the order of seconds at near-melting temperatures, while such large defects were observed after laser-annealing times of just hundreds of nanoseconds [129].

The small sizes and short timescales involved in ultrafast laser annealing make the direct experimental study of the anomalous generation of {001} loops very difficult. But precisely at the nanoscale, atomistic simulation techniques can become very useful to get some physical insight. In particular, CMD is very adequate to address this issue, since with the use of empirical potentials, parallel codes running on average cluster computers allow the dynamic simulation of systems with sizes of nanometers for times of the order of nanoseconds, which are the typical size and timescales involved in ultrafast laser annealing experiments.

As a first approximation, ultrafast laser annealing can be simulated within CMD assuming a pure heat transfer from the laser to the lattice atoms. Several CMD studies have used this approximation [233, 247]. Initial samples consist of a pure Si crystal where a given amount of extra self-I atoms are introduced in random positions. Samples are then annealed at different temperatures to analyze how self-Is move and interact to give rise to more complex defects. Defects obtained are then analyzed using different computational techniques to extract their particular configurations and thermodynamic properties [234, 241, 248].

CMD simulations show that at relatively low temperatures (corresponding to conventional annealing techniques), self-I aggregation is consistent with the Ostwald ripening process, with the dominance of stable ordered structures of relatively small size that exchange free self-Is (see Fig. 3.16). Several families of self-I clusters are observed: {111} rod-like defects, ⟨011⟩ chains (considered

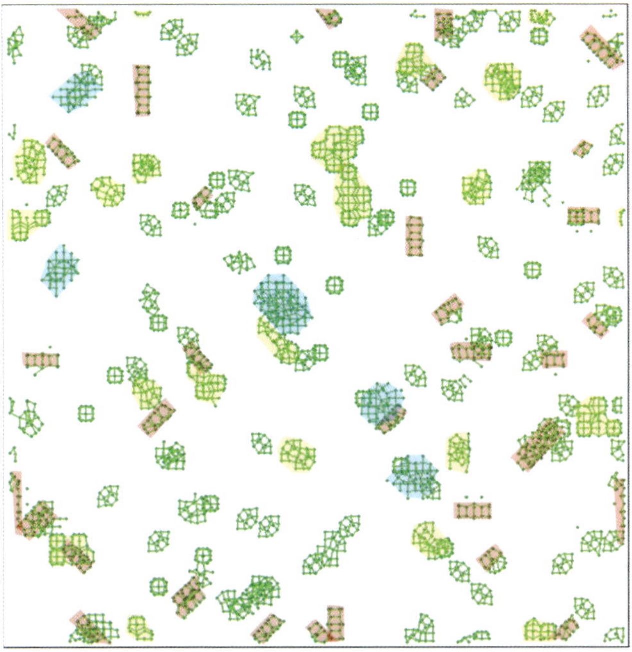

**FIG. 3.16** [001] projection of self-I clusters formed in a CMD simulation of I aggregation at 850 K after 170 ns annealing. Clusters shadowed in blue are {111} defects, in red are (110) chains, and in yellow small agglomerates of {001} Arai tetra-Is. Remaining defects are isolated Arai tetra-Is. *(Extracted from L.A. Marqués, M. Aboy, M. Ruiz, I. Santos, P. López, L. Pelaz, Molecular dynamics simulation of the early stages of self-interstitial clustering in silicon, Mater. Sci. Semicond. Process. 42 (2016) 235–238.)*

as precursors of {113} defects), compact {001} Arai tetra-Is, and {001} agglomerates of them [33]. For long annealing times, the dominant self-I clusters will result from the competition among the different defect families. In principle, the Ostwald ripening mechanism will favor the growth of clusters with lower energies. For {111} defects and (011) chains, formation energies decrease monotonously, but in the case of {001} Arai type of defects, formation energies show minima for configurations whose number of Is is a multiple of 4 [129, 249]. Interestingly enough, these are the "magic numbers" found in the inverse modeling of experimental measurements done by Cowern et al. [204]. This is because only in these multiple-of-four configurations, all {001} defect atoms are tetra-coordinated and, therefore, strongly bounded. Consequently, while {111} defects and (011) chains can grow by capturing free Is through the Ostwald ripening mechanism, this is not the case for {001} Arai clusters, which in order to grow they must capture four Is simultaneously, and such an event is highly unlikely at low annealing temperatures. This explains why no {001} extended defects are observed in Si after conventional annealing conditions.

In the CMD simulations at higher temperatures, closer to the melting point and therefore typical of laser annealing conditions, the process of self-I defect growth is completely different [128, 233]. During the simulations, large agglomerates, containing hundreds of Is, are formed. Unlike extended defects found in conventional annealing conditions, these agglomerates have no regular crystallographic structures. Another unexpected observation is that, in spite of their large size, they are highly mobile. A careful analysis of the atomic structure of the disordered agglomerates determined that they are of liquid nature, which justifies the high mobility they show in the simulations [128, 250]. The existence of I-rich liquid droplets within a solid semiconductor is possible due to the strong compressive stress induced by the surrounding crystal. After several nanoseconds, liquid agglomerates spontaneously transform into large {001} loops. At the end of the simulations, formed defects are much larger than in the previous case of low-temperature annealing. Most of them are {001} loops, but some {111} dislocations can also be observed (see Fig. 3.17). In this regime, as annealing temperature is increased, formed clusters are larger and the amount of Is stored in {111} defects decreases with respect to {001} loops, in very good agreement with experimental observations [251]. (011) chains are not

FIG. 3.17 [001] projection of self-I clusters formed in a CMD simulation of I aggregation at 1300 K after 30 ns annealing. Clusters shadowed in blue are {111} dislocations, and in yellow, {001} loops. Remaining defects are isolated Arai tetra-Is. *(Extracted from L.A. Marqués, M. Aboy, M. Ruiz, I. Santos, P. López, L. Pelaz, Molecular dynamics simulation of the early stages of self-interstitial clustering in silicon, Mater. Sci. Semicond. Process. 42 (2016) 235–238.)*

stable at temperatures close to the melting point, and consequently, no {113} extended defects are formed in this regime.

CMD techniques have also been used to study how I-rich liquid agglomerates grow and transform into {001} loops [129, 250]. As it was previously mentioned, these liquid droplets move through the Si lattice. They adopt a quasispherical shape to minimize the surface-volume ratio and consequently their interfacial energy. When two droplets approximate, a narrow neck connecting them is formed, allowing the atoms to redistribute and recover the spherical shape in a final larger droplet. This process is known as *coalescence*. While responsible for defect growth in metals [252], coalescence had not been reported in Si before, as Ostwald ripening has been for decades the only growth mechanism considered for defect coarsening in semiconductors. CMD simulations have shown that, as I-rich liquid droplets coalesce and grow, they tend to occupy a larger volume within the lattice, with a consequent decrease of their compressive stress [128]. At a given size depending on annealing temperature (larger for higher temperatures, in agreement with experimental observations), liquid agglomerates become undercooled and spontaneously transform into {001} loops. This has been identified as a first-order transition that starts from the nucleation of {001} Arai tetra-Is at droplet edges in contact with the surrounding lattice. It is worth noting that the Arai tetra-I is the only ordered atomic structure stable at temperatures close to the melting point, due to its high vibrational entropy [241]. Once nucleated, Arai tetra-Is act as a template that promotes the placement of excess I atoms within the agglomerates along {001} planes. It has been demonstrated that this particular arrangement is caused by the tensile and compressive stress fields that develop concurrently around the forming loop [250]. {001} loops, once generated, can further grow by capturing moving liquid agglomerates. When they reach a certain size, they spontaneously unfault into {001} dislocations.

Atomistic simulations have also served to characterize {001} loops [247, 250]. The atomic structure observed in the core of generated {001} loops, fully consistent with experimental TEM images (see Fig. 3.18), differs from the one formerly proposed for such defects in Ge [246]. Other information extracted from the simulations is the formation energy per I atom (0.68 eV) and I density (13.6 $nm^2$) of {001} loops. This information is relevant from a practical point of view, because from the measurable size distribution and I density of {001} loops, it is possible to evaluate the total amount of self-Is stored in such defects. They are also key parameters for predictive KMC or continuum modeling of ultrafast laser annealing treatments or other extremely high-temperature processing steps.

## 3.5 Conclusions

The complexity of laser-related phenomena for semiconductor technology requires a multiscale modeling approach that captures from the detailed excitation of the electronic subsystem to the multiple atomic dopant-defect

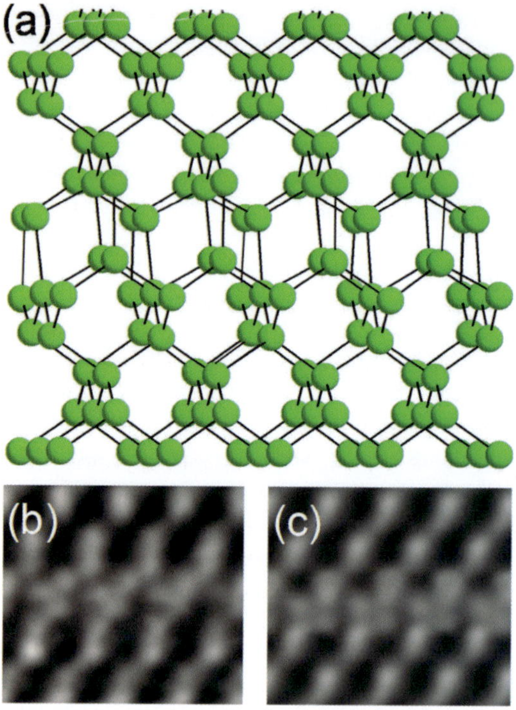


FIG. 3.18 (A) Atomic structure of the {001} loops as observed in CMD simulations, (B) experimental HTREM image from Qiu et al. [242], and (C) simulated TEM image of the structure shown in (A). *(Extracted from L.A. Marqués, M. Aboy, M. Ruiz, I. Santos, P. López, L. Pelaz, {001} loops in silicon unraveled, Acta Mater. 166 (2019) 192–201.)*

interactions. Therefore, computational tools that resolve at atomic and femtosecond timescales must be combined with simulations that can provide a relevant description of semiconductor technology issues at macroscopic scales. Ab initio and tight binding methods can describe the laser-induced electron excitation and the ultrafast heat transfer from electrons to the lattice, but they are limited to systems with a few hundreds of atoms and very short times. Nevertheless, these methods can provide clues for physics-based approximations used in higher level models. Large-scale CMD and KMC simulations can provide detailed insight into the atomic-level dopant-defect dynamics involved in laser processing.

Modeling of Si laser annealing has evolved with the improvement of computing capabilities as more sophisticated approaches could be applied. The physics of the interaction between a laser beam and matter is complex and it involves the excitation of a fraction of electrons from occupied (valence electrons) to unoccupied electronic levels (conduction band in semiconductors), bond softening, and modification of phonon spectrum. Simplifications frequently used for near-equilibrium systems are no longer valid in laser heating.

More elaborated models have been developed to provide us with a better understanding of how laser photons interact with electrons, how excited electrons eventually relax, and how energy is transferred to the lattice. This process is relatively fast (within 100 femtoseconds to tens of picoseconds) compared to the characteristic timescale of the collective movement of atoms responsible for the structural and phase transformations triggered by the laser excitation. For laser pulses longer than several nanoseconds (much larger than the electron thermal relaxation time), it is often assumed that the energy has already been transferred to the lattice, and modeling focuses on the evolution of the heating profiles resulting after laser irradiation and the atomic displacement of dopant and defects. However, ultrafast (pico- and femtosecond) laser pulses have a duration shorter than or comparable to the relaxation time of electrons to phonons and detailed modeling of the coupling between electrons and phonons needs to be simulated as well as the effect of electron excitation on interatomic forces.

Due to the intrinsic features of laser annealing treatments, i.e., localized irradiation in the nanometer and nanosecond scale, pulsed laser irradiation of a damaged solid has a high potential for nanotechnology applications. This process promotes ultrafast nonequilibrium kinetics leading to microscopic modifications of the material for which atomistic modeling methods are well suited. Techniques such as CMD are particularly suitable for systems with sizes of nanometers for times of the order of nanoseconds, which are the typical size and timescales involved in ultrafast laser annealing experiments. Improved models such as TTM or specifically designed empirical potentials allow to describe the physics of the electron-nuclei interactions, and at the same time to follow the dynamics of defects and the phase transformations without the need of fitting parameters. Models that focus on dopant-defect interactions, such as KMC, have been adapted to capture nonuniform fast-varying thermal fields along with dopant-defect kinetics with parameters valid in a wide range of temperatures and considering eventual phase transitions that include defect generation or anomalous impurity redistribution.

Atomistic simulation techniques are key to get physical insight at small sizes and short timescales involved in fast laser annealing processes. They do not only help us understand the physics of the system but they can also provide technological clues for the improvement of semiconductor processing. The future in this field of research appears to be very promising, thanks to the continuously increasing power of computers on each new technology node. This improvement in computing power is allowing the treatment of larger and more realistic systems in ab initio and tight binding simulations, the use of sophisticated potentials with near-quantum accuracy in CMD simulations with billions of atoms, and the effective scaling to the modeling of microstructure evolution with the concurrent use of KMC and PF methods. These tools, together in carefully designed multiscale approaches, can serve to obtain more realistic fundamental understanding and increasingly quantitative predictions of semiconductor material behavior and processing.

## References

[1] L. Pelaz, L.A. Marqués, M. Aboy, P. López, I. Santos, Front-end process modeling in silicon, Eur. Phys. J. B 72 (2009) 323–359.

[2] S.F. Lombardo, S. Boninelli, F. Cristiano, G. Fisicaro, G. Fortunato, M. Grimaldi, G. Impellizzeri, M. Italia, A. Marino, R. Milazzo, E. Napolitani, V. Privitera, A. La Magna, Laser annealing in Si and Ge: anomalous physical aspects and modeling approaches, Mater. Sci. Semicond. Process. 62 (2017) 80–91.

[3] G. Fisicaro, K. Huet, R. Negru, M. Hackenberg, P. Pichler, N. Taleb, A. La Magna, Anomalous impurity segregation and local bonding fluctuation in *l*-Si, Phys. Rev. Lett. 110 (11) (2013), 117801.

[4] E.V. Monakhov, B.G. Svensson, M.K. Linnarsson, A. La Magna, M. Italia, V. Privitera, G. Fortunato, M. Cuscunà, L. Mariucci, The effect of excimer laser pretreatment on diffusion and activation of boron implanted in silicon, Appl. Phys. Lett. 87 (2005) 192109.

[5] K.K. Ong, K.L. Pey, P.S. Lee, A.T.S. Wee, X.C. Wang, Y.F. Chong, Dopant distribution in the recrystallization transient at the maximum melt depth induced by laser annealing, Appl. Phys. Lett. 89 (2006), 172111.

[6] A.R. Leach, Molecular Modelling: Principles and Applications, second ed., Pearson Education, Harlow, England, 2001.

[7] P. Hohenberg, W. Kohn, Inhomogeneous electron gas, Phys. Rev. 136 (3B) (1964) B864–B871.

[8] H. Jónsson, G. Mills, K.W. Jacobsen, Nudged elastic band method for finding minimum energy transitions, in: B.J. Berne, G. Ciccotti, D.F. Coker (Eds.), Classical and Quantum Dynamics in Condensed Phase Simulations, World Scientific, Singapore, 1998, pp. 385–404.

[9] O.K. Al-Mushadani, R.J. Needs, Free-energy calculations of intrinsic point defects in silicon, Phys. Rev. B 68 (2003), 235205.

[10] F. Cargnoni, C. Gatti, L. Colombo, Formation and annihilation of a bond defect in silicon: an ab initio quantum-mechanical characterization, Phys. Rev. B 57 (1) (1998) 170–177.

[11] S.J. Clark, G.J. Ackland, Ab initio calculations of the self-interstitial in silicon, Phys. Rev. B 56 (1) (1997) 47–50.

[12] J.L. Mercer, J.S. Nelson, A.F. Wright, E.B. Stechel, Ab initio calculations of the energetics of the neutral Si vacancy defect, Model. Simul. Mater. Sci. Eng. 6 (1) (1998) 1–8.

[13] P.E. Blöchl, E. Smargiassi, R. Car, D.B. Laks, W. Andreoni, S.T. Pantelides, First-principles calculations of self-diffusion constants in silicon, Phys. Rev. Lett. 70 (16) (1993) 2435–2438.

[14] B.J. Coomer, J.P. Goss, R. Jones, S. Oberg, P.R. Briddon, Identification of the tetra-interstitial in silicon, J. Phys. Condens. Matter 13 (1) (2001) L1–L7.

[15] Y.A. Du, R.G. Hennig, T. Lenosky, J.W. Wilkins, From compact point defects to extended structures in silicon, Europhys. J. B 57 (2007) 229–234.

[16] M. Kohyama, S. Takeda, First-principles calculations of the self-interstitial cluster I4 in Si, Phys. Rev. B 60 (11) (1999) 8075–8080.

[17] D.V. Makhov, L.J. Lewis, Stable fourfold configurations for small vacancy clusters in silicon from ab initio calculations, Phys. Rev. Lett. 92 (25) (2004), 255504.

[18] C.S. Nichols, C.G. Van de Walle, S.T. Pantelides, Mechanisms of dopant impurity diffusion in silicon, Phys. Rev. B 40 (8) (1989) 5484–5496.

[19] M. Ramamoorthy, S.T. Pantelides, Complex dynamical phenomena in heavily arsenic doped silicon, Phys. Rev. Lett. 76 (25) (1996) 4753–4756.

[20] B. Sadigh, T.J. Lenosky, S.K. Theiss, M.J. Caturla, Mechanism of boron diffusion in silicon: an ab initio and kinetic Monte Carlo study, Phys. Rev. Lett. 83 (21) (1999) 4341–4344.
[21] W. Windl, M.M. Bunea, R. Stumpf, S.T. Dunham, M.P. Masquelier, First-principles study of boron diffusion in silicon, Phys. Rev. Lett. 83 (21) (1999) 4345–4348.
[22] J. Zhu, T.D. de la Rubia, L.H. Yang, C. Mailhiot, G.H. Gilmer, Ab initio pseudopotential calculations of B diffusion and pairing in Si, Phys. Rev. B 54 (7) (1996) 4741–4747.
[23] T.J. Lenosky, B. Sadigh, S.K. Theiss, M.J. Caturla, T.D. de la Rubia, Ab initio energetics of boron-interstitial clusters in crystalline Si, Appl. Phys. Lett. 77 (12) (2000) 1834–1836.
[24] X.Y. Liu, W. Windl, M.P. Masquelier, Ab initio modeling of boron clustering in silicon, Appl. Phys. Lett. 77 (13) (2000) 2018–2020.
[25] F. Bloch, Uber die Quantenmechanik der Elektronen in Kristallgittern, Z. Phys. 52 (1928) 555–600.
[26] L. Colombo, Tight-binding molecular dynamics: a primer, La Riv. Nuovo Cimento 28 (10) (2005) 1–59.
[27] M. Cogoni, B.P. Uberuaga, A.F. Voter, L. Colombo, Diffusion of small self-interstitial clusters in silicon: temperature-accelerated tight-binding molecular dynamics simulations, Phys. Rev. B 71 (2005). 121203(R).
[28] L. Colombo, Tight-binding theory of native point defects in silicon, Annu. Rev. Mater. Res. 32 (2002) 271–295.
[29] A. Jääskeläinen, L. Colombo, R. Nieminen, Silicon self-diffusion constants by tight-binding molecular dynamics, Phys. Rev. B 64 (2001), 233203.
[30] L.J. Munro, D.J. Wales, Defect migration in crystalline silicon, Phys. Rev. B 59 (6) (1999) 3969–3980.
[31] M. Tang, L. Colombo, J. Zhu, T. Diaz de la Rubia, Intrinsic point defects in crystalline silicon: tight-binding molecular dynamics studies of self-diffusion, interstitial-vacancy recombination, and formation volumes, Phys. Rev. B 55 (21) (1997) 14279–14289.
[32] M.T. Zawadzki, W. Luo, P. Clancy, Tight-binding molecular dynamics study of vacancy-interstitial annihilation in silicon, Phys. Rev. B 63 (2001), 205205.
[33] N. Arai, S. Takeda, M. Kohyama, Self-interstitial clustering in crystalline silicon, Phys. Rev. Lett. 78 (22) (1997) 4265–4268.
[34] D.A. Richie, J. Kim, S.A. Barr, K.R.A. Hazzard, R. Hennig, J.W. Wilkins, Complexity of small silicon self-interstitial defects, Phys. Rev. Lett. 92 (4) (2004), 045501.
[35] E.G. Song, E. Kim, Y.H. Lee, Y.G. Hwang, Fully relaxed point defects in crystalline silicon, Phys. Rev. B 48 (3) (1993) 1486–1489.
[36] C.Z. Wang, C.T. Chan, K.M. Ho, Tight-binding molecular-dynamics study of defects in silicon, Phys. Rev. Lett. 66 (2) (1991) 189–192.
[37] P. Alippi, L. Colombo, Lattice-strain field induced by {311} self-interstitial defects in silicon, Phys. Rev. B 62 (3) (2000) 1815–1820.
[38] J. Kim, J.W. Wilkins, F.S. Khan, A. Canning, Extended Si {311} defects, Phys. Rev. B 55 (24) (1997) 16186–16197.
[39] M. Kohyama, S. Takeda, Tight-binding study of the {113} planar interstitial defects in Si, Phys. Rev. B 51 (19) (1995) 13111–13116.
[40] W. Luo, P.B. Rasband, P. Clancy, B.W. Roberts, Tight-binding studies of the tendency for boron to cluster in c-Si. II. Interaction of dopants and defects in boron-doped Si, J. Appl. Phys. 84 (5) (1998) 2476–2486.
[41] W. Luo, P. Clancy, Identification of stable boron clusters in c-Si using tight-binding statics, J. Appl. Phys. 89 (3) (2001) 1596–1604.

[42] B.J. Garrison, D. Srivastava, Potential-energy surfaces for chemical-reactions at solid-surfaces, Annu. Rev. Phys. Chem. 46 (1995) 373–394.

[43] S.B. Sinnott, D.W. Brenner, Three decades of many-body potentials in materials research, MRS Bull. 37 (5) (2012) 469–473.

[44] A.F. Voter, Interatomic potentials for atomistic simulations, MRS Bull. 21 (2) (1996) 17–19.

[45] F.H. Stillinger, T.A. Weber, Computer-simulation of local order in condensed phases of silicon, Phys. Rev. B 31 (1985) 5262–5271.

[46] J. Tersoff, New empirical-approach for the structure and energy of covalent systems, Phys. Rev. B 37 (1988) 6991–7000.

[47] A.P. Bartok, R. Kondor, G. Csanyi, Gaussian approximation potentials: the accuracy of quantum mechanics, without the electrons, Phys. Rev. Lett. 104 (2010), 136403.

[48] A.P. Bartok, M.C. Payne, R. Kondor, G. Csanyi, On representing chemical environments, Phys. Rev. B 87 (2013) 184115.

[49] J. Behler, Perspective: machine learning potentials for atomistic simulations, J. Chem. Phys. 145 (2016) 170901.

[50] K. Kadau, T.C. Germann, P.S. Lomdahl, Molecular dynamics comes of age: 320 billion atom simulation on BlueGene/L, Int. J. Mod. Phys. C 17 (2006) 1755–1761.

[51] Stanford University n.d., Folding@Home. Available from: https://foldingathome.org/ (14 July 2020).

[52] S. Plimpton, Fast parallel algorithms for short-range molecular dynamics, J. Comput. Phys. 117 (1) (1995) 1–19.

[53] Sandia National Laboratories, n.d., LAMMPS Molecular Dynamics Simulator. Available from: https://lammps.sandia.gov (14 July 2020).

[54] M.P. Allen, D.J. Tildesley, Computer Simulation of Liquids, Oxford University Press, New York, 1989.

[55] D. Frenkel, B. Smit, Understanding Molecular Simulation: From Algorithms to Applications, Academic Press, San Diego, 1996.

[56] J.M. Haile, Molecular Dynamics Simulations: Elementary Methods, John Wiley & Sons, New York, 1992.

[57] D.C. Rapaport, The Art of Molecular Dynamics Simulation, second ed., Cambridge University Press, Cambridge, UK, 2004.

[58] R. Darkins, D.M. Duffy, Modelling radiation effects in solids with two-temperature molecular dynamics, Comput. Mater. Sci. 147 (2018) 145–153.

[59] B. Rethfeld, D.S. Ivanov, M.E. Garcia, S.I. Anisimov, Modelling ultrafast laser ablation, J. Phys. D Appl. Phys. 50 (19) (2017), 193001.

[60] C. Wu, E.T. Karim, A.N. Volkov, L.V. Zhigilei, Atomic movies of laser-induced structural and phase transformations from molecular dynamics simulations, in: M. Castillejo, P.M. Ossi, L. Zhigilei (Eds.), Lasers in Materials Science, vol. 191, Springer, New York, 2014, pp. 67–100.

[61] L.V. Zhigilei, Z. Lin, D.S. Ivanov, E. Leveugle, D.H. Duff, D. Thomas, C. Sevilla, S.J. Guy, Atomic/molecular-level simulations of laser-materials interactions, in: A. Miotello, P.M. Ossi (Eds.), Laser-Surface Interactions for New Materials Production: Tailoring Structure and Properties, vol. 130, Springer, New York, 2010, pp. 43–79.

[62] L. Pelaz, M. Jaraiz, G.H. Gilmer, H.-J. Gossmann, C.S. Rafferty, D.J. Eaglesham, J.M. Poate, B diffusion and clustering in ion implanted Si: the role of B cluster precursors, Appl. Phys. Lett. 70 (1997) 2285.

[63] M. Strobel, A. La Magna, S. Coffa, A kinetic lattice Monte-Carlo approach to the evolution of boron in silicon, Nucl. Instrum. Methods Phys. Res., Sect. B 186 (1–4) (2002) 339–343.

[64] M. Aboy, L. Pelaz, E. Bruno, S. Mirabella, S. Boninelli, Kinetics of large B clusters in crystalline and preamorphized silicon, J. Appl. Phys. 110 (2011) 073524.
[65] G. Fisicaro, C. Bongiorno, I. Deretzis, F. Giannazzo, F. La Via, F. Roccaforte, M. Zielinski, M. Zimbone, A. La Magna1, Genesis and evolution of extended defects: the role of evolving interface instabilities in cubic SiC, Appl. Phys. Rev. 7 (021402) (2020) 1–14.
[66] P. López, L. Pelaz, L.A. Marqués, I. Santos, M. Aboy, J. Barbolla, Atomistic modeling of defect evolution in Si for amorphizing and subamorphizing implants, Mater. Sci. Eng. B 114-115 (2004) 82–87.
[67] I. Martin-Bragado, V. Moroz, Facet formation during solid phase epitaxy regrowth: a lattice kinetic Monte Carlo model, Appl. Phys. Lett. 95 (12) (2009), 123123.
[68] I. Martín-Bragado, R. Borges, J.P. Balbuena, M. Jaraiz, Kinetic Monte Carlo simulation for semiconductor processing: a review, Prog. Mater. Sci. 92 (2018) 1–32.
[69] P.L. Silvestrelli, A. Alavi, M. Parrinello, D. Frenkel, Ab initio molecular dynamics simulation of laser melting of silicon, Phys. Rev. Lett. 77 (15) (1996) 3149–3152.
[70] P.L. Silvestrelli, A. Alavi, M. Parrinello, D. Frenkel, Structural, dynamical, electronic, and bonding properties of laser-heated silicon: an ab initio molecular-dynamics study, Phys. Rev. B 56 (7) (1997) 3806–3812.
[71] A. Alavi, J. Kohanoff, M. Parrinello, D. Frenkel, Ab initio molecular dynamics with excited electrons, Phys. Rev. Lett. 73 (19) (1994) 2599–2602.
[72] R.O. Jones, Density functional theory: its origins, rise to prominence, and future, Rev. Mod. Phys. 87 (3) (2015) 897–923.
[73] W. Kohn, L.J. Sham, Self-consistent equations including exchange and correlation effects, Phys. Rev. 140 (4A) (1965) A1133–A1138.
[74] H.M. Van Driel, Kinetics of high-density plasmas generated in Si by 1.06- and 0.53-μm picosecond laser pulses, Phys. Rev. B 35 (15) (1987) 8166–8176.
[75] J.A. Van Vechten, R. Tsu, F.W. Saris, D. Hoonhout, Reasons to believe pulsed laser annealing of Si does not involve simple thermal melting, Phys. Lett. A 47 (6) (1979) 417–421.
[76] J.A. Van Vechten, R. Tsu, F.W. Saris, D. Hoonhout, Nonthermal pulsed laser annealing of Si; plasma annealing, Phys. Lett. A 47 (6) (1979) 422–426.
[77] N.S. Grigoryan, T. Zier, M.E. Garcia, E.S. Zijlstra, Ultrafast structural phenomena: theory of phonon frequency changes and simulations with code for highly excited valence electron systems, J. Opt. Soc. Am. B 31 (11) (2014) C22–C27.
[78] T. Zier, E.S. Zijlstra, M.E. Garcia, Silicon before the bonds break, Appl. Phys. A 117 (1) (2014) 1–5.
[79] T. Zier, E.S. Zijlstra, M.E. Garcia, Quasimomentum-space image for ultrafast melting of silicon, Phys. Rev. Lett. 116 (15) (2016), 153901.
[80] T. Zier, E.S. Zijlstra, S. Krylow, M.E. Garcia, Simulations of laser-induced dynamics in free-standing thin silicon films, Appl. Phys. A 123 (10) (2017) 625.
[81] E.S. Zijlstra, A. Kalitsov, T. Zier, M.E. Garcia, Squeezed thermal phonons precurse nonthermal melting of silicon as a function of fluence, Phys. Rev. X 3 (1) (2013), 011005.
[82] E.S. Zijlstra, F.C. Kabeer, B. Bauerhenne, T. Zier, N. Grigoryan, M.E. Garcia, Modeling of material properties after ultrashort laser and XUV excitation, Appl. Phys. A 110 (3) (2013) 519–528.
[83] E.S. Zijlstra, A. Kalitsov, T. Zier, M.E. Garcia, Fractional diffusion in silicon, Adv. Mater. 25 (39) (2013) 5605–5608.
[84] E.S. Zijlstra, T. Zier, B. Bauerhenne, S. Krylow, P.M. Geiger, M.E. Garcia, Femtosecond-laser-induced bond breaking and structural modifications in silicon, $TiO_2$, and defective graphene: an ab initio molecular dynamics study, Appl. Phys. A 114 (1) (2014) 1–9.

[85] R. Biswas, V. Ambegaokar, Phonon spectrum of a model of electronically excited silicon, Phys. Rev. B 26 (4) (1982) 1980–1988.
[86] M.T. Dove, Structure and Dynamics: An Atomic View of Materials, Oxford Master Series in Condensed Matter Physics, Oxford University Press, New York, 2003.
[87] S. Baroni, S. de Gironcoli, A. Dal Corso, P. Giannozzi, Phonons and related crystal properties from density-functional perturbation theory, Rev. Mod. Phys. 73 (2) (2001) 515–562.
[88] X. Gonze, First-principles responses of solids to atomic displacements and homogeneous electric fields: implementation of a conjugate-gradient algorithm, Phys. Rev. B 55 (16) (1997) 10337–10354.
[89] V. Recoules, J. Clérouin, G. Zérah, P.M. Anglade, S. Mazevet, Effect of intense laser irradiation on the lattice stability of semiconductors and metals, Phys. Rev. Lett. 96 (5) (2006), 055503.
[90] G.Q. Yan, X.L. Cheng, H. Zhang, Z.Y. Zhu, D.H. Ren, Different effects of electronic excitation on metals and semiconductors, Phys. Rev. B 93 (21) (2016) 214302.
[91] M.A.L. Marques, E.K.U. Gross, Time-dependent density functional theory, in: C. Fiolhais, F. Nogueira, M.A.L. Marques (Eds.), A Primer in Density Functional Theory, Springer, Berlin, Heidelberg, 2003, pp. 144–184.
[92] M.A.L. Marques, C.A. Ullrich, F. Nogueira, A. Rubio, K. Burke, E.K.U. Gross (Eds.), Time Dependent Density Functional Theory, Lecture Notes in Physics 706, Springer, Berlin Heidelberg, 2006.
[93] G. Onida, L. Reining, A. Rubio, Electronic excitations: density-functional versus many-body Green's-function approaches, Rev. Mod. Phys. 47 (2) (2002) 601–659.
[94] S.A. Sato, Y. Shinohara, T. Otobe, K. Yabana, Dielectric response of laser-excited silicon at finite electron temperature, Phys. Rev. B 90 (17) (2014), 174303.
[95] C. Lian, S.B. Zhang, S. Meng, Ab initio evidence for nonthermal characteristics in ultrafast laser melting, Phys. Rev. B 94 (18) (2016), 184310.
[96] C. Lian, S.B. Zhang, S. Meng, Reply to "Comment on "Ab initio evidence for nonthermal characteristics in ultrafast laser melting"", Phys. Rev. B 99 (6) (2019), 066302.
[97] T. Zier, M.E. Garcia, P.L. Silvestrelli, Comment on "ab initio evidence for nonthermal characteristics in ultrafast laser melting", Phys. Rev. B 99 (6) (2019), 066301.
[98] S.A. Sato, K. Yabana, Y. Shinohara, T. Otobe, G.F. Bertsch, Numerical pump-probe experiments of laser-excited silicon in nonequilibrium phase, Phys. Rev. B 89 (6) (2014), 064304.
[99] M. Schultze, K. Ramasesha, C.D. Pemmaraju, S.A. Sato, D. Whitmore, A. Gandman, J.S. Prell, L.J. Borja, D. Prendergast, K. Yabana, D.M. Neumark, S.R. Leone, Attosecond band-gap dynamics in silicon, Science 346 (6215) (2014) 1348–1352.
[100] Y. Shinohara, K. Yabana, Y. Kawashita, J.I. Iwata, T. Otobe, G.F. Bertsch, Coherent phonon generation in time-dependent density functional theory, Phys. Rev. B 82 (15) (2010), 155110.
[101] K. Uchida, Y. Suzuki, K. Watanabe, Thermal effects on laser-assisted field evaporation from a Si surface: a real-time first-principles study, J. Appl. Phys. 126 (3) (2019), 034302.
[102] C. Covington, D. Kidd, H. Buckner, H. Appel, K. Varga, Time propagation of the coupled Maxwell and Kohn-Sham equations using the Riemann-Silberstein formalism, Phys. Rev. E 100 (5) (2019), 053301.
[103] M. Noda, S.A. Sato, Y. Hirokawa, M. Uemoto, T. Takeuchi, S. Yamada, A. Yamada, Y. Shinohara, M. Yamaguchi, K. Iida, I. Floss, T. Otobe, K.M. Lee, K. Ishimura, T. Boku, G.F. Bertsch, K. Nobusada, K. Yabana, SALMON: scalable ab-initio light–matter simulator for optics and nanoscience, Comput. Phys. Commun. 235 (2019) 356–365.
[104] SALMON software, n.d., Scalable Ab-Initio Light-Matter Simulator for Optics and Nanoscience, Available from: https://salmon-tddft.jp/index.html (14 July 2020).

[105] K. Yabana, T. Sugiyama, Y. Shinohara, T. Otobe, G.F. Bertsch, Time-dependent density functional theory for strong electromagnetic fields in crystalline solids, Phys. Rev. B 85 (4) (2012), 045134.

[106] S. Yamada, M. Noda, K. Nobusada, K. Yabana, Time-dependent density functional theory for interaction of ultrashort light pulse with thin materials, Phys. Rev. B 98 (24) (2018), 245147.

[107] A. Yamada, K. Yabana, Energy transfer from intense laser pulse to dielectrics in time-dependent density functional theory, Eur. Phys. J. D 73 (5) (2019) 87.

[108] A. Yamada, K. Yabana, Multiscale time-dependent density functional theory for a unified description of ultrafast dynamics: pulsed light, electron, and lattice motions in crystalline solids, Phys. Rev. B 99 (24) (2019), 245103.

[109] P. Stampfli, K.H. Bennemann, Theory for the instability of the diamond structure of Si, Ge, and C induced by a dense electron-hole plasma, Phys. Rev. B 42 (11) (1990) 7163–7173.

[110] P. Stampfli, K.H. Bennemann, Dynamical theory of the laser-induced lattice instability of silicon, Phys. Rev. B 46 (17) (1992) 10686–10692.

[111] A. Gambirasio, M. Bernasconi, L. Colombo, Laser-induced melting of silicon: a tight-binding molecular dynamics simulation, Phys. Rev. B 61 (12) (2000) 8233–8237.

[112] H.O. Jeschke, M.E. Garcia, K.H. Bennemann, Microscopic analysis of the laser-induced femtosecond graphitization of diamond, Phys. Rev. B 60 (6) (1999) R3701–R3704.

[113] H.O. Jeschke, M.E. Garcia, K.H. Bennemann, Theory for the ultrafast ablation of graphite films, Phys. Rev. Lett. 87 (1) (2001), 015003.

[114] H.O. Jeschke, M.S. Diakhate, M.E. Garcia, Molecular dynamics simulations of laser-induced damage of nanostructures and solids, Appl. Phys. A 96 (1) (2009) 33–42.

[115] N. Medvedev, H.O. Jeschke, B. Ziaja, Nonthermal phase transitions in semiconductors induced by a femtosecond extreme ultraviolet laser pulse, New J. Phys. 15 (1) (2013), 015016.

[116] N. Medvedev, Z. Li, B. Ziaja, Thermal and nonthermal melting of silicon under femtosecond X-ray irradiation, Phys. Rev. B 91 (5) (2015), 054113.

[117] N. Medvedev, Z. Li, V. Tkachenko, B. Ziaja, Electron-ion coupling in semiconductors beyond Fermi's golden rule, Phys. Rev. B 95 (1) (2017), 014309.

[118] N. Medvedev, V. Tkachenko, V. Lipp, Z. Li, B. Ziaja, Various damage mechanisms in carbon and silicon materials under femtosecond X-ray irradiation, 4open 1 (2018) 3. Available from: https://doi.org/10.1051/fopen/2018003. (14 July 2020).

[119] N. Medvedev, M. Kopecky, J. Chalupsky, L. Juha, Femtosecond X-ray diffraction can discern nonthermal from thermal melting, Phys. Rev. B 99 (10) (2019). 100303(R).

[120] N. Medvedev, Z. Fang, C. Xia, Z. Li, Thermal and nonthermal melting of III-V compound semiconductors, Phys. Rev. B 99 (174) (2019), 144101.

[121] B. Ziaja, N. Medvedev, Modelling ultrafast transitions within laser-irradiated solids, High Energy Density Phys. 8 (1) (2012) 18–29.

[122] B.M. Lee, S. Munetoh, T. Motooka, Molecular dynamics study of velocity distribution and local temperature change during rapid cooling processes in excimer-laser annealed silicon, Comput. Mater. Sci. 37 (2006) 198–202.

[123] B.M. Lee, H.K. Baik, B.S. Seong, S. Munetoh, T. Motooka, Molecular-dynamics analysis of the nucleation and crystallization process of Si, Physica B 392 (2007) 266–271.

[124] L.A. Marqués, L. Pelaz, M. Aboy, J. Barbolla, The laser annealing induced phase transition in silicon: a molecular dynamics study, Nucl. Inst. Methods Phys. Res. B 216 (2004) 57–61.

[125] S. Munetoh, X.Y. Ping, T. Ogata, T. Motooka, R. Teranishi, Excimer laser crystallization processes of amorphous silicon thin films by using molecular-dynamics simulations, ISIJ Int. 50 (12) (2010) 1925–1928.

[126] B.M. Lee, T. Motooka, S. Munetoh, Molecular-dynamics simulations of nucleation and crystallization processes of laser crystallized poly-Si, J. Phys. Condens. Matter 20 (2008), 055205.

[127] E.J. Albenze, M.O. Thompson, P. Clancy, Molecular dynamics study of explosive crystallization of SiGe and boron-doped SiGe alloys, Ind. Eng. Chem. Res. 45 (16) (2006) 5628–5639.

[128] L.A. Marqués, M. Aboy, I. Santos, P. López, F. Cristiano, A. La Magna, K. Huet, T. Tabata, L. Pelaz, Ultrafast generation of unconventional {001} loops in Si, Phys. Rev. Lett. 119 (2017), 205503.

[129] L.A. Marqués, M. Aboy, M. Ruiz, I. Santos, P. López, L. Pelaz, {001} loops in silicon unraveled, Acta Mater. 166 (2019) 192–201.

[130] C. Yang, Y. Wang, X. Xu, Molecular dynamics studies of ultrafast laser-induced phase and structural change in crystalline Si, Int. J. Heat Mass Transf. 55 (2012) 6060–6066.

[131] Q.Y. Liu, J. Zhou, Y.W. Zhao, L.C. Xiong, T.L. Shi, Y.H. Long, On the dynamic behaviors of silicon single crystal under nanosecond laser irradiation, Comput. Mater. Sci. 160 (2019) 420–429.

[132] D.K. Chokappa, S.J. Cook, P. Clancy, Nonequilibrium simulation method for the study of directed thermal processing, Phys. Rev. B 39 (14) (1989) 10075–10087.

[133] L. Wang, M.O. Thompson, P. Clancy, Non-equilibrium molecular dynamic simulations of laser thermal processing of heavily B-doped Si, in: Proceedings of the Rapid Thermal and Other Short-Time Processing Technologies II, Electrochemical Society Proceedings, 2001-9, 2001, pp. 345–352.

[134] L. Wang, P. Clancy, M.O. Thompson, C.S. Murthy, Thermodynamic and kinetic studies of laser thermal processing of heavily boron-doped amorphous silicon using molecular dynamics, J. Appl. Phys. 92 (5) (2002) 2412–2419.

[135] Q. Yu, P. Clancy, Molecular dynamics simulation of crystal growth in $Si_{1-x}Ge_x$/Si (100) heterostructures, J. Cryst. Growth 149 (1995) 45–58.

[136] M.I. Kaganov, I.M. Lifshitz, L.V. Tanatarov, Relaxation between electrons and the crystalline lattice, Sov. Phys. JETP 4 (2) (1957) 173–178.

[137] S.I. Anisimov, B.L. Kapeliovich, T.L. Perelman, Electron-emission from surface of metals induced by ultrashort laser pulses, Zh. Eksp. Teor. Fiz. 66 (2) (1974) 776–781.

[138] B.N. Chichkov, C. Momma, S. Nolte, F. von Alvensleben, A. Tünnermann, Femtosecond, picosecond and nanosecond laser ablation of solids, Appl. Phys. A Mater. Sci. Process. 63 (2) (1996) 109–115.

[139] J.G. Fujimoto, J.M. Liu, E.P. Ippen, N. Bloembergen, Femtosecond laser interaction with metallic tungsten and nonequilibrium electron and lattice temperatures, Phys. Rev. Lett. 53 (1984) 1837–1840.

[140] J. Hohlfeld, S.S. Wellershoff, J. Gudde, U. Conrad, V. Jahnke, E. Matthias, Electron and lattice dynamics following optical excitation of metals, Chem. Phys. 251 (2000) 237–258.

[141] J.K. Chen, D.Y. Tzou, J.E. Beraun, Numerical investigation of ultrashort laser damage in semiconductors, Int. J. Heat Mass Transf. 48 (2005) 501–509.

[142] S.L. Daraszewicz, D.M. Duffy, Hybrid continuum—atomistic modelling of swift heavy ion radiation damage in germanium, Nucl. Inst. Methods Phys. Res. B 303 (2013) 112–115.

[143] A. Rämer, O. Osmani, B. Rethfeld, Laser damage in silicon: energy absorption, relaxation, and transport, J. Appl. Phys. 116 (2014), 053508.

[144] H. Häkkinen, U. Landman, Superheating, melting, and annealing of copper surfaces, Phys. Rev. Lett. 71 (7) (1993) 1023–1026.

[145] D.S. Ivanov, L.V. Zhigilei, Combined atomistic-continuum modeling of short-pulse laser melting and disintegration of metal films, Phys. Rev. B 68 (2003), 064114.

[146] D.S. Ivanov, L.V. Zhigilei, Effect of pressure relaxation on the mechanisms of short-pulse laser melting, Phys. Rev. Lett. 91 (2003), 105701.

[147] C. Schäfer, H.M. Urbassek, L.V. Zhigilei, Metal ablation by picosecond laser pulses: a hybrid simulation, Phys. Rev. B 66 (2002), 115404.

[148] L.V. Zhigilei, Z. Lin, D.S. Ivanov, Atomistic modeling of short pulse laser ablation of metals: connections between melting, spallation, and phase explosion, J. Phys. Chem. C 113 (2009) 11892–11906.

[149] A. Duvenbeck, F. Sroubek, Z. Sroubek, A. Wucher, Nucl. Instrum. Methods Phys. Res. Sect. B 225 (2004) 464–477.

[150] A. Duvenbeck, Z. Sroubek, A. Wucher, Electronic excitation in atomic collision cascades, Nucl. Instrum. Methods Phys. Res. Sect. B 228 (2005) 325–329.

[151] D.M. Duffy, A.M. Rutherford, Including the effects of electronic stopping and electron–ion interactions in radiation damage simulations, J. Phys. Condens. Matter 19 (2007) 016207–016218.

[152] A.M. Rutherford, D.M. Duffy, The effect of electron-ion interactions on radiation damage simulations, J. Phys. Condens. Matter 19 (2007) 496201–496210.

[153] V.P. Lipp, B. Rethfeld, M.E. Garcia, D.S. Ivanov, Atomistic-continuum modeling of short laser pulse melting of Si targets, Phys. Rev. B 90 (2014), 245306.

[154] Y. Gan, J.K. Chen, Combined continuum-atomistic modelling of ultrashort-pulsed laser irradiation of silicon, Appl. Phys. A 105 (2011) 427–437.

[155] Y. Gan, J.K. Chen, A hybrid method for integrated atomistic-continuum simulation of ultrashort-pulsed laser interaction with semiconductors, Comput. Phys. Commun. 183 (2012) 278–284.

[156] Y. Gan, J.K. Chen, Atomic-level modeling of a silicon nanofilm irradiated by ultrashort-pulsed laser bursts, Int. J. Therm. Sci. 65 (2013) 1–8.

[157] A. Rousse, C. Rischel, S. Fourmaux, I. Uschmann, S. Sebban, G. Grillon, P. Balcou, E. Förster, J.P. Geindre, P. Audebert, J.C. Gauthier, D. Hulin, Non-thermal melting in semiconductors measured at femtosecond resolution, Nature (London) 410 (2001) 65–68.

[158] V.P. Lipp, D.S. Ivanov, B. Rethfeld, M.E. Garcia, On the interatomic interaction potential that describes bond weakening in classical molecular-dynamic modelling, J. Opt. Technol. 81 (2014) 254–255.

[159] S. Khakshouri, D. Alfè, D.M. Duffy, Development of an electron-temperature-dependent interatomic potential for molecular dynamics simulation of tungsten under electronic excitation, Phys. Rev. B 78 (2008), 224304.

[160] S.T. Murphy, S.L. Daraszewicz, Y. Giret, M. Watkins, A.L. Shluger, K. Tanimura, D.M. Duffy, Dynamical simulations of an electronically induced solid-solid phase transformation in tungsten, Phys. Rev. B 92 (2015), 134110.

[161] J.A. Moriarty, R.Q. Hood, L.H. Yang, Quantum-mechanical interatomic potentials with electron temperature for strong-coupling transition metals, Phys. Rev. Lett. 108 (2012), 036401.

[162] G.E. Norman, S.V. Starikov, V.V. Stegailov, Atomistic simulation of laser ablation of gold: effect of pressure relaxation, J. Exp. Theor. Phys. 114 (2012) 792–800.

[163] L. Shokeen, P.K. Schelling, An empirical potential for silicon under conditions of strong electronic excitation, Appl. Phys. Lett. 97 (2010), 151907.

[164] T. Kumagai, S. Izumi, S. Hara, S. Sakai, Development of bond-order potentials that can reproduce the elastic constants and melting point of silicon for classical molecular dynamics simulation, Comput. Mater. Sci. 39 (2007) 457–464.

[165] L. Shokeen, P.K. Schelling, Thermodynamics and kinetics of silicon under conditions of strong electronic excitation, J. Appl. Phys. 109 (2011), 073503.

[166] L. Shokeen, P.K. Schelling, Role of electronic-excitation effects in the melting and ablation of laser-excited silicon, Comput. Mater. Sci. 67 (2013) 316–328.

[167] R. Darkins, P.-W. Ma, S.T. Murphy, D.M. Duffy, Simulating electronically driven structural changes in silicon with two-temperature molecular dynamics, Phys. Rev. B 98 (2018), 024304.

[168] B. Bauerhenne, V.P. Lipp, T. Zier, E.S. Zijlstra, M.E. Garcia, Self-learning method for construction of analytical interatomic potentials to describe laser-excited materials, Phys. Rev. Lett. 124 (8) (2020), 085501.

[169] B. Bauerhenne, M.E. Garcia, Performance of state-of-the-art force fields for atomistic simulations of silicon at high electronic temperatures, Eur. Phys. J. Spec. Top. 227 (2019) 1615–1629.

[170] J. Tersoff, New empirical model for the structural properties of silicon, Phys. Rev. Lett. 56 (1986) 632–635.

[171] B.W. Dodson, Development of a many-body Tersoff-type potential for silicon, Phys. Rev. B 35 (1987) 2795–2798.

[172] M. Ishimaru, S. Munetoh, T. Motooka, K. Moriguchi, A. Shintani, Molecular-dynamics studies on defect-formation processes during crystal growth of silicon from melt, Phys. Rev. B 58 (19) (1998) 12583–12586.

[173] U. Landman, W.D. Luedtke, M.W. Ribarsky, R.N. Barnett, C.L. Cleveland, Molecular-dynamics simulations of epitaxial crystal growth from the melt. I. Si(100), Phys. Rev. B 37 (9) (1988) 4637–4646.

[174] W.D. Luedtke, U. Landman, M.W. Ribarsky, R.N. Barnett, C.L. Cleveland, Molecular-dynamics simulations of epitaxial crystal growth from the melt. II. Si(111), Phys. Rev. B 37 (9) (1988) 4647–4655.

[175] T. Motooka, K. Nishihira, R. Oshima, H. Nishizawa, F. Hori, Atomic diffusion at solid/liquid interface of silicon: transition layer and defect formation, Phys. Rev. B 65 (2002), 081304.

[176] K. Nishihira, T. Motooka, Molecular-dynamics simulations of crystal growth from melted Si: self-interstitial formation and migration, Phys. Rev. B 66 (2002), 233310.

[177] A.G. Cullis, N.G. Chew, H.C. Webber, D.J. Smith, Orientation dependence of high speed silicon crystal growth from the melt, J. Cryst. Growth 68 (1984) 624–638.

[178] R. Tsu, R.T. Hodgson, T.Y. Yan, J.E.E. Baglin, Order-disorder transition in single-crystal silicon induced by pulsed UV laser irradiation, Phys. Rev. Lett. 42 (20) (1979) 1356–1358.

[179] A.G. Cullis, H.C. Webber, N.G. Chew, J.M. Poate, P. Baeri, Transitions to defective crystal and the amorphous state induced in elemental Si by laser quenching, Phys. Rev. Lett. 49 (3) (1982) 219–222.

[180] E. Dornberger, W. Von Ammon, The dependence of ring-like distributed stacking faults on the axial temperature gradient of growing Czochralski silicon crystals, J. Electrochem. Soc. 143 (5) (1996) 1648–1653.

[181] W. Von Ammon, E. Dornberger, H. Oelkrug, H. Weider, The dependence of bulk defects on the axial temperature gradient of silicon crystals during Czochralski growth, J. Cryst. Growth 151 (1995) 273–277.

[182] A. La Magna, V. Privitera, G. Fortunato, M. Cuscunà, B.G. Svensson, E. Monakhov, K. Kuitunen, J. Slotte, F. Tuomisto, Vacancy generation in liquid phase epitaxy of Si, Phys. Rev. B 75 (2007), 235201.

[183] L.M. Swanson, J.R. Parsons, C.W. Howlke, Damaged regions in neutron-irradiated and ion-bombarded Ge and Si, Radiat. Eff. 9 (3) (1971) 249–256.

[184] H. Cerva, G. Hobler, Comparison of transmission electron microscope cross sections of amorphous regions in ion implanted silicon with point-defect density calculations, J. Electrochem. Soc. 139 (12) (1992) 3631–3638.

[185] L.A. Marqués, L. Pelaz, J. Hernández, J. Barbolla, G.H. Gilmer, Stability of defects in crystalline silicon and their role in amorphization, Phys. Rev. B 64 (4) (2001), 045114.

[186] L. Pelaz, L.A. Marqués, J. Barbolla, G.H. Gilmer, Atomistic modeling of amorphization and recrystallization in silicon, Appl. Phys. Lett. 82 (13) (2003) 2038–2040.

[187] L.A. Marqués, L. Pelaz, L. Aboy, L. Enriquez, J. Barbolla, Microscopic description of the irradiation-induced Amorphization in silicon, Phys. Rev. Lett. 91 (13) (2003), 135504.

[188] K.R.C. Mok, M. Jaraiz, I. Martin-Bragado, J.E. Rubio, P. Castrillo, R. Pinacho, J. Barbolla, M.P. Srinivasan, Ion-beam amorphization of semiconductors: a physical model based on the amorphous pocket population, J. Appl. Phys. 98 (4) (2005), 046104.

[189] L. Pelaz, L.A. Marqués, J. Barbolla, Ion-beam-induced amorphization and recrystallization in silicon, J. Appl. Phys. 96 (11) (2004) 5947–5976.

[190] I. Martin-Bragado, {111} local configurations: the main source of silicon defects during solid phase epitaxial regrowth modeled by lattice kinetic Monte Carlo, Appl. Phys. Lett. 98 (23) (2011), 233109.

[191] K.A. Jackson, G.H. Gilmer, D.E. Temkin, Monte Carlo simulation of the rapid crystallization of bismuth-doped silicon, Phys. Rev. Lett. 75 (1995) 2530.

[192] D.L. Woodraska, J.A. Jaszczak, A Monte Carlo simulation method for {111} surfaces of silicon and other diamond-cubic materials, Surf. Sci. 374 (1–3) (1997) 319–332.

[193] S.R. Stiffler, P.V. Evans, A.L. Greer, Interfacial transport kinetics during the solidification of silicon, Acta Metall. Mater. 40 (7) (1992) 1617–1622.

[194] G. Fisicaro, L. Pelaz, P. Lopez, A. La Magna, Kinetic Monte Carlo simulations for transient thermal fields: computational methodology and application to the submicrosecond laser processes in implanted silicon, Phys. Rev. E 86 (3) (2012), 036705.

[195] G. Fisicaro, M. Italia, V. Privitera, G. Piccitto, K. Huet, J. Venturini, A. La Magna, Dopant activation and damage evolution in implanted silicon after excimer laser annealing, Phys. Status Solidi C 8 (3) (2011) 940–943.

[196] G. Fisicaro, M. Italia, V. Privitera, G. Piccitto, K. Huet, J. Venturini, A. La Magna, Solid phase phosphorous activation in implanted silicon by excimer laser irradiation, J. Appl. Phys. 109 (11) (2011), 113513.

[197] K. Huet, G. Fisicaro, J. Venturini, H. Besaucèle, A. La Magna, Defect kinetics and dopant activation in submicrosecond laser thermal processes, Appl. Phys. Lett. 95 (23) (2009), 231901.

[198] A. La Magna, P. Alippi, V. Privitera, G. Fortunato, M. Camalleri, B.G. Svensson, A phase-field approach to the simulation of the excimer laser annealing process in Si, J. Appl. Phys. 95 (2004) 4806.

[199] A. Karma, W.J. Rappel, Quantitative phase-field modeling of dendritic growth in two and three dimensions, Phys. Rev. E 57 (4) (1998) 4323.

[200] K.S. Jones, J. Chen, S. Bharatan, J. Jackson, L. Rubin, M. Puga-Lambers, D. Venables, The effect of dose rate and implant temperature on transient enhanced diffusion in boron implanted silicon, J. Electron. Mater. 26 (11) (1997) 1361–1364.

[201] R. Falster, V.V. Voronkov, F. Quast, On the properties of the intrinsic point defects in silicon: a perspective from crystal growth and wafer processing, Phys. Status Solidi B 222 (1) (2000) 219–244.

[202] M. Jaraiz, L. Pelaz, E. Rubio, J. Barbolla, G.H. Gilmer, D.J. Eaglesham, H.-J. Gossmann, J.M. Poate, Atomistic modeling of point and extended defects in crystalline materials, Mater. Res. Soc. Proc. 532 (1998) 43.

[203] P.A. Stolk, H.-J. Gossmann, D.J. Eaglesham, D.C. Jacobson, C.S. Rafferty, G.H. Gilmer, M. Jaraíz, J.M. Poate, H.S. Luftman, T.E. Haynes, Physical mechanisms of transient enhanced dopant diffusion in ion-implanted silicon, J. Appl. Phys. 81 (1997) 6031.

[204] N.E.B. Cowern, G. Mannino, P.A. Stolk, F. Roozeboom, H.G.A. Huizing, J.G.M. van Berkum, F. Cristiano, A. Claverie, M. Jaraíz, Energetics of self-interstitial clusters in Si, Phys. Rev. Lett. 82 (1999) 4460–4463.

[205] F. Cristiano, J. Grisolia, B. Colombeau, M. Omri, B. de Mauduit, A. Claverie, L.F. Giles, N.E. B. Cowern, Formation energies and relative stability of perfect and faulted dislocation loops in silicon, J. Appl. Phys. 87 (2000) 8420–8428.

[206] A. Bongiorno, L. Colombo, T.D. De la Rubia, Structural and binding properties of vacancy clusters in silicon, Europhys. Lett. 43 (1998) 695.

[207] M. Jaraiz, G.H. Gilmer, J.M. Poate, T.D. de la Rubia, Atomistic calculations of ion implantation in Si: point defect and transient enhanced diffusion phenomena, Appl. Phys. Lett. 68 (1996) 409–411.

[208] G. Fisicaro, L. Pelaz, M. Aboy, P. Lopez, M. Italia, K. Huet, F. Cristiano, Z. Essa, Q. Yang, E. Bedel-Pereira, M. Quillec, A. La Magna, Kinetic Monte Carlo simulations of boron activation in implanted Si under laser thermal annealing, Appl. Phys. Express 7 (2) (2014), 021301.

[209] G. Fisicaro, A. La Magna, G. Piccitto, V. Privitera, Laser annealing of SiGe and Ge based devices, Microelectron. Eng. 88 (4) (2011) 488–491.

[210] G. Fisicaro, A. La Magna, Modeling of laser annealing, J. Comput. Electron. 13 (1) (2014) 70–94.

[211] K.M. Beatty, K.A. Jackson, Orientation dependence of the distribution coefficient obtained from a spin-1 Ising model, J. Cryst. Growth 174 (1–4) (1997) 28–34.

[212] K.M. Beatty, K.A. Jackson, Monte Carlo modeling of dopant segregation, J. Cryst. Growth 271 (3–4) (2004) 495–512.

[213] M.J. Aziz, Model for solute redistribution during rapid solidification, J. Appl. Phys. 53 (2) (1982) 1158.

[214] M.J. Aziz, J.Y. Tsao, M.O. Thompson, P.S. Peercy, C.W. White, Solute trapping: comparison of theory with experiment, Phys. Rev. Lett. 56 (1986) 2489.

[215] P.G. Sanders, M.J. Aziz, Self-diffusivity of liquid silicon measured by pulsed laser melting, J. Appl. Phys. 86 (8) (1999) 4258.

[216] K.A. Jackson, G.H. Gilmer, D.E. Temkin, K.M. Beatty, Microsegregation far from equilibrium, J. Cryst. Growth 163 (4) (1996) 461–469.

[217] K.A. Jackson, K.M. Beatty, M.V. Minke, Monte Carlo simulation of non-equilibrium segregation during crystal growth, J. Non-Cryst. Solids 219 (1997) 100–109.

[218] W. Bergholz, Defect engineering in silicon materials, in: Y. Yoshida, G. Langouche (Eds.), Defects and Impurities in Silicon Materials, Lecture Notes in Physics, vol. 916, Springer, Tokyo, 2015.

[219] M. Ieong, B. Doris, J. Kedzierski, K. Rim, M. Yang, Silicon device scaling to the sub-10-nm regime, Science 306 (2004) 2057–2060.

[220] D.J. Eaglesham, P.A. Stolk, H.-J. Gossmann, J.M. Poate, Implantation and transient B diffusion in Si: the source of the interstitials, Appl. Phys. Lett. 65 (1994) 2305–2307.

[221] H.-C. Liao, J.-C. Lin, R.-D. Chang, Time evolution of boron deactivation with carbon coimplantation in preamorphized silicon, Jpn. J. Appl. Phys. 57 (2018), 081301.

[222] F. Panciera, P.F. Fazzini, M. Collet, J. Boucher, E. Bedel, F. Cristiano, End-of-range defects in germanium and their role in boron deactivation, Appl. Phys. Lett. 97 (2010), 012105.

[223] C.-P. Chou, C.-Y. Chen, K.-Y. Chen, S.-C. Teng, Y.-H. Wu, Improved leakage current and device uniformity for sub-20 nm N-FinFETs by cryogenic Ge preamorphization implant in contact, Microelectron. Eng. 178 (2017) 137–140.

[224] R. Duffy, A. Heringa, V.C. Venezia, J. Loo, A.M. Verheijen, M.J.P. Hopstaken, K. van der Tak, M. de Potter, J.C. Hooker, P. Meunier-Beillard, R. Delhougne, Quantitative prediction of junction leakage in bulk-technology CMOS devices, Solid State Electron. 54 (2010) 243–251.

[225] M.A. Green, S.P. Bremner, Energy conversion approaches and materials for high-efficiency photovoltaics, Nat. Mater. 16 (2017) 23–34.

[226] C. Leroy, P.G. Rancoita, Particle interaction and displacement damage in silicon devices operated in radiation environments, Rep. Prog. Phys. 70 (2007) 493–625.

[227] I. Pintilie, G. Lindstroem, A. Junkes, E. Fretwurst, Radiation-induced point-and cluster-related defects with strong impact on damage properties of silicon detectors, Nucl. Inst. Methods Phys. Res. A 611 (2009) 52–68.

[228] J.R. Srour, J.W. Palko, Displacement damage effects in irradiated semiconductor devices, IEEE Trans. Nucl. Sci. 60 (2013) 1740–1766.

[229] I. Avci, M.E. Law, E. Kuryliw, A.F. Saavedra, K.S. Jones, Modeling extended defect ({311} and dislocation) nucleation and evolution in silicon, J. Appl. Phys. 95 (2004) 2452–2460.

[230] A. Claverie, B. Colombeau, F. Cristiano, A. Altibelli, C. Bonafos, Modeling of the Ostwald ripening of extrinsic defects and transient enhanced diffusion in silicon, Nucl. Inst. Methods Phys. Res. B 186 (2002) 281–286.

[231] A. Claverie, B. Colombeau, B. de Mauduit, C. Bonafos, X. Hebras, G. Ben Assayag, F. Cristiano, Extended defects in shallow implants, Appl. Phys. A 76 (2003) 1025–1033.

[232] S. Coffa, S. Libertino, C. Spinella, Transition from small interstitial clusters to extended {311} defects in ion-implanted Si, Appl. Phys. Lett. 76 (2000) 321–323.

[233] S.S. Kapur, T. Sinno, Detailed microscopic analysis of self-interstitial aggregation in silicon. I. Direct molecular dynamics simulations of aggregation, Phys. Rev. B 82 (2010), 045205.

[234] S.S. Kapur, A.M. Nieves, T. Sinno, Detailed microscopic analysis of self-interstitial aggregation in silicon. II. Thermodynamic analysis of single clusters, Phys. Rev. B 82 (2010), 045206.

[235] S. Lee, G.S. Hwang, Structure and stability of small compact self-interstitial clusters in crystalline silicon, Phys. Rev. B 77 (2008), 085210.

[236] S. Lee, G.S. Hwang, Growth and shape transition of small silicon self-interstitial clusters, Phys. Rev. B 78 (2008), 045204.

[237] J. Li, K.S. Jones, {311} defects in silicon: the source of the loops, Appl. Phys. Lett. 73 (1998) 3748–3750.

[238] S. Libertino, S. Coffa, J.L. Benton, Formation, evolution, and annihilation of interstitial clusters in ion-implanted Si, Phys. Rev. B 63 (2001), 195206.

[239] N. Zographos, C. Zechner, I. Avci, Efficient TCAD model for the evolution of interstitial clusters, {311} defects, and dislocation loops in silicon, Mater. Res. Soc. Symp. Proc. 994 (2007) 297–306.

[240] B. Colombeau, F. Cristiano, A. Altibelli, C. Bonafos, G. Ben Assayag, A. Claverie, Atomistic simulations of extrinsic defects evolution and transient enhanced diffusion in silicon, Appl. Phys. Lett. 78 (2001) 940–942.

[241] S.S. Kapur, T. Sinno, Entropic origins of stability in silicon interstitial clusters, Appl. Phys. Lett. 93 (2008), 221911.

[242] Y. Qiu, F. Cristiano, K. Huet, F. Mazzamuto, G. Fisicaro, A. La Magna, M. Quillec, N. Cherkashin, H. Wang, S. Duguay, D.C. Blavette, Extended defects formation in nanosecond laser-annealed ion implanted silicon, Nano Lett. 14 (2014) 1769–1775.

[243] P. Batude, M. Vinet, B. Previtali, C. Tabone, C. Xu, J. Mazurier, O. Weber, F. Andrieu, L. Tosti, L. Brevard, B. Sklenard, P. Coudrain, S. Bobba, H. Ben Jamaa, P.-E. Gaillardon, A. Pouydebasque, O. Thomas, C. Le Royer, J.-M. Hartmann, L. Sanchez, L. Baud, V. Carron, L. Clavelier, G. De Micheli, S. Deleonibus, O. Faynot, T. Poiroux, Advances, challenges and opportunities in 3D CMOS sequential integration, in: Proceedings of the 2011 International Electron Devices Meeting, 2011, pp. 151–154.

[244] B. Rajendran, R.S. Shenoy, D.J. White, N.S. Chokshi, R.L. De Leon, G.S. Tompa, R. Fabian, Low thermal budget processing for sequential 3-D IC fabrication, IEEE Trans. Electron Devices 54 (2007) 707–714.

[245] A. Vandooren, J. Franco, B. Parvais, Z. Wu, L. Witters, A. Walke, W. Li, L. Peng, V. Deshpande, F.M. Bufler, N. Rassoul, G. Hellings, G. Jamieson, F. Inoue, G. Verbinnen, K. Devriendt, L. Teugels, N. Heylen, E. Vecchio, T. Zheng, E. Rosseel, W. Vanherle, A. Hikavyy, B. T. Chan, R. Ritzenthaler, G. Besnard, W. Schwarzenbach, G. Gaudin, I. Radu, B.-Y. Nguyen, N. Waldron, V. De Heyn, D. Mocuta, N. Collaert, 3-D sequential stacked planar devices featuring low-temperature replacement metal gate junctionless top devices with improved reliability, IEEE Trans. Electron Devices 65 (2018) 5165–5171.

[246] S. Takeda, Structure analysis of defects in nanometer space inside a crystal: creation and agglomeration of point defects in Si and Ge revealed by high-resolution electron microscopy, Microsc. Res. Tech. 40 (1998) 313–335.

[247] L.A. Marqués, M. Aboy, M. Ruiz, I. Santos, P. López, L. Pelaz, Molecular dynamics simulation of the early stages of self-interstitial clustering in silicon, Mater. Sci. Semicond. Process. 42 (2016) 235–238.

[248] C.Y. Chuang, A. Sattler, T. Sinno, Thermodynamic and morphological analysis of large silicon self-interstitial clusters using atomistic simulations, J. Appl. Phys. 117 (2015), 135706.

[249] S. Birner, J. Kim, D.A. Richie, J.W. Wilkins, A.F. Voter, T. Lenosky, Accelerated dynamics simulations of interstitial-cluster growth, Solid State Commun. 120 (2001) 279–282.

[250] L.A. Marqués, M. Aboy, I. Santos, P. López, F. Cristiano, A. La Magna, K. Huet, T. Tabata, L. Pelaz, On the anomalous generation of {001} loops during laser annealing of ion implanted silicon, Nucl. Inst. Methods Phys. Res. B 458 (2019) 179–183.

[251] F. Cristiano, M. Shayesteh, R. Duffy, K. Huet, F. Mazzamuto, Y. Qiu, M. Quillec, H.H. Henrichsen, P.F. Nielsen, D.H. Petersen, A. La Magna, G. Caruso, S. Boninelli, Defect evolution and dopant activation in laser annealed Si and Ge, Mater. Sci. Semicond. Process. 42 (2016) 188–195.

[252] A. Weck, D.S. Wilkinson, E. Maire, H. Toda, Visualization by X-ray tomography of void growth and coalescence leading to fracture in model materials, Acta Mater. 56 (2008) 2919–2928.

Chapter 4

# Laser annealing applications for semiconductor devices manufacturing

Karim Huet
*Laser Systems and Solutions of Europe, SCREEN Semiconductor Solutions, Co., Ltd, Gennevilliers, France*

## 4.1 Introduction

In the history of semiconductor device fabrication, annealing is one of the critical process steps used to achieve a targeted device performance. Thermal budget requirements in most semiconductor manufacturing process steps have always followed a downward trend through either a reduction of the time spent at high temperature or a lowering of the peak temperature [1]. This was driven at first by the traditional shrinking of structure dimensions, which naturally required reducing species diffusion. From the last decade, performance improvements started to be driven essentially by the introduction of new materials (e.g., SiGe alloy, high-*k*/metal gate stack) and device architecture changes (e.g., 2D planar to 3D FinFET). The current development trend is leading toward 3D integrated circuits to answer the growing demand for increased functionality and performance. Such new drivers impact the whole supply chain, and low thermal budget solutions are now key to manage local strain and stress; limit species diffusion and interdiffusion; and control layer deposition, etch, and quality. By nature, it is seen as a key enabler for any 3D-IC such as sensors and vertical power devices, which were indeed among the early adopters of laser technologies for annealing purposes [2, 3].

As depicted in Fig. 4.1, the idea is to reduce the annealing timescale to achieve higher anneal temperatures, thus allowing a better control of the thermal budget and improving material properties. Lowering the time scale and, consequently, the depth scale, comes with its own set of challenges, especially concerning the temperature variation scale [4], highlighting the process variability from wafer to wafer (batch furnace) to within wafer (single wafer RTP and spike) to within die (millisecond anneal) and finally to within layout

Laser Annealing Processes in Semiconductor Technology. https://doi.org/10.1016/B978-0-12-820255-5.00008-8

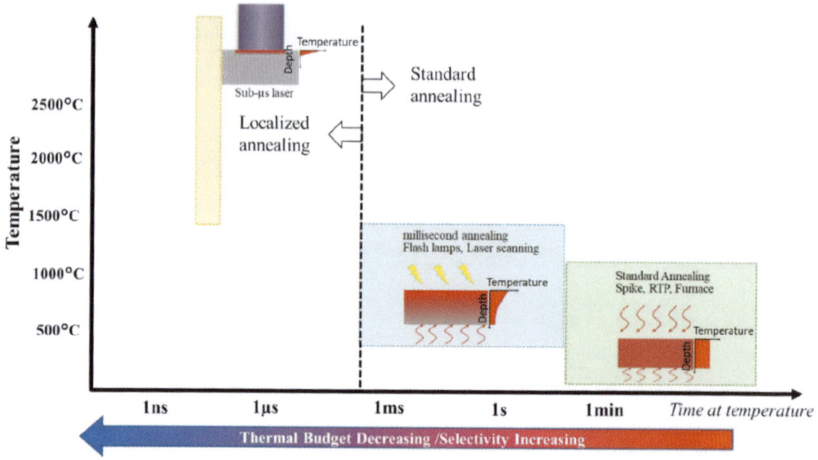


**FIG. 4.1** Illustration of thermal budget of different annealing approaches.

(sub-μs anneal). Many of these challenges have been overcome over the years for millisecond and longer anneal schemes by improving the equipment (e.g., spatial distribution of backside heating for spike), adjusting the process flow (e.g., combination of RTP or spike with millisecond anneal), or adapting the layout design [5–7], and they are still being actively worked on for sub-μs annealing.

For a wide variety of applications, laser exhibits superior performance and a more cost-effective solution than nonlaser technologies. Moreover, laser provides flexible, higher throughput to process various materials. As stated in Ref. [8], trends regarding laser annealing are expected to experience fast growth, driven by 3D integration applications.

Specifically, pulsed laser annealing of semiconductor materials at submicrosecond timescales has been investigated for many years ([9, 10] and references therein), especially for front end of line (FEOL) applications. Initially, the most obvious interest was on the shallow junction formation in the melting regime with huge potential benefits such as highly localized dopant activation above the solid-phase equilibrium limit [11] and near-perfect liquid-phase-induced recrystallization [9]. However, adoption for high-volume manufacturing was never achieved in CMOS due to integration issues such as dopant deactivation induced by the thermal budget associated with subsequent annealing steps in the process flow [12] and pattern variations at the transistor level. As such, the key drivers for allowing the integration of sub-μs laser annealing for 3D devices and architectures are (i) the minimization of pattern variation effects, e.g., by using dual beams [13, 14] or by having the layout pattern buried deeper than the thermal diffusion length, and (ii) the reduction of the overall thermal budget of process steps following laser anneal in order to keep its outstanding benefits.

In this chapter, with such requirements in mind, we will review the most promising applications for laser annealing from the recent published literature with a focus on realistic candidates for integration in a high-volume semiconductor manufacturing flow, based on electrically relevant results. The first section will be dedicated to power devices, the second section to CMOS logic and 3D integration, and the final section to memory applications.

## 4.2 Power devices

Historically [3], laser anneal for semiconductor devices was first integrated in high-volume manufacturing for the backside dopant activation in power devices (anode formation), mainly isolated gate bipolar transistors (IGBT) and diodes. In this section, recent work to extend the approach to Si-based and wide bandgap devices is reported.

### 4.2.1 Si power devices

For Si-based vertical power devices, such as insulated gate bipolar transistors (IGBTs) and diodes, the strategy for reducing on-state voltage, minimizing switching losses, and improving heat dissipation is to thin the wafer thicknesses. Today, the thinnest wafer thicknesses in production are below 100 μm. The mainstream processes to accommodate thinning of the device and keep breakdown characteristics on target is double-implant activation on the backside of Si-IGBT devices, where a p-collector (B-doped) and n-buffer layer (P-doped) are formed. To ensure the high breakdown voltage and minimize the yield loss, thick (μm-scale) defect-free and highly active n-buffer layers are required. The most cost-effective and productive approach is a single backside annealing step. In this case, heating of the frontside should be limited to less than 400°C (or even 200°C, depending on frontside limitations) with carefully controlled and repeatable dopant profiles. To satisfy such requirements, pulsed laser annealing in the melting regime is the adopted approach in the industry [3, 15–21]. Indeed, thanks to such a low thermal budget and surface-selective annealing process, the thermal gradient in the Si wafer is very abrupt, avoiding any undesirable thermal effects on the frontside. For instance, as illustrated in Fig. 4.2 [21], a very deep P activation in Si over several micrometers having a near-perfect dopant activation is reported.

### 4.2.2 Wide bandgap material power devices

Thanks to their unique electrical and thermal properties, silicon carbide (SiC) and gallium nitride (GaN) are considered as viable materials to overcome silicon limitations. Among the many challenges for the integration of these compound materials in industrial devices, two are identified as being addressable by laser annealing: crystal modification and contact formation. Indeed, recent

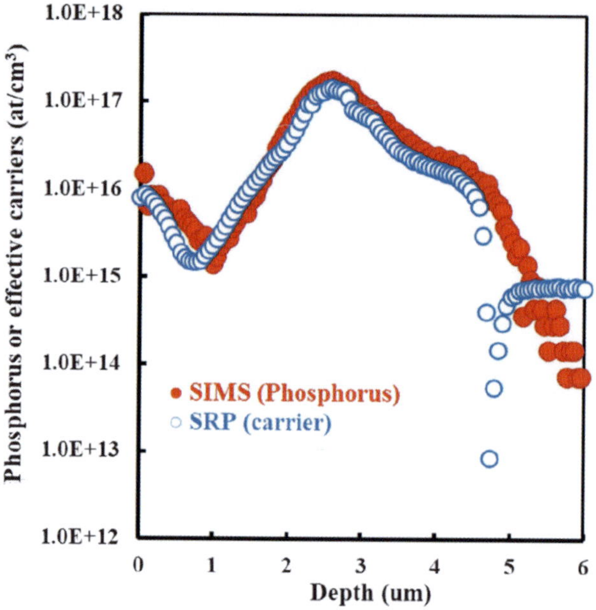


**FIG. 4.2** Phosphorus chemical (SIMS) and active (SRP) profiles in deep implanted samples after single pulse laser annealing. *(Extracted from T. Tabata, S. Halty, I. Toque-Tresonne, F. Mazzamuto, K. Huet, Y. Mori, UV excimer laser annealing for next generation power electronics, in: 2016 21st International Conference on Ion Implantation Technology (IIT), Tainan, 2016, pp. 1–4, https://doi.org/10.1109/IIT.2016.7882917.)*

works have focused on substitutional doping, defect engineering, crystallization, and ohmic contact formation with lower thermal budgets, identified as the current bottleneck of these materials. In this regard, an important progress has been made by using laser annealing to overcome traditional high-temperature annealing process limitations. For these wide bandgap materials, the effective wavelength spectrum is quite limited since the band gap is too large to emit phonons by photon excitation with wavelengths above the UV range; thus, mostly UV lasers have been considered in the literature.

### *4.2.2.1 SiC annealing*

Manufacturing of SiC-based power devices such as MOS transistors requires the formation of doped areas involving dopant implantation (traditionally phosphorus and aluminum for N- and P-type regions, respectively). Due to the very low diffusivity of the species in SiC, multiple dopant implantation must be used and traditional furnace diffusion cannot be considered. Moreover, extremely high temperatures (above 1500°C) are mandatory to achieve a sufficient, although not perfect, recovery of implantation defects and high enough dopant activation [22]. Such high thermal budgets come with their set of integration issues such as surface degradation from Si evaporation, uncontrolled species interdiffusion, not to mention the complexity of the equipment. In this framework, laser annealing has been applied for defect curing and dopant activation in SiC with promising results [23–29].

Crystal modification induced by sub-μs laser annealing was studied from low temperature up to temperatures supposed to completely break all lattice bonding. The latter condition, studied in Ref. [27], corresponds to the liquid phase of the material, even though theoretical studies have shown that such liquid-phase-driven SiC growth would require temperatures and pressures exceeding ~3200°C and $\sim 10^5$ bar in equilibrium conditions [30]. Such conditions could be considered as unrealistic for device manufacturing from an industrial point of view, although the possibility of melting SiC by laser anneal to recrystallize amorphous SiC to 3C polytype was already demonstrated [31]. In chemical equilibrium conditions, the dissociation of SiC bonding should not happen below 3000 K. This is possible only because the UV laser process can localize the heating within the first 200 nm in sub-μs time scale, thus enabling out-of-equilibrium anneal conditions from the chemical standpoint. However, laser annealing process is fast enough so that the Si atoms do not have time to out diffuse from the SiC surface. In summary, despite the local SiC dissociation observed, this is a promising solution to enable the recovery of SiC lattices strongly damaged by ion beam implantation.

On the lower temperature range [25, 28], i.e., below the surface melting threshold, as shown in Fig. 4.3, significant crystal recovery and activation of

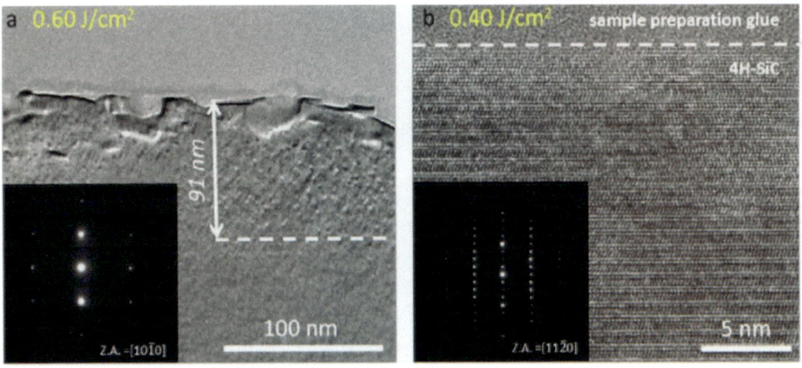


**FIG. 4.3** Structural analysis of laser-annealed phosphorus-implanted 4H-SiC showing the crystal quality achieved in the submelt regime with multiple short pulses (1000 times). (A) XTEM bright field image of the 0.60 J/cm$^2$ laser-annealed source implant. The inset shows the 4H-SiC diffraction pattern under $[11\bar{1}0]$ zone axis. (B) HRTEM of the source implant sample subjected to 0.40 J/cm$^2$ laser fluence showing a periodic ABAC stacking sequence acquired in correspondence of the surface under the $[11\bar{2}0]$ zone axis. The inset reports the related 4H-SiC diffraction pattern. *(Reproduced with permission C. Calabretta, M. Agati, M. Zimbone, S. Boninelli, A. Castiello, A. Pecora, G. Fortunato, L. Calcagno, L. Torrisi, F. La Via, Laser annealing of P and Al implanted 4H-SiC epitaxial layers, Materials, 12 (20) (2019) 3362, https://doi.org/10.3390/ma12203362.)*

various implanted dopants could be obtained with nonamorphizing implants (e.g., by hot implantation at 500°C) and multiple pulse irradiation with substrate heating (up to 800°C). Less degradation of surface morphology could also be obtained compared with traditional furnace. Such results show a promise to overcome the limitations from high-temperature furnace. However, the need of a relatively high substrate temperature and significant number of pulses (several thousands) may be an issue for high-volume manufacturing, especially regarding productivity matters. Nonetheless, moving forward toward industry interest for electrical device evaluations, a very recent study [29] investigated the impact of laser for postimplant backside dopant activation for junction barrier Schottky (JBS) diodes on thinned SiC. The result showed a small but significant improvement in voltage drop (thus better on-state resistance) with respect to the standard annealing case.

#### *4.2.2.2 GaN annealing*

Similar to SiC, the complexity of the GaN compound limits the possibility for defect curing and junction engineering using traditional furnace anneals. Especially, doping is limited by the GaN dissociation and subsequent nitrogen evaporation, leading to the formation of voids and gallium-rich regions. In the case of sub-μs laser annealing, it may be expected that, similar to the SiC case described in the previous section, N and Ga would dissociate in the near-surface region (N being evaporated and Ga segregating near the surface); however, the GaN stoichiometry would remain intact in the deeper region, which would still be improved. For device fabrication, this usually involves additional process steps such as capping the surface with a dielectric [32] to avoid surface degradation from N evaporation. Early works studying laser anneal on GaN showed the promising results in terms of junction formation and defect curing. Concerning dopant activation, in Ref. [33], the effects of sub-μs laser annealing on the properties of magnesium (Mg)-doped GaN epitaxial film grown on sapphire substrate were studied. Photoluminescence (PL) spectra showed the capability to improve Mg activation, resulting in a significant improvement of I–V characteristics of the PN junction. Concerning defect curing, in Ref. [34], the possibility to combine laser annealing with RTP to specifically cure surface defects and reduce the roughness induced by the RTP step in beryllium (Be)-implanted GaN was studied. Furthermore, in Ref. [35], a $N_xAs_{1-x}$ thin film is formed by heavily $N^+$-implanted GaAs molten and recrystallized by excimer laser anneal, thus demonstrating the potential to regrow lattice from melting phase by laser annealing. Recrystallization from melting phase showed clear benefits in terms of defect curing and activation rate of the implanted species. This possibility opens to promising approaches for junction formation and band gap engineering in GaN.

#### *4.2.2.3 Ohmic contact formation*

SiC and GaN need very high temperatures to lower the contact resistance [36]; however, such high thermal budget induces process nonuniformity and important segregation and clustering issues. Specifically in the SiC case, similar to conventional Si-based devices discussed above, the trend is to improve performance by thinning the substrate [37, 38], thus requiring again a low thermal budget for the backside contact formation step to avoid alternating between wafer frontside and backside processing as schematized in Fig. 4.4. Again, laser annealing has shown great potential for such requirements thanks to its low thermal budget and surface-selective annealing process [27, 37, 39–41].

The possibility of forming nickel silicide for backside contact was studied in Ref. [27] on thinned 4H-SiC and sputtered with a 100-nm-thick nickel (Ni) layer. From sheet resistance, XRD and cross-sectional TEM measurements, a low resistance nickel silicide contact was formed, with the formation of a uniform carbon nanocluster layer at the interface and the recovery of thinning-induced defects. As stated in Ref. [39], extending the study to industrial-grade power SiC diodes, the fine-grained carbon layer structure (highlighted on the TEM reported in Fig. 4.5) leads to an excellent integrity of the contact stack compared with traditional annealing by minimizing the risk of contact delamination at the following steps in the manufacturing flow. Moreover, although the understanding related to the origin of such ohmic contact is still unclear [42], such clusters may be beneficial for the formation of a low ohmic contact to SiC; therefore, such a uniform layer would be ideal for a homogenous current flow through the contact area. More recently [40], it was found that the laser

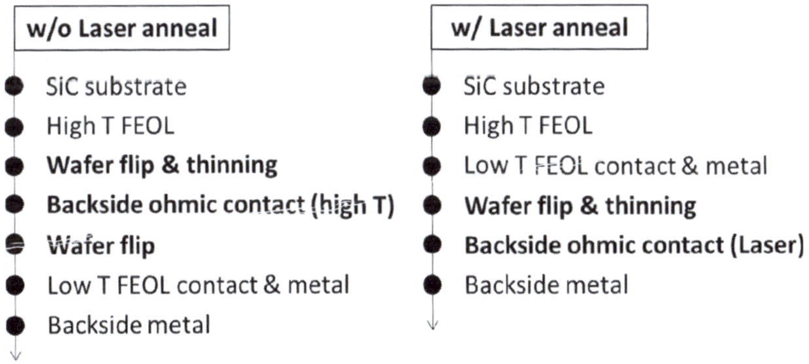


**FIG. 4.4** Junction Barrier Schottky diode schematic process flow without and with laser anneal for backside ohmic contact formation. The key advantage is avoiding two times wafer flip thanks to compatibility of laser anneal with frontside metal being present. *(Adapted from S. Rascunà, P. Badalà, C. Tringali, C. Bongiorno, E. Smecca, A. Alberti, S. Di Franco, F. Giannazzo, G. Greco, F. Roccaforte, M. Saggio, Morphological and electrical properties of nickel based Ohmic contacts formed by laser annealing process on n-type 4H-SiC, Mater. Sci. Semicond. Process. 97 (2019) 62–66, https://doi.org/10.1016/j.mssp.2019.02.031.)*

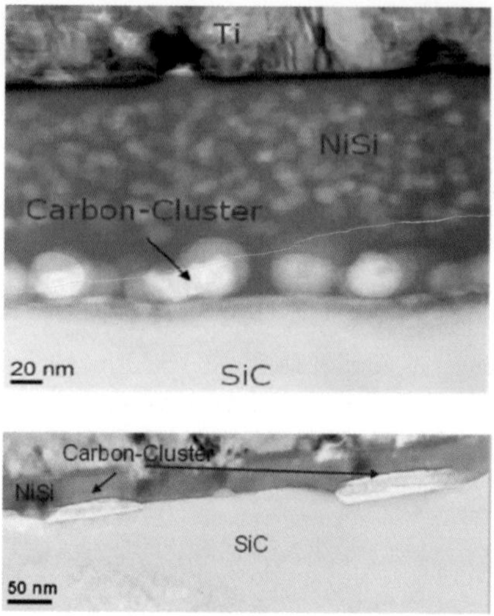


**FIG. 4.5** Cross-sectional TEM of the contact formed by standard annealing *(bottom)* and UV laser annealing *(top)* for SiC. *(Reproduced with permission R. Rupp, R. Kern, R. Gerlach, Laser backside contact annealing of SiC power devices: a prerequisite for SiC thin wafer technology, in: 2013 25th International Symposium on Power Semiconductor Devices & IC's (ISPSD), Kanazawa, 2013, pp. 51–54, https://doi.org/10.1109/ISPSD.2013.6694396.)*

annealing conditions and initial metal layer thickness could be tuned to obtain an optimal stoichiometry distribution of the silicide layer (carbon-free with an Ni content spanning from Ni-rich surface to Si-rich at the SiC interface) while preserving the layer integrity to prevent delamination and achieve better device performance than with the standard thermal approach.

A similar, although more prospective achievement was obtained on GaN [41] using Ti/Al and Cr/Au as metal stacks for n- and p-type contacts, respectively. The resistivity improvement was measured on TLM structures. In the case of n-type contacts, this was attributed to vacancy generation from GaN dissociation, while in the case of p-type contacts, the resistivity improvement was suggested to be related to an increase of Mg activation thanks to the dissociation of Mg—H bonds during the short pulse annealing. While still being far from an industrial case, such results show the potential to apply sub-μs laser annealing to GaN contact formation.

### 4.2.3 Conclusion

In conclusion, laser annealing is shown to be scalable to the next-generation power devices, by extending the Si-based devices with μm-scale deep buffer activation and allowing chemical out-of-equilibrium processing of wide band-gap materials such as SiC and GaN for defect curing, lattice modification, and contact formation. While laser annealing is already adopted in volume manufacturing for Si- and SiC-based devices, the maturity and interest levels for application to GaN are not sufficient yet, although the limited research results available today show good potential.

## 4.3 CMOS logic and 3D integration

CMOS manufacturing is the most demanding field in terms of performance and yield requirements and as such, is on the frontline to adopt new technologies to meet the growing demands of increased functionality, performance, miniaturization, and cost reduction. Indeed, the initial adoption of laser annealing for semiconductor device manufacturing started in the mid-2000s with millisecond range timescales (from several hundred μs to 10 ms) [4 and references therein]. Millisecond annealing showed the improved activation and low species diffusion over high-temperature rapid thermal annealing (RTA), which added a most useful process knob for junction engineering by lowering the RTA thermal budget for diffusion-related needs while using laser to achieve high dopant activation rates. Since then, many applications for laser annealing have been studied. Following the millisecond anneal, the approach that exhibits the most outstanding results is the sub-μs range pulsed laser annealing [4, 9, 10], based on its strong potential for performance improvement and very low thermal budget. Indeed, the thermal budget restrictions are further pushed down with the current development trend toward 3D integrated circuit architecture.

An illustration shown in Fig. 4.6A is a simplified process flow depicting key process modules for CMOS FinFET manufacturing from front end of line (FEOL) to back end of line (BEOL), through middle end of line (MEOL), together with an illustration of 3D sequential integration with its remaining process issues in Fig. 4.6B. In such framework, for the top-tier manufacturing part, the target here is to achieve performance equivalent to traditional high-temperature approaches but with thermal budgets compatible with back end of line (BEOL) limitations (typically 500°C, 2 h at most [43, 44]). Laser annealing compatibility with such restrictions was demonstrated in the sub-μs regime [45, 46]. In this section, we will review results regarding promising applications of laser annealing for the traditional CMOS process modules, extending to the new opportunities opened by the development of 3D architectures for which laser annealing can help releasing the thermal budget process constraints.

### 4.3.1 Top-tier active area formation

A straightforward approach for obtaining a high-quality crystal layer as active area for the top-tier devices is to consider ultra-thin silicon-on-insulator (SOI) wafers and layer transfer by bonding as described in Refs. [43, 47–49]. Another approach is to crystallize a deposited amorphous layer by laser annealing. Recent work demonstrated the viability of laser annealing for active layer crystallization. In Refs. [50–52], the location grain crystallization approach [53] is used to localize grain growth to the regions of interest to avoid the variability induced by the presence of randomly located grain boundaries in the active devices. As described in Ref [51] and reproduced on Fig. 4.7, the key steps are (i) via hole etching into the interlayer dielectric (ILD) to form selective cooling and crystallization seeding points, (ii) conformal a-Si deposition, (iii) sub-μs

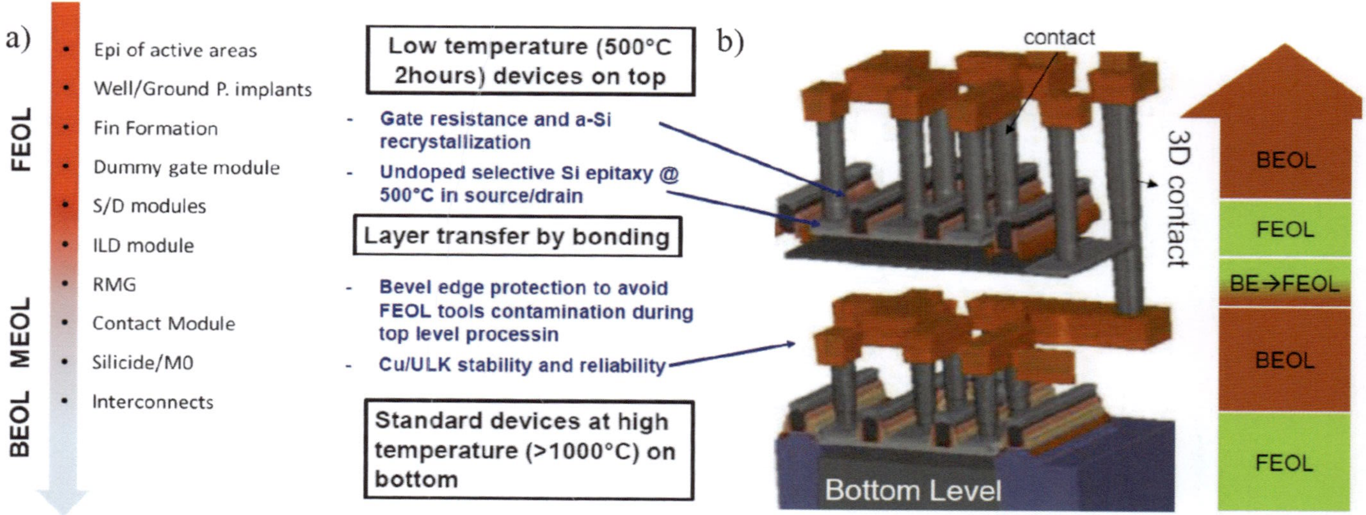


**FIG. 4.6** (A) CMOS manufacturing flow: key process modules, (B) 3D sequential CoolCube integration and remaining process issues, where S/D stands for source/drain, RMG for replacement metal gate, ILD for interlayer dielectric (ILD), and ULK for ultra-low *k*. *(Reproduced with permission L. Brunet, et al., Breakthroughs in 3D sequential technology, in: 2018 IEEE International Electron Devices Meeting (IEDM), San Francisco, CA, 2018, pp. 7.2.1–7.2.4, https://doi.org/10.1109/IEDM.2018.8614653.)*

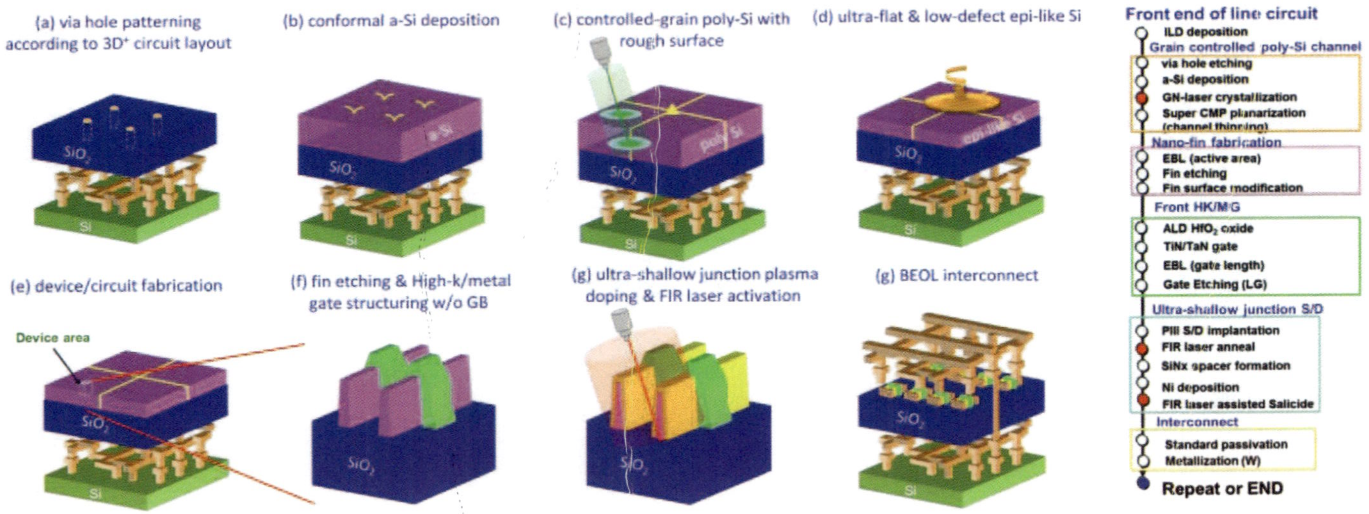


**FIG. 4.7** 3D-IC Active area formation process module steps with location grain control, where GN laser stands for green nanosecond laser, FIR laser for far-infrared laser, GB for grain boundaries, CMP for chemical mechanical planarization, EBL for etch back layer, ALD for atomic layer deposition. *(Reproduced with permission C. Yang, et al., Location-controlled-grain technique for monolithic 3D BEOL FinFET circuits, in: 2018 IEEE International Electron Devices Meeting (IEDM), San Francisco, CA, 2018, pp. 11.3.1–11.3.4, https://doi.org/10.1109/IEDM.2018.8614708.)*

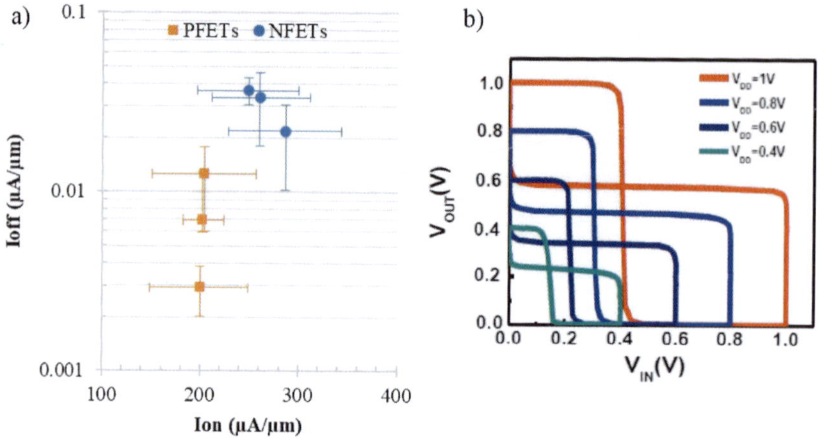


**FIG. 4.8** (A) Ioff vs Ion cloud chart for stackable single-grain n/p GAA-MOSFETs. (B) Corresponding 6T-SRAM cell butterfly curve performance operated at various drain voltages. *((A) Adapted from F. Hsueh, et al., Monolithic 3D SRAM-CIM macro fabricated with BEOL gate-all-around MOSFETs, in: 2019 IEEE International Electron Devices Meeting (IEDM), San Francisco, CA, USA, 2019, pp. 3.3.1–3.3.4, https://doi.org/10.1109/IEDM19573.2019.8993628. (B) Reproduced with permission F. Hsueh, et al., Monolithic 3D SRAM-CIM macro fabricated with BEOL gate-all-around MOSFETs, in: 2019 IEEE International Electron Devices Meeting (IEDM), San Francisco, CA, USA, 2019, pp. 3.3.1–3.3.4, https://doi.org/10.1109/IEDM19573.2019.8993628.)*

laser crystallization, and (iv) chemical-mechanical planarization (CMP). Devices can then be created within the single grain areas, and the design will have to take into account layout coordination between circuits and grain-boundary patterns. As shown in Fig. 4.8A, the normalized drive current achieved is relatively low (about twice lower compared with SOI performance [51]), essentially due to the poorer crystal quality. Nevertheless, as shown in Fig. 4.8B, functional SRAM-based circuits (e.g., XNOR) [50] could be achieved with potential applications for low-power in-memory and near-memory computing circuits, foreseen to be used for future artificial intelligence chips.

### 4.3.2 Polygate formation

With a similar concept, pulsed laser anneal has been successfully used for poly-gate crystallization of the top layer in a gate-first scheme [44]. In this case, an in situ doped amorphous layer was crystallized and dopants activated by laser annealing before the gate patterning steps. After fine tuning of the laser parameters, a high activation rate and full recrystallization of the layer with few hundred nm size range grains were obtained, without impacting BEOL integrity. As a result, the melt laser process could be integrated in the gate formation module for 3D sequential CoolCube. Again, the key point for a successful integration was to address the surface roughness induced by local grain boundaries after crystallization using an optimized CMP step. Very recently [54], high-performance CMOS devices manufactured with such a low thermal budget process flow integrating sub-μs laser anneal for the polygate formation step were integrated in fully functional electrical circuits (ring oscillators and SRAM cells).

### 4.3.3 Dopant activation for extensions, S/D, and contact areas

Dopant activation has been a field of interest for laser annealing since the early days in the 1970s [55]. With the introduction of new materials and 3D structures and dimensions of doped areas close to several atomic layers, such interest is still very much alive. A review of fundamental studies involving advanced characterization and numerical simulations will be, respectively, detailed in Chapters 5 and 6 of this book. In this section, after summarizing the key factors for junction formation process, we will review the latest results on electrical structures which show a promise for integration in high-volume manufacturing.

#### *4.3.3.1 Shallow junction formation key factors*

In the case of actual device manufacturing, the requirement is to activate dopants in highly confined regions due to scaled 3D architectures such as ultra-thin body SOI (UTB-SOI), FinFET, or gate all-around (GAA) such as nanosheet or nanowire devices.

A picture of the promising performance in terms of laser anneal for dopant activation is illustrated in Fig. 4.9 [56], comparing the active dopant concentration for a highly doped 40-nm Si:P epitaxial layer (4.6% P content, i.e., $2.3 \times 10^{21}\ cm^{-3}$) for various annealing approaches, namely, epi (as reference without anneal), spike annealing, ms laser anneal (with and without combination to spike), and sub-μs melt anneal (pulsed excimer laser), the latter showing close to 100% activation. Diffusion was kept under control in all laser cases (<3–5 nm). Such performance was further confirmed by contact resistivity results with about two times lower values for sub-μs melt anneal compared with the best millisecond laser anneal case.

In the full device process flow, such activation has to be obtained without any degrading side effects such as dopant overdiffusion or species interdiffusion. Indeed, species diffusion depth should be tightly controlled to avoid any impact, either on short channel effects for dopants or on local strain relaxation for alloys. In a more general way, the profiles of the various species involved should be controlled to achieve the adequate tradeoff between thermal budget and performance. For example, in SiGe based p-type contacts, a higher Ge fraction near surface has been shown to be beneficial to improve contact resistivity [57, 58] thanks to the Schottky barrier height lowering (from the narrower bandgap) and smaller hole effective mass [59].

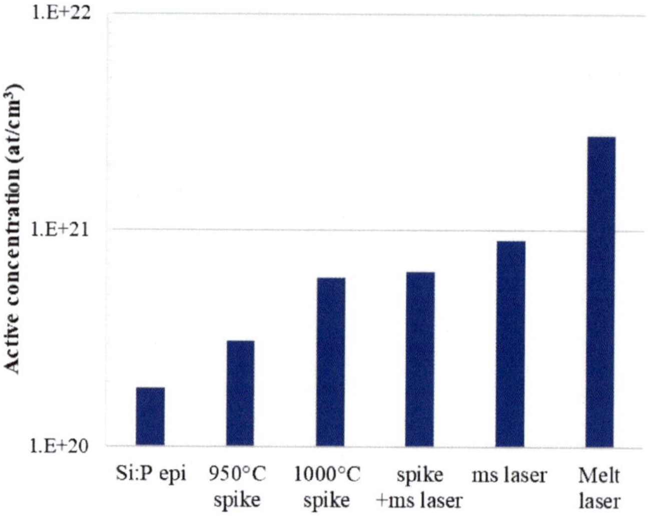


**FIG. 4.9** Active dopant concentration of in situ doped Si:P epitaxial layer (4.6% P content) for different annealing conditions. *(Adapted from E. Rosseel, et al., Selective epitaxial growth of high-P Si:P for source/drain formation in advanced Si nFETs, ECS Trans. 75 (8) (2016) 347, https://doi.org/10.1149/07508.0347ecst.)*

Thermal budget limitations to avoid species interdiffusion depend on the process step of interest. For example, for presilicide contact annealing, when the metal gate stack is present, the thermal budget should be low enough to avoid interdiffusion of the species in the gate stack, which are carefully engineered to control the threshold voltages. In this case, sub-μs laser annealing is naturally the most suitable approach. On the other side, for active area doping, located early in the flow without any fragile structures, the thermal budget margin is higher and accessible to millisecond range laser annealing, although its peak temperature will have to be limited in the case of 3D sequential integration (typically 800°C [43]).

### 4.3.3.2 Application to electrical device structures

Obtaining meaningful figures of merit on electrical structures, namely, contact resistivity, is best achieved by contact chain structures [60] or transfer length method (TLM), especially in its multiring circular form (MR-CTLM) [57, 61] to access the sub-$10^{-9}\ \Omega\,cm^2$ range required by the industry for coming technology nodes.

The corresponding process flows for contact chains and MR-CTLM are reproduced in Figs. 4.10 and 4.11, respectively. Such structures, mimicking the contact unit process module, allow to extract meaningful trends to fine tune the process windows for laser annealing in terms of tradeoffs between diffusion control, dopant activation, and potential deactivation from additional thermal

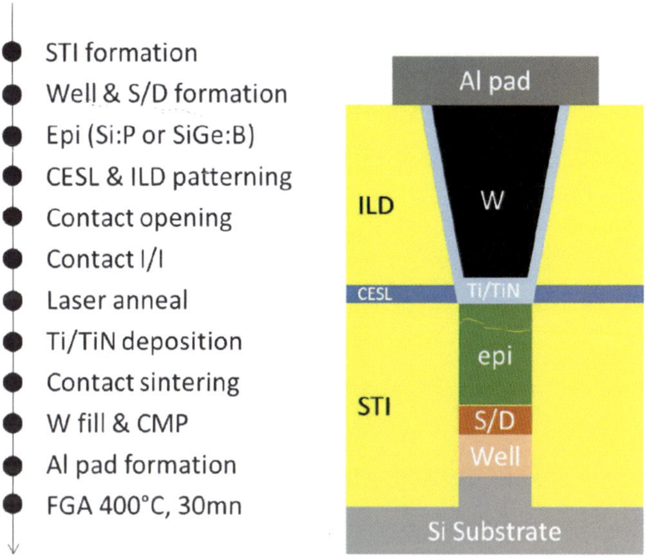


**FIG. 4.10** Contact chain schematic structure for resistivity extraction with its simplified process flow. *(Adapted from R. Hung, et al., Novel solutions to enable contact resistivity <1E-9 Ω-cm2 for 5nm node and beyond, in: 2018 International Symposium on VLSI Technology, Systems and Application (VLSI-TSA), Hsinchu, 2018, pp. 1–2, https://doi.org/10.1109/VLSI-TSA.2018.8403817.)*

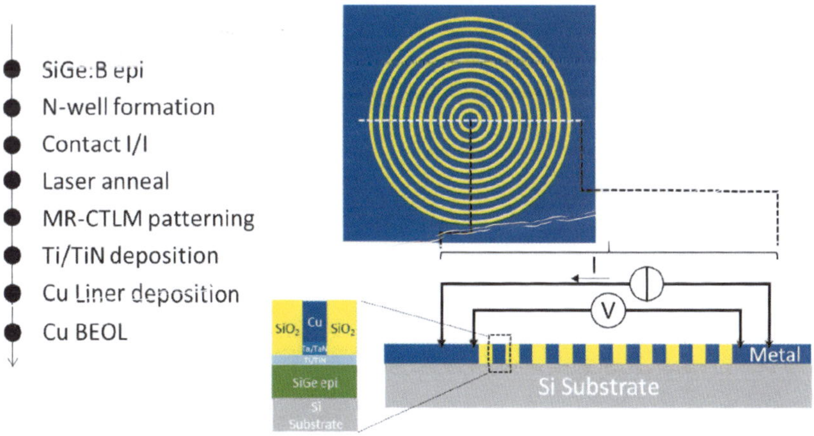


**FIG. 4.11** MR-CTLM schematic structure with its simplified process flow. Extracted contact resistivity measurement error is estimated to be $2.10^{-10}$ Ω $cm^2$. *(Adapted from J. Everaert, et al., Sub-$10^9$ Ω·$cm^2$ contact resistivity on p-SiGe achieved by Ga doping and nanosecond laser activation, in: 2017 Symposium on VLSI Technology, Kyoto, 2017, pp. T214–T215, https://doi.org/10.23919/VLSIT.2017.7998176.)*

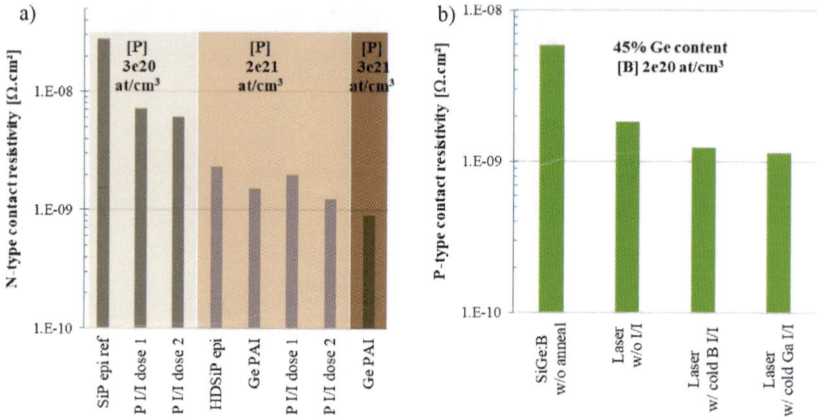


**FIG. 4.12** Contact resistivity results from contact chain structures with sub-μs laser melt annealing illustrating the studied optimization pathways (A) for N-type (B) for P-type. *(Adapted from R. Hung, et al., Novel solutions to enable contact resistivity <$10^{-9}$ Ω-cm2 for 5nm node and beyond, in: 2018 International Symposium on VLSI Technology, Systems and Application (VLSI-TSA), Hsinchu, 2018, pp. 1–2, https://doi.org/10.1109/VLSI-TSA.2018.8403817.)*

budgets later in the flow (e.g., contact anneal, Metal 1 layer deposition, and sintering). As summarized in Fig. 4.12, the latest results [60] show that the industry targets can be achieved with a proper tuning of the alloy profile (in the P-type contact case), doping approach (e.g., cold Ga implant for P-type, high concentration in situ doped Si:P for N-type), and sub-μs laser annealing conditions. An interesting finding for P-type contacts [57, 58] is that the best performance (i.e., less than $10^{-9}$ Ω cm$^2$) is achieved when dopants can be segregated close to the contact interface during laser anneal. Such segregation effects can be achieved thanks to the peculiar dynamics of the melting/recrystallization process induced by sub-μs laser annealing in the proper regime as studied in Refs. [62, 63], allowing for a near-complete solute trapping of the dopants related to relatively high solidification velocities (1–2 m/s).

Unfortunately, with such structures, there can be only very limited learnings relative to the impact of pattern density and metal gate stack integrity, for which a much more costly and complex full device flow is required, ideally up to metal 1 or metal 2 layer. In this regard, only a few literature studies are available showing actual device results, essentially on p-type transistors. Indeed, considering the use of low melting point SiGe material may allow for an improved selectivity over Si and other fragile layers and lower overall thermal budget, thus opening a wider process window for integration. Indeed, promising results were obtained on FinFET based transistor devices cumulating multiple annealing techniques, including millisecond and nanosecond (melt) laser annealing [64, 65]. In Ref. [64], in accordance with fin resistor TLM structure resistivity trends, the best results were achieved by a combination of plasma doping and sub-μs melt laser annealing, as shown in Fig. 4.13A. They were explained by the

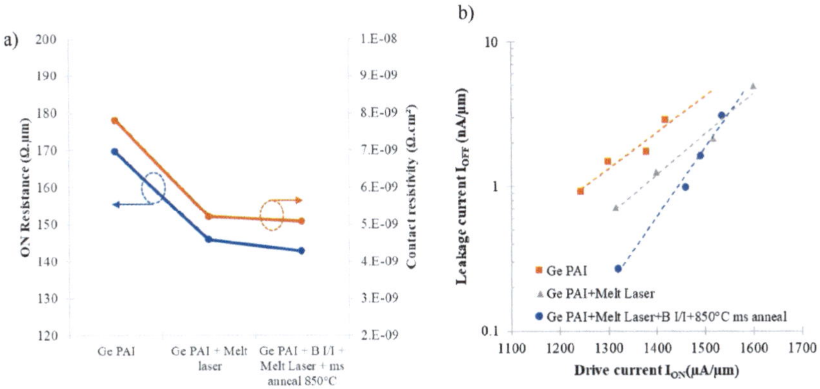


**FIG. 4.13** (A) Device on-resistance compared with fin resistor TLM contact resistivity showing the optimization pathway, (B) corresponding $I_{OFF}$ vs $I_{ON}$ cloud chart comparing different process flow approaches. *(Adapted from C.I. Li, et al., P-type MOSFET contact resistance improvement by conformal plasma doping and nanosecond laser annealing, in: IEEE Electron Device Lett. 40 (2) (2019) 307–309, https://doi.org/10.1109/LED.2019.2890950.)*

good conformality and high dose of the plasma doping approach, combined with the shallow diffusion and high activation benefits of the melt laser process. The device characteristics, reproduced in Fig. 4.13B, show good electrical performance in terms of drive current and off-state leakage. As it can be seen, although the additional B implant and millisecond anneal steps do not show a significant improvement in contact resistivity, the final device performance is still improved, especially in terms of leakage in the lower range of drive current.

A similar study [65] reports the effect of sub-µs laser melt annealing on external parasitic resistance and short channel effects in pFinFETs, determining the hierarchy of melting thresholds and their effect onto external resistance improvement and CMOS process window, identifying key process window dependencies on gate contact pitch and number of fins. Compared with traditional ms laser anneal reference, sub-µs laser annealing allowed the improvement of ON resistance of about 10% and external resistance of about 40% without degrading of short channel performance for optimized energy density and substrate preheating conditions, for which the melting was restricted to the source/drain regions. Device performance degradation and failure at higher energy density was explained by channel or even sub-fin melting (for the extreme case), leading to excessive dopant diffusion and Ge intermixing in the fin structures. The impact of pattern variations, estimated from the effect of different contacted gate pitch (CPP) and number of fins, showed that while the process window is not significantly sensitive to CPP variations, it is shifted down by ~50°C for the smallest number of fins, due to the higher thermal confinement (lower average thermal conductivity) in such isolated structures (thus reaching higher temperatures compared with higher number of fins at a given laser energy density).

The integration of sub-μs laser annealing process was further confirmed in recent work [66] on lateral nanowire Ge-channel based n-type transistors, one of the most promising architectures for the coming technology nodes. This time, the source/drain contact regions are formed using an implant-free approach by optimized multilayer epitaxy with heavy in situ doping above the solid solubility limit. After the contact opening step, the dopants were activated using sub-μs laser annealing before the metallization step. The implant-free approach avoids the usual drawbacks of ion implantation such as end-of-range defects, which would potentially require deeper melt to fully recover, with the associated risk of too much in-depth diffusion in the contact fins. This scheme was first applied to MR-CTLM structures to evaluate its potential for contact resistivity improvement. By optimizing the epitaxy and laser conditions, very low resistivities, down to $1.6 \times 10^{-9}\ \Omega\ \mathrm{cm}^2$ (reported in Fig. 4.14A), could be achieved, setting a record for contacts on n-Ge at the time of publication. The optimized process was then successfully integrated in the Ge single nanowire transistor process flow. As a result, a 9-nm-diameter NW, 70 nm gate length with scaled contacts was demonstrated showing the well-behaved transfer characteristics as reported in Fig. 4.14B, thus setting a first for a functional scaled n-type Ge-based nanowire field effect transistor manufactured in a high-volume manufacturing CMOS compatible process flow.

These studies show that sub-μs laser anneal could be applied to complete process flow (up to Metal 1) and gives a promise that pattern effects could be overcome with the correct combination of process steps and optimization of laser annealing conditions.

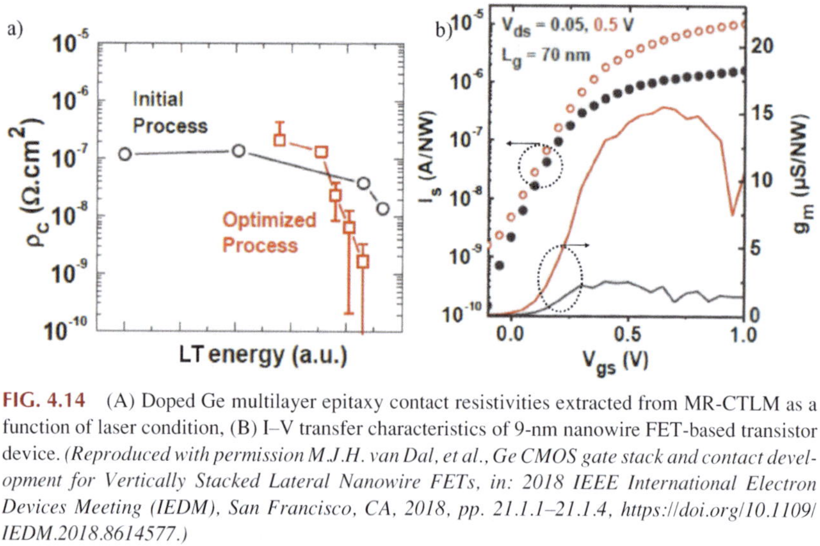


**FIG. 4.14** (A) Doped Ge multilayer epitaxy contact resistivities extracted from MR-CTLM as a function of laser condition, (B) I–V transfer characteristics of 9-nm nanowire FET-based transistor device. *(Reproduced with permission M.J.H. van Dal, et al., Ge CMOS gate stack and contact development for Vertically Stacked Lateral Nanowire FETs, in: 2018 IEEE International Electron Devices Meeting (IEDM), San Francisco, CA, 2018, pp. 21.1.1–21.1.4, https://doi.org/10.1109/IEDM.2018.8614577.)*

Very recently [49], sub-μs laser annealing for junction extension dopant activation was successfully integrated in a 3D sequential device manufacturing process flow with the layer transfer approach. As stated earlier, the target is to achieve top-tier device performance at least as good as their equivalent with traditional "high-temperature" process flow. As described in Fig. 4.15A, laser annealing was applied after junction extension implants and nitride spacer deposition but before offset spacer etch, in order to open the process window and strengthen the patterned structure integrity (an approach that was already demonstrated in Ref. [47]). The "low-temperature" process flow was then pursued until Metal 1. As shown in Fig. 4.15, the resulting device characteristics are matching the performance of the high-temperature reference devices, with a slight degradation of gate-induced drain leakage (GIDL) from the sharper junction profiles, while an improvement of short channel effects such as drain-induced barrier lowering (DIBL) was achieved thanks to the better control of dopant diffusion over the high-temperature spike anneal counterpart.

Such an achievement highlights the renewed potential of laser annealing enabled by the development of 3D-based architectures, for which the essential benefits do not lie on the specifics of nonequilibrium annealing (especially metastable activation by melting, which may disappear by deactivation from the following steps), but on its selectivity and effectiveness in relation to the low thermal budget involved in the process, compatible with the need to keep the integrity of fragile buried layers.

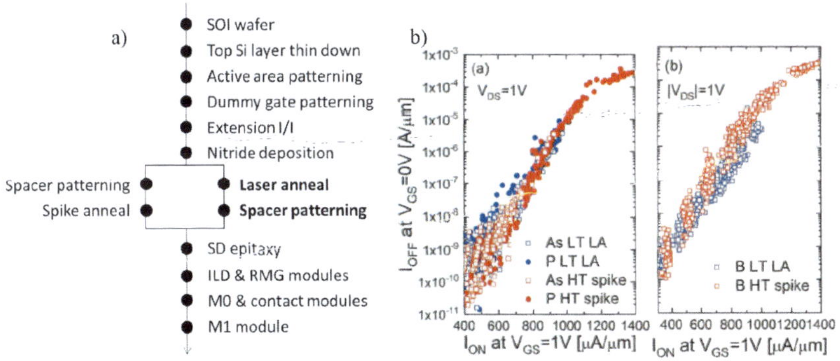


**FIG. 4.15** (A) Low-temperature process flow for top-tier planar FD-SOI devices with the split sequence for the standard high-temperature flow. (B) Corresponding $I_{OFF}$ vs $I_{ON}$ cloud chart comparing laser-annealed devices with spike anneal reference for NMOS *(left)* and PMOS *(right)* devices. *((A) Adapted from A. Vandooren, et al., 3D sequential low temperature top tier devices using dopant activation with excimer laser anneal and strained silicon as performance boosters, in: 2020 IEEE Symposium on VLSI Technology, Honolulu, HI, USA, 2020, pp. 1–2, https://doi.org/10.1109/VLSITechnology18217.2020.9265026. (B) Reproduced with permission A. Vandooren, et al., 3D sequential low temperature top tier devices using dopant activation with excimer laser anneal and strained silicon as performance boosters, in: 2020 IEEE Symposium on VLSI Technology, Honolulu, HI, USA, 2020, pp. 1–2, https://doi.org/10.1109/VLSITechnology18217.2020.9265026.)*

### 4.3.4 Silicide formation

Silicide formation represents a well-known application of laser annealing in the millisecond regime, typically applied as a replacement of the second RTA step after the removal of the unreacted metal from the first RTA step. As initially demonstrated by Ref. [67] for nickel-based silicides, the main advantages of ms laser anneal, which has driven its adoption in high-volume manufacturing, are device yield improvement by the reduction Ni piping-related defects (also highlighted by e-beam scan and brightness contrast voltage in Ref. [68]) and the lower silicide diffusion from the low thermal budget, allowing for the use of shallower S/D junctions. Its applicability to titanium-based silicides could be further confirmed by Ref. [69]. Relatively recent work [70] demonstrated its applicability to 28-nm SRAM cells, additionally studying in detail mechanisms for yield degradation (shown on Fig. 4.16 in terms of SRAM bit failures) when too high peak temperatures are used (>800°C). Such degradation would essentially be related to stress build-up during ms anneal and its relaxation during the subsequent annealing steps, leading to defect generation and, eventually, to the diffusion of Ni into the substrate.

Moving down to lower thermal budget approaches, sub-μs laser annealing was recently investigated for Ti-based contacts on 3D-integrated backside-illuminated CMOS image sensors [71]. Compared with the traditional two-step RTA silicidation process, UV sub-μs laser annealing was performed in this work as a replacement of the first RTA step to promote the formation of the low-resistivity C54-$TiSi_2$ phase at lower temperature than the one used for the second RTA step. It was determined that the C40-$TiSi_2$ formed by laser annealing prior to RTA allowed to reduce the subsequent formation temperature of the C54 phase from 800°C to 650°C. Phase nucleation was found to be the controlling mechanism (rather than the silicide growth kinetics) thanks to the higher driving force (compared with starting from the C49 phase) and template effect of the C40 phase. Such approach is a step toward another "low-temperature" process module, which could be applied to 3D architectures.

As an additional hint for development trends relative to silicide formation, recent work on germanides [72] shows the potential of laser anneal in the sub-μs timescale for germanide formation with reasonably low (but still quite far from industry requirements) contact resistivities in the $10^{-8}\,\Omega\,cm^2$ range, as extracted from traditional TLM structures.

### 4.3.5 Front end and back end of line dielectrics

Another important module of the CMOS manufacturing flow involves the process of dielectrics, with a critical figure of merit being their relative permittivity ($k$).

In the FEOL, hafnium oxide ($HfO_2$) is well known as a gate dielectric material with high $k$ value, which is currently obtained in its cubic phase

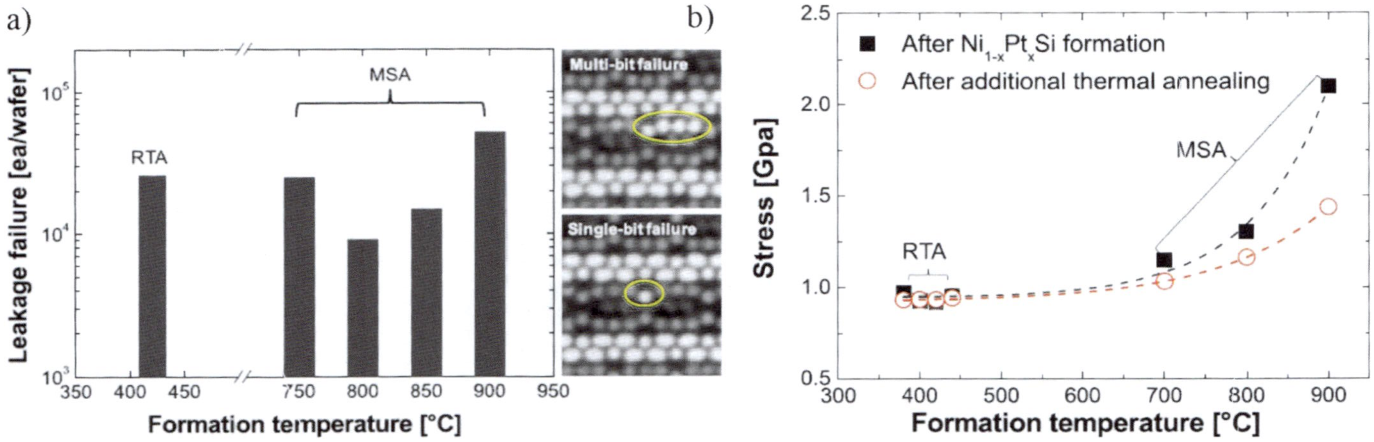


**FIG. 4.16** (A) Leakage failure extracted by electron beam inspection on SRAM cells as a function of annealing condition. MSA stands for millisecond annealing (in this case laser-based), (B) silicide film mechanical stress measured from wafer curvature with and without post-anneal. *(Reproduced with permission J. Kim, et al., Effect of millisecond annealing temperature of Ni1-xPtx Si formation on leakage current characteristics of static random- access memory cells, IEEE Trans. Electron Devices, 66 (1) (2019) 389–394, https://doi.org/10.1109/TED.2018.2878871.)*

(theoretically up to $k \sim 29$) [73] in the usual manufacturing flows. However, achieving its tetragonal phase would enable access to much higher $k$-value (up to ~70 theoretically) [73] and thus improve device performance in terms of parasitic delay and gate leakage [74]. In Ref. [75], the nucleation and crystal growth in nondoped $HfO_2$ thin films have been investigated using sub-μs laser annealing. Using metal insulator metal (MIM) capacitors to extract $k$ values (reported on Fig. 4.17), it has been demonstrated that the crystallinity of the laser-annealed $HfO_2$ films can be cumulatively improved on the TiN substrate by repeating the laser pulse irradiation while maintaining the identical cubic-like phase. The extremely fast heating and cooling during the laser annealing process are therefore supposed to prevent the nucleation and crystal growth of the monoclinic phase in the films. It has been also suggested that the selective nucleation and crystal growth of a specific $HfO_2$ phase might be realized by adjusting the process parameters of the nanosecond laser annealing. These results suggest the possible nucleation and crystal growth of an intentionally selected $HfO_2$ phase by controlling the thermal budget of the single laser irradiation, which would enable the application of sub-μs laser annealing on high-$k$ and Hf-based dielectrics engineering. As an illustration, very recently [76], orthorhombic phase creation by sub-μs laser anneal was achieved in 10-nm-thin Si-implanted $HfO_2$ and used for integrated ferroelectric capacitors in BEOL with excellent performance characteristics in terms of remanent polarization, retention, and endurance compared with the reference low-temperature approach.

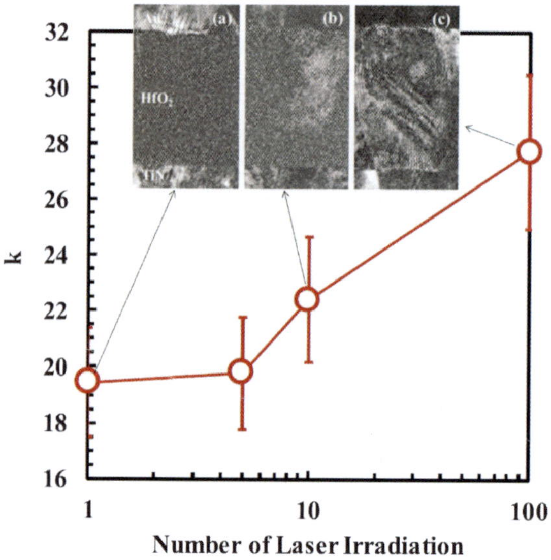


**FIG. 4.17** Relative permittivity ($k$) extracted from C–V measurements as a function of laser conditions for MIM capacitor with 50-nm-thick $HfO_2$ layer. Inset: corresponding cross-sectional dark field TEM images. *(Reproduced with permission T. Tabata, Nucleation and crystal growth in $HfO_2$ thin films by UV nanosecond pulsed laser annealing, Appl. Phys. Express, 13 (1) (2019) 015509, https://doi.org/10.7567/1882-0786/ab5ce2. Copyright (2019) The Japan Society of Applied Physics.)*

Moving to the other side of the manufacturing flow, a major figure of merit in BEOL at current and future technology nodes is the interconnect capacitance, dominated by the interlayer dielectric (ILD) performance. In this framework, not only must its permittivity ($k$) be as low as possible, but it must be as robust as possible to all the BEOL process steps such as process-induced damage (PID) during via etch. In the work recently reported in Ref. [77], significant toughening of ILD layers was achieved using a 7-nm node BEOL flow using laser annealing. The idea is to enable the fully aligned via (FAV) scheme, which allows to improve time-dependent dielectric breakdown (TDDB) and use larger vias with reduced resistance. The key requirement is to achieve an ILD layer with a good tradeoff between robustness, $k$ value, and gapfill capability.

High-porosity flowable CVD (FCVD) and plasma-enhanced CVD (PECVD) SiCOH low-$k$ films ($k \sim 2.55$) were subjected to UV cure and laser anneal and compared with a reference low-porosity PECVD film ($k \sim 2.77$) in terms of TDDB, via profile, line resistance vs lateral capacitance (RC), and $k$ value. Comparing "laser last" (laser after UV cure) with "laser first" (laser before UV cure) integration scheme, it was found that the best robustness-performance tradeoff was obtained using the "laser first" approach, which allows to set a lower porosity compared with UV cure with a minor negative impact on the $k$ value. Remarkably, as shown in Fig. 4.18, PECVD SiCOH subjected to laser anneal followed by UV cure resulted in a $\sim 2.5\times$ TDDB

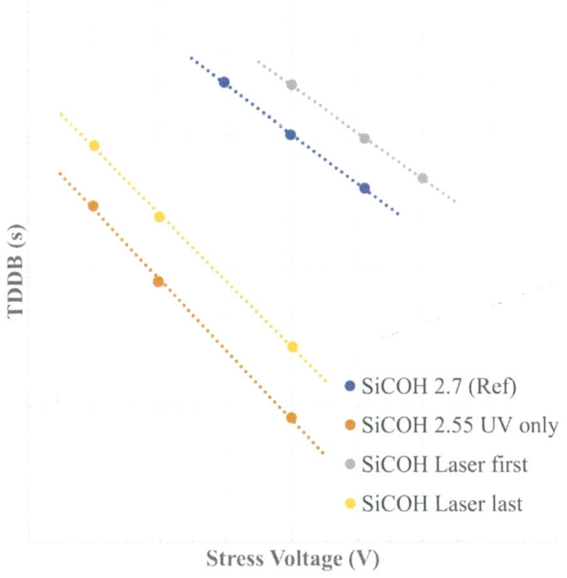


**FIG. 4.18** TDDB performance of SiCOH 2.55 with and without laser anneal compared with SiCOH 2.7 reference. *(Adapted from D. Sil, et al., Ion implantation and laser annealing for toughening low-k dielectric in scaled-down Interconnects, in: 2019 IEEE International Interconnect Technology Conference (IITC), 2019.)*

improvement together with 6% RC advantage over the $k = 2.77$ reference, also partly thanks to the observed improvement in via profile shape. Such a toughened low-$k$ material is an excellent candidate for FAV interconnects in the future technology nodes.

### 4.3.6 Back end of line copper interconnects

BEOL contact resistance is expected to become the bottleneck for circuit performance in future technology nodes. Cu line resistance is known to increase dramatically for narrow interconnect lines as linewidths become smaller than the electron mean free path in copper. As introduced in [4 and references therein], melt laser anneal is a promising path to improving copper microstructure, electrical properties, and reliability performance while keeping the thermal budget low enough to avoid damaging the fragile surrounding interfaces and low-$k$ dielectrics. In Ref. [4], large grains and ~25% line resistance improvement were shown for linewidths down to 40 nm. The relative improvement trend reproduced in Fig. 4.19 remains high even for smaller linewidths and is thus believed to be due not only to a grain size enlargement, but also to a reduction in interface scattering from impurities.

This result was confirmed more recently [78] using a 14-nm node finFET technology with 40-nm metal pitch. In this work, M1 interconnects were formed using the process flowchart described by Fig. 4.20.

The key optimization point in the process flow was the removal of the electroplated Cu overburden by chemical-mechanical planarization (CMP)

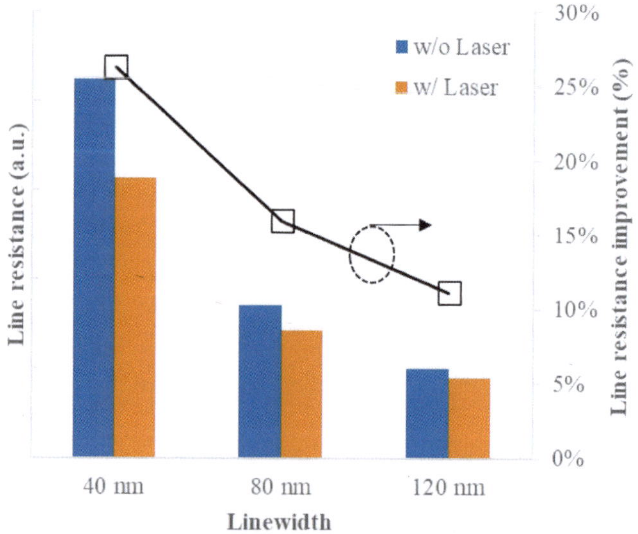


**FIG. 4.19** Line resistance and its relative improvement after sub-μs laser anneal as a function of linewidth. *(Adapted from O. Gluschenkov, H. Jagannathan, Laser annealing in CMOS manufacturing, ECS Trans. 85 (6) (2018) 11–23, https://doi.org/10.1149/08506.0011ecst.)*

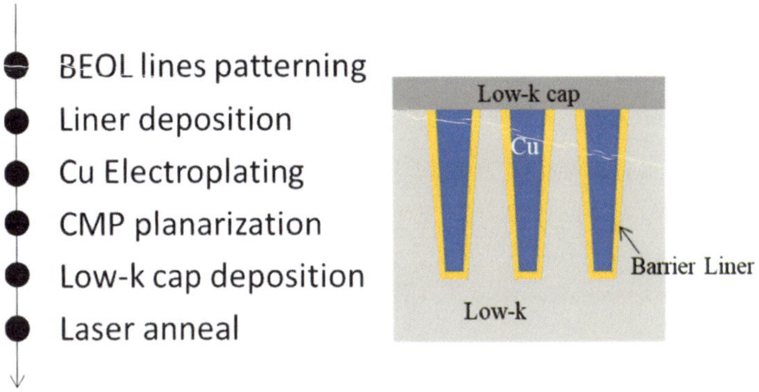


**FIG. 4.20** Simplified BEOL process flow for M1 interconnects with laser anneal with corresponding schematic structure. *(Adapted from R.T.P. Lee, et al., Nanosecond laser anneal for BEOL performance boost in advanced FinFETs, in: 2018 IEEE Symposium on VLSI Technology, Honolulu, HI, 2018, pp. 61–62, https://doi.org/10.1109/VLSIT.2018.8510651.)*

followed by the deposition of a back-end compatible low-temperature confinement layer before applying the melt laser anneal. The laser process parameters (especially the melt laser peak temperature and wafer base temperature) were then optimized to find the best tradeoff between line resistance improvement (up to +35%) and capacitance degradation (~9%, likely due to some impact on low-*k* material) for an overall RC delay improvement of ~15%. The line resistance vs capacitance tradeoff is reported in Fig. 4.21 for various laser

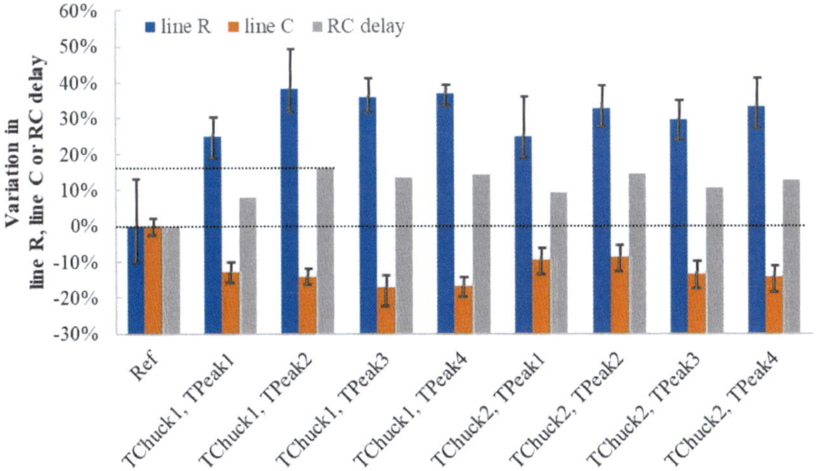


**FIG. 4.21** Line resistance improvement, line capacitance degradation, and estimated RC delay improvement for various laser annealing conditions (chuck preheat temperature and process peak temperature). *(Adapted from R.T.P. Lee et al., Nanosecond laser anneal for BEOL performance boost in advanced FinFETs, in: 2018 IEEE Symposium on VLSI Technology, Honolulu, HI, 2018, pp. 61–62, https://doi.org/10.1109/VLSIT.2018.8510651.)*

anneal conditions. The resulting improvement in final device performance after completing the full 11 metal layers was significant (2%–5%). Combined with excellent yield characteristics, the robustness of the approach is clearly demonstrated and hints that this could be further improved by applying the laser anneal to more interconnect layers in future technology nodes, in which linewidths at higher metal levels will also get smaller. However, for smaller metal pitch technologies, the tradeoff may be more difficult to find, as the resistivity improvement may not be sufficient and solutions to minimize the capacitance degradation will be required.

### 4.3.7 Conclusion

To summarize, for the CMOS manufacturing flow, since the last couple of decades, there has been a continuous effort to demonstrate the potential of laser annealing for various applications from the front end to the back end of the line. In the last decade, clear trends can be seen in the literature and the industry toward a real integration effort in actual device structures and circuits. Indeed, the interest has been renewed with the development of 3D architectures for which laser annealing is clearly identified as a mean to relieve some of the constraints related to the low thermal budget requirements.

## 4.4 Memory

In the memory ecosystem, the three well-known markets could be addressed by laser annealing, namely, NAND Flash, DRAM, and emerging memories (EM). In the NAND space, 3D NAND channel annealing for crystallization is not easily accessible to sub-μs laser annealing, due to the most challenging tradeoff required to crystallize several μm depth at high temperature while keeping fragile charge trap memory layers undamaged. However, vertical diodes and vertical channel transistors are interesting candidates as steering elements (or access devices) for EM devices considered for storage class memory (SCM) products such as phase change random access memory (PCRAM) or resistive RAM (RRAM) [79–81] and vertical DRAM [82–84], respectively. As illustrations, Fig. 4.22A shows a typical case of a 3D memory array circuit [85], highlighting the steering element, while a typical unit memory cell is schematized in Fig. 4.22B [86] and an ultimately scaled vertical DRAM structure is reported in Fig. 4.22C [82]. Here, similarly to 3D logic discussed earlier, the requirements for laser annealing are to achieve at least similar performance as the high-temperature process flows, without impacting the buried layers. In this section, we will review laser annealing potential for vertical access devices (also called steering elements) and vertical channel transistors.

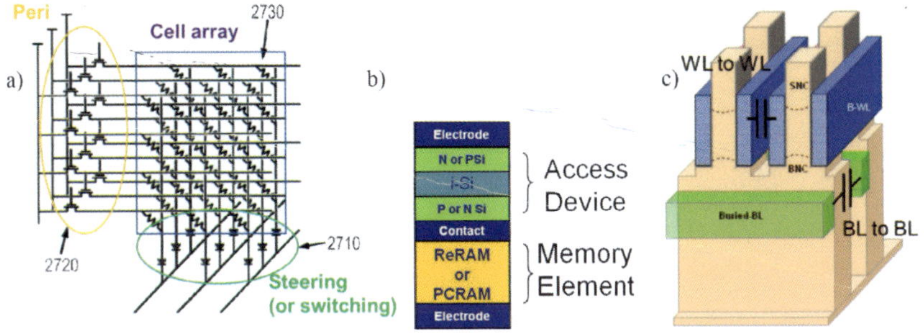


**FIG. 4.22** (A) Example of 3D memory array circuit with key function highlighted. (B) typical silicon diode steering element layer stack on top of a memory cell, (C) 3D DRAM with vertical pillar transistor structure. *((A) Adapted from D.R. Shepard, Vertical Switch Three-Dimensional Memory Array. U.S. Patent No. US 2010/0232200 A1, September 16, 2010. (B, C) Adapted from D.S. Woo, DRAM: its challenging history and future, in: 2017 IEEE International Electron Devices Meeting (IEDM), San Francisco, CA, 2017, Short course 2.)*

### 4.4.1 Vertical access devices

Access devices for emerging memories are typically unipolar or bipolar diodes. The thickness of such diodes should be scalable down from few hundred nm to a few tens of nm with high current density (few $MA/cm^2$, driving the memory cell) and high breakdown voltage (typically over 3 V, to minimize the leakage). To achieve such performance in Si-based devices, the polysilicon quality should be as good as possible. Moreover, such devices need a proper control of junction depths to keep a large enough intrinsic region. Traditional low-pressure chemical vapor deposition (LPCVD) with in situ doping operates at temperatures higher than 500°C. In this range, the crystallinity of the poly-Si diode is poor and the dopants may not be fully activated. As a consequence, the current may not be enough to achieve proper switching and the breakdown voltage will be too low for a reliable device operation. Any further activation or crystallization anneal will induce further dopant diffusion and impact the device reliability. Moreover, in the framework of cell stacking, the thermal budget applied to the memory element during access device formation must be low enough to avoid damaging it. For example, considering traditional RTP, the GeSbTe (GST) alloy material used as memory layer in PCRAM should not be heated at temperatures higher than 300°C to 600°C, depending on composition.

Regarding vertical diode devices, the potential of sub-μs laser annealing was evaluated on PIN diodes with an in situ doped layer stack deposited by low-temperature LPCVD (<550°C) [86]. As shown in Fig. 4.23A and B, a process window could be found to achieve the challenging tradeoff between good layer crystallinity (for high drive currents), diffusion control (for high breakdown voltage and low leakage), and buried memory layer integrity (in that case GST).

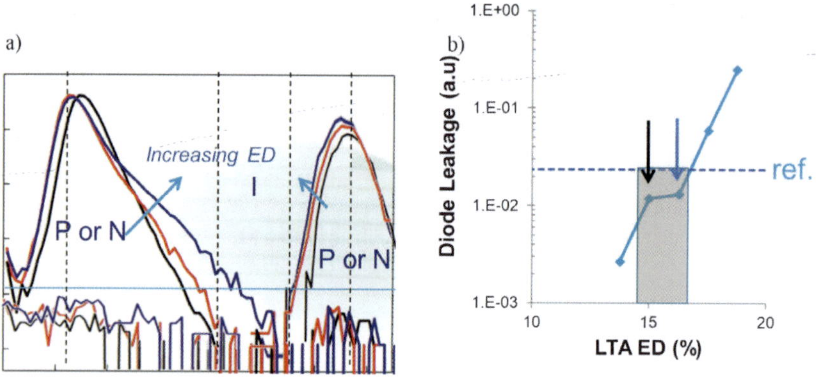


**FIG. 4.23** (A) PIN diode layer stack SIMS profiles for various laser annealing conditions, (B) diode leakage as a function of laser condition (energy density ED) compared with high-temperature reference, highlighting the process window region. *(Reproduced with permission K. Huet, C. Boniface, R. Negru, J. Venturini, Ultra low thermal budget anneals for 3D memories: access device formation, AIP Conf. Proc. 1496 (2012) 135, https://doi.org/10.1063/1.4766508.)*

## 4.4.2 Vertical channel transistors

Regarding vertical transistor devices, the extensive work done on vertical memory cells with sub-μm channel length [87–89] gives a useful insight on the potential for laser annealing to integrate such 3D memory process flows. A typical vertical transistor process flow is reproduced in Fig. 4.24 [87], in which the

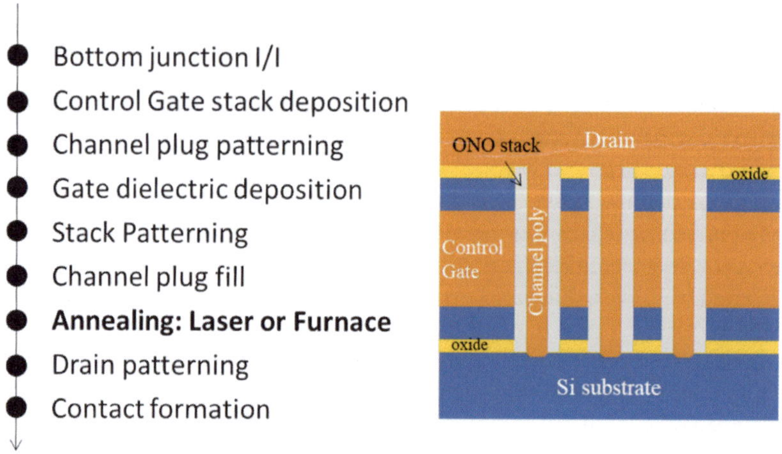


**FIG. 4.24** Simplified vertical channel NAND flash memory (3D SONOS) process flow and corresponding schematic structure. *(Adapted from J.G. Lisoni, et al., Laser thermal anneal of polysilicon channel to boost 3D memory performance, in: 2014 Symposium on VLSI Technology (VLSI-Technology): Digest of Technical Papers, Honolulu, HI, 2014, pp. 1–2, https://doi.org/10.1109/VLSIT.2014.6894346.)*

channel crystallization step is achieved by laser annealing after the memory plug patterning and channel filling steps. An important point is that the polysilicon overburden (several tens of nm thick) from the filling step is recessed only after laser annealing. Such layer helps in mitigating optical pattern effects and allows for proper Si reflow in the structure. The resulting polysilicon grain size in the vertical channel is much improved as shown by TEM analysis reproduced in Fig. 4.25. The resulting improvement in transistor performance compared with the traditional furnace annealing reference is significant with over 5X improvement in drive current and over 2X improvement in subthreshold swing, as reported in Fig. 4.26. Such improvement is believed to be related not only to conductivity enhancement from crystal quality, but also to a reduction of defects at the gate/oxide interfaces, which are even more critical in the so-called Macaroni-type vertical memory cells [88] with very thin channels and additional oxide fillers.

### 4.4.3 Conclusion

To summarize, laser annealing has been less studied for memory applications compared with other fields such as CMOS and power devices, essentially due to the expected integration challenges. Nevertheless, with the trend toward high aspect ratio features and 3D devices, some of its unique advantages may find their space. For instance, DRAM polysilicon contact plug annealing [10] could be a promising application for which laser annealing may be considered to remove voids induced from the imperfect filling of high aspect ratio contact plugs, as demonstrated in the work on vertical 3D NAND transistors.

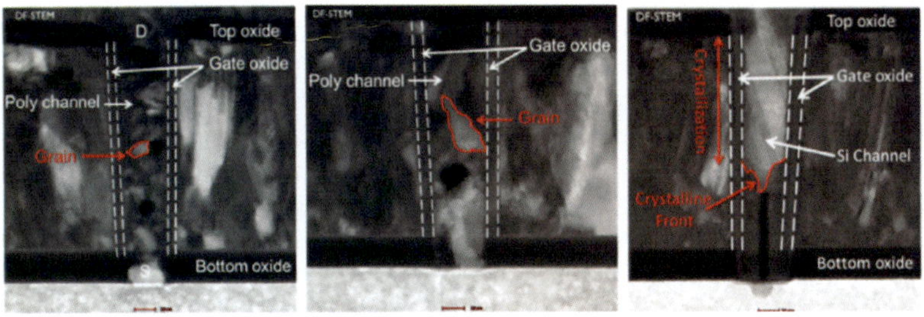


**FIG. 4.25** Channel polycrystallization with sub-μs laser annealing. Dark field STEM of *(left)* poly, *(center)* LTA (2.0 J/cm$^2$) and *(right)* LTA 1.6 J/cm$^2$. *(Reproduced with permission J.G. Lisoni, et al., Laser thermal anneal of polysilicon channel to boost 3D memory performance, in: 2014 Symposium on VLSI Technology (VLSI-Technology): Digest of Technical Papers, Honolulu, HI, 2014, pp. 1–2, https://doi.org/10.1109/VLSIT.2014.6894346.)*

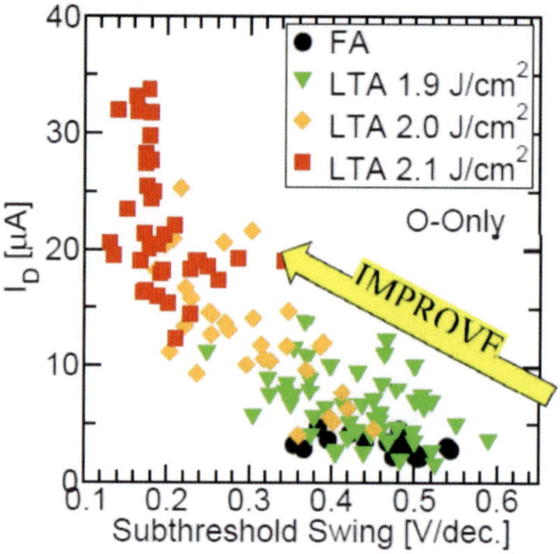


**FIG. 4.26** Performance improvement of vertical transistor with laser annealing relative to traditional furnace-annealed polysilicon. *(Reproduced with permission J.G. Lisoni, et al., Laser thermal anneal of polysilicon channel to boost 3D memory performance, in: 2014 Symposium on VLSI Technology (VLSI-Technology): Digest of Technical Papers, Honolulu, HI, 2014, pp. 1–2, https://doi.org/10.1109/VLSIT.2014.6894346.)*

## 4.5 Conclusion

In this chapter, we have reviewed the applications of laser annealing for major semiconductor industry fields, namely, power devices, CMOS, and memory.

For power devices, laser annealing adoption is already well established, especially for the Si-based devices. It has very good potential for growth, driven by the automotive industry and the recent switch to more efficient and better-performing devices enabled by SiC-based technologies, for which laser annealing is already adopted for contact formation and shows a high promise for dopant activation.

For traditional CMOS logic, laser annealing was only adopted in the millisecond regime for niche applications such as silicide formation. Unfortunately, lower thermal budget approaches such as sub-μs annealing do not show a good enough tradeoff between the expected benefit and the complexity of its integration in highly complex process flows for which many modules are interdependent. As a result, at the device or circuit level, key target figures of merit can be achieved by optimizing several process steps in different unit modules without the hurdle of integrating laser annealing. However, the renewed interest for 3D architectures opens new opportunities since the associated process flows come with severely limited thermal budgets constraints and only require at least as good performance as the corresponding high thermal budget counterpart. In this case, leveraging the optimization of existing process steps is way more complex and laser annealing thus becomes an attractive tool for key applications such as active area formation, dielectrics postdeposition annealing, silicide formation, and contact resistivity improvement.

Finally, for memory, laser annealing was never adopted in mainstream devices manufacturing flows, although its applicability to vertical devices with tight thermal budget constraints was extensively studied. However, similarly to CMOS logic, the trend toward 3D architectures to complement or, in the long term, replace existing technologies opens new opportunities, essentially related to material modifications such as crystallization and defect curing.

## References


[1] P. Loewenhardt, T.S. Lee, J.-C. Wang, L. Dorf, S. Rauf, A. Athyde, Atomic Level Precision Materials Engineering, Semicon Taiwan, Taipei, Taiwan, 2015.

[2] S.G. Wuu, C.C. Wang, D.N. Yaung, et al., A manufacturable Back-Side Illumination Technology using bulk-Si substrate for advanced CMOS Image Sensor, in: IEEE International Image Sensor Workshop, Bergen, Norway, 2009.

[3] T. Gutt, H. Schulze, T. Rupp, J. Venturini, Laser thermal annealing for power field effect transistor by using deep melt activation, in: 2006 14th IEEE International Conference on Advanced Thermal Processing of Semiconductors, Kyoto, 2006, pp. 193–197, https://doi.org/10.1109/RTP.2006.367999.

[4] O. Gluschenkov, H. Jagannathan, Laser annealing in CMOS manufacturing, ECS Trans. 85 (6) (2018) 11–23, https://doi.org/10.1149/08506.0011ecst.

[5] L.M. Feng, Y. Wang, D.A. Markle, Minimizing pattern Dependency in millisecond annealing, in: 2006 International Workshop on Junction Technology, Shanghai, 2006, pp. 25–30, https://doi.org/10.1109/IWJT.2006.220853.

[6] J.C. Scott, O. Gluschenkov, B. Goplen, et al., Reduction of RTA-driven intra-die variation via model-based layout optimization, in: IEEE International Symposium on VLSI Technology, Honolulu, HI, 2009, pp. 152–153.

[7] T. Miyashita, T. Kubo, Y.S. Kim, et al., A study on millisecond annealing (MSA) induced layout dependence for flash lamp annealing (FLA) and laser spike annealing (LSA) in multiple MSA scheme with 45 nm high-performance technology, in: 2009 IEEE International Electron Devices Meeting (IEDM), Baltimore, MD, 2009, pp. 1–4, https://doi.org/10.1109/IEDM.2009.5424429.

[8] A. Pizzagalli, Laser technologies for semiconductor manufacturing, Yole Developpement, 2017. Report.

[9] K. Huet, F. Mazzamuto, T. Tabata, I. Toqué-Tresonne, Y. Mori, Doping of semiconductor devices by laser thermal annealing, Mater. Sci. Semicond. Process. 62 (2017) 92–102, https://doi.org/10.1016/j.mssp.2016.11.008.

[10] K. Huet, T. Tabata, J. Aubin, F. Rozé, L. Thuries, S. Halty, B. Curvers, F. Mazzamuto, J. Liu, Y. Mori, Laser thermal annealing for low thermal budget applications: from contact formation to material modification, ECS Trans. 89 (3) (2019) 137, https://doi.org/10.1149/08903.0137ecst.

[11] C.W. White, B.R. Appleton, S.R. Wilson, in: J.M. Poate, J.W. Mayer (Eds.), Laser Annealing of Semiconductors, Academic Press, 1982, pp. 111–146.

[12] R. Murto, K. Jones, M. Rendon, S. Talwar, Activation and deactivation studies of laser thermal annealed boron, arsenic, phosphorus, and antimony ultra-shallow abrupt junctions, in: 2000

International Conference on Ion Implantation Technology Proceedings. Ion Implantation Technology - 2000 (Cat. No.00EX432), Alpbach, 2000, pp. 155–158, https://doi.org/10.1109/IIT.2000.924113.

[13] Y. Wang, S. Chen, M. Shen, X. Wang, S. Zhou, J. Hebb, D. Owen, Dual beam laser spike annealing technology, in: 2010 International Workshop on Junction Technology Extended Abstracts, Shanghai, 2010, pp. 1–6, https://doi.org/10.1109/IWJT.2010.5474998.

[14] Z. Liu, O. Gluschenkov, H. Niimi, B. Liu, J. Li, J. Demarest, S. Mochizuki, P. Adusumilli, M. Raymond, A. Carr, S. Chen, Dual beam laser annealing for contact resistance reduction and its impact on VLSI integrated circuit variability, in: 2017 Symposium on VLSI Technology, Kyoto, 2017, pp. T212–T213, https://doi.org/10.23919/VLSIT.2017.7998175.

[15] T. Gutt, H. Schulze, Deep melt activation using laser thermal annealing for IGBT thin wafer technology, in: 2010 22nd International Symposium on Power Semiconductor Devices & IC's (ISPSD), Hiroshima, 2010, pp. 29–32.

[16] D. Friedrich, H. Bernt, H. Hanssen, P. Oesterlin, H. Schmidt, Laser annealing of power devices, in: 2007 15th International Conference on Advanced Thermal Processing of Semiconductors, Catania, Sicily, 2007, pp. 263–269, https://doi.org/10.1109/RTP.2007.4383852.

[17] K. Ukawa, Y. Kanda, T. Sameshima, N. Sano, M. Naito, N. Hamamoto, Activation of silicon implanted with phosphorus and boron atoms by infrared semiconductor laser rapid annealing, Jpn. J. Appl. Phys. 49 (7R) (2010), https://doi.org/10.1143/JJAP.49.076503, 076503.

[18] Y. Arai, T. Milsuda, T. Kudo, K. Sano, H. Schmidt, P. Oestcrlin, Backside-activation technique of power device IGBTs by a microsecond-pulsed green laser, in: 2009 17th International Conference on Advanced Thermal Processing of Semiconductors, Albany, NY, 2009, pp. 1–4, https://doi.org/10.1109/RTP.2009.5373455.

[19] T. Seino, Y. Arai, N. Kobayashi, T. Kudo, K. Sano, Backside activation of power device IGBTs by microsecond-pulsed green laser annealing thermally assisted with CW diode laser, in: 2010 18th International Conference on Advanced Thermal Processing of Semiconductors (RTP), Gainesville, FL, 2010, pp. 140–143, https://doi.org/10.1109/RTP.2010.5623799.

[20] M. Rahimo, C. Corvasce, J. Vobecky, Y. Otani, K. Huet, Thin-wafer silicon IGBT with advanced laser annealing and sintering process, IEEE Electron Device Lett. 33 (11) (2012) 1601–1603, https://doi.org/10.1109/LED.2012.2215304.

[21] T. Tabata, S. Halty, I. Toque-Tresonne, F. Mazzamuto, K. Huet, Y. Mori, UV excimer laser annealing for next generation power electronics, in: 2016 21st International Conference on Ion Implantation Technology (IIT), Tainan, 2016, pp. 1–4, https://doi.org/10.1109/IIT.2016.7882917.

[22] Y. Negoro, T. Kimoto, H. Matsunami, F. Schmid, G. Pensl, Electrical activation of high-concentration aluminum implanted in 4H-SiC, J. Appl. Phys. 96 (2004) 4916, https://doi.org/10.1063/1.1796518.

[23] S.Y. Chou, Y. Chang, K.H. Weiner, T. Sigmon, J.D. Parsons, Annealing of implantation damage and redistribution of impurities in SiC using a pulsed excimer laser, Appl. Phys. Lett. 56 (1990) 530, https://doi.org/10.1063/1.102735.

[24] C. Dutto, E. Fogarassy, D. Mathiot, D. Muller, P. Kern, D. Ballutaud, Long-pulse duration excimer laser annealing of Al+ ion implanted 4H-SiC for pn junction formation, Appl. Surf. Sci. 208–209 (2003) 292–297, https://doi.org/10.1016/S0169-4332(02)01357-0.

[25] Y. Tanaka, H. Tanoue, K. Arai, Electrical activation of the ion-implanted phosphorus in 4H-SiC by excimer laser annealing, J. Appl. Phys. 93 (10) (2003) 5934, https://doi.org/10.1063/1.1565190.

[26] Y. Tanaka, H. Tanoue, K. Arai, Electrical activation of ion-implanted nitrogen and aluminum in 4H-SiC by excimer laser annealing, Mater. Sci. Forum 433–436 (2003) 605–608, https://doi.org/10.4028/www.scientific.net/msf.433-436.605.
[27] F. Mazzamuto, S. Halty, Y. Mori, Silicon carbide recrystallization mechanism by non-equilibrium melting laser anneal, Mater. Sci. Forum 858 (2016) 540–543, https://doi.org/10.4028/www.scientific.net/msf.858.540.
[28] C. Calabretta, M. Agati, M. Zimbone, S. Boninelli, A. Castiello, A. Pecora, G. Fortunato, L. Calcagno, L. Torrisi, F. La Via, Laser annealing of P and Al implanted 4H-SiC epitaxial layers, Materials 12 (20) (2019) 3362, https://doi.org/10.3390/ma12203362.
[29] C. Hellinger, O. Rusch, M. Rommel, A.J. Bauer, T. Erlbacher, Low-resistance ohmic contact formation by laser annealing of N-implanted 4H-SiC, Mater. Sci. Forum 1004 (2020) 718–724, https://doi.org/10.4028/www.scientific.net/msf.1004.718.
[30] V.S. Tsvetkov, S.T. Allen, H.S. Kong, C.H. Carter, Recent progress in SiC crystal growth, Inst. Phys. Conf. Ser. 142 (1996) 17–22. https://ci.nii.ac.jp/naid/80009054308/en/.
[31] A. Hedler, S. Urban, F. Falk, H. Hobert, W. Wesch, Excimer laser crystallization of amorphous silicon carbide produced by ion implantation, Appl. Surf. Sci. 205 (1–4) (2003) 240–248, https://doi.org/10.1016/S0169-4332(02)01071-1.
[32] V.C. Sorg, Laser Annealing and Dopant Activation in III-V Materials, Cornell University, 2017, https://doi.org/10.7298/X4ZW1J2T.
[33] X.C. Wang, G.C. Lim, W. Liu, C.B. Soh, S.J. Chua, Effects of 248nm excimer laser irradiation on the properties of Mg-doped GaN, Appl. Surf. Sci. 252 (5) (2005) 2071–2077, https://doi.org/10.1016/j.apsusc.2005.03.195.
[34] H.T. Wang, L.S. Tan, E.F. Chor, Pulsed laser annealing of Be-implanted GaN, J. Appl. Phys. 98 (2005), https://doi.org/10.1063/1.2120893, 094901.
[35] K.M. Yu, W. Walukiewicz, Synthesis of $GaN_xAs_{1-x}$ thin films by pulsed laser melting and rapid thermal annealing of N+-implanted GaAs, J. Appl. Phys. 94 (2003) 1043, https://doi.org/10.1063/1.1582393.
[36] S.Y. Han, K.H. Kim, J.K. Kim, H.W. Jang, K.H. Lee, N.K. Kim, J.L. Lee, Ohmic contact formation mechanism of Ni on n-type 4H–SiC, Appl. Phys. Lett. 79 (12) (2001) 1816–1818, https://doi.org/10.1063/1.1404998.
[37] S. Rascunà, P. Badalà, C. Tringali, C. Bongiorno, E. Smecca, A. Alberti, S. Di Franco, F. Giannazzo, G. Greco, F. Roccaforte, M. Saggio, Morphological and electrical properties of nickel based Ohmic contacts formed by laser annealing process on n-type 4H-SiC, Mater. Sci. Semicond. Process. 97 (2019) 62–66, https://doi.org/10.1016/j.mssp.2019.02.031.
[38] O. Rusch, C. Hellinger, J. Moult, Y. Corcoran, T. Erlbacher, Reducing on-resistance for SiC diodes by thin wafer and laser anneal technology, Mater. Sci. Forum 1004 (2020) 155–160, https://doi.org/10.4028/www.scientific.net/msf.1004.155.
[39] R. Rupp, R. Kern, R. Gerlach, Laser backside contact annealing of SiC power devices: a prerequisite for SiC thin wafer technology, in: 2013 25th International Symposium on Power Semiconductor Devices & IC's (ISPSD), Kanazawa, 2013, pp. 51–54, https://doi.org/10.1109/ISPSD.2013.6694396.
[40] P. Badalà, S. Rascunà, B. Cafra, A. Bassi, E. Smecca, M. Zimbone, C. Bongiorno, C. Calabretta, F. La Via, F. Roccaforte, M. Saggio, G. Franco, A. Messina, A. La Magna, A. Alberti, Ni/4H-SiC interaction and silicide formation under excimer laser annealing for ohmic contact, Materialia 9 (2020) 100528, https://doi.org/10.1016/j.mtla.2019.100528.
[41] G.H. Wang, T.-C. Wong, X. Wang, H.-Y. Zheng, T.-K. Chan, T. Osipowicz, Y.-L. Foo, S. Tripathy, Reduced contact resistance and improved surface morphology of ohmic contacts

on GaN employing KrF laser irradiation, Jpn. J. Appl. Phys. 50 (2011) 4S, https://doi.org/10.1143/JJAP.50.04DF06.

[42] Y. Cheng, W. Lu, T. Wang, Z. Chen, Fabrication of Ohmic contact on semi-insulating 4H-SiC substrate by laser thermal annealing, J. Appl. Phys. 119 (2016), https://doi.org/10.1063/1.4953778, 225705.

[43] P. Batude, et al., 3DVLSI with CoolCube process: an alternative path to scaling, in: 2015 Symposium on VLSI Technology (VLSI Technology), Kyoto, 2015, pp. T48–T49, https://doi.org/10.1109/VLSIT.2015.7223698.

[44] L. Brunet, et al., Breakthroughs in 3D sequential technology, in: 2018 IEEE International Electron Devices Meeting (IEDM), San Francisco, CA, 2018, pp. 7.2.1–7.2.4, https://doi.org/10.1109/IEDM.2018.8614653.

[45] C. Fenouillet-Beranger, et al., Ns laser annealing for junction activation preserving inter-tier interconnections stability within a 3D sequential integration, in: 2016 IEEE SOI-3D-Subthreshold Microelectronics Technology Unified Conference (S3S), Burlingame, CA, 2016, pp. 1–2, https://doi.org/10.1109/S3S.2016.7804375.

[46] C. Cavalcante, et al., 28nm FDSOI CMOS technology (FEOL and BEOL) thermal stability for 3D Sequential Integration: yield and reliability analysis, in: 2020 IEEE Symposium on VLSI Technology, Honolulu, HI, USA, 2020, pp. 1–2, https://doi.org/10.1109/VLSITechnology18217.2020.9265075.

[47] C. Fenouillet-Beranger, et al., New insights on bottom layer thermal stability and laser annealing promises for high performance 3D VLSI, in: 2014 IEEE International Electron Devices Meeting, San Francisco, CA, 2014, pp. 27.5.1–27.5.4, https://doi.org/10.1109/IEDM.2014.7047121.

[48] A. Vandooren, et al., Key challenges and opportunities for 3D sequential integration, in: 2018 IEEE SOI-3D-Subthreshold Microelectronics Technology Unified Conference (S3S), Burlingame, CA, USA, 2018, pp. 1–4, https://doi.org/10.1109/S3S.2018.8640203.

[49] A. Vandooren, et al., 3D sequential low temperature top tier devices using dopant activation with excimer laser anneal and strained silicon as performance boosters, in: 2020 IEEE Symposium on VLSI Technology, Honolulu, HI, USA, 2020, pp. 1–2, https://doi.org/10.1109/VLSITechnology18217.2020.9265026.

[50] F. Hsueh, et al., Monolithic 3D SRAM-CIM macro fabricated with BEOL gate-all-around MOSFETs, in: 2019 IEEE International Electron Devices Meeting (IEDM), San Francisco, CA, USA, 2019, pp. 3.3.1–3.3.4, https://doi.org/10.1109/IEDM19573.2019.8993628.

[51] C. Yang, et al., Location-controlled-grain technique for monolithic 3D BEOL FinFET circuits, in: 2018 IEEE International Electron Devices Meeting (IEDM), San Francisco, CA, 2018, pp. 11.3.1–11.3.4, https://doi.org/10.1109/IEDM.2018.8614708.

[52] S. Jung, H. Lim, K.H. Kwak, K. Kim, A 500-MHz DDR high-performance 72-Mb 3-D SRAM fabricated with laser-induced epitaxial c-Si growth technology for a stand-alone and embedded memory application, IEEE Trans. Electron Devices 57 (2) (2010) 474–481, https://doi.org/10.1109/TED.2009.2037368.

[53] R. Ishihara, P.C. van der Wilt, B.D. van Dijk, A. Burtsev, J.W. Metselaar, C.I.M. Beenakker, Advanced excimer-laser crystallization process for single-crystalline thin film transistors, Thin Solid Films 427 (1–2) (2003) 77–85, https://doi.org/10.1016/S0040-6090(02)01250-6.

[54] C. Fenouillet, et al., First demonstration of low temperature ($\leq$500°C) CMOS devices featuring functional RO and SRAM bitcells toward 3D VLSI integration, in: 2020 Symposium on VLSI Technology (VLSI Technology), 2020.

[55] J.M. Poate, W.L. Brown, Laser annealing of silicon, Phys. Today 35 (6) (1982) 24, https://doi.org/10.1063/1.2915125.

[56] E. Rosseel, et al., Selective epitaxial growth of high-P Si:P for source/drain formation in advanced Si nFETs, ECS Trans. 75 (8) (2016) 347, https://doi.org/10.1149/07508.0347ecst.

[57] J. Everaert, et al., Sub-10 − 9 Ω·cm2 contact resistivity on p-SiGe achieved by Ga doping and nanosecond laser activation, in: 2017 Symposium on VLSI Technology, Kyoto, 2017, pp. T214–T215, https://doi.org/10.23919/VLSIT.2017.7998176.

[58] L. Wang, et al., Comprehensive study of Ga activation in Si, SiGe and Ge with 5 × 10 − 10 Ω·cm2 contact resistivity achieved on Ga doped Ge using nanosecond laser activation, in: 2017 IEEE International Electron Devices Meeting (IEDM), San Francisco, CA, 2017, pp. 22.4.1–22.4.4, https://doi.org/10.1109/IEDM.2017.8268441.

[59] P. Sarangapani, C. Weber, J. Chang, S. Cea, M. Povolotskyi, G. Klimeck, T. Kubis, Atomistic tight-binding study of contact resistivity in Si/SiGe PMOS Schottky contacts, IEEE Trans. Nanotechnol. 17 (5) (2018) 968–973, https://doi.org/10.1109/TNANO.2018.2840836.

[60] R. Hung, et al., Novel solutions to enable contact resistivity <1E-9 Ω-cm2 for 5nm node and beyond, in: 2018 International Symposium on VLSI Technology, Systems and Application (VLSI-TSA), Hsinchu, 2018, pp. 1–2, https://doi.org/10.1109/VLSI-TSA.2018.8403817.

[61] H. Yu, et al., Multiring circular transmission line model for ultralow contact resistivity extraction, IEEE Electron Device Lett. 36 (6) (2015) 600–602, https://doi.org/10.1109/LED.2015.2425792.

[62] T. Tabata, J. Aubin, K. Huet, F. Mazzamuto, Segregation and activation of Ga in high Ge content SiGe by UV melt laser anneal, J. Appl. Phys. 125 (2019) 215702, https://doi.org/10.1063/1.5096889.

[63] T. Tabata, P.-E. Raynal, K. Huet, J.-L. Everaert, Segregation and activation of Sb implanted in Si by UV nanosecond-laser-anneal-induced non-equilibrium solidification, J. Appl. Phys. 127 (135701) (2020), https://doi.org/10.1063/5.0005176.

[64] C.I. Li, et al., P-type MOSFET contact resistance improvement by conformal plasma doping and nanosecond laser annealing, IEEE Electron Device Lett. 40 (2) (2019) 307–309, https://doi.org/10.1109/LED.2019.2890950.

[65] O. Gluschenkov, et al., External resistance reduction by nanosecond laser anneal in Si/SiGe CMOS technology, in: 2018 IEEE International Electron Devices Meeting (IEDM), San Francisco, CA, 2018, pp. 35.3.1–35.3.4, https://doi.org/10.1109/IEDM.2018.8614628.

[66] M.J.H. van Dal, et al., Ge CMOS gate stack and contact development for Vertically Stacked Lateral Nanowire FETs, in: 2018 IEEE International Electron Devices Meeting (IEDM), San Francisco, CA, 2018, pp. 21.1.1–21.1.4, https://doi.org/10.1109/IEDM.2018.8614577.

[67] C. Ortolland, et al., Silicide yield improvement with NiPtSi formation by laser anneal for advanced low power platform CMOS technology, in: 2009 IEEE International Electron Devices Meeting (IEDM), Baltimore, MD, 2009, pp. 1–4, https://doi.org/10.1109/IEDM.2009.5424428.

[68] Y. He, et al., Investigation of Ni-based silicide formation by different dynamic surface annealing approaches, in: 2010 18th International Conference on Advanced Thermal Processing of Semiconductors (RTP), Gainesville, FL, 2010, pp. 76–78, https://doi.org/10.1109/RTP.2010.5623785.

[69] R. Beneyton, P. Morin, S. Muthukrishnan, D. Larmagnac, A. Mayur, C.T. Richard, Formation of titanium silicide by millisecond anneal, in: 2010 18th International Conference on Advanced Thermal Processing of Semiconductors (RTP), Gainesville, FL, 2010, pp. 80–85, https://doi.org/10.1109/RTP.2010.5623790.

[70] J. Kim, et al., Effect of millisecond annealing temperature of $Ni_{1-x}Pt_x$ Si formation on leakage current characteristics of static random- access memory cells, IEEE Trans. Electron Devices 66 (1) (2019) 389–394, https://doi.org/10.1109/TED.2018.2878871.

[71] L. Esposito, S. Kerdilès, M. Gregoire, P. Benigni, K. Dabertrand, J.-G. Mattei, D. Mangelinck, Impact of nanosecond laser energy density on the C40-TiSi2 formation and C54-TiSi2 transformation temperature, J. Appl. Phys. 128 (2020), https://doi.org/10.1063/5.0016091, 085305.
[72] S. Teng, Z. Liang, C. Chou, Y. Tsai, P. Chiu, Y. Wu, Nearly epitaxial low-resistive co Germanide formed by atomic layer deposited cobalt and laser thermal annealing, IEEE Electron Device Lett. 41 (2) (2020) 272–275, https://doi.org/10.1109/LED.2019.2961382.
[73] X. Zhao, D. Vanderbilt, First-principles study of structural, vibrational, and lattice dielectric properties of hafnium oxide, Phys. Rev. B 65 (2002), https://doi.org/10.1103/PhysRevB.65.233106, 233106.
[74] S. Migita, et al., Design and demonstration of very high-k (k∼50) HfO2 for ultra-scaled Si CMOS, in: 2008 Symposium on VLSI Technology, Honolulu, HI, 2008, pp. 152–153, https://doi.org/10.1109/VLSIT.2008.4588599.
[75] T. Tabata, Nucleation and crystal growth in HfO2 thin films by UV nanosecond pulsed laser annealing, Appl. Phys. Express 13 (1) (2019), https://doi.org/10.7567/1882-0786/ab5ce2, 015509.
[76] L. Grenouillet, et al., Performance assessment of BEOL-integrated HfO2-based ferroelectric capacitors for FeRAM memory arrays, in: 2020 Symposium on VLSI Technology (VLSI Technology), 2020.
[77] D. Sil, et al., Ion implantation and laser annealing for toughening low-k dielectric in scaled-down Interconnects, in: 2019 IEEE International Interconnect Technology Conference (IITC), 2019.
[78] R.T.P. Lee, et al., Nanosecond laser anneal for BEOL performance boost in advanced FinFETs, in: 2018 IEEE Symposium on VLSI Technology, Honolulu, HI, 2018, pp. 61–62, https://doi.org/10.1109/VLSIT.2018.8510651.
[79] M.-J. Lee, et al., Stack friendly all-oxide 3D RRAM using GaInZnO peripheral TFT realized over glass substrates, in: 2008 IEEE International Electron Devices Meeting, San Francisco, CA, 2008, pp. 1–4, https://doi.org/10.1109/IEDM.2008.4796620.
[80] M.J. Marinella, V.V. Zhirnov, Emerging memory devices, assessment and benchmarking, in: A. Chen, J. Hutchby, V. Zhirnov, G. Bourianov (Eds.), Emerging Nanoelectronic Devices, first ed., John Wiley and Sons, 2015, pp. 246–276, https://doi.org/10.1002/9781118958254.ch13.
[81] A. Chen, Memory selector devices and crossbar array design: a modeling-based assessment, J. Comput. Electron. 16 (2017) 1186–1200, https://doi.org/10.1007/s10825-017-1059-7.
[82] D.S. Woo, DRAM: its challenging history and future, in: 2017 IEEE International Electron Devices Meeting (IEDM), San Francisco, CA, 2017. Short course 2.
[83] H. Chung, et al., Novel 4F2 DRAM cell with Vertical Pillar Transistor (VPT), in: 2011 Proceedings of the European Solid-State Device Research Conference (ESSDERC), Helsinki, 2011, pp. 211–214, https://doi.org/10.1109/ESSDERC.2011.6044197.
[84] Y. Cho, et al., Suppression of the floating-body effect of vertical-cell DRAM with the buried body engineering method, IEEE Trans. Electron Devices 65 (8) (2018) 3237–3242, https://doi.org/10.1109/TED.2018.2849106.
[85] D. R. Shepard, Vertical Switch Three-Dimensional Memory Array. U.S. Patent No. US 2010/0232200 A1, September 16, 2010.
[86] K. Huet, C. Boniface, R. Negru, J. Venturini, Ultra low thermal budget anneals for 3D memories: access device formation, AIP Conf. Proc. 1496 (2012) 135, https://doi.org/10.1063/1.4766508.
[87] J.G. Lisoni, et al., Laser thermal anneal of polysilicon channel to boost 3D memory performance, in: 2014 Symposium on VLSI Technology (VLSI-Technology): Digest of Technical Papers, Honolulu, HI, 2014, pp. 1–2, https://doi.org/10.1109/VLSIT.2014.6894346.

[88] G. Congedo, et al., Analysis of performance/variability trade-off in Macaroni-type 3-D NAND memory, in: 2014 IEEE 6th International Memory Workshop (IMW), Taipei, 2014, pp. 1–4, https://doi.org/10.1109/IMW.2014.6849381.

[89] K. Huet, I. Toqué-Tresonne, F. Mazzamuto, T. Emeraud, H. Besaucèle, Laser thermal annealing: a low thermal budget solution for advanced structures and new materials, in: 2014 International Workshop on Junction Technology (IWJT), Shanghai, 2014, pp. 1–6, https://doi.org/10.1109/IWJT.2014.6842020.

Chapter 5

# Materials science issues related to the fabrication of highly doped junctions by laser annealing of Group IV semiconductors

Ray Duffy[a], Enrico Napolitani[b], and Fuccio Cristiano[c]
[a]*Tyndall National Institute, University College Cork, Cork, Ireland,* [b]*Department of Physics and Astronomy, University of Padua, Padua, Italy,* [c]*LAAS-CNRS, University of Toulouse, Toulouse, France*

## 5.1 Introduction

The search for improved doping solutions aiming at achieving highly doped and shallow source and drain (S/D) regions in silicon-based MOS field-effect transistors has accompanied the evolution of the semiconductor industry since its emergence in the 1960s. This has supported the constant development of increasingly fast annealing methods able to remove the process-induced damage and limit dopant diffusion while achieving high temperatures and hence high activation rates. With the end of the "pure miniaturization era" that occurred during the last decade, mainly thanks to the appearance of multigate 3D device geometries such as FinFETs or nanowire-based transistors, the stringent requirements on the reduction of S/D thickness have been somehow relieved. Yet, the need for extremely high active dopant concentration has remained as crucial as before, especially for the optimization of the S/D contact resistivity.

Within this context, the use of pulsed laser annealing in the nanosecond regime, initially investigated in the late 1970s, has therefore regained a strong momentum in both the research and industry communities. This concerns in particular the investigation of the peculiar phenomena occurring in the melt annealing regime during the ultra-fast solidification of laser-induced liquid layers. Several of them were already precisely described in the exhaustive

Laser Annealing Processes in Semiconductor Technology https://doi.org/10.1016/B978-0-12-820255-5.00007-6

1985 review article by Cullis [1], while others were further investigated, understood, or discovered in the following years, such as the anomalous formation of extended defects during nanosecond laser anneals at temperatures just below the melt temperature (also discussed in Chapter 3 of this book). In contrast, longer laser annealing processes, typically in the millisecond regime, achieved, for instance, by continuous wave laser beams result in nonmelt annealing conditions under which the damage recovery and/or the formation of residual defects occur in the solid phase [2], similar to other "non-laser" millisecond or conventional RTA/furnace anneals (cf. [3] and references therein).

For these reasons, this chapter will be mainly devoted to the review of the phenomena occurring when highly doped Si, Ge, and SiGe layers are fabricated by nanosecond laser annealing in the melt regime. Those related to the laser kinetics (melt duration, melt and regrowth velocity, solute trapping...) or to the morphology, crystalline nature, and residual damage of the annealed structures following recrystallization will be reviewed in Section 5.2. The detailed review of the specific issues related to the achievement of highly doped junctions (dopant activation, redistribution and segregation, thermal stability...) will be discussed in the following sections, with a focus on the laser doping of both silicon (Section 5.3) and germanium (Section 5.4) substrates.

## 5.2 Structural investigations

This section is divided into two parts. In Section 5.2.1, the current understanding of the main phenomena related to laser kinetics in the melt regime will be first reviewed, including the determination of the reduced melt temperature of the amorphous phases (Section 5.2.1.1), the melt front propagation (Section 5.2.1.2), and the solute trapping phenomenon (Section 5.2.1.3).

The final morphology, crystalline nature, and residual damage of Si, Ge, and SiGe materials subjected to nanosecond laser anneals in the melt regime will be discussed in Section 5.2.2. The specific case of the onset of surface melting when the threshold energy for melting is reached will be first discussed in Section 5.2.2.1, while the partial and full melt-related phenomena will be presented in Section 5.2.2.2 with a focus on three specific material systems, namely, ion-implanted crystalline materials, amorphous layers, and strained SiGe alloys.

### 5.2.1 Melting and recrystallization kinetics

Historically, the use of preamorphization in conjunction with melt laser anneal was based on the lower melting temperature of amorphous Si (a-Si) or germanium (a-Ge) compared with its crystalline counterpart. In this way, it was possible to achieve an enhanced dopant activation only in a well-controlled melted layer, whose thickness was only determined by the preamorphizing implant conditions. Melting temperature determination is discussed in Section 5.2.1.1.

Whether occurring in amorphous or crystalline material, melting usually starts at the surface where the laser energy is more strongly absorbed. A melt front separating the surface liquid and the solid substrate forms and propagates into the solid down to a maximum depth that depends on both the material and the laser properties. Following the laser switch-off, thermal diffusion into the substrate induces a temperature reduction of the melt and eventually triggers the resolidification process. Melt front propagation, melt duration, and recrystallization velocity are the important parameters from both theoretical and applied points of view. They will be discussed in Section 5.2.1.2.

Finally, complex impurity redistribution is systematically observed after solidification of the melt. This is due to the nonequilibrium impurity segregation (solute trapping) occurring at the fast-moving liquid/solid interface during solidification. Some relevant experimental studies that evidenced this phenomenon will be reviewed in Section 5.2.1.3, with a specific reference to Ge segregation in laser-annealed SiGe alloys.

#### *5.2.1.1 Melting temperature (amorphous vs crystal)*

In early theoretical works, the melting temperature difference between amorphous and crystalline Si (or Ge) was inferred from thermodynamics arguments. Spaepen and Turnbull [4] and Bagley and Chen [5] independently investigated the growth kinetics of semiconductor crystals from either the rapid quenching of its melt phase or the heating of its amorphous phase. To this end, they calculated the temperature dependence of the free energy of the amorphous phase (with respect to the crystalline one) and compared it with the corresponding free energy curve of the liquid phase (again with respect to the crystalline one), so that the intersection between the two curves corresponds to the amorphous-phase melt temperature. This is schematically shown in the original drawing from Bagley and Chen (cf. Fig. 5.1A).

One of the critical parameters necessary for the calculation of the free energy curves is the temperature variation of the crystallization enthalpy of the amorphous phase, $\Delta H_{ac}$. This triggered a considerable amount of experiments aimed at the evaluation of $\Delta H_{ac}$. In a series of experiments based on differential scanning calorimetry (DSC) measurements [6–8], Donovan et al. provided the precise values of $\Delta H_{ac}$ for both Si and Ge amorphous materials, allowing reliable calculations of their respective melt temperatures. The value found for a-Ge (965 K) was very close to those initially estimated in Spaepen and Turnbull [4] (960–975 K) or in Bagley and Chen [5] (969 K) and about 20% lower than the melt temperature of c-Ge (1210 K [9]). As for silicon, the melt temperature of the amorphous phase was found at 1420 K, i.e., 16% lower than the melt temperature of c-Si (1687 K, cf [10] and refs. therein). The calculated phase diagram for silicon is reported in Fig. 5.1B.

Alternative methods for the determination of the a-Si melt temperature were also investigated. Thompson et al. [11] used transient conductance to directly

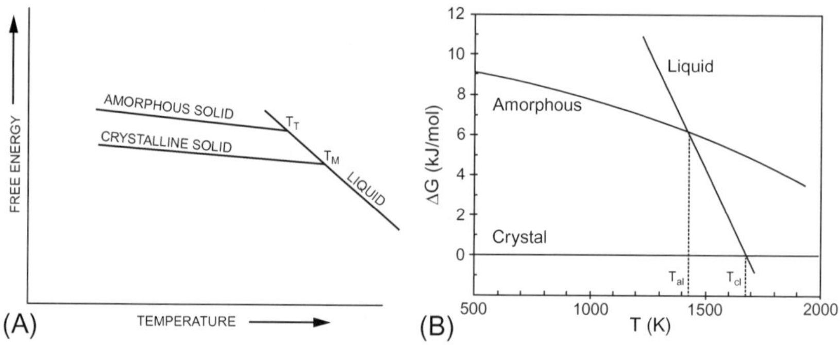


**FIG. 5.1** Phase diagrams of group IV semiconductors. The Gibbs free energy relationships between the solid crystalline, solid amorphous, and metallic liquid phases allow the determination of the melting temperature of both the amorphous and the crystalline solid phases. (A) Schematic representation from the original work of Bagley and Chen [5]. (B) Calculated values for silicon. *(From L. Pelaz, L.A. Marqués, J. Barbolla, Ion-beam-induced amorphization and recrystallization in silicon, J. Appl. Phys. 96 (2004) 5947–5976. https://doi.org/10.1063/1.1808484. Original data from E.P. Donovan, F. Spaepen, D. Turnbull, J.M. Poate, D.C. Jacobson, Calorimetric studies of crystallization and relaxation of amorphous Si and Ge prepared by ion implantation, J. Appl. Phys. 57 (1985) 1795–1804. https://doi.org/10.1063/1.334406.)*

measure the melt depth, in conjunction with time-resolved reflectivity (TRR) to follow the kinetics of the amorphous-liquid transition. The a-Si melt temperature found in this way was slightly higher than the one determined by DSC. In contrast, Baeri et al. [12] used electron beam irradiation in combination with RBS and TEM measurements and found a significantly lower value of the a-Si melt temperature. Finally, on the basis of TRR measurements, Kokorowski et al. [13] reported a melt temperature for a-Si almost identical to that of c-Si.

The extensive investigations carried out in the 1980s also evidenced the existence of different states for a-Si, whose properties strongly depend on its thermal history and/or preparation conditions. In particular, "unrelaxed" a-Si (typically formed by ion implantation) "relaxes" towards a lower free energy configuration during annealing, due to the rearrangement of the atomic network, leading to a decrease in the tetrahedral bond-angle distortion and a consequent short-range ordering [14,15]. The structural relaxation process is sufficiently fast to occur during the temperature ramp up of typical ns laser anneals. Only when ps anneals are carried out, unrelaxed a-Si can be brought up to its melt temperature, which is ~150 K lower than that of relaxed a-Si (~1270 K [16]). Comprehensive reviews of a-Si properties can be found in Spinella [15] and Pelaz [17].

As for SiGe alloys, the melt temperature of the crystalline alloy is intermediate between those of pure Si and pure Ge, as established from the Ge-Si phase diagram [18,19]. Similar to the pure elements case discussed above, the melt temperature of the amorphous SiGe phase is therefore expected to be lower than that of crystalline one. However, only few qualitative evidences of this behavior are available in the literature, which are based on the decrease in the laser energy threshold for the melt onset of a-SiGe layers. Examples are available for melting experiments investigated by transient conductance [20] or time-resolved reflectivity [21].

Finally, it is important to note how, in spite of the limited amount of experimental data available at that time, the early speculative works of Spaepen and Turnbull [4] and of Bagley and Chen [5] not only provided reliable estimations of both a-Ge and a-Si melt temperatures, but were also able to predict most of the main "practical" implications that need to be accounted for when applying ns laser anneal for dopant engineering. These included (i) the modification of the surface morphology at the onset of surface melt, (ii) the formation and displacement of a planar amorphous-liquid interface during melt and resolidification, (iii) the possibility to induce secondary melt through the latent heat which is released during solidification, and (iv) the nature of the solid phase (amorphous, single-, or polycrystalline), which is formed during quenching, depending on the solidification temperature and the cooling rate.

### *5.2.1.2 Melt front propagation, melt duration, and resolidification velocity*

Since the late 1970s, it was possible to predict the main features of melt dynamics on the basis of simple heat-flow calculations [22,23]. Fig. 5.2A gives an example of such calculations by tracing the melt depth as a function of time during (and after) a pulsed laser irradiation [22] (see Chapters 3 and 6 of this book for a review of the most advanced atomistic and continuum physical models applied to laser annealing). The combined effect of two competing

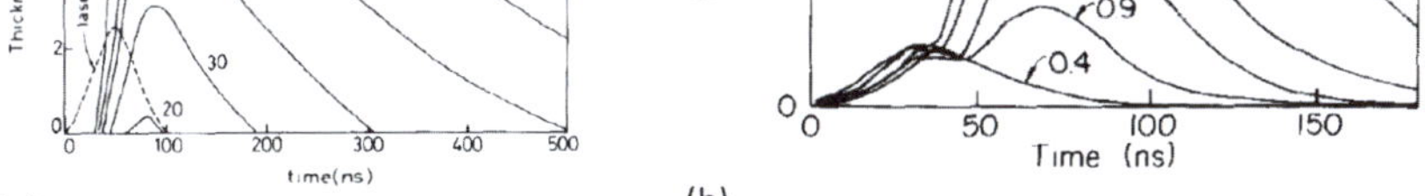


(a) (b)

**FIG. 5.2** Melt kinetics investigations: melt depth vs time obtained by (A) heat-flow calculations for amorphous silicon irradiated with a ruby laser ($\lambda = 694$ nm) with 50 nsec pulse and different power densities [22] and (B) measurements by transient conductance for crystalline silicon irradiated with a ruby laser ($\lambda = 694$ nm) with 20-nsec pulse and different energy densities [24]. The small common peak at $\sim$35 ns corresponds to the photoconductive current and is not related to melting.

processes (surface heating induced by laser light absorption and temperature decrease due to heat diffusion into the substrate) results in the fast increase of the melt depth at the beginning of the anneal, followed by a relatively slower decrease during solidification. Several important parameters for the understanding and the optimization of the laser annealing process can be directly extracted from such curves, including melt-in speed, maximum melt depth (MMD), melt duration, and solidification velocity.

The experimental investigation of melt dynamics is necessarily based on the real-time measurement techniques, which are eventually supported by ex situ measurements (such as Rutherford backscattering spectrometry, RBS, or transmission electron microscopy, TEM) for calibration purposes or in order to provide additional information.

The earliest comprehensive investigation of melt and crystallization kinetics was carried out by Auston and co-workers in the late 1970s [2] (see also [25] for a more general review). Their works were mainly based on TRR measurements, where the higher reflectivity of liquid Si compared with the solid phase allows distinguishing between the two phases during the anneal. Auston et al. [26] precisely measured the melt duration as a function of laser energy density during ns laser anneal of amorphous silicon. This provided a conclusive evidence for the melting of the surface of ion-implanted semiconductors during ns laser annealing (under debate at that time), in contrast to continuous wave (CW) laser anneals [27], where regrowth was found to occur, without melting, through solid-phase epitaxy (SPE). These works were later extended to other materials (including Ge and GaAs) [28], where the melt duration was measured for different laser wavelengths and energy densities.

Considerable improvements in the TRR technique were later introduced, mainly applied to the investigation of poly-Si formation from thin amorphous films (such studies were driven by the considerable progress of the excimer laser technology for applications in the display industry, see Chapter 1 of this book). Examples of these improvements can be found in Boneberg and Leiderer [29] or Voogt [30], where two different lasers were used to monitor the TRR signal at both the surface and the interface between the amorphous film and the substrate. Alternatively, a single laser was used to simultaneously monitor the reflected and the transmitted signal during the laser anneal [31]. Nevertheless, due to the limited penetration depth of the laser radiation used for TRR measurements, the measurement of the melt depth was inaccessible for this technique, while the estimation of melt-in and solidification velocities remained unreliable.

A significant advance in the real-time investigation of melt and recrystallization kinetics was provided by Galvin et al. [32,33] who developed a time-resolved electrical conductance technique (also referred to as transient conductance, TC). Thanks to the increased electrical conductance of liquid (metallic) silicon compared with the solid phase, the measurements of the electrical conductance of the sample during laser annealing provide a measure of the volume of molten material as a function of time. For a 30-ns pulse obtained with a Q-switched ruby laser, solidification velocities in the range 2 to 3 m/s were reliably measured for the first time on (111) oriented silicon substrates. In addition, the regrowth velocity was found to decrease with increasing laser energy, in agreement with the decrease in the thermal gradients induced by thermal diffusion in the late stages of the anneal.

Subsequent work from the same group [24,34] introduced an improved version of the TC technique, where Au-doped (111) Si bulk wafers were replaced by ~0.5-μm-thick (100) SOS (silicon-on-sapphire) substrates, in order to decrease the carrier lifetime and hence reduce the strong photoconductive current signal appearing at the beginning of the anneal. In such a way, the evolution of the melt depth during anneal was experimentally accessible over the (almost) entire time range, as shown in Fig. 5.2B, where the melt-related signal is affected by the small photoconductive peak only in the first ~40 ns of the anneal. The theoretical picture described by simple heat-flow calculations (cf. Fig. 5.2A) could therefore be confirmed. For a 25-ns pulse obtained with a Q-switched ruby laser, the melt-in speed was found to scale linearly with incident laser energy, with measured values in the interval 5–13 m/s, compared with relatively constant (and lower) solidification velocities of ~3 m/s [24]. A similar behavior was observed with shorter laser pulses (2.5 to 10 ns) [34], but with much higher melt-in speeds (up to 40 m/s). In fact, by minimizing heat losses during light absorption, melt-in speeds as high as 200 m/s were measured. The corresponding solidification velocities were also found to increase compared with longer laser pulses, thanks to the increased thermal gradients associated with extremely short pulse durations. Solidification velocities values up to 10 m/s were recorded for a 10-ns pulse, and up to ~20 m/s for a 2.5-ns pulse. The impact of crystal orientation on the regrowth velocity was investigated in detail by Cullis and co-workers [1]. A discussion about this effect, including a comparison with related atomistic simulations, is given in Chapter 3 (Section 4.1).

The transient conductance technique has also been utilized to probe the melting and solidification of amorphous and polycrystalline SiGe layers deposited on $SiO_2$ substrates under UV laser irradiation (20-ns pulse duration) [20]. Measured solidification velocities were reported in the 0.5–2.5 m/s range, i.e., below the maximum theoretical value for SiGe (~10 m/s), indicating that solidification occurred in moderately undercooled liquid. In addition, in the partial-melt regime (melt thickness lower than that of the initial SiGe layer), the solidification velocity was found to increase with laser energy, due to a progressive increase of the degree of supercooling of the liquid. In contrast, in the full-melt regime, the solidification velocity exhibited the expected reduction with increasing laser energy, due to the increased thermal diffusion and the consequent reduction of temperature gradients.

Finally, the progressive development of dynamic TEM (DTEM) methods has made it possible to further investigate laser annealing-related phenomena in real time, by simultaneously providing structural observations and crystallization kinetics. For instance, this technique has been extensively applied to the investigation of the explosive crystallization (cf. Section "Partial melt of amorphous layers: Explosive crystallization" of this chapter) in Ge since the 1980s [35–38]. In a recent work from Santala [39], TEM images were recorded at 95-ns time intervals following a laser irradiation of thin Ge amorphous layers obtained with a frequency-doubled Nd:YAG laser (532 nm) and a 12-ns pulse duration. Such ultra-fast imaging rate allowed measuring "explosive" crystallization velocities in the range 6–13 m/s.

### 5.2.1.3 Dopant segregation

Since the very first studies published in this field, it was clear that the enhanced dopant activation achievable by melt laser annealing, well in excess of the solid solubility, was accompanied by a complex impurity redistribution, which is at the origin of the final shape of the dopant depth profiles and often results in a considerable dopant enrichment at the silicon surface. This redistribution was rapidly explained in terms of a nonequilibrium impurity segregation (solute trapping) occurring at the fast-moving liquid/solid interface during solidification (see [40] and references therein). It is important to note that, in the work of Leamy et al., the demonstration of the nonequilibrium nature of the solute trapping mechanism was supported by the fact that concentrations in excess of the maximum solid solubility were obtained for dopants exhibiting a retrograde solubility behavior (Ga, In, Bi, Sb, As…). An example is given in Fig. 5.3A, showing that a concentration of substitutional Gallium as high as $\sim 4 \times 10^{20}\ cm^{-3}$ (i.e., a decade more than the maximum solid solubility) was obtained following melt laser annealing (Nd:YAG). Similar results were obtained by Cullis et al. [41] in the case of platinum, who registered an effective solubility enhancement of 3 orders of magnitude compared with the equilibrium solubility value.

Based on these early works, it was therefore proposed that the interfacial segregation coefficient, $k = C_S/C_L$, defined as the ratio of the solute concentration in the solid to that in the liquid at the moving interface, differs from the equilibrium value, $k_0$, and varies as a function of the liquid/solid interface velocity. For typical $k_0$ values lower than unity, the effective segregation coefficient therefore increases as a function of the l/s interface velocity. A detailed experimental investigation of this effect was later provided by Baeri et al. [42], who studied the trapping and segregation of In in silicon for different solidification velocities obtained by changing the temperature of the substrate during irradiation and/or the laser pulse duration (15 to 50 ns, using a ruby laser). Within the entire explored solidification velocity range (1.8 to 5.2 m/s), the authors demonstrated that each of the measured in-depth profiles could be fitted with a unique, velocity-dependent, interfacial segregation coefficient. In particular, for an interface velocity increase from 2 to 5 m/s, the In substitutional fraction

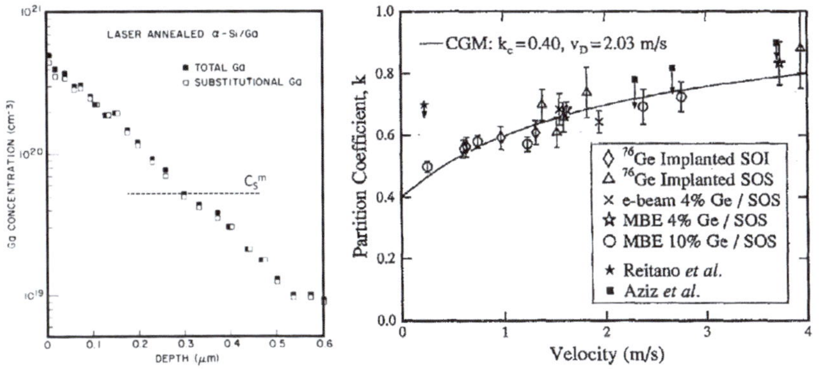


(a) (b)

**FIG. 5.3** (A) Gallium depth distribution in silicon obtained following multiple melt laser irradiations of a Ga-doped a-Si layer deposited on a c-Si substrate. The totality of the in-diffused Ga atoms is in substitutional position, as determined by RBS. In the near-surface region, the Ga concentration largely exceeds the maximum solid solubility, $C_S^m$. (B) germanium partition coefficient as a function of solidification velocity: comparison between various experimentally determined values *(symbols)* and a fit curve *(solid line)* based on the dilute continuous growth model. *(A: Taken from H.J. Leamy, J.C. Bean, J.M. Poate, G.K. Celler, Nonequilibrium incorporation of impurities during rapid solidification, J. Cryst. Growth. 48 (1980) 379–382. https://doi.org/10.1016/0022-0248(80)90032-9. B: Taken from D.P. Brunco, M.O. Thompson, D.E. Hoglund, M.J. Aziz, H.J. Gossmann, Germanium partitioning in silicon during rapid solidification, J. Appl. Phys. 78 (1995) 1575–1582. https://doi.org/10.1063/1.360251.)*

was found to increase from less than 5% to ~60%. The continuous growth model (CGM) was later proposed by Aziz et al. [43,44], to predict the velocity dependence of the segregation coefficient, while an extension of this model [45] allowed to account for the additional dependence on the crystallographic orientation of the substrate. The latter effect was experimentally investigated by Reitano et al., for several impurities from Groups III, IV, and V in silicon [46,47], who validated the theoretical predictions (nonequilibrium segregation modeling is discussed in more detail in Chapters 3 and 6 of this book).

Additional segregation-related results concerning relevant dopant impurities for both Si and Ge are discussed in detail in Sections 5.3 and 5.4 of this chapter, while an equally important case is that of germanium segregation during melt laser annealing of SiGe alloys. Indeed, the strong Ge redistribution occurring during the solidification step critically modifies the local alloy composition. This has an impact on the final physical properties of the alloy (including carrier mobility) and may eventually lead to its elastic relaxation. The occurrence of Ge segregation during the solidification of laser-melted SiGe layers was experimentally observed by several groups in the early 1990s, e.g., Lombardo et al. [48,49] for SiGe alloys with up to 40% at. Ge content and fabricated by laser-assisted e-beam evaporation (XeCl laser, 30-ns pulse duration); Berti et al. [50] for alloys with up to 18% at. Ge content and fabricated by high-dose Ge implantation (ruby laser, 20 ns pulse or XeCl laser, ~45 ns pulse); Repplinger [51] for a similar experiment with alloys with up to ~20% at. Ge content, also fabricated by high-dose Ge implantation (ArF laser, 20-ns pulse). Subsequently, a comprehensive picture of the variation of the Ge segregation coefficient $k$ as a function of solidification velocity was provided by Brunco et al. [52], who used UV excimer irradiation (XeCl, 30-ns pulse duration) for their study. An example is shown in Fig. 5.3B, where several values of $k$ are plotted as a function of the corresponding solidification velocity for SiGe alloys with a Ge content up to 10% at. and fabricated using different methods (ion implantation, MBE, e-beam evaporation). The segregation coefficient values, extracted using the continuous growth model in the dilute approximation, appear to be independent from the used fabrication method. In addition, the extrapolation of the $k$ curve to zero velocity allows an estimation of the equilibrium segregation coefficient $k_e$. On the basis of several fitting tests done using different physical models, a value of $k_e = 0.45 \pm 0.02$ was proposed in that work, i.e., slightly different from the one reported in the particular case shown in Fig. 5.3B, while, for higher Ge contents (above 20% at.), the nondilute CGM predicted a slight increase of the $k(v)$ values. Finally, Weizman et al. [20] observed that, similarly to what is found for low-solubility impurities (cf. Section 5.3.2.3 of this chapter), liquid/solid interface instability can develop during fast recrystallization of SiGe, leading to lateral Ge segregation within the moving interface and eventually to the formation of surface ripples in the recrystallized material. The formation of segregation-related surface ripples was only observed in SiGe alloys with the initial Ge content in the range 30%–70% at., with lateral variations of up to 15% at. in SiGe alloys with an initial Ge content of 35% at.

### 5.2.2 Laser-induced damage

The final morphology, crystalline nature, and residual damage of Si, Ge, and SiGe materials submitted to nanosecond laser anneals in the melt regime depend on the combination of several factors, including both the material fabrication history (deposition on crystalline or glass substrates, in situ or ion implantation doping, amorphization…) and the laser anneal conditions (laser energy, pulse duration, partial- or full-melt regime…).

The specific case of the localized surface melting when the threshold energy for melting is reached will be first discussed in Section 5.2.2.1. For materials systems in which a continuous liquid layer is formed instead upon laser anneal, the nature and crystalline quality of the regrown material also depends on whether or not the layer under investigation is partially or entirely melted by the laser irradiation. This will be discussed in Section 5.2.2.2 with a focus on three specific material systems, namely, ion-implanted crystalline materials, amorphous layers, and strained SiGe alloys. We note that, for all the case here discussed, laser-induced melt initiates at the surface, where the laser energy is more strongly absorbed.

#### *5.2.2.1 Melt threshold: Localized surface melt*

The melting kinetics process is usually described assuming the simplified scheme that when the melt temperature is reached at the surface, a continuous thin liquid layer forms there. However, early investigations carried out with $CO_2$ CW lasers on both Ge [53] and Si [54] already indicated that, at the melting threshold, solid and liquid phases coexist at the surface before a continuous liquid layer is formed. This was later confirmed by Young et al. [55] who conducted a detailed investigation of the formation of periodic structures on Ge surface, induced by 20-ns laser pulses (Q-switched Nd:YAG, $\lambda = 1.06$-μm). They provided several experimental evidences that random surface local melting occurs at the melt onset, including (i) TRR measurements showing that, at the melt threshold, the surface reflectivity never reached the characteristic value of a uniformly molten surface, or (ii) SEM images, such as the one presented in Fig. 5.4A, where a "pebble-like" structure is formed at the surface for laser conditions close to the melt threshold. It was proposed that each of these "pebbles" or "hillocks" is generated by the resolidification of a locally molten area, which has the net effect of raising its central portion above the original surface level. The coexistence of solid and liquid regions at the melt onset was further confirmed by Jellison et al. [56] on the basis of both TRR and ellipsometry (TRE) measurements. Similar results were also obtained in works where the surface melt was induced by non-laser-based processes, such as electron beam irradiation of Si or Ge [57], or flash lamp annealing (FLA) of silicon [58,59], silicon covered by a thin SiC layer [60], or $SiO_2$/Ge stacks [61].

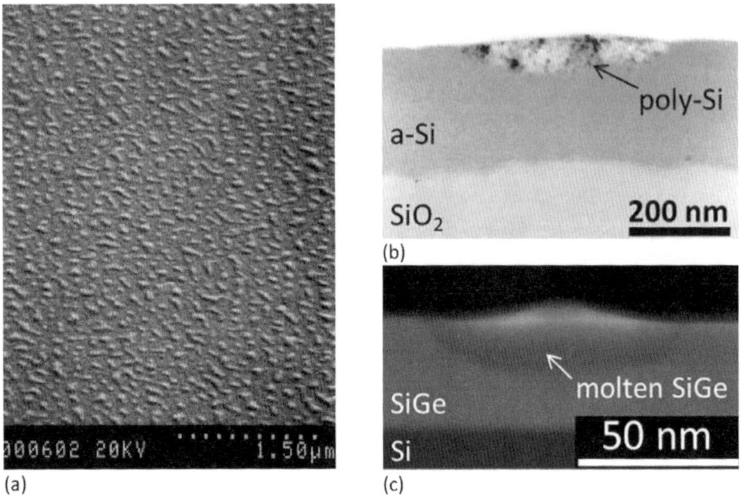


**FIG. 5.4** Illustrations of random surface melting obtained by electron microscopy techniques in different material systems. (A) Crystalline silicon (SEM) [55]. (B) Amorphous silicon (Bright Field TEM) [30]. (C) Silicon-Germanium alloys (STEM-HAADF) [62].

Several factors have been invoked to explain the origin of the random surface melt mechanism, including inhomogeneities in the spatial distribution of the laser radiation [56], or the nucleation process inherent to the first-order solid-liquid phase transition, in which liquid "seeds" first nucleate and subsequently grow until a continuous liquid layer is formed [60]. The latter process is further supported by the fact that the free surface may provide sites for heterogeneous nucleation to occur, which locally reduces the temperature needed to initiate the nucleation process [59,63]. Indeed, recent molecular dynamics simulations [64] take this mechanism into account to predict the surface melting of laser-irradiated crystalline silicon.

All the works considered so far were conducted in homogeneous crystalline Si or Ge substrates. In this case, the random nature of the surface melt process had to be inferred from investigations of the surface morphology or from time-resolved measurements recorded during the anneal. In contrast, material systems where the surface layer undergoing the laser anneal has different properties compared with the substrate (i.e., crystallographic phase or chemical composition) may provide a direct evidence that the melting process starts with the formation of isolated liquid islands thanks to TEM analysis. Two examples (both obtained with a 308-nm XeCl laser) are shown in Fig. 5.4. In the case of a thick Si amorphous layer [30] (cf. Fig. 5.4B), the surface random melt (induced by a 56-ns laser pulse) results in the formation of isolated poly-Si islands, which are easily distinguished from the surrounding un-melted amorphous material in bright field TEM images. In the case of a SiGe layer grown on a Si substrate [62] (cf. Fig. 5.4C), the Ge redistribution (cf. Section 5.2.1.3) occurring during resolidification of an isolated melted SiGe region obtained with a 146-ns laser pulse can be perfectly visualized by STEM-HAADF images, whose contrast is proportional to the atomic number of observed material. Indeed, both the 2D profile of the liquid/solid interface at the MMD and the Ge-rich raised tip at the center of the solidified area can easily be identified in such images. In particular, it appears that, even at the onset of the surface melt, the thickness of an isolated liquid droplet is $\sim$10 nm, i.e., the same order as the extremely thin layers which are currently investigated for the state-of-the-art and future semiconductor technology (including UT-SOI, FinFETs, and NW-based devices). In view of developing reliable melt laser processes for their application to such thin materials, it is therefore important to understand the random surface melt mechanism and identify solutions for limiting its impact. In fact, in the case of 3D devices, a more "deterministic" scenario could emerge due to the 3D temperature fields, which could have strong peaks in the interesting device zone (see Chapter 6 for a more detailed discussion).

### 5.2.2.2 *Partial- and full-melt regimes*

For laser conditions that largely exceed the threshold melt condition, it can be safely assumed that a surface continuous liquid layer is formed, which subsequently propagates into the solid down to a maximum depth. The nature and crystalline quality of the regrown material depends on several laser- and -material-related parameters. In practice, it is convenient to classify the different cases according to the material system under investigation.

The simplest case is obviously that of an initially undamaged elemental (Si or Ge) single-crystal substrate. In this case, once the continuous liquid layer reaches the MMD, resolidification starts from a perfect substrate and typically occurs through the liquid-phase epitaxy (LPE), therefore leading to defect-free crystalline materials [34]. A deviation from this behavior is observed only when regrowth occurs at extremely high speeds. In such situations, it is expected that the high interface velocities exceed the rate at which atoms can rearrange as a crystal at the solidifying interface [25].

Early evidence of crystalline silicon regrowing into amorphous phase following the fast melt solidification was supplied by experiments run with both picosecond laser pulses (30 psec with wavelengths of 532 nm or 266 nm [65]) and nanosecond laser pulses (10 nsec with wavelengths of 266 nm [66]). However, in these works, the expected fast solidification velocities were only inferred from the estimated cooling rate achievable with such lasers pulses ($10^{10}$–$10^{11}$°C/s in the ns regime or rates up to $10^{14}$°C/s in the ps regime).

More complete studies were later conducted when reliable methods for the measurement of the solidification velocity were established (cf. Section 5.2.1.2). In particular, using a frequency-doubled Q-switched ruby laser ($\lambda = 347$ nm), Thompson et al. [34] showed that for short laser pulses (10 ns and less), high solidification velocities were found compared with longer laser pulses, with measured values up to 10 m/s for a 10-ns pulse, and up to ~20 m/s for a 2.5-ns pulse (compared with ~3 m/s for 25-ns laser pulses) [34]. In particular, for solidification velocities higher than ~15 m/s, TRR profiles indicated that the surface region of the solidified liquid becomes amorphous. This result was interpreted on the basis of a strong undercooling of the melt below the a-Si melt temperature just before solidification [34], while the wavy nature of the amorphous-crystalline interface ([1,67] for a more extensive discussion) suggested that the formation of the amorphous phase is initiated by interface breakdown at high velocities.

In the next subsections, we will consider material systems where the surface layer undergoing the laser anneal has different properties compared with the substrate (i.e., different damage level, crystallographic phase, or chemical composition). In this case, the characteristics of the regrown material strongly depend on whether or not the surface layer under investigation is partially or entirely melted by the laser irradiation.

## Extended defects in ion-implanted damaged layers

Let us consider first a silicon matrix containing a surface damage layer induced by ion implantation. For implant conditions below the amorphization threshold, the material is essentially in the crystalline phase, with a damage region typically centered around the mean projected range of the implanted species and consisting of high concentrations of point defects, defect complexes, and locally amorphous regions. Early investigations carried out on such materials using Q-switched ruby lasers ($\lambda = 1094$ nm) [68–70] already revealed the formation of various types of extended defects upon irradiation. The presence of elongated dislocations [69] was associated with a defective regrowth mechanism occurring when the melt front at the deepest penetration intersects the implantation-induced damage layer. In contrast, the observation of dislocation loops localized at the depth of maximum implant damage [70] was ascribed to a defect generation mechanism occurring in a "quasimolten" state, i.e., just below the melt temperature. Similar loops were also observed by Earles et al. [71], who specifically investigated the defect formation in $B^+$ ion-implanted silicon following laser anneal in nonmelt conditions obtained using a XeCl (308-nm) pulsed laser (15 ns) with energy densities below or equal to 0.6 $J/cm^2$. Following a single laser pulse, the presence of "small defects" was inferred from the nonuniform contrast appearing in dark field TEM images. In contrast, for multiple laser shots (ten pulses at 0.55 $J/cm^2$), dislocation loops were directly observed in TEM images.

More recently, a detailed investigation about the nature of the loops formed in such "quasimolten state" was performed by Qiu et al. [72,73] (XeCl laser, $\lambda = 308$ nm, pulse duration $<200$ ns). In particular, the implant conditions were chosen in order to create relatively wide damaged regions (below the amorphization threshold) extending over $\sim$150 nm in depth, either by $B^+$ or by $Si^+$ implants. By varying the laser energy density and the corresponding melt depth, it was therefore possible to extensively investigate the formation of defects in the nonmelted regions just below the l/s interface (the melted regions being of excellent crystalline quality and defect-free upon solidification). An example is shown in Fig. 5.5A, taken from a $Si^+$-implanted sample melted to a depth of $\sim$100 nm. Defects are formed only below the depth corresponding to the thickness of the melted region. As shown in the inset, they exhibit the typical contrast of dislocation loops. In addition, their size (up to $\sim$10 nm in the upper limit of the defect layer) decreases with increasing depth, in agreement with the rapid decrease of the substrate temperature below the melted region. Further TEM analysis of these defects based on high-resolution imaging (cf. Fig. 5.5B and C) and geometrical phase analysis (GPA) unambiguously allowed to establish that the loops have a (001) habit plane (parallel to the surface) and a Burgers vector equal to $\sim$1/4 [001].

Such "unconventional" (001) loops (with respect to the {111} dislocation loops or {311} rodlike defects typically formed in ion implanted silicon) were equally observed in both $Si^+$- and $B^+$-implanted samples, therefore excluding any role of Boron-related clustering in their formation. Further investigation of their formation in SOI substrates [74] also excluded "thermal stress"-induced modification of the defect formation energy as a possible explanation of their

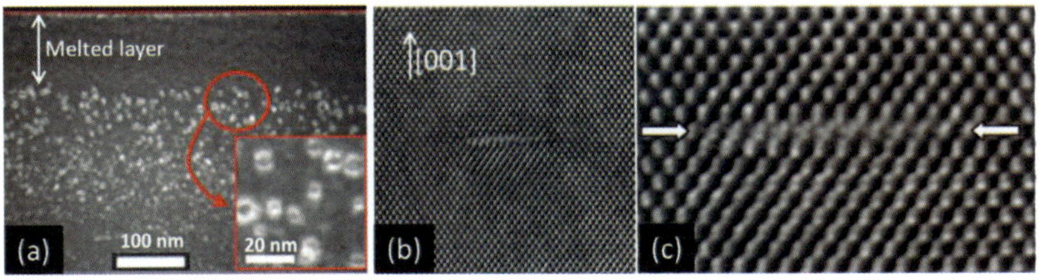


**FIG. 5.5** Cross-sectional TEM analysis of $Si^{+}$ and $B^{+}$-implanted silicon samples (non-amorphizing conditions) following a 1-pulse laser anneal at 2.6 J/cm$^2$ showing the formation of (001) dislocation loops. (A) Weak beam dark field analysis of the $Si^{+}$-implanted sample. Surface is indicated by the *red dashed line*. (B) HRTEM image in [-110] orientation of the $B^{+}$-implanted sample; (C) enlarged view of the dislocation imaged in (B). The habit plane is (001). *(From Y. Qiu, F. Cristiano, K. Huet, F. Mazzamuto, G. Fisicaro, A. La Magna, M. Quillec, N. Cherkashin, H. Wang, S. Duguay, D. Blavette, Extended defects formation in nanosecond laser-annealed ion implanted silicon, Nano Lett. 14 (2014) 1769–1775. https://doi.org/10.1021/nl4042438.)*

formation. On the basis of extended molecular dynamics calculations (discussed in detail in Chapter 3 of this book), Marqués et al. [74,75] finally demonstrated that at temperatures close to the melting point, self-interstitial rich silicon is driven into dense liquid-like droplets that are highly mobile within the solid crystalline Si matrix. These liquid droplets grow by a coalescence mechanism and eventually transform into {001} loops through a liquid-to-solid-phase transition in the nanosecond timescale.

For highly damaged ion-implanted materials, the generated point defects, defect clusters, and amorphous pockets can progressively accumulate until the highly damaged crystalline material becomes unstable and transforms into the amorphous phase. The most common case is that of a continuous surface amorphous layer separated from the perfect crystalline substrate by a thin damaged (but still crystalline) "end-of-range" layer located just below the amorphous/crystalline (*a*/*c*) interface.

The "end-of-range" (EOR) layer is conceptually similar to the damaged layers created by non-amorphizing implants and discussed in previous paragraphs. Hence, when a laser anneal is performed in melt regime and the l/s interface at MMD falls inside the EOR layer, the layer regrows in the crystalline phase, but defects are expected to be formed upon solidification, similar to those discussed above, i.e., elongated threading dislocations and small loops. However, most of the investigations in amorphized materials mainly focused only on the elongated dislocations formed when the solidification starts from the *a*/c interface [41,76]. In this case, their formation was associated with a defective regrowth mechanism caused by the roughness of the *a*/*c* interface [77,78]. In particular, for a Si layer self-amorphized with $Si^{+}$ ions and annealed with a XeCl UV laser ($\lambda = 308$ nm, 15-nsec pulse), Kuryliw [79] showed that a thermal treatment at very low temperature (450°C) prior to laser anneal resulted in the reduction of both the *a*/*c* interface roughness and the dislocation density.

## Partial melt of amorphous layers: Explosive crystallization

The resolidification of an amorphous layer occurs differently when the layer is only partially melted by a laser anneal. In this case, no crystalline seeds are available at the l/s interface at the onset of resolidification, which therefore proceeds by random nucleation of crystalline grains, leading to the formation of a continuous surface polycrystalline layer, separated from the crystalline substrate by the remaining nonmelted amorphous layer. In fact, this scenario was observed only for low laser energies [41]. Instead, in most cases, the thickness of the polycrystalline layer was found to rapidly reach the original *a*/*c* interface when increasing the laser energy density and remained unchanged upon further energy increase until the threshold for single-crystal regrowth was reached [41,80].

This behavior was initially ascribed to the lower values of the melt temperature and the thermal conductivity of the amorphous phase compared with the crystalline one [81]. However, all these early works assumed that the thickness of the polycrystalline material (determined by TEM) always corresponded to the laser-induced melt depth. On the contrary, later observations carried out by Thompson et al. [11] clearly demonstrated that the "primary" melt depth (as measured by transient conductance) induced by laser irradiation (Q-switched ruby laser, pulse duration: 28 ns) of a partially amorphized silicon layer is systematically smaller than the corresponding polycrystalline thickness (measured by TEM), which instead results from the recrystallization of a "secondary" buried molten layer, also visible from transient conductance measurements (cf. Fig. 5.6A), while the two quantities converge towards the *a*/*c*/ interface depth at higher energies. The same result was found by Sinke and Saris [82] (cf. Fig. 5.6B) from RBS measurements of Cu segregation in laser-melted Si amorphous layers (Q-switched ruby laser, pulse duration: 20 ns). Based on these results, Thompson and co-workers unambiguously explained the solidification mechanism of a partially melted amorphous layer in terms of an explosive crystallization (EC) process, mediated by a thin undercooled liquid layer propagating within the recrystallizing material at a velocity larger than ~10 m/s. In such a process, the latent heat released from the solidification of the primary melt into poly-Si is sufficient to melt a thin underlying layer of amorphous material. This undercooled liquid rapidly solidifies by liberating in turn the corresponding latent heat. A self-sustained EC process therefore continues until the heat losses in the substrate inhibit the temperature of the remaining amorphous material to raise above the melt threshold. The formation and the motion of a buried molted liquid propagating away from the surface during EC was later confirmed by the rapid oscillations of the TRR profiles, originating from the interfering reflections from the sample surface and from the moving buried layer [83,84]. The corresponding crystallization velocities were in the range 13–14 m/s, in agreement with those estimated in Thompson [11]. In addition, Z-contrast TEM analysis of Cu impurities "transported" by the buried layer during EC allowed estimating an upper limit for its thickness (~15 nm) [84].

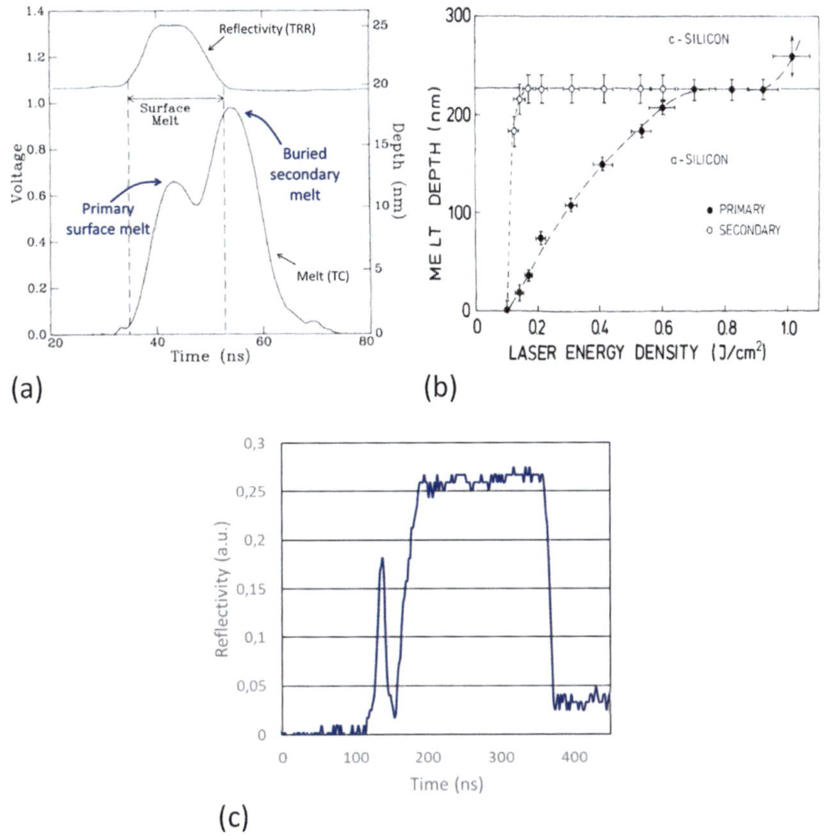


**FIG. 5.6** Experimental evidence of the occurrence of explosive crystallization (EC) in different material systems using different characterization methods. (A) Evolution for the time-resolved surface reflectivity, TRR *(top)* and transient conductance, TC *(bottom)* of an a-Si layer submitted to a 28-ns Q-switched ruby laser pulse. The occurrence of EC is evidenced by the second peak of the TC signal (buried secondary melt). (B) Depths of the primary and secondary EC-related melts in a-Si samples formed by $Cu^+$ implantation as a function of the laser energy density (20-ns Q-switched ruby laser pulse). The occurrence of EC is inferred from the observation of a double-peak structure in the depth distribution of Cu impurities after laser anneal, as measured by Rutherford backscattering spectrometry (RBS). (C) TRR signal evolution during laser irradiation (~160-ns XeCl laser pulse) of an a-Ge layer formed by $P^+$ implantation. The occurrence of EC is evidenced by the observation of a sharp reflectivity peak to the laser-induced second melt of the recrystallized layer. *(A: Taken from M.O. Thompson, G.J. Galvin, J.W. Mayer, P.S. Peercy, J.M. Poate, D.C. Jacobson, A.G. Cullis, N.G. Chew, Melting temperature and explosive crystallization of amorphous silicon during pulsed laser irradiation, Phys. Rev. Lett. 52 (1984) 2360–2363. https://doi.org/10.1103/PhysRevLett.52.2360. B: Taken from W. Sinke, F.W. Saris, Evidence for a self-propagating melt in amorphous silicon upon pulsed-laser irradiation, Phys. Rev. Lett. 53 (1984) 2121–2124. https://doi.org/10.1103/PhysRevLett.53.2121. C: Adapted from S.F. Lombardo, S. Boninelli, F. Cristiano, I. Deretzis, M.G. Grimaldi, K. Huet, E. Napolitani, A. La Magna, Phase field model of the nanoscale evolution during the explosive crystallization phenomenon, J. Appl. Phys. 123 (2018) 105105. https://doi.org/10.1063/1.5008362.)*

The same conclusions were drawn for similar experiments in which the starting silicon amorphous material is deposited on a $SiO_2$ substrate by CVD instead of being formed by ion implantation into a bulk crystalline material [85–87]. In particular, Voogt et al. [30] provided a TRR-based evidence of the formation of a buried liquid formation during EC in such layers ("secondary" melt). In addition, they showed that if the whole EC process is completed while the laser irradiation is still active, a second melt process of the recrystallized layer can occur, this time at a higher temperature that is close to the melting point of c-Si. The time separation between the first and the second melt events was found to strongly depend on the laser pulse duration. Indeed, for excessively short laser pulses, the primary melt might be so rapid that the entire amorphous layer is melted prior to the onset of EC (an example of this behavior can be found in Bruines [88] where EC was not observed following laser irradiation with 7.5-ns pulses). Finally, the investigation of EC in CVD-grown amorphous layers strongly contributed to the uncovering of the super lateral growth (SLG) mechanism, by which extremely large crystal grains can be obtained in the solidified material [86]. This laser-induced mechanism is discussed in detail in Chapter 1 of this book and is at the heart of the application of poly-Si layers in TFT technology. As for the predictive simulations of this phenomenon, a state-of-the-art phase-field model is discussed in Chapter 6 of this book [89], which is applied to the case of germanium samples amorphized by $P^+$ implants and annealed with a XeCl laser ($\lambda = 308$ nm, pulse duration $\sim$150 ns). In this case, the occurrence of EC is observed by TRR measurements, as shown in Fig. 5.6C, where the sharp reflectivity peak corresponds to the EC step prior to the laser-induced second melt of the recrystallized layer.

Concerning the investigation of EC in SiGe alloys, very few reports are available in the literature. The results of an unpublished work from Yu [90] are briefly discussed in two theoretical papers [91,92] in which molecular dynamics (MD) simulations were implemented to estimate the recrystallization speed during EC of SiGe alloys. In her original work [90], Yu investigated vertical EC in Si-rich $Si_{1-x}Ge_x$ alloys, with x ranging from 0 to 0.35, and found that the EC speed exhibits a strong nonlinear decrease when increasing Ge content. A similar behavior was observed in Ge-rich $Si_{1-x}Ge_x$ alloys when the Ge fraction is slightly decreased from 1 to 0.975 and 0.95. In contrast, the experiments were unable to observe EC in the composition range $0.90 < x < 0.95$. In a qualitative agreement with the experiments, the MD calculations of Albenze et al. [91] also predicted a strong nonlinear decrease in the crystallization speed as a function of the alloy composition, although the EC was expected to occur in the whole composition range. Indeed, the occurrence of EC in laser-annealed Ge-rich amorphous SiGe layers deposited on a glass substrate was inferred in more recent experiments based on TEM analysis ([93], $x_{Ge} = 0.67$) or a combination of SEM and transient conductance measurements ([20], $x_{Ge} = 0.50$). In particular, both coarse and fine polycrystalline

SiGe grains were observed in Sedky [93] following EC induced by a KrF 248-nm laser with a pulse duration of 24 ns. Finally, TRR measurements of $Si_{0.5}Ge_{0.5}$ alloys amorphized by $Ga^+$ ion implantation and laser-annealed by a XeCl 308-nm laser with a pulse duration of 160 ns [21] unambiguously showed a sharp EC-related peak, similar to the one shown in Fig. 5.6C.

As already mentioned in Section 5.2.1.2, further insight into the EC phenomenon has been achieved in recent years, thanks to the development of the dynamic TEM (DTEM) technique, allowing electron microscopic images to be taken in situ (with exposures in the nanosecond time scale) during laser annealing of amorphous materials. These works have provided, in addition to a detailed investigation of the microstructure evolution, a quantitative analysis of the time/position-dependent speed of the crystallization front during EC, although the reliability of the extracted growth rate values might depend on the ability to generate multiple closely time-spaced images from a single crystallization event [39], rather than collecting single images, taken at different times, from several equivalent crystallization events. Based on these results, in one of these works [38], the authors proposed a reconsideration of the currently accepted description of the EC phenomenon, whereby the formation of a "liquid-like" intermediate material, considered to be the driving mechanism for the ultra-fast recrystallization process, might be replaced by a solid-state transition in which the chemical bonds in the amorphous solid layer directly heated by the latent heat release transform from covalent to metallic type.

As a final comment, it is interesting to highlight the critical role played by laser annealing in the investigation of the EC mechanism. Indeed, EC had been known for a very long time, including reports dating back to the mid-19$^{th}$ century about electro-deposited Sb layers that, when being "*gently strucked*" provoked "*an explosion…with a small cloud of white vapour…sufficient to burn one's fingers*" [94], as well as similar experiments where EC was induced on amorphous Ge layers by a local pressure induced by a stylus [95–97]. However, a full understanding of this phenomenon was gained when laser irradiation started to be used as an additional process to induce EC of an amorphous Ge layer. This included the role of the latent heat in the crystallization dynamics [98] and the predicted existence of a thin liquid layer between the resolidified polycrystalline material and the non-melted solid amorphous [99,100], which was later inferred also from impurity diffusion experiments in laser-annealed a-Ge materials [101].

## Strain relaxation in melted SiGe layers

Pseudomorphic SiGe/Si stacks are widely used in semiconductor technologies. The elastic stress generated by the deformation of the SiGe layer, necessary to accommodate for its lattice mismatch with respect to the Si substrate, induces a band structure modification in the neighboring silicon regions that may result in the increase in carrier mobility and a consequent improvement of the final device performances.

Laser annealing can be used as a means of fabrication of pseudomorphic SiGe/Si stacks. In this method, known as pulsed laser-induced epitaxy (PLIE), a thin amorphous layer made of SiGe [49] or pure germanium [102–106], is first deposited on a Si substrate; then it undergoes a laser-induced melt down to the Si substrate, leading to the formation of a graded SiGe alloy layer upon resolidification. Alternatively, laser annealing can be used to electrically activate dopants in SiGe/Si stacks previously grown pseudomorphically by CVD [107]. In both cases, the strain level of the final structure strongly depends on the laser anneal conditions. In the worst case, relaxation occurs, which is accompanied by the formation of extended defects. Several works were therefore dedicated to the understanding of the relaxation mechanisms and to the identification of the best laser anneal conditions allowing a perfectly strained structure to be achieved, including fundamental studies in which undoped CVD-grown strained SiGe layers were submitted to laser anneal under different conditions [62,108,109].

Concerning the PLIE fabrication technique, strained SiGe layers fabricated with XeCl lasers ($\lambda = 308$ nm) were initially demonstrated by Abelson et al. [102], with Ge concentrations up to 17% at. (and a corresponding thickness of ~110 nm) and later confirmed by Lombardo et al. [49] ($x_{Ge} = 7\%$ at., thickness = 500 nm) and Kociniewski et al. [104] ($x_{Ge} = 19\%$ at., estimated thickness = 170 nm). Similar results were obtained in the case of SiGe layers fabricated by KrF laser irradiation ($\lambda = 248$ nm), with maximum Ge concentrations in the strained layer up to 12% at. [105]. The presence of extended defects [49] and more specifically misfit dislocations [103] was reported as a consequence of partial relaxation of the SiGe layer.

In the case of CVD-grown strained SiGe, the impact of a KrF laser irradiation was investigated in Ong [109] for 50-nm-thick layers with a single composition ($x_{Ge} = 24\%$ at.), showing that a fully strained configuration was obtained following a partial melt (~30 nm) of the initial structure. The Ge segregation that occurred during solidification resulted in a maximum Ge concentration of ~36% at. at the surface of the recrystallized material. Similar results were reported in Luong [107] from a 20-nm-thick $Si_{0.64}Ge_{0.36}$ layer, who reported a fully strained regrown material after the partial melt (~15 nm) induced by a KrF laser.

More recently, a detailed investigation of the impact of UV laser annealing on strained SiGe epitaxial layers of different Ge content in the range 10–40% at. was reported in Dagault [62], in which the authors proposed a unified understanding of the laser-induced SiGe relaxation mechanism, taking into account the annealing regime, the impact of Ge concentration, and the initial layer characteristics. In particular, it was shown that, when the l/s interface at MMD (i.e., at the beginning of solidification) is not flat but exhibits a considerable roughness, the solidification results in the formation of strain-relieving extended defects throughout the whole SiGe layer, independently of the Ge content. An example is shown in Fig. 5.7A for a $Si_{0.8}Ge_{0.2}$ layer annealed with a XeCl

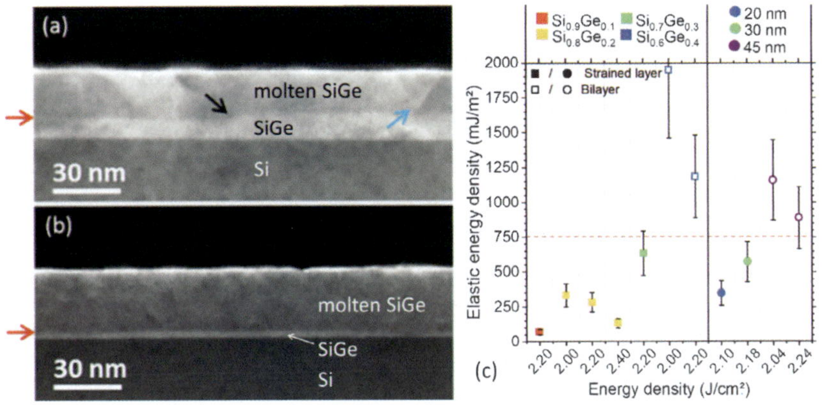


**FIG. 5.7** (A) STEM-HAADF micrograph of a SiGe layer annealed at 1.80 J/cm$^2$. The sudden contrast variation indicated by the *red arrow* corresponds to the position of the l/s interface at the maximum melt depth (i/e/ at the beginning of resolidification). The rough l/s interface, as indicated by the *black arrow*, leads to the formation of extended defects *(blue arrow)* and to partial strain relaxation. (B) Same as (A) for a laser energy density of 2.00 J/cm$^2$. In this case, the flat l/s interface at MMD leads to a perfectly crystalline and strained SiGe layer. (C) Total elastic energy density calculated for 30-nm-thick SiGe layers with different Ge content *(left)* and SiGe layers with 30% Ge content and different thicknesses *(right)* after laser annealing with different energy densities. Full symbols represent perfectly strained layers, while open symbols correspond to partially/fully relaxed layers. A threshold value for relaxation is observed close to 750 mJ/m$^2$. *(A and B: Taken from L. Dagault, P. Acosta-Alba, S. Kerdilès, J.P. Barnes, J.M. Hartmann, P. Gergaud, T.T. Nguyen, A. Grenier, A.M. Papon, N. Bernier, V. Delaye, J. Aubin, F. Cristiano, Impact of UV nanosecond laser annealing on composition and strain of Undoped Si 0.8 Ge 0.2 epitaxial layers, ECS J. Solid State Sci. Technol. 8 (2019) P202–P208. https://doi.org/10.1149/2.0191903jss. C: Taken from L. Dagault, S. Kerdilès, P. Acosta Alba, J.M. Hartmann, J.P. Barnes, P. Gergaud, E. Scheid, F. Cristiano, Investigation of recrystallization and stress relaxation in nanosecond laser annealed Si1−xGex/Si epilayers, Appl. Surf. Sci. 527 (2020) 146752. https://doi.org/10.1016/j.apsusc.2020.146752.)*

laser at an energy density of 1.8 J/cm$^2$ [108]. In this STEM-HAADF image, the sudden contrast variation indicated by the red arrow corresponds to the position of the l/s interface at the beginning of resolidification. Faceting of the l/s interface, with few nm-thick steps, is clearly visible (cf. black arrow), while extended defects nucleated close to the l/s interface are highlighted by the blue arrow in the same figure. Such interface morphology is consistent with the considerable stress relaxation (~25%) measured by XRD and RSM in the same sample.

In contrast, when the resolidification starts from a virtually flat l/s interface (such as the one shown in Fig. 5.7B for a $Si_{0.8}Ge_{0.2}$ layer annealed with a XeCl laser at an energy density of 2.0 J/cm$^2$), the regrowth behavior is determined by the elastic energy stored in the whole SiGe layer, similar to the case of low-temperature solid-phase epitaxial regrowth (LT-SPER) of amorphized SiGe/Si epilayers [110]. In this case, a threshold elastic energy value for strain relaxation exists, below which the resolidification results in a pseudomorphic defect-free single-crystal material. When the elastic energy stored in the regrown layer exceeds a critical value l/s interface, roughening is expected to occur, leading to the formation of strain-relieving defects in the upper resolidified, which are easily observed by TEM and are consistent with the stress relaxation levels detected by XRD analysis. By calculating the elastic energy of all the observed samples obeying the "elastic energy-driven" relaxation mechanism (cf. in Fig. 5.7C), the authors were finally able to estimate the threshold value ($\sim$750 mJ/m$^2$) for strain relaxation. While being well above the value previously determined for ion beam synthesized SiGe layers ($\sim$300 mJ/m$^2$ [111]), the critical elastic energy for NLA SiGe layers is lower than the one achieved for CVD-grown SiGe layer with a constant Ge concentration ($\sim$1000 mJ/cm$^2$ [112]). However, NLA offers more versatility by allowing the formation of steep Ge gradients with concentration levels at the surface as high as 80% at., which are beneficial for several applications, including source/drain contact engineering.

## 5.3 Dopant redistribution and activation in Si

### 5.3.1 Laser annealing for impurity activation

#### *5.3.1.1 Introduction*

High dopant activation in crystalline Si in order to reduce sheet resistance, and low dopant diffusivity to facilitate device scaling are aspirational metrics for an annealing methodology. Equilibrium solid solubility is not sufficient to meet the aggressive device access resistance targets at advanced device dimensions; thus, the above-equilibrium metastable solubility is desired. Furthermore, thermal budget restraints in terms of time versus temperature have to be controlled with subsecond anneals in order to control dopant diffusion. With the advent of 3-dimensional above-IC integration, the advantage of pulsed laser anneal is in that thermal profiles and gradients are spatially confined to the top surfaces of the exposed wafer. However, the formation of above-equilibrium metastable solubility requires consideration during subsequent processing, because supply of thermal energy, e.g., by back-end processing, causes the metastable condition to degenerate back to the lower equilibrium state.

A key factor in this approach is the concept of amorphization and recrystallization of the Si substrate. Ion implantation is considered as the most common approach to locally introduce impurities into a semiconductor material. These highly energetic ions damage the semiconductor material as they strike it, impacting on target atoms and displacing many of them from lattice positions in the Si crystal. If the concentration of these damage events is high enough, the crystalline nature of the Si is broken down, and it is amorphized [17,113]. Amorphous Si is considered to be a continuous random network with short-range order similar to that in the crystal, but without the long-range order. However, in certain circumstances, a process known as dynamic annealing can reduce the target damage during the implant process so it has either dissipated or recombined within the Si. Higher wafer temperatures (above room temperature) suppress amorphization as damage anneals out during the implant process itself. Colder wafer temperatures (lower than room temperature) promote amorphization as dynamic annealing is reduced, and thus, damage build-up is promoted.

Heating amorphous Si by thermal annealing, such as by laser annealing, provokes a process known as recrystallization. This is related to the fact that the free energy of crystalline Si is lower than the amorphous state, due to more disordered atom arrangements and distorted bonds in terms of length and angle. Distorted bonds are thermodynamically less stable than regularly arranged bonds; as a result, atoms prefer to rearrange themselves into the form of a regular crystal. In the presence of a template, the amorphous region will recrystallize into monocrystalline Si, with the same crystalline orientation as the template. For example, an amorphous Si region on top of a (100) oriented crystal seed will recrystallize with a (100) orientation.

In this section of this chapter, the concepts of above-equilibrium solubility and thermal stability are first discussed in general, with pulsed laser anneal in mind, applied to Si. Subsequently a review of the most relevant literature will be presented, structured in terms of Group III elements (p-type dopants), Group V and Group VI elements (n-type dopants).

#### *5.3.1.2 Metastable solubility*

Above-equilibrium levels of solubility allow advanced device scaling and greater freedom for process optimization. By the late 1970s and the early 1980s, impurity solubilities due to solid-phase-epitaxial-regrowth (SPER) [114,115] and full-melt laser anneals [116] were reported. In fact, as will be discussed later, many of the works in that era were performed on impurity species that were larger in size than Si (e.g., As, Sb, In) as it was common practice in that time to use Rutherford backscattering channeling spectrometry (RBS) to profile dopant impurities in terms of diffusion and solubility. It is required in RBS that the impurity atom under study is larger than the host substrate atoms, and so as As, Sb, and In are larger than Si, they were chosen as impurity subjects.

Many junction formation techniques historically involved or required amorphization of the Si substrate, either in the form of self-amorphization during the implantation of large-sized dopant impurities, or in the form of preamorphization to prevent channeling of the small-sized dopant impurities such as B or P. An important parameter to consider when studying impurities in the post-recrystallization regime is the metastable solubility limit. Metastable solubility is higher than the maximum equilibrium solubility, as under the extremely nonequilibrium conditions just after recrystallization the Si lattice can substitutionally incorporate a supersaturation of impurity atoms.

As an example of a dopant metastable solubility, Group III and V impurity solubilities were extracted and compared for the solid-phase epitaxial regrowth (SPER), flash lamp, and submelt laser anneals [117]. RBS channeling analysis was used to determine the substitutional impurity depth profiles and residual amorphous Si layer thicknesses. The metastable solubilities produced by the different anneals are largely similar relative to the maximum equilibrium values. This was a remarkable result as there is a 600°C difference between the SPER anneal and the other two anneals. This temperature independence of metastable solubility is in contrast to the temperature dependence of equilibrium solubility.

Moreover, there was a correlation between metastable solubility and the equilibrium distribution coefficient (k), which is defined as the relative tendency of various impurities to dissolve in solid Si. Trumbore reported k values for many impurities in Si, in 1960 [118]. In terms of concentration, $k = C_{solid}/C_{liquid}$ at the melting point of Si. Thus, if equilibrium solubility is high, k is close to 1. If equilibrium solubility is low, k is significantly lower than 1. In Fig. 5.8,

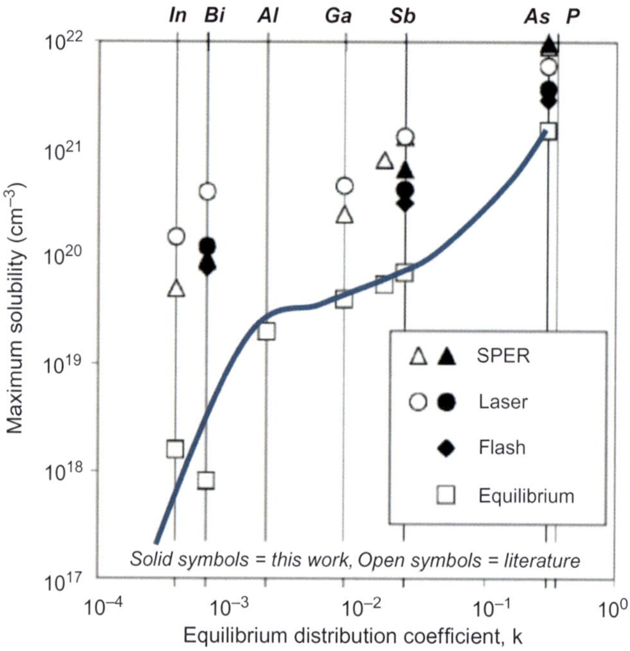


**FIG. 5.8** Maximum solubility vs equilibrium distribution coefficient for Group III and V impurities in Si. The *solid symbols* represent the experimental data for SPER, flash lamp, and submelt laser anneals. The *open symbols* represent data for equilibrium conditions and SPER and full-melt laser anneals. *(R. Duffy, T. Dao, Y. Tamminga, K. van der Tak, F. Roozeboom, and E. Augendre, Appl. Phys. Lett. 89 (2006) 071915, and references therein. Copyright, 2006 American Institute of Physics.)*

the maximum solubility data are plotted versus k. Note that solubility does not guarantee electrical activity, but it is generally understood that non-substitutional impurities will not contribute to the electrical behavior of the device. Literature data for equilibrium conditions and SPER and full-melt laser anneals are also incorporated in Fig. 5.8. Metastable solubility is significantly higher than the corresponding equilibrium value. The benefit of switching to advanced anneals is greater for larger-sized impurities such as In and Bi.

#### *5.3.1.3 Thermal stability*

A significant factor to consider when laser annealing processes is the region of damaged Si just beyond the amorphous-crystalline interface. This region was damaged during implant, but not to a high enough level to transform to amorphous material. As the recrystallized amorphous Si region regrows essentially defect-free, there exists a buried layer rich in point defects, such as interstitial point defects. The annealing of damaged regions proceeds with an evolution of defect types from small to large. Simple point defects may agglomerate and form larger extended defects, typically rich in interstitials, or line-like defects known as {311}s, or agglomerate even further to form larger disk-like defects known as dislocation loops. These {311}s and dislocation loops are problematic as they act as sources of interstitials and can enhance local point defect populations and lead to unwanted physical effects. Specifically with respect to laser annealing, even though there has been growth in popularity, one drawback is that it can leave residual damage in the Si substrate [71,119,120]. These disturbances create localized energy levels in the Si band gap, and if located in or close to the space charge depletion region, they may act as centers for carrier generation or recombination events [121]. deteriorating carrier lifetimes, and nonideal leakage currents [122,123], which ultimately degrade electron device performance.

Metastable solubility and activation can be hindered by dopant-defect cluster formation. In an intricate fashion, dopant-defect clusters may be in a state of formation or dissolution depending on the thermal budget of the post-laser anneal, the local concentration of the point defects, and of the dopants. Different dopants may interact preferentially with different interstitials or vacancies to form immobile electrically inactive clusters. As mentioned earlier, one major drawback of using a low-thermal-budget anneal like laser anneal is that the end-of-range (EOR) Si interstitial defect band, just beyond the amorphous-crystalline interface, is still present after the anneal. During any subsequent thermal step, such as in back-end processing, EOR defects dissolve and emit Si interstitials, which subsequently diffuse and then interact with the active dopant profile.

For example, B is the most commonly used p-type dopant in Si technologies and interacts predominantly with interstitial point defects [123] to form boron-interstitial clusters (BICs) [124–127]. Growth of BICs, and corresponding degradation of the active B profile, depends on the local concentrations of B and Si interstitials [128]. Several studies have linked the evolution of the EOR defect band with B deactivation via the change of sheet resistance versus time or temperature of post-anneals [129,130].

For n-type dopants in Si, the clustering paths are potentially more complicated as common dopants may interact with both interstitials and vacancies [123]. As stated previously, interstitials coming from the EOR defect band can raise the local interstitial population, while vacancies, which do not have an equivalent obvious defect-band source, may be present in large numbers if created by means of vacancy-engineering or injected from surfaces. For P, $P_nV_m$ and $P_nI_m$ clusters have been studied by several authors, and the physics needs to be controlled for an optimized P active doping profile [131–134]. Growth and dissolution of these clusters depend on the local concentrations of P and point defects. Most reports of clustering behavior relate to $As_nV_m$ clusters [135,136]; however, $As_nI_m$ clustering studies are also available in the literature [137,138].

In terms of technological relevance, dopant deactivation is a major concern in the fabrication of advanced electronic devices, as there is a significant thermal budget involved in back-end processing, performed post-laser anneal. Achievement of highly activated doped layers with the above-equilibrium solubility is only a benefit to the electrical performance of working devices if deactivation can be suppressed.

## 5.3.2 Laser annealing of Group III impurities

In the following sections, a review of the historical literature from the past 40 years will be presented. Within this, notes will be given on the different laser anneal systems and conditions used in many of the works. It should be stated from the outset that the optimized energy densities and pulse durations are specific to that particular laser, and that those parameters don't necessarily transfer from one laser tool to another. As an example of this, the reader is pointed to the work of Ong et al. where calculations of melting threshold energies of crystalline and amorphous materials due to pulsed laser irradiation were characterized [139]. It was demonstrated that the melting threshold energy was different for different lasers, and so depending on the laser tool, different conditions would have to be used in order to produce the same result in the Si samples. That study compared a variety of lasers with different wavelengths, namely, Nd:YAG double frequency 0.532 μm; Ruby 0.693 μm; Alexandrite 0.750 μm; Nd:YAG 1.06 μm. They observed that threshold melting energy increased with laser wavelength.

### *5.3.2.1 Boron (B)*

Work on laser anneal studies for improved B activation in Si started to appear in literature in the late 1970s. One of the first works was that by Young et al. who used a Q-switched ruby laser as their annealing source and compared with results obtained by conventional thermal annealing (see Fig. 5.9). It was found that laser annealing of B implanted layer resulted in significantly increased electrical activity, thus proving the concept of the above-equilibrium solubility and activation of Group III impurity dopants in Si. In their study, the B implant dose range was $4.2 \times 10^{14}$–$2.5 \times 10^{16}$ atoms/cm$^2$ [140].

Also in 1978, Narayan et al. wrote a number of seminal papers. The first one described again a comparative study of laser and thermal annealing of B-implanted Si. The concepts of implant damage annihilation and B precipitation were explored. Few, if no, defects in the form of dislocations, loops, and stacking faults were observed in samples annealed with a Q-switched ruby laser, with a pulse energy of 1.5–1.7 J cm$^{-2}$. Furthermore, the concept of metastable solubility was discussed as precipitation of B was observed if the implanted B dose was above the equilibrium solubility limit and if the laser-treated specimens were subsequently thermally annealed. It was concluded that B in the precipitates was electrically inactive [141]. The same group

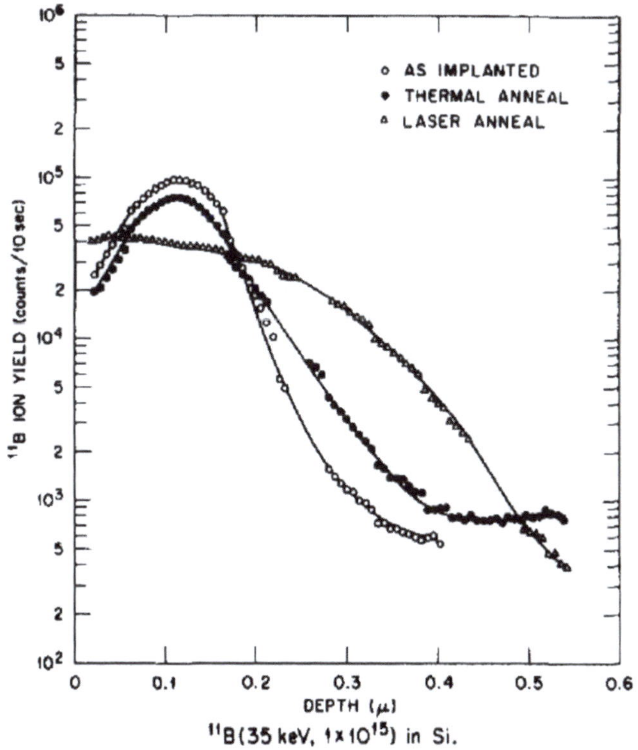


**FIG. 5.9** Depth profiles for B+ (35 keV, $1 \times 10^{15}$ cm$^{-2}$)-implanted silicon. Flattening of the profiles at counting rates ≤ 600/10 sec is believed related to insufficient aperturing and crater wall effects. *(R.T. Young, C.W. White, G.J. Clark, J. Narayan, W.H. Christie, M. Murakami, P.W. King, S.D. Kramer, Appl. Phys. Lett. 32 (1978) 139.)*

followed up that work by forming p-n junctions in B-deposited Si by laser-induced diffusion. In that parallel study, B was deposited rather than ion-implanted. The dopant profiles were very shallow and had a graduated shape when profiled, namely, continuously decreasing without any plateau. Good diode characteristics were demonstrated, with low off-state leakage [142].

Also at that time, a number of groups studied the area using other characterization techniques to explore the problem. X-ray Bragg reflection, Raman scattering, and infrared reflectivity and transmissivity were all used in that period for metrology. Larson et al. implemented X-ray Bragg reflection in order to study the lattice strain in B-implanted laser-annealed Si. The shape of the depth distribution of the strain was found to be essentially the same as that for the B distribution after laser annealing [143,144]. Engstrom et al. applied Raman scattering and infrared reflectivity and transmissivity, in their works. In terms of the laser conditions employed, a single pulse from a Q-switched ruby laser ($\lambda = 693.4$ nm) was used, with a pulse energy and duration of 1.7 J/cm$^2$ and 40 ns, respectively [145,146].

By the early 1980s, work had moved on to more electron device-oriented analysis. Masetti et al. modeled carrier mobility versus carrier concentration B-doped Si, as well as for P and As. Primarily, this was a study on carrier mobility at supersaturated active concentrations and how this might impact device performance. B-doped samples were activated by a pulsed ruby laser with an energy of 1.9 J/cm$^2$. In conclusion, the carrier mobility continuously decreased as the carrier concentration increased, even for carrier densities one order of magnitude higher than the solid solubility [147].

In the mid-1980s, the fashion was to combine F with B to control diffusion. This might have been a separate F implant, or more commonly at that time, using $BF_2$ as an implant source. Narayan et al. once again featured in the literature, but in this case they used a pulsed excimer laser anneal ($\lambda = 0.308$ μm, $\tau = 50 \times 10^{-9}$ sec). The benefit of laser anneal was equally present for $BF_2$ as for B in terms of improved active concentration and reduced resistivity, but the implant damage control was more difficult due to the heavier implanted species of course. F was seen to outgas from the samples during the anneal, which was also observed in other annealing methodologies [148].

The problem of B precipitation and deactivation with ultra-high implant doses was still on the landscape in these times, and into the 1990s. Solmi et al. developed a model to explain and describe the precipitation phenomena of high-concentration B diffusion in Si.

The experimental part of that work included the characterization data of samples annealed by a pulsed ruby laser [149].

By the turn of the millennium, the field had moved into the era of preamorphizing ion implants in combination with B implants to help with junction scaling and diffusion control. The difference for laser annealing is that amorphous Si has a different melting threshold than crystalline Si, so that had to be considered when optimizing the processes. For example, Jones et al. studied a 1 keV

$4 \times 10^{15}$ atoms/cm$^2$ B implant into preamorphized Si with a 308-nm XeCl excimer laser. The laser power was varied between 0.625 and 0.825 J/cm$^2$. Local melting of the amorphous Si was observed. Secondary ion mass spectrometry (SIMS) profiling showed a significant pile-up of B in the end of range region of the implant, as well as B dose loss from the samples [150]. Later, the same group explored nonmelt laser annealing of 1 and 5 keV $10^{15}$ ions/cm$^2$ B-implanted Si, where Earles et al. used a 308-nm XeCl excimer laser again. With a 15-ns pulse at energy densities ranging from 0.4 to 0.6 J/cm$^2$, dopant diffusion could be decreased while supplying enough energy to the surface to produce dopant activation [71].

In parallel, Whelan et al. also used an XeCl excimer laser, $\lambda \sim 308$ nm, but with a 28-ns pulse duration. That work undertook lateral and vertical dopant profiling of B-implanted Si using selective etching and cross-sectional TEM [151].

Finally, in the 2000s, the field had progressed to the study of using repetitive pulses or multiple scans for laser annealing B-implanted Si. The research of Poon et al. explored multiple-pulse laser annealing of preamorphized Si for ultra-shallow B junction formation with a KrF excimer laser using a wavelength of 248 nm. The laser pulses had a repetition frequency of 1 Hz and a pulse duration of 23 ns. The laser energy density was chosen to melt the amorphous Si layer but not crystalline Si, and thus control the junction depth by defining the preamorphized layer depth [152]. In another work, Sharp et al. studied B activation and post-laser anneal deactivation as a function of nonmelt laser anneal scans. Under annealing conditions where a high density of {113} defects were produced after the laser annealing, further processing at 700–1000°C leads to strong deactivation, with B decorating these end of range defects. However, with an increased number of scans, the {113} defects evolved into loops, and deactivation was strongly suppressed during post annealing [130].

### 5.3.2.2 *Aluminum (Al)*

Al is not an ideal p-type dopant compared with B, due to relatively poor solubility and activation, so there are few works on Al-doped Si, especially in combination with laser anneal. Nonetheless, for completeness, works involving Al or Al-based materials in combination with laser anneal will be briefly summarized.

Jung et al. used a $Nd^{3+}$ YAG laser with a wavelength of 532 nm, for selective annealing of the $Al_2O_3$/Si interface for electron device applications [153]. Near-ideal shallow-junction diode formation was achieved in combination with an Al masking layer during excimer laser annealing by Gonda et al. [154].

Lu et al. worked on optimizing Si solar cell applications, where they used a 532-nm wavelength laser with a Gaussian beam distribution, applied to B-doped anodic aluminum oxide layers as a p-type dopant source. Their laser doping processes formed localized $p^+$ regions on the Si surfaces [155].

Al-induced or Al-assisted crystallization has been studied in the past as a means of lowering processing thermal budgets. For example, Nast et al. used a rapid thermal anneal for Al-induced crystallization of Si on glass for thin-film solar cells [156]. Liu et al. demonstrated that Al-assisted crystallization reaction times could be reduced from minutes to milliseconds when using laser anneal. A diode laser with a wavelength of 808 nm and a beam size of 12 mm × 270 μm was applied to deposit Al films. The conclusion was the diode laser could replace conventional furnace process, which would have been the conventional approach [157,158].

Finally, Tabata et al. did study Al doping in combination with laser anneal, although that was on SiGe. The impact of solidification velocity by UV melt laser anneal on activation of Ga, In, and Al in high Ge content SiGe was explored on $Si_{0.5}Ge_{0.5}$ substrates with $1 \times 10^{16}$ atoms/cm$^2$ ion implants of Ga, Al, and In. The boost in the above-equilibrium activation was characterized as a function of melt versus nonmelt laser annealing conditions with the best results observed for largest size element (In) in the partial-melt regime, where a 400 × activation increase was recorded [159].

#### 5.3.2.3 Gallium (Ga)

As with Al, the lack of work on Ga in Si is possibly related to the relative advantages of B as a p-type dopant. There have been a number of proposals of Ga as a p-type dopant in Ge devices, but not so many in Si technology.

The key publication with respect to Ga in combination with laser anneal is the work by White et al. [116]; see Fig. 5.10, for example, of supersaturation Ga results. In summary, supersaturated substitutional alloys of Group III and Group V dopants in Si were formed by ion implantation and pulsed laser annealing. More specifically, Group III (Ga, In) and Group V (As, Sb, Bi) dopants were implanted at doses ranging from $1 \times 10^{15}$ to $1 \times 10^{17}$ atoms/cm$^2$. The samples were annealed with a Q-switched ruby laser (energy density was 1.5 J/cm$^2$, and pulse duration was $15 \times 10^{-9}$ sec), and the substitutional concentrations were characterized by RBS channeling. This was an early and important work in the field demonstrating clearly that laser annealing could boost dopant activation and solubility well above the equilibrium levels.

The interested reader can learn more about Ga doping in Si from the significant works of Tsai et al. [160], Arora et al. [161], Elliman and Carter [162], and Tamura et al. [163], carried out in the early to mid-1980s.

#### 5.3.2.4 Indium (In)

As we travel down Group III of the periodic table, we next come to In, which is larger than B, Al, and Ga. It has the worst solubility of this group of elements, but it has found applications in doping electronic devices, as it does not diffuse as quickly compared with the other three. The premise is that one can implant In

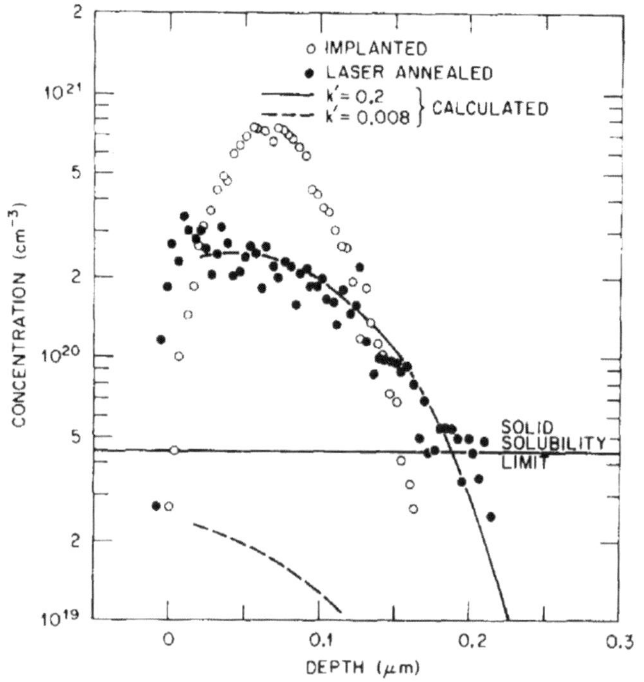


**FIG. 5.10** Dopant profiles for Ga (100 keV, $6.17 \times 10^{15}$ atoms/cm$^2$) in silicon compared with calculations. After annealing, the integrated Ga concentration was $3.2 \times 10^{15}$/cm$^2$. *(C.W. White, S.R. Wilson, B.R. Appleton, F.W. Young Jr., J. Appl. Phys. 51 (1980) 738.)*

where you want it, and it does not move much during the activation anneal [164].

As mentioned in the previous section, the work by White et al. in 1980 demonstrated how In, and other dopants, can benefit from laser anneal processing.

Around that time, Kurnaev et al. showed that a single pulse from a ruby laser can produce a significant regrowth in the amorphous region of In-implanted Si. In that work, RBS measurements demonstrated that ~60% of implanted In atoms occupy regular lattice positions in the crystal. There was also a redistribution of the indium atoms during recrystallization, causing the maximum concentration to be at the surface [165]. Similarly, Baeri and Poate studied the segregation phenomena of In-implanted following the melting and epitaxial regrowth of Si surface layers by pulsed ruby laser irradiation. The velocity of the liquid-solid interface during recrystallization was varied from 1.8 to 5.2 m/s by the laser conditions. In was observed to be trapped on substitutional sites, above equilibrium, or segregated in a surface layer [42]. This effect of segregation has also been shown to occur during rapid thermal annealing [166].

Kemerink et al., in their study involving In, concluded that the vacancy concentration in laser-annealed Si amounts to at least $10^{16}$–$10^{17}$ atoms/cm$^3$, and explored the annealing behavior of the In-vacancy interaction. In the experiments reported in that paper, the samples were annealed with a Q-switched ruby laser with a pulse width of 25 ns and an energy density of 1.6 $J/cm^2$ per pulse [167].

Finally, Clark et al. explored the strain compensation with B combined with In, using 308-nm wavelength XeCl excimer irradiation source, with a pulse duration of 18 ns. Compensation of the strain corresponded to a reduction in the deactivation during post-laser anneals [168].

### 5.3.3 Laser annealing of Group V impurities

As with the previous section, this section on n-type dopants will be presented in the form of a historical review and perspective. The early work relevant to Group V impurities for doping applications in Si, like the Group III case, started to appear regularly in the literature in the late 1970s.

#### *5.3.3.1 Phosphorus (P)*

P is of a similar size to Si, so it is not possible to perform the same RBS channeling analysis and profiling of it, in contrast to the larger species As, Sb, In, and Bi. Thus, the early works on P characterized the laser-annealed samples in different ways. For example in 1978, Battaglin studied the crystal damage by transmission electron microscopy (TEM), in order to understand the lattice disorder after laser anneal of P-implanted Si. Many types of imperfections were observed from small defects, to twins, to dislocation loops. Laser irradiation was performed in air at room temperature with a Q-switched ruby laser pulse with ~15 ns duration. The energy density range was 1.0–3.0 $J/cm^2$ [169].

Carrier concentration, carrier mobility, and sheet resistivity determined by van der Pauw measurements were undertaken by Young et al. again to explore crystal damage associated with P-implanted Si after laser anneal. In that study, the laser treatment was a Q-switched ruby laser ($\lambda = 0.694$ μm, pulse duration time = 50 ns) with an energy density of 1.5–1.8 J $cm^2$/pulse. Diffusion-induced precipitates and dislocations were removed by the pulse laser treatment in melt mode. The laser process also had the benefit of a reduction in sheet resistivity [170]. Also in 1978, Affolter et al. achieved P doping of Si by covering the surface with a layer of dopant and melting locally with pulses from either an Nd: YAG or a $CO_2$ laser [171].

Around the late 1970s, Hitachi Ltd., Tokyo, Japan, were very active in the literature combining P ion implant with laser annealing. Miyao and Tamura et al. published the nice work comparing the annealing behaviors of P-implanted Si irradiated by several lasers of different wavelength; namely, $CO_2$, Nd-YAG, and Ruby lasers were compared. The absorption coefficients were different depending on the laser wavelength, while the reflectivity of the surface was tuned by cap, and by varying the $SiO_2$ thickness, selective annealing was possible [172,173]. A year later, the same group followed up this work by performing a P deactivation study using furnace annealing post-laser anneal. Characteristics of supersaturated solutions produced by P+ implantation (50 keV, 8 × $10^{15}$ atoms/$cm^{-2}$) and Q-switched pulse ruby laser annealing (0.3–1.5 J/$cm^2$) were investigated [174]. Surface carrier concentrations, electron mobilities, and carrier concentration profiles were determined by van der Pauw test structures and the anodic oxidation layer removal method. Defects were examined by TEM and photoluminescence characterization. Subsequently, Nakashima et al. performed a dedicated photoluminescence study on the effects of laser annealing on P-implanted and unimplanted Si, with the dependence of crystal-quality recovery on laser annealing energy determined [175].

Inada et al. also used van der Pauw profiling techniques to demonstrate ~2 × $10^{20}$ atoms/$cm^3$ active P after pulsed electron beam annealing. Highly doped n-type Si layers were formed using a single shot of a pulsed electron beam, with pulse width approximately 50 ns, and incident energy density of 0.92 or 1.17 J/$cm^2$ [176].

The study of the deactivation phenomenon once again appeared in the literature by the early 1980s. Finetti et al. explored the thermal stability of the supersaturated P-doped Si, determined by isochronal and isothermal annealing. It was determined that deactivation and precipitation were strongly dependent on the supersaturation. Even short processes at low temperature could produce a significant precipitation of dopant in the more heavily doped samples. The inactive P was proposed to be present in the form of small precipitates, with densities and mean diameter depending on supersaturation [177]. In more detail, Nobili et al. carried out laser annealing by a 20-ns single pulse delivered by a Q-switched ruby oscillator laser with an energy density of 1.8 J/$cm^2$. Electrical activation was shown to be as high as 5 × $10^{21}$ atoms/$cm^3$, but strong deactivation was seen after 850°C for 30 min post-anneal [178].

Into the 2000s, Takamura et al. did a very comprehensive dopant deactivation and thermal stability of dopants in laser-annealed silicon across a range of concentrations and annealing conditions. P, As, B, and Sb were all studied and compared, as shown in Fig. 5.11. Laser annealing was performed on a frequency-doubled Nd:YAG laser operating at 532 nm, with laser energy densities of 0.73, 0.80, or 0.90 J/$cm^2$, with ten shots applied at each location. The resulting melt depths were 110, 140, and 180 nm, respectively, which repaired the damage from the implants, as well as the end of range damage. In conclusion, during post-laser anneal, P and As samples deactivated quickly at low temperatures, while B and Sb remained fully active until 700–800°C, giving insight into the interaction of supersaturated dopant profiles with local point defects, as well as the clustering and precipitation dynamics [132].

Grun and Fischer performed athermal annealing of P-ion-implanted Si using a 1.06-mm wavelength laser pulse, with 0–30 J, and 5–40 ns duration. It was described as athermal because under the tool configuration the intensity was

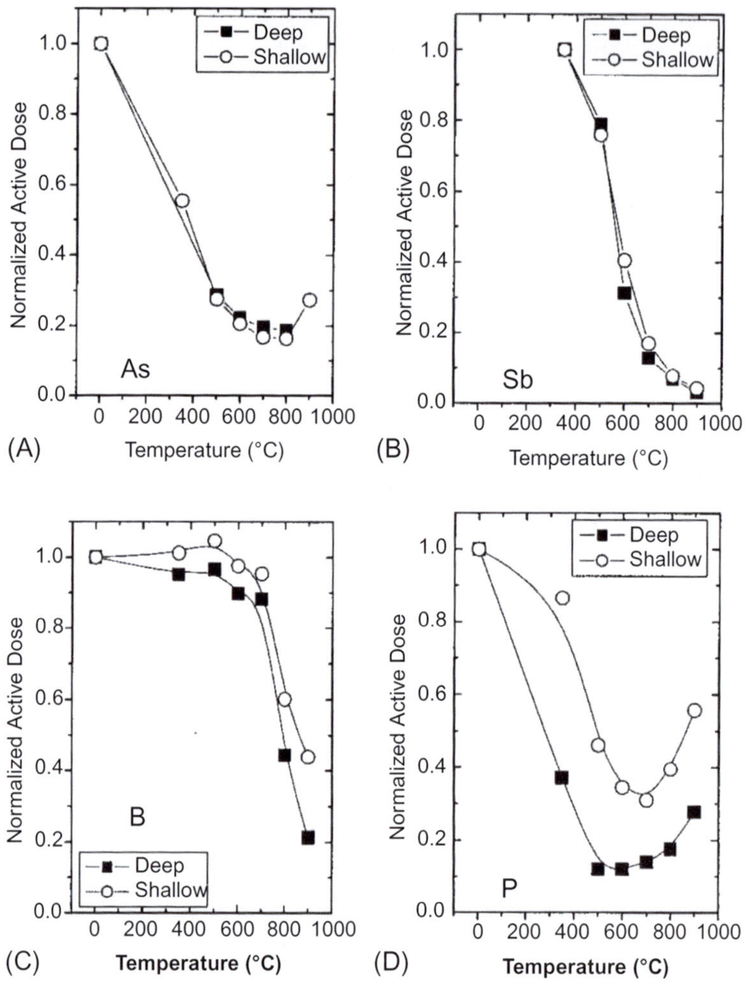


**FIG. 5.11** Normalized active dose as a function of anneal temperature for (A) As-doped; (B) Sb-doped; (C) B-doped; and (D) P-doped samples for both shallow and deep junctions. Only in the case of the P-doped sample is there a noticeable reduction in deactivation for the shallow samples. The concentration is the dominant factor determining the deactivation behavior. *(Y. Takamura, S.H. Jain, P.B. Griffin, J.D. Plummer, J. Appl. Phys. 92 (2002) 230.)*

sufficiently high to produce a hot plasma that streams away from the irradiated spot at high velocity, while still activating the P ion-implanted region. The effect was proposed to be related to multimode acoustical stress waves and shock waves in the wafer [179].

Finally, laser anneal was explored by Loubet et al. for resistivity reduction of in situ P-doped Si and SiC epitaxy for source/drain formation in field-effect transistor devices. This was successfully achieved using a 1250°C millisecond laser anneal resulting in a 30% and 42% resistivity reduction for SiP and $SiC_{2.1\%}P$, respectively [180].

### 5.3.3.2 *Arsenic (As)*

As stated above, RBS channeling analysis was a common characterization technique in the initial studies of laser anneal of ion-implanted Si, particularly for impurity species atoms that were larger than the Si atom. A very early report by Krynicki et al. in 1977 showed exactly how RBS channeling could profile both the As profile and the residual implant damage in {111} oriented Si substrates, based on the comparison between the random and aligned spectra. In that work, they used a Q-switched ruby laser ($\lambda = 0.6943$ μm) to irradiate $1 \times 10^{15}$ atoms/cm$^2$ implanted As in Si [181].

In those days, the experiments focused mainly on melt laser, where the energy densities were relatively high. Baeri et al. performed work on Si implanted with 400 keV As to a dose of $5 \times 10^{15}$ atoms/cm$^2$ and irradiated by a 50-ns Q-switched ruby laser pulse in the energy range of 1.0–3.0 J/cm$^2$. The broadening of an ion-implanted profile was assumed to correspond to the diffusion of the impurity atoms mainly in the melted region [182]. In a similar theme, Venkatesan et al. discovered that increased laser energy absorption causes the liquid layer to be formed sooner, resulting in more total absorption, and that a high As concentration could increase that absorption. Their samples were irradiated with an Nd: YAG laser using 110-nsec pulses with a peak power density of 57 MW/cm$^2$. The authors also noted that if the laser radiation energy was greater than the semiconductor band gap, the substrate material would have a substantial absorption coefficient [183].

The choice of laser source was quite varied between different research groups in the late 1970s. Auston et al. studied CW argon laser annealing of As-implanted Si where they concluded that the recrystallization mechanism is an accelerated solid-state epitaxial regrowth process similar to thermal annealing [27]. In parallel, Takai et al. used a CW $CO_2$-laser annealing tool, applied to 150 keV As ions implanted in Si wafers to a dose of $10^{16}$ atoms/cm$^2$. Laser beams with a power density ranged from 100 to 640 W cm$^{-2}$. It was seen that lattice disorder induced during implantation was completely removed with a power density of 500 Wcm$^{-2}$. Furthermore, about 90% of As atoms were placed on the lattice sites by laser annealing for >2 s with a power density of >100 W cm$^{-2}$ [184]. Delfino et al. and Tsien et al. also used $CO_2$ lasers around this time [185,186]. In the latter case, it was seen that at high annealing temperatures, complete electrical activation was achieved, but after a prolonged annealing time, a relaxation of metastable concentrations took place.

As was the case for other dopants by the early 1980s, the concern around metastable activation was being examined in dedicated studies. Chu et al. used sheet resistance measurements to track the stability of the As metastable state produced by laser irradiation as a function of post-laser heat treatment between 300°C and 900°C [187].

A very significant and extensive work was carried out by Williams et al. using an Nd:YAG laser source that was applied to As implants as a function of implant dose, substrate orientation, laser power, and laser pulse length. The work is very complete, and an example of their data is shown in Fig. 5.12. The main conclusions were that increased As concentration within the near surface resulted in an enhanced laser absorption, but only for the Nd:YAG laser source not for the ruby source. Substitutional As concentrations in the bulk exceeded the maximum equilibrium values, but substitutional solubility and segregation effects near surface suggest that supersaturation close to the surface may drop off from bulk values. Within 5 nm of the surface, the solubility was below the maximum equilibrium substitutionality even when the bulk value greatly exceeded this value. This has a significant implication for thin body Si devices. The higher As concentrations coming from the surface segregation may stimulate precipitate nuclei for the excess As [188]. Also using a Nd:YAG laser processing, Erbil et al. looked into lattice distortion studies using X-ray-absorption fine-structure measurements for 0.1, 0.7, and 7 at. % As impurities in single-crystal silicon [189].

Laser annealing was also applied to polycrystalline Si films. Solmi et al performed a novel study and analysis where electrical properties of laser-annealed heavily n-type doped polycrystalline Si films were explored. Poly-Si film grain sizes and carrier mobilities were extracted [190].

Finally, by the 1990s, the advantage of supersaturation created by laser annealing was well established and the focus returned to the thermal stability and electrical deactivation problem. Rousseau et al. showed that the deactivation of As after laser anneal was accompanied by an injection of point defects into the substrate. Specifically, the vacancies required for As deactivation were provided by a Frenkel pair generation, thus releasing large numbers of interstitials [191]. This was followed up by a very complete analysis by the same group in 1998, where a time, temperature, and concentration study of As deactivation was presented [192].

### 5.3.3.3 Antimony (Sb)

Sb was an ideal candidate for studying the benefits of laser anneal on impurity activation in Si due to its large size, and thus ease of profiling in RBS, and it's relatively poor maximum equilibrium solubility, which is $<10^{20}$ atoms/cm$^3$, and so the boost from laser annealing would be greater than other dopants. Hence, the literature is quite rich for laser-assisted activation of Sb in Si. Before the historical review, the reader should note a few excellent review articles by J. Williams in the area of metastable activation of dopant in Si, namely [115,193].

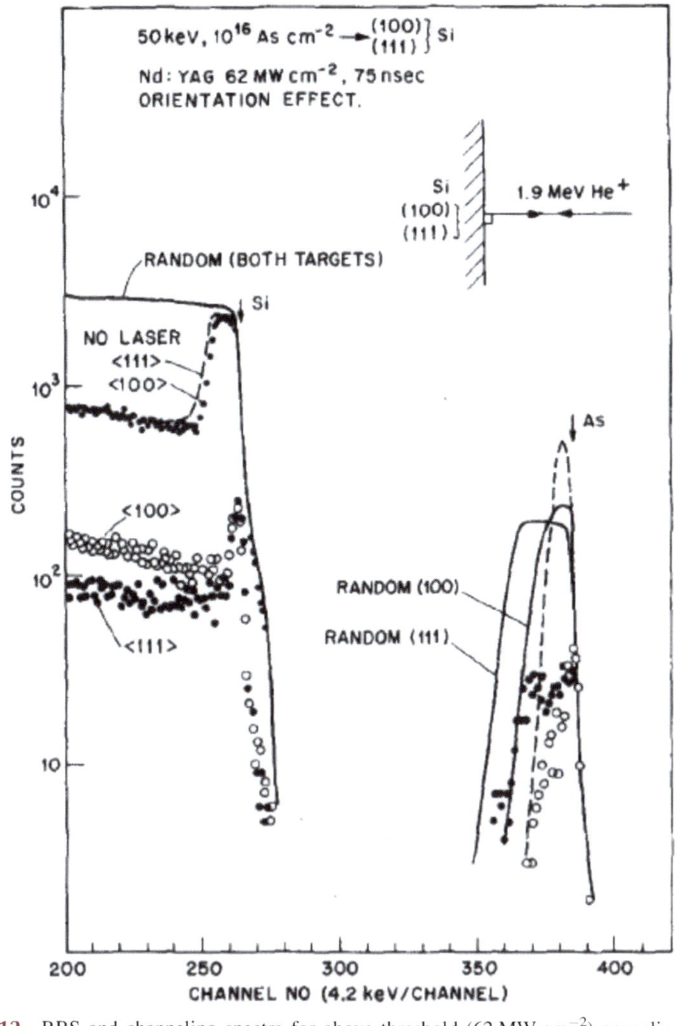


**FIG. 5.12** RBS and channeling spectra for above threshold (62 MW cm$^{-2}$) annealing of both (100) and (111) oriented silicon indicating good regrowth in both cases. *(J.S. Williams, W.L. Brown, G.K. Celler, H.J. Leamy, J.M. Poate, G.A. Rozgonyi, T.T. Sheng, J. Appl. Phys. 52 (1981) 1038.)*

In 1978, White et al. studied the redistribution of dopants in ion-implanted Si due to pulsed-laser annealing. The authors studied B, P, As, and Sb in single-crystal Si and after laser anneal with a Q-switched ruby of energies in the range 1.5–1.7 J/cm$^2$. Analysis was carried out by secondary-ion mass spectrometry and ion backscattering [194]. In parallel, Williams et al. explored Sb-implanted Si wafers annealed by a CW Ar ion laser and determined that the mechanism for recrystallization was thermal solid-phase regrowth from the underlying substrate template [195].

White et al. followed up their work by looking into pulsed ruby laser annealing on As and Sb ion-implanted Si in the dose range of $1 \times 10^{15}$ to $2 \times 10^{16}$ atoms/cm$^2$. The laser pulses were from a Q-switched source having 50-nsec durations and power densities of 1.5–1.7 J/cm$^2$. It was noted that impurity segregation was observed most noticeable for the high-dose Sb implants, resulting in a partially precipitated region in the first 40 nm close to the surface [196]. The problem of instability caused by supersaturated Si was also observed in the work of Narayan [197]. Rapid solidification of the solid-liquid interface caused instability during directional solidification in In-, Ga-, Fe-, and Sb-implanted Si after pulsed laser annealing. Interface instability results in lateral segregation of impurity, which leads to the formation of cellular structures that can be detected in cross section in TEM or top-down in scanning electron microscopy.

Also in the early 1980s, the application to Si solar cells was demonstrated by Fogarassy et al. A technique for solar cell fabrication was shown, based on the deposition of thin films of dopants on Si, followed by irradiation with a high-energy ruby laser with energy densities in the range 1.1–2.0 J/cm$^2$. Several dopants like P, Sb, Bi, Al, Ga, and In were investigated in that study, and by using deposited thin films of dopants, the above-equilibrium solubility levels were achieved [198,199].

As in the previous section on As, by the early 1980s, the community had moved onto the problem of sustaining the metastable solubility of Sb during post-laser thermal treatments, or at least understanding it. Pogany et al. looked at metastable solid solutions of Sb in (100) Si and determined that Sb precipitation and relief of lattice strain were associated with the movement of Si interstitials, which precipitate and agglomerate as TEM-visible crystal defects [200]. Further studies by Pennycook et al. [201,202], and Allain et al. [203] added to the knowledge of precipitation, deactivation, and diffusion of metastable Sb in Si.

In the mid-1990s, things became a little more varied as Desmur et al. presented a novel application of highly doped Si in combination with laser anneal, as they proposed pulsed laser etching of Si for dopant profile modification and dopant desorption induced by surface melting. A 308-nm excimer laser anneal was used, and the desorption efficacy of several dopants including Sb was tracked during melt laser anneal in $Cl_2$ pressurized ambient [204].

By 2000, Sb was being studied alongside the more commonly used dopants of B, P, and As for transistor source-drain evaluation. Murto et al. used a 532-nm laser pulse of approximately 20-ns duration, with a focused beam size of 1 cm × 1 cm. Laser energies were varied from 0.30 to 0.68 J/cm$^2$ with post-laser anneal rapid-thermal-anneals in the range of 700–1000°C. Extensive Hall effect and SIMS analysis were carried out and presented for B, P, As, and Sb in parallel [205].

We should again mention the extensive thermal stability of dopants study by Takamura here in relation to Sb, as shown in Fig. 5.11 [132]. Laser annealing was performed on a frequency-doubled Nd:YAG laser operating at 532 nm, with laser energy densities of 0.73–0.90 J/cm$^2$. Post-laser anneal B and Sb remained fully active until 700–800°C. For Sb specifically, the same group followed up on the defect annihilation study where diffuse X-ray scattering and TEM study of defects in Sb-implanted Si was executed [206].

Finally, very recently, Tabata et al. studied the segregation and activation of Sb-implanted Si by 308-nm UV excimer nanosecond-laser-anneal-induced solidification. Sb atoms strongly segregated toward the surface when the laser melted the Sb-implanted Si layer (see Fig. 5.13). The active dopant level after laser anneal showed a significant improvement ($\sim$2.6 × 10$^{20}$ atoms/cm$^3$), far higher than the equilibrium solid solubility limit [207].

### 5.3.4 Laser annealing of Group VI impurities

Group VI impurities are not commonly used in Si technology for doping, although they may have other uses. The problem is that the double donors do not readily give up their free electrons, making their applicability somewhat limited. Indeed, as a guide to Group VI donor states, Schaub et al. wrote in 1984 that Te in Si has one double donor and several monovalent donor species. There is a deep level with an electrical activation energy of 200 meV that was determined from the saturation of the free electron concentration. This level represents the first ionization stage of the Te double donor. The second ionization stage is estimated to have an even higher activation energy of 440 meV [208].

This scenario has been somehow confirmed by most of the experiments published on Group VI impurities in combination with laser anneal until the early 2000s, as briefly summarized below.

Sheehy et al. did an extensive study of chalcogen doping of Si via intense femtosecond-laser irradiation using S, Se, and Te powders (see Fig. 5.14). Approximately 2 mg of the desired dopant in powder form was manually dispersed across the Si surface. Samples were irradiated with a 1-kHz train of 100-fs, 800-nm laser pulses with an energy density of 10 kJ m$^{-2}$ focused to a spot size of 150 μm in diameter. The Se- and Te-doped samples showed a promise for use in infrared photodetectors as they retained their infrared absorptance even after thermal annealing, correlated with a supersaturated solution of trapped chalcogen dopants and point defects in the Si, which modified the electronic structure of the surface layer [209].

An early study of Te in Si, by Foti et al., involved a lattice location study of impurity after Q-switched ruby laser irradiation. It was determined that the laser processing can reorder the damaged region and to put implanted Te ions in substitutional lattice sites with better results than in thermal annealing [210]. Also in the early 1980s, Kemerink et al. published a number of papers on Te in Si. For example, after pulsed ruby laser annealing of Te-implanted Si, considerable impurity redistribution was observed with little surface segregation. In the bulk, substitutional fractions of 85-90% were determined [211]. Furthermore, the same group saw that the fraction of substitutional Te atoms in Si, deduced from

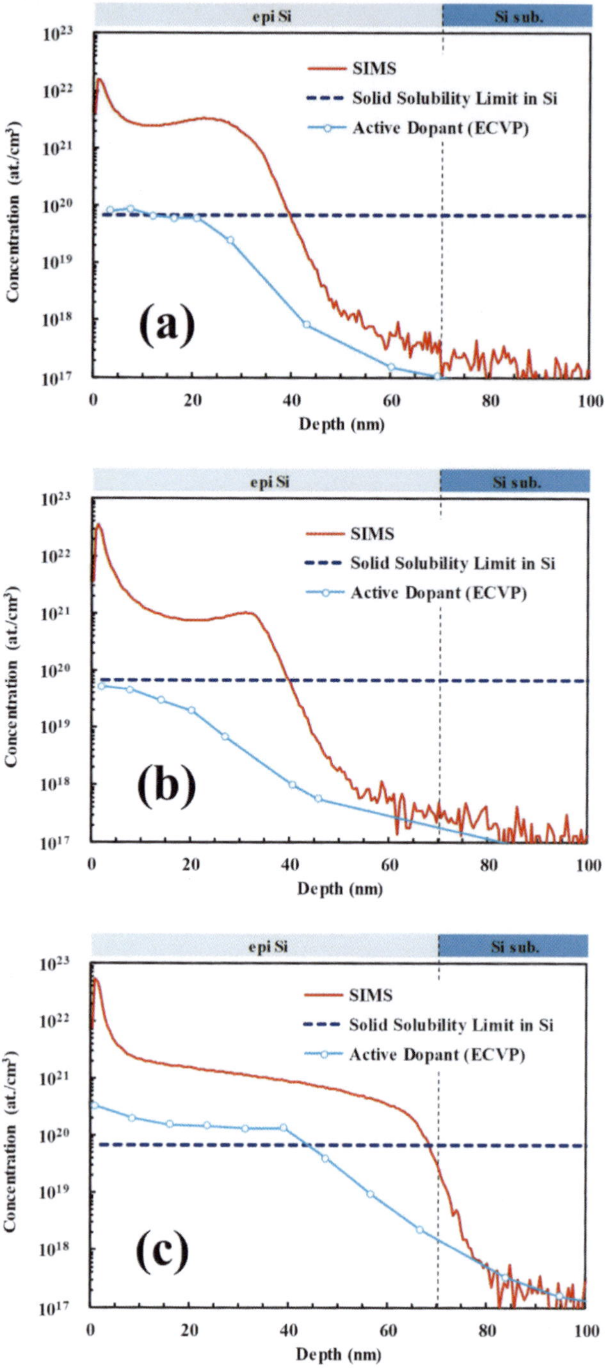


**FIG. 5.13** Electrochemical capacitance-voltage profiling measured on the samples annealed at (A) 1.50 J/cm$^2$, (B) 1.80 J/cm$^2$, and (C) 2.20 J/cm$^2$. The Sb SIMS profiles *(red solid lines)* and the equilibrium solid solubility limit of Sb in Si given by the literature *(blue dotted lines)* are also shown. *(T. Tabata, P.-E. Raynal, K. Huet, J.-L. Everaert, J. Appl. Phys. 127 (2020) 135701.)*

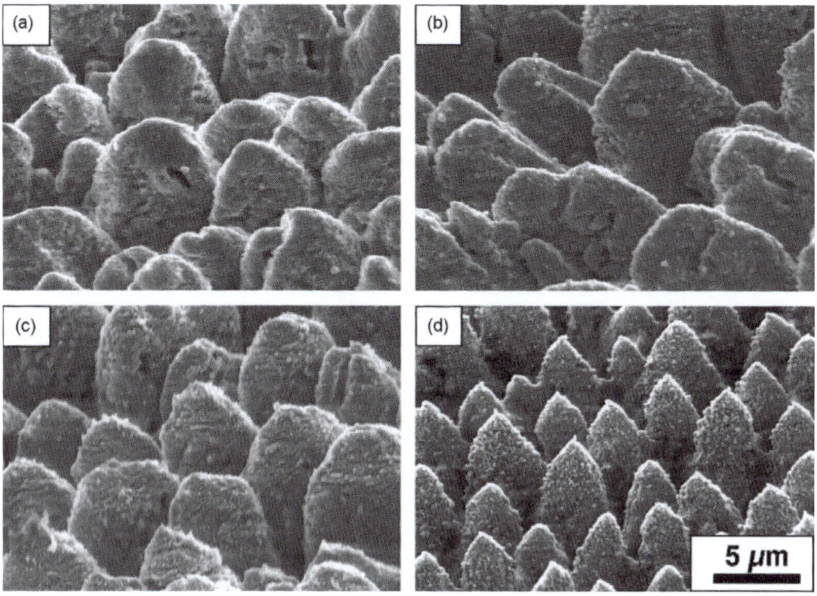


**FIG. 5.14** Scanning electron microscopic images of irradiated surfaces using (A) sulfur powder, (B) selenium powder, (C) tellurium powder, and (D) sulfur hexafluoride gas as a dopant source. In each image, the surface is at a 45 degree angle and the long axis of the structures' elliptical base is parallel to the plane of the image. *(M.A. Sheehy, B.R. Tull, C.M. Friend, E. Mazur, Mater. Sci. Eng. B 137 (2007) 289. (D) Taken from M.A. Sheehy, L. Winston, J.E. Carey, C.M. Friend, E. Mazur, Role of the background gas in the morphology and optical properties of laser-microstructured silicon. Chem. Mater. 17 (2005) 3582–3586. https://doi.org/10.1021/cm049029i.)*

temperature-dependent Hall measurements, is comparable with that derived from channeling experiments. The donor activation energy was measured to be a relatively deep at 0.19 eV. Approximately 10% of the Te atoms produce a shallow donor level at 0.06 eV below the conduction band, and were hypothesized to be associated with a defect [212,213].

In contrast, more promising results were recently reported by Wang et al. [214,215], who investigated the silicon n-type doping by Te ion implantation and melt laser annealing (XeCl laser, $\lambda = 308$ nm, pulse duration 28 ns). By combining Hall effect measurements with RBS analysis in channeling geometry (including angular axial scans around different crystallographic directions) and ab initio calculations, the authors demonstrated that when the implanted Te dose increases, the interstitial Te fraction decreases, while substitutional Te dimers become the dominant configuration as effective double donors, leading to free carrier concentrations as high as $8.3 \times 10^{20}$ cm$^{-3}$.

## 5.4 Dopant redistribution and activation in Ge

### 5.4.1 Introduction

Germanium, after the ground-breaking studies performed between the 1940s and early 1960s, which contributed to the foundation of the modern microelectronic industry, experienced decades of oblivion. In the last years (since more than one decade), the research community returned instead to investigate Ge with more and more increasing interest [216], mainly as a possible high-mobility replacement of Si in nanolectronic devices [217], but also for applications in a huge number of other fields, mostly in IR photonics such as photodetectors [218], lasers [219], LEDs [220], waveguides [221], plasmonics [221,222], as well as in photovoltaics [223], spintronics [224], and quantum computing [225].

Besides the interesting characteristics in terms of electronic and optical properties, Ge presents several significant differences with respect to Si also in terms of junction formation technology [226–229]. For example, Ge presents a significantly higher tendency to accumulate damage upon ion implantation [230,231] which, in the case of heavy ion implantation, can even bring to the formation of detrimental surface honeycomb structures [232]. Furthermore, a comparison of maximum solid solubilities in Ge reported in Fig. 5.15 (solid squares) with solid solubilities in Si reported previously in this chapter in Fig. 5.8 shows that maximum solid solubilities for donors in Ge are significantly lower than in Si, with values of 20, 8, and $1 \times 10^{19}$ $cm^{-3}$ for P, As and Sb, respectively, which makes it quite challenging to achieve high donor concentrations. Solubilities of B, Al, and Ga acceptors are instead relatively high, with values around $5 \times 10^{20} cm^{-3}$, even if B presents severe activation issues (in Si, the values are close to $1 \times 10^{19} cm^{-3}$ for Al and Ga, and $5 \times 10^{20} cm^{-3}$ for B—not reported in Fig. 5.8).

Another significant difference is that, in contrast to Si where both self-interstitial (I) and vacancy (V) point defects play a major role, Ge is dominated by V's. The equilibrium concentration of V's is considerably much higher than I's, and V's are present in the charged states neutral $V^0$ and doubly negatively charged $V^=$ [226]. This determines several peculiar effects. The first is that the ion implantation damage, which has a main interstitial nature, dissolves relatively quickly, and no significant transient enhanced diffusion phenomena are generally observed after ion implantation [253,254]. Second, diffusion mechanisms are mediated by V's for all the dopants, except for B that shows a weak, I-mediated, diffusion. Third, donor-$V^=$ pairing causes both extremely high donor diffusivities at high concentrations due to extrinsic effects and also significant dopant clustering, further limiting the achievement of high active donor concentrations. This is summarized in Fig. 5.15, where the maximum active concentrations are reported after different doping processes. Those obtained for donors after equilibrium processes are systematically lower than the corresponding solubility values.

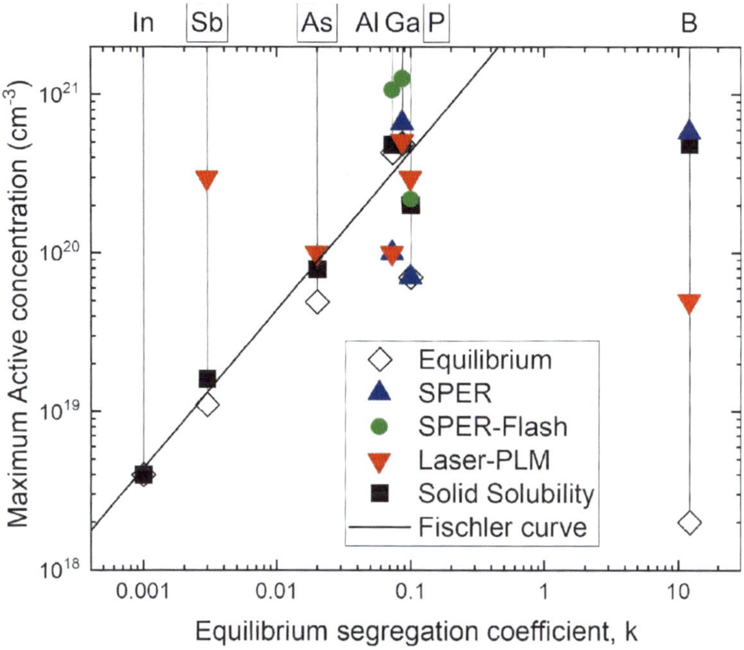


**FIG. 5.15** Maximum active concentration vs equilibrium distribution coefficient, compared with maximum equilibrium solid solubilities *(solid squares)*, for Group III and V *(squared labels)* impurities in Ge, obtained with different processing methods: equilibrium processes *(diamonds)*, solid-phase epitaxy regrowth (SPER) *(up triangles)*, SPER followed by flash lamp annealing (circles) and pulsed laser melting *(down triangles)*. Fischler curve is also reported for comparison [233]. All the equilibrium data are from the reviews by Olelsinski et al. [234–239] and Trumbore [118]. SPER data are from Impellizzeri et al. [240–242] and Posselt et al. [243], Flash lamp annealing data are from Prucnal et al. [244,245], PLM data are from Impellizzeri et al. [246], Mazzocchi et al. [247], Milazzo et al. [248–250], Huang et al. [251], Carraro et al. [252].

The above solubility, diffusion, and activation issues have made it particularly challenging in Ge to meet the requirements for future CMOS technology in terms of dopant activation and confinement, in particular for donors [228]. As a consequence, different strategies for junction formation have been explored during the last years, such as multiple implants, implants at high temperatures, co-implantation, solid-phase epitaxial regrowth (SPER), SPER induced by flash lamp annealing, and laser annealing in the melting regime, here referred to as pulsed laser melting (PLM). Fig. 5.15 includes the maximum active concentrations obtained with the most successful among the above processes, namely, SPER, SPER associated with flash annealing, and PLM. It is clear from the figure that the panorama is significantly different and considerably more complex than in Si: first, metastable solubility is quite hard to achieve in Ge; second, improvements in the maximum active concentration always require phase transitions, namely, crystallization of Ge from solid (SPER) or liquid (PLM) phase;

third, PLM is clearly outperforming in some cases (e.g., for Sb), but seems to be not always the best option in terms of maximum active concentration even if, as it will be underlined in the following sections, PLM still keeps space for improvement for most of the dopants.

The solubility and activation issues depicted previously represent also a major concern for the thermal stability where the above-equilibrium metastable active concentrations are achieved. Therefore, thermal stability must be carefully evaluated, in particular concerning n-type dopants, due to their extrinsic diffusivity and enhanced dopant-V pairing.

In the following sections, a review of the most relevant literature on pulsed laser melting in Ge will be presented, structured in terms of Group III elements (p-type dopants) and Group V elements (n-type dopants). The interest in laser processing of semiconductors (dated in the 1970s, as stated previously in this book, cf. Chapter 1) raised after the beginning of the Ge oblivion era; thus, the neglect of Ge also included the laser processes. The Ge resurgence that started in the 2000s promoted also the investigation of laser processes in such material as a possible method to solve the dopant activation issues reported above. Thus, the literature in this section will be considerably more recent than the one in the previous section on Si. Data will be obtained by different lasers, mostly excimer 308 nm and 248 nm with 20- to 25-ns pulse duration, and Nd-Yag 355 nm with 7-ns pulse duration. Dopants will be both deposited, by sputter deposition or by monolayer doping, and ion-implanted.

### 5.4.2 Laser annealing of Group III impurities in Ge

#### 5.4.2.1 Boron (B)

PLM on Ge implanted with B have been first explored by Mazzocchi et al. [247]. They implanted B in Ge with and without PAI, and performed PLM at 308 nm with a relatively long 180-ns pulse. They observed both a significant out-diffusion and a partial activation, with the best results obtained in non-PAI samples with a maximum active concentration of $2.4 \times 10^{19}$ cm$^{-3}$. They suggested the observed increase in the O contamination as the possible cause of the partial B electrical activation.

A detailed study of the behavior of B in Ge during pulsed laser melting has been then performed by Impellizzeri et al. [246] (see Fig. 5.16). They implanted B at 20 keV, $1 \times 10^{15}$cm$^{-2}$ in Ge (100), and performed PLM with a XeCl excimer laser ($\lambda = 308$ nm, 30-ns pulse duration) at 630 mJ/cm$^2$ with 1, 3, or 10 pulses, in vacuum. The implant was below the amorphization threshold.

A significant redistribution of B is observed (Fig. 5.16A) down to the maximum melt depth (MMD), which is at about 150 nm depth, where a pile-up is also clearly visible. A marked out-diffusion causes a significant dose loss, up to 35% after 10 laser pulses. The profiles have been also successfully simulated by

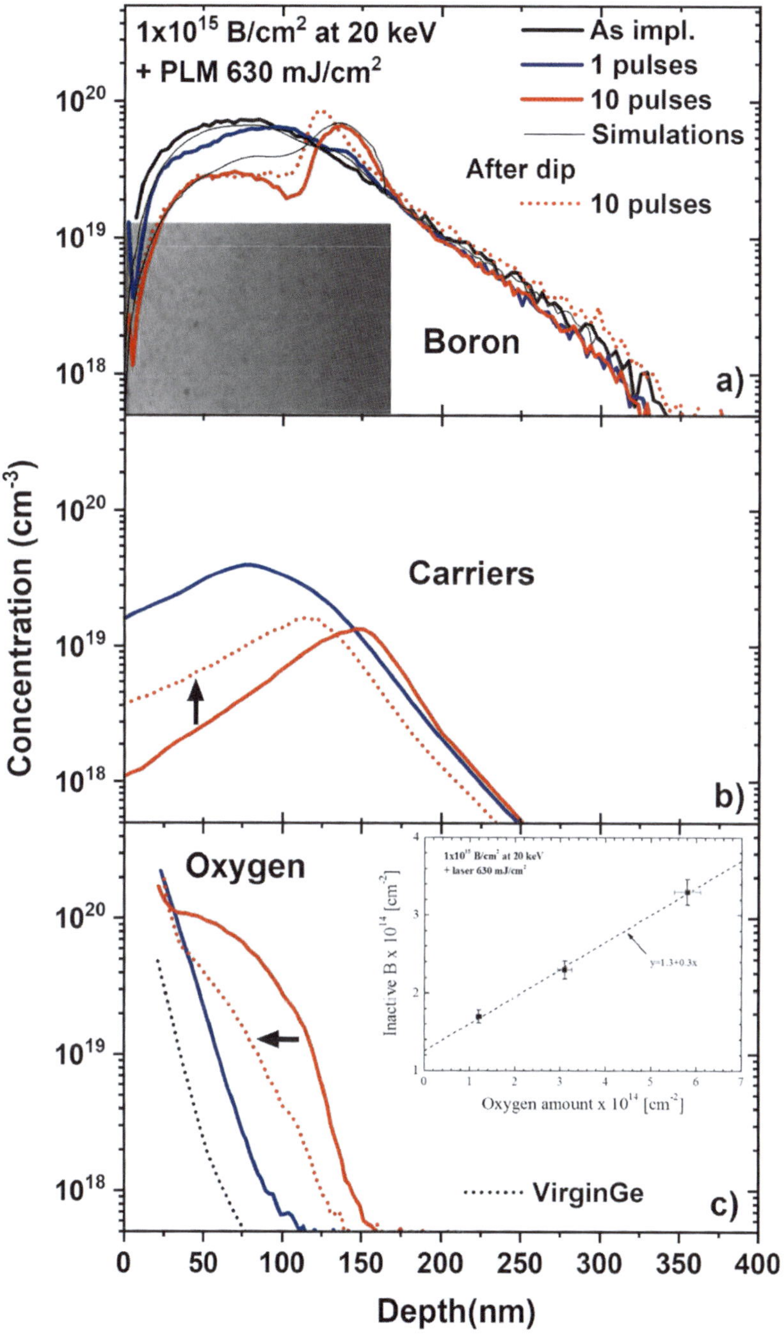


**FIG. 5.16** (A) Boron redistribution in Ge after PLM, compared with (B) corresponding carrier depth profiles and (C) oxygen depth profiles. If PLM is done after native oxide removal by dipping in water at 50°C, oxygen in-diffusion reduces and carrier concentration increases. The inset shows the correlation of the inactive B with the O amount. The X-TEM in (A) refers to the 10 pulses sample without dip and shows no sign of defects or precipitates. *(Readapted from G. Impellizzeri, E. Napolitani, S. Boninelli, G. Fisicaro, M. Cuscuna, R. Milazzo, A. La Magna, G. Fortunato, F. Priolo, V. Privitera, B-doping in Ge by excimer laser annealing, J. Appl. Phys. 113 (2013) 113505. https://doi.org/10.1063/1.4795268.)*

using a detailed phase field model (see [255]; see also Chapter 6 for further details), which considers all the phenomena of nonequilibrium kinetics, including the non-Fickian impurity diffusion in liquid-Ge due to its mixed (covalent/metallic) bonding state [256], which is responsible for the observed pile-up.

The carrier depth profiles (Fig. 5.16B) are lower than the chemical ones with active fraction decreasing by increasing the number of pulses. Best result is obtained after 1 pulse with 50% active fraction and the maximum active concentration of 4 x $10^{19}$ $cm^{-3}$. Such limited activation of B in Ge after PLM has been always confirmed in the literature, including the aforementioned paper Mazzocchi et al. [247], as well as the subsequent paper by Wang et al. [257].

Interestingly, Impellizzeri et al. [246] demonstrate (Fig. 5.16C) that the B activation is well correlated with the oxygen in-diffusion from the surface (see the inset, where the inactive B increases linearly with the O amount). The origin of O is mainly attributed to the native oxide, as oxide removal prior to PLM by rinsing in warm water at 50°C causes a reduction of O diffusion and correspondingly an increase in B activation.

The behavior of O in Ge during PLM is therefore relevant for the activation of B (and also for other dopants, see later in this chapter), and it has been further investigated by Impellizzeri et al. [258] and Milazzo et al. [259–261], showing that the diffusivity of O in pure Ge during PLM is about $6 \times 10^{-5}$ $cm^2/s$ and is significantly reduced in the presence of B to $1 \times 10^{-5}$ $cm^2/s$. This suggests a B—O interaction in the liquid phase leading to the formation of inactive B—O clusters, further confirming O in-diffusion as the main cause of B deactivation.

Ge lattice modifications induced by inactive B after PLM have been investigated by Milazzo et al. [260], indicating that different B clusters form closer to the surface, where B interacts with O, and at larger depth, i.e., at the pile-up, where B—B interaction is mainly responsible for the B deactivation.

Metastability of the B-doped layers has been investigated by Impellizzeri et al. [246], showing that no sign of deactivation and negligible diffusion occurs at least up to 1 h at 600°C.

#### 5.4.2.2 Aluminum (Al)

PLM of Ge doped with Al has been investigated only recently by Milazzo et al. [249]. They implanted Ge (001) with 25 keV Al at two doses, $6.4 \times 10^{14}$ $Al/cm^2$ and $4.2 \times 10^{15}$ $Al/cm^2$, and performed PLM in air atmosphere with an Nd:Yag laser apparatus emitting 7-ns pulses at 355-nm wavelength. They used a fixed energy density of 400 $mJ/cm^2$, corresponding to a MMD of about 150 nm.

The Al diffusion reported in Fig. 5.17 presents differences and similarities with respect to that shown previously for B in Fig. 5.16. The diffusion extent is smaller, due to the shorter laser pulse, and consequently, Al does not reach the MMD and there is no sign of the related pile-up. On the contrary, a clear pile-up forms close to the surface, being well correlated with an O pile-up. O is induced by the ion implantation (high dose, Fig. 5.17C) or in-diffused from the surface (low dose, panel 3d). It is nice also to note, as reported in the paper, that O diffusion is retarded by the presence of Al with respect to unimplanted Ge, similar to what reported previously for B. It seems that Al has a strong tendency, even more than B, to interact with O in the liquid phase. The pile-up is similar to the one observed on Al implanted in Ge and annealed at temperatures above 700°C [242]. Integration of the Al profiles indicates a significant dopant loss of 48% and 27% for the low dose and high dose, respectively.

Electrical activation is partial, being limited to $1.2 \times 10^{20}\ cm^{-3}$ for higher-dose samples, i.e., close to what is achieved by SPER [242]. The limited active

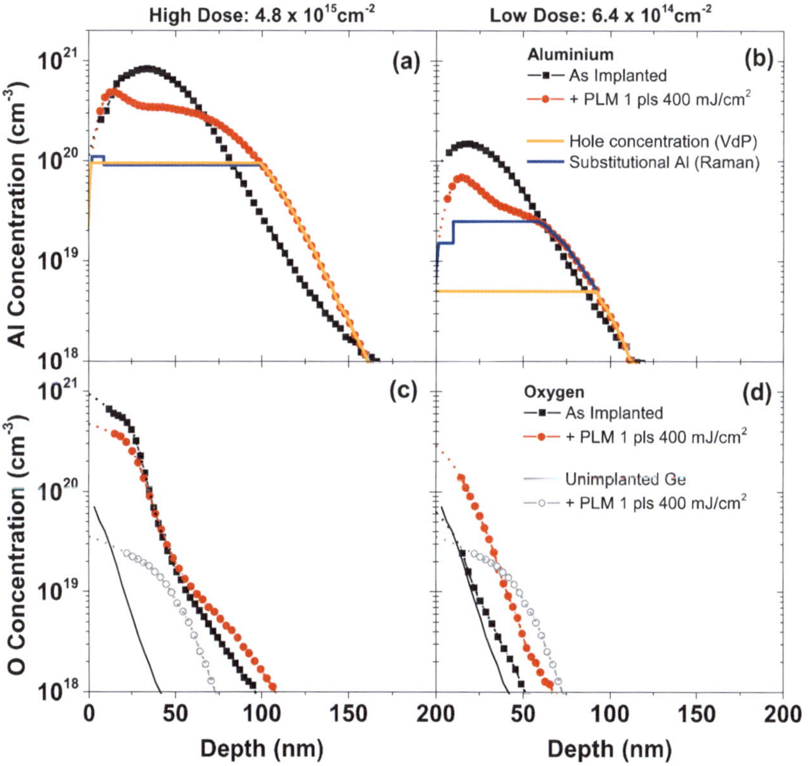


**FIG. 5.17** *Top panels*: Al concentration profile as measured by SIMS for the high-dose (A) and low-dose (B) samples as implanted *(squares)* and after 1 PLM pulse at 400 mJ/cm$^2$ *(circles)*. Active levels as calculated from VdP and Raman measurements are also reported in orange and blue line, respectively. *Bottom panels*: O concentration profiles for the high-dose (C) and low-dose (D) samples as implanted *(squares)* and after 1 PLM pulse at 400 mJ/cm$^2$ *(circles)*. As reference, the O profile measured on unimplanted Ge before *(black line)* and after *(gray open squares)* the same PLM irradiation. The profiles are extrapolated *(dashed lines)* in the first 10 nm below the surface, where the technique is known to exhibit artifacts [249].

fraction at the higher dose is ascribed to Al-Al interaction occurring during the recrystallization process, rather than Al-O interaction, which instead takes place close to the surface at the lower dose. However, it would be interesting to confirm this aspect by investigating Al activation at high doses during PLM without any O contamination induced by the ion implantation. Therefore, Al doping in Ge by PLM might still have space for improvement, through a better control of the process conditions.

#### *5.4.2.3 Gallium (Ga)*

Ga doping of Ge by PLM has been investigated only by Wang et al. [257]. They implanted 5 keV Ga at a dose of $5 \times 10^{15}$ cm$^{-2}$. No details on the implant temperature are reported, as well as no information is present on the possible onset of surface honeycomb structures after ion implantation. They performed PLM with long and short pulses, different energy densities and single and multipulse mode, although no details are given on the exact numbers used for the laser processing conditions, as well as on the processing atmosphere.

Anyhow, they reported a remarkable active concentration of $5.1 \times 10^{20}$ cm$^{-3}$ for the long pulse with multipulse processes at higher energies, despite the observation of a significant out-diffusion (up to 90% of the implanted dose). Unfortunately, no information is reported on the maximum active concentration obtained with the other conditions explored.

Even if scarce of details, the results obtained by Wang et al. are quite relevant and highly promising. By considering also the high active concentrations obtained by other nonmelting techniques (see Fig. 5.15 and [240,241,245]), it can be expected that PLM of Ge doped with Ga has large margins for improvement. This demands for future detailed explorations of different processing conditions in terms of implantation doses, short pulse duration, different number of pulses, and laser processing atmosphere.

### 5.4.3 Laser annealing of Group V impurities in Ge

Given the major diffusion and activation issues experienced by donors in Ge, as outlined in Section 5.4.1, PLM has been investigated more deeply than for acceptors, as an alternative annealing technique to boost the active fraction and confine donor diffusion. This section will review the work done on Group V impurities, starting with the early works appeared at the mid-2000s.

#### *5.4.3.1 Phosphorus (P)*

Phosphorus in Ge during PLM has been first studied by Huang et al. [262]. They implanted 20 keV P at a dose of $1 \times 10^{15}$ cm$^{-2}$ and subsequently performed PLM with a KrF laser at 248-nm wavelength and 23-ns pulses in air atmosphere. They observed a clear improvement compared with other annealing methods of the sheet resistance $R_s$ after PLM, with an increase with the number of pulses, suggesting the presence of P out-diffusion. They did not extract the maximum active concentration, but from the lowest $R_s$ value they reported (50 Ohm/square) and the corresponding P depth profile thickness (around 80 nm). a maximum active concentration around $1 \times 10^{20}$ cm$^{-3}$ could be inferred, i.e., in line with other results reported in the literature afterward.

Interestingly, they reported also an increase of the MMD by increasing the number of pulses and interpreted this as a result of the decrease in reflectivity related to the increase in surface roughness induced by the PLM. Indeed, they observed a progressive increase in the root-mean-square surface roughness with the number of pulses from 0.4 nm before PLM to 7 nm after 10 pulses. This increase of MMD with the number of pulses was previously observed also in Si [152], where a significant reduction of the phenomenon is obtained by reducing the O concentration in the processing atmosphere [263].

Heo et al. [264] investigated the effect of 1 pulse 0.3–0.5 J/cm$^2$ KrF excimer laser at 248-nm wavelength on Ge (001) doped by $PH_3$ plasma doping (PLAD) at 1 keV, $1 \times 10^{16}$ P/cm$^2$. They reported a box-like, shallow profile with relatively low Rs at 0.5 J/cm$^2$, and a Hall dose pointing to a huge Hall carrier concentration above $1 \times 10^{21}$ cm$^{-3}$. However, it should be pointed out that, even if none of the several other studies reported in the literature on PLM on Ge doped with P employed PLAD as doping method, all of them stuck at maximum active around few $1 \times 10^{20}$ cm$^{-3}$, i.e., one order of magnitude below the Heo's value. Therefore, we believe that such an extremely high value, not in line with the relatively vast available literature, should be confirmed by other independent studies, possibly trying to reproduce the Heo's experiment and investigate the role of PLAD in the P activation, before being considered as a benchmark for the maximum active P concentration achievable in Ge by PLM (cf. Fig. 5.15).

Tsouroutas et al. [265–267] subsequently performed the experiments similar to those reported above by Huang et al. [262] but with a shorter pulse laser, i.e., 4-ns pulses performed with a Nd-Yag laser operating at 355-nm wavelength in the range 0.15–0.4 J/cm$^2$. They performed 5 pulses per each spot. Prior to PLM, they implanted 30 keV P $1 \times 10^{15}$ cm$^{-2}$ at room temperature, which produced an amorphous layer of 65 nm thickness. Melting beyond the amorphous/crystalline interface was reached for energy densities equal to and above 0.3 J/cm$^2$ (about 80 nm MMD), and box-like profiles were always reached thanks to the use of multipulse processes. The dose reduction during PLM was relatively limited, i.e., below 20%, thanks to the short pulses.

Mazzocchi et al. [247] and Shayesteh et al. [268]—the latter have been reviewed also by Huet et al. [269]—made quite similar experiments. Both groups implanted P at 15 keV, $1 \times 10^{15}$ cm$^{-2}$ at room temperature producing a 30- to 35-nm-thick amorphous layer and used the same long pulse (180 ns) XeCl excimer laser emitting at 308-nm wavelength by LASSE-SCREEN, i.e., a significantly much longer pulse than those used in the previous literature studies. They investigated a wide range of energy densities, from submelt to melt well beyond the a/c interface. Mazzocchi et al. used a 1.5-μm-thick Ge layers deposited on Si, whereas Shayesteh et al. used bulk p-type Ge as a substrate. They obtained similar results in terms of dopant diffusion and activation, structural modification, and surface roughness. The samples by Shayesteh et al. received also an extensive electrical characterization and have been the subject of further investigations in subsequent papers.

Representative results are reported in Fig. 5.18. Thanks to the long pulses used, a significant redistribution of P during PLM is observed already with single pulses, with a dose loss up to 50% and box-like profiles for the higher energy densities. P is almost fully active, reaching a maximum carrier concentration of $1.65 \times 10^{20}$ cm$^{-3}$, as shown by the ECV results reported in Fig. 5.18 and also confirmed by Hall measurements.

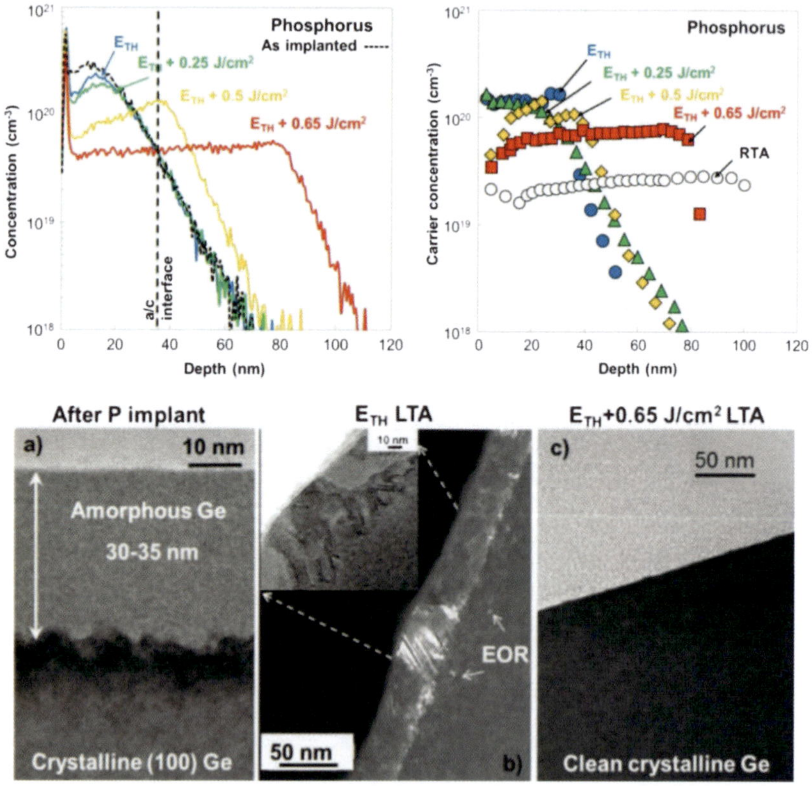


**FIG. 5.18** P profiles obtained by SIMS *(top-left)* and carrier profiles obtained by ECV of P-implanted samples subjected to PLM at energies ranging from $E_{TH}$ to $E_{TH}$ + 0.65 J/cm$^2$. $E_{TH}$ represents the threshold for surface melting and, in this experiment, is $E_{TH}$ = 0.55 J/cm$^2$. Bottom panels are TEM images of (A) amorphous and crystalline interface in Ge with 30–35 nm amorphous depth after P implant, (B) defective crystalline Ge after $E_{TH}$ PLM(LTA), and (C) crystalline Ge with no visible defects after $E_{TH}$ + 0.65 J/cm$^2$ PLM [268].

It's nice to note that ECV profiles follow quite well the SIMS ones, indicating a uniform 100% active fraction, and also suggesting that an O in-diffusion from the surface, if present, has no role on the P activation. We remind that B experiences instead a marked depletion of the active concentration close to the surface due to a strong B—O interaction (see Fig. 5.16).

Interestingly (Fig. 5.18), no defects are observed by TEM at the higher ED, where the Ge is melted beyond the a/c interface, whereas at intermediate ED, where melting arrived just in correspondence of the a/c interface, end-of-range defects form as well as defects or stacking faults in the regrown zone [73,247,268]. A similar effect has been reported also by Tsouroutas et al. [266]. Melting beyond the a/c interface is important also to control the surface morphology, as RMS significantly increases at intermediate ED, due to a defective Ge recrystallization, then decreases again at higher ED, reaching values only slightly above the as-implanted ones [247,268].

Shayesteh et al. [268] also investigated the leakage currents of the above n+/p junctions. The carrier lifetimes on the same samples were subsequently investigated by Filippone et al. [270]. It was found that both leakage currents and generation lifetimes improved with increasing laser anneal energy density, coinciding with better crystal quality, however without outperforming reference samples processed by RTP.

Lombardo et al. [89] successfully modeled P depth profiles extracted from the same samples of the study by Shayesteh et al. [268], fully taking into account the phenomenon of explosive crystallization. Still on the same samples, Boninelli et al. [271] made an extensive structural characterization involving CBED, HRXRD, and SIMS, and obtained for the first time a precise measurement of the lattice parameter modification induced by substitutional electrically active P atoms. They found that, as predicted by the theory including the "electronic contribution" to the lattice parameter modification, P doping results in a very small lattice expansion, despite the small atomic radius of P compared with Ge.

In 2015, both Borland et al. [272] and Huang et al. [251] explored a P concentration regime higher than most of the previous studies. Borland et al. implanted P at 4 keV, $5 \times 10^{15} cm^{-2}$ in Ge layers deposited on Si, reaching as-implanted peak concentrations well above $1 \times 10^{21} cm^{-3}$. Huang et al. deposited Ge doped in situ with $7 \times 10^{20} cm^{-3}$ P on SOI substrates. Despite the much different starting samples and laser conditions used, they obtained the similar maximum active concentrations of $3 \times 10^{20} cm^{-3}$, suggesting that this might represent the metastable solubility limit for P in Ge during PLM.

Remarkably, Huang et al. [251] evidenced for the first time that PLM can induce a modification of the strain of Ge layers deposited on Si. In their samples, the biaxial in-plane strain changed from compressive −0.17% before PLM, to tensile +0.35% after PLM, and they attributed this effect to the thermal strain induced by the different expansion coefficients of Ge and Si.

Frigerio et al. [273], similar to Huang et al. [251], performed PLM on Ge/Si epitaxial layers doped with P in situ at a lower concentration of $1.1 \times 10^{20}$ cm$^{-3}$. They obtained an active concentration of about $9 \times 10^{19}$ cm$^{-3}$, estimated by reflectivity measurements, and obtained tensile strain after PLM of +0.38%. More importantly, they performed for the first time an optical characterization of a Ge-doped layer after PLM obtaining that, as a result of the high active P concentration and the high tensile strain resulting from the PLM, a plasma wavelength reduction to about 5.4 μm, a clear bandgap narrowing, and an increased ratio of direct to indirect bandgap emission had been achieved. These results make PLM quite attractive for several Ge applications, not only in nanoelectronics, but also in photonics and plasmonics.

Pastor et al. [274] explored 4-ns, 355-nm PLM on P+F co-implanted 1-μm-thick Ge layers grown on Si. F shows a beneficial role in reducing the surface deterioration which is instead observed, presumably due to the very short pulse used, in samples implanted only with P. They obtained a maximum active concentration of $1 \times 10^{20}$cm$^{-3}$ and observed that the as-grown tensile strain of +0.16% is stable against PLM, at variance with previous results, presumably due to the relatively thick Ge layer and the short pulse.

Milazzo et al. [250] investigated the possible role of threading dislocations (TD) in the modifications induced by 7-ns, 355-nm PLM performed on 500-nm-thick Ge layer grown on Si and implanted with 30 keV P $1 \times 10^{15}$ cm$^{-2}$. They compared two different Ge/Si samples with high and low TD density with a reference bulk Ge sample, confirming that neither TD nor the Si substrate plays any significant role in dopant diffusion and activation during PLM (with a maximum active concentration of about $1.5 \times 10^{20}$ cm$^{-3}$).

Jiménez et al. [223] recently demonstrated that PLM with a KrF excimer laser emitting 22-ns pulses at 248 nm can be successfully exploited to dope Ge with P by a spin-on-dopant source. They deposited on Ge a SoD layer 180 nm thick with a P concentration of $5 \times 10^{20}$ cm$^{-3}$ by spin coating and cured it at 200°C. Then, they performed PLM in the 0.65 to 0.95 J/cm$^2$ range, in both single- and multipulse mode. In comparison with other doping methods described previously, SoD+PLM suffers from the limited availability of P able to diffuse into Ge from the surface SoD source, and from out-diffusion at high ED values in multipulse mode. Furthermore, P partially activates at the highest concentrations and tends to saturate to an active concentration of $1.5 \times 10^{19}$cm$^{-3}$. This partial activation is at variance with all the observations reported previously regarding P introduced in situ or by ion implantation and seems to be related to the presence of a SoD surface layer during PLM.

### 5.4.3.2 *Arsenic (As)*

PLM in Ge implanted with As has been investigated by Milazzo et al. [248] and Shayesteh et al. [268], obtaining results in agreement with each other.

Shayesteh et al. implanted As at 28 keV, 1 × $10^{15}$ $cm^{-2}$ forming an amorphous layer of 35–40 nm thickness, and used a XeCl excimer laser with 308-nm wavelength and 160-ns pulse duration. Milazzo et al. implanted 40 keV at two doses, i.e., 2 × $10^{14}$ $cm^{-2}$ (amorphous Ge of 48 nm thickness) and 3 × $10^{15}$ $cm^{-2}$ (amorphous Ge of 61 nm thickness) and used a XeCl excimer laser with a shorter duration of 28 ns. Both papers report that, if the energy density is high enough to melt beyond the a/c interface, excellent crystallization is achieved after PLM with no evidence of extended defects in X-TEM analyses.

As shown in Fig. 5.19, which reports representative results by Milazzo et al. [248], As redistributes within the melted region, reaching box-like profiles in the case of multipulse processes. Differently than the case for other dopants in Ge reported previously in this chapter, no dose loss has been detected after PLM for As in Ge. The chemical profiles have been successfully simulated by using a phase field model (see [255]; and also Chapter 6 for further details), which considers all the phenomena of non-equilibrium kinetics, including the non-Fickian impurity diffusion in liquid-Ge due to its mixed (covalent/metallic) bonding state [256]. In the case of As, its low equilibrium segregation coefficient $k = 0.02$ prevented the formation of a pile-up at the MMD.

As shown by the bottom panel and the inset in Fig. 5.19, the electrical activation is only partial, reaching a maximum active fraction of 50% at low As concentrations, and saturating to a maximum active value of 1 × $10^{20}$ $cm^{-3}$ at the highest concentrations. The origin of the partial and saturating As activation and the structural modification occurring during PLM of Ge implanted by As has been subsequently deeply investigated [73,259,260,275,276]. Even if O actually in-diffuses at appreciable concentrations in As-implanted Ge during PLM [259], its role in the activation of As has been definitely excluded, thanks to the absence of any correlation between the O dose present in the sample and the fraction of electrically active As. Furthermore, neither As diffusion nor the O diffusion is influenced by the coexistence of the two elements, excluding any As-O interaction in the liquid phase during PLM [259], in contrast to what is instead observed for B. Extensive characterization of those samples by HRXRD, APT, and PAS [260,275] indicated that a high concentration of very small clusters containing both vacancies and As atoms are formed during PLM, and suggested that the mechanism of As partial activation might rely on the incorporation of a fraction of As atoms in the crystal at the liquid/solid interface in non-substitutional sites, in the form of AsV complexes. Such complexes could subsequently migrate and/or dissociate during lattice cooling by interacting with As atoms forming inactive $As_2V$. Interestingly, PLM in undoped Ge seems not to introduce vacancy clusters in the matrix [275], at variance with what instead observed in silicon [277,278]. In Ge, vacancy introduction can be triggered instead by the presence of donors, as will be also discussed later in this chapter in the antimony section.

The concept of metastable dopant solubility of PLM processes has been frequently underlined within this chapter (see, for example, Section 5.3.1). The

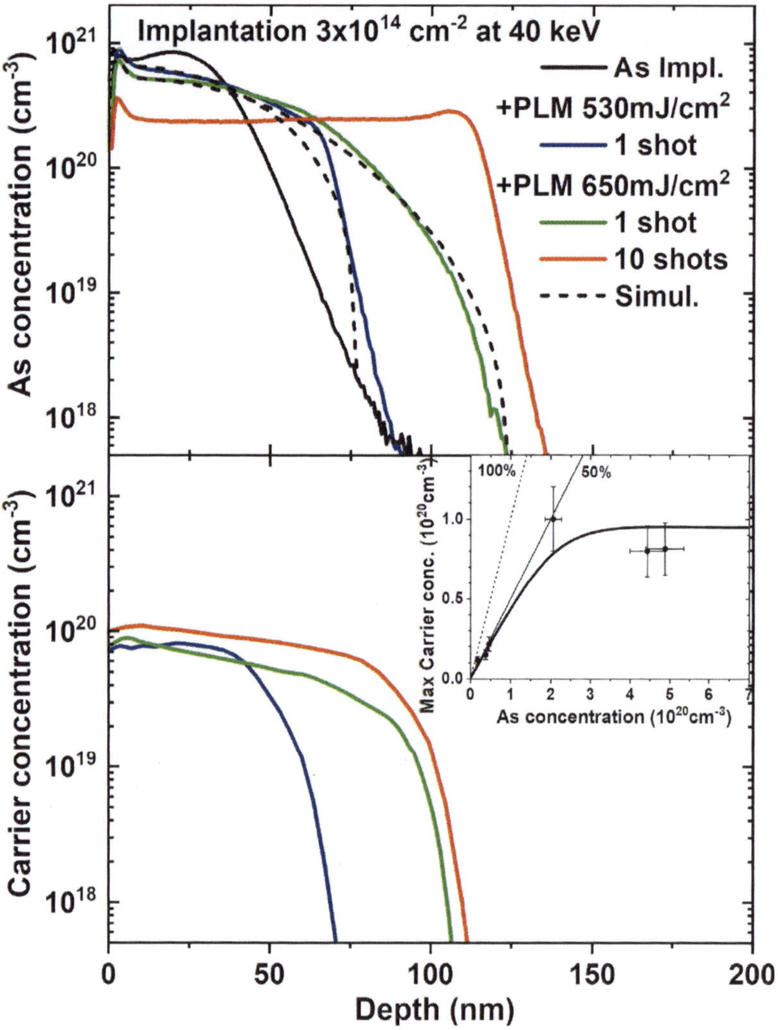


**FIG. 5.19** Arsenic *(top panel)* and carrier *(bottom panel)* concentration depth profiles of Ge implanted with As and annealed by pulsed laser melting at different energy densities and number of pulses. Simulations based on a phase field model are also reported. The inset shows the partial and saturating behavior of the As active concentration as a function of the As chemical concentration. *(Readapted from R. Milazzo, E. Napolitani, G. Impellizzeri, G. Fisicaro, S. Boninelli, M., Cuscunà, D. De Salvador, M. Mastromatteo, M. Italia, A. La Magna, G. Fortunato, F. Priolo, V. Privitera, A. Carnera, N-type doping of Ge by As implantation and excimer laser annealing, J. Appl. Phys. 115 (2014) 053501. https://doi.org/10.1063/1.4863779.)*

paper by Milazzo et al. [276] reports a detailed investigation of the thermal stability of the same above samples highly doped with As by ion implantation and PLM. They have shown that the active As concentration starts to progressively decrease at temperatures of 350°C and above (see Fig. 5.20 left panel).

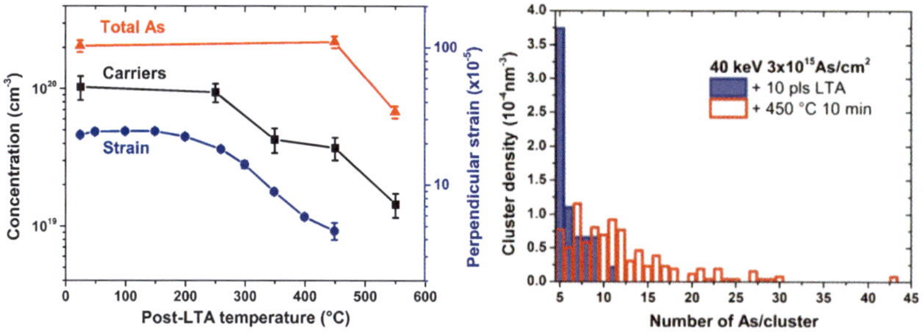


**FIG. 5.20** *Left panel* shows the deactivation of Ge doped with As by PLM as a result of the post-PLM annealing temperature. *Right panel* shows the corresponding increase of the As cluster size distribution, obtained by atom probe tomography, indicative of an Ostwald ripening mechanism occurring during post-LTA annealing at 450°C [276].

This is accompanied with a decrease of the lattice strain, as substitutional As induces a significant lattice expansion, and the lattice parameter reduces as As atoms progressively deactivate and move out of substitutional sites. Detailed analysis of HRXRD data and APT characterization of cluster size distribution (see Fig. 5.20 right panel) suggested that As deactivation upon post-LTA annealing evolves through a cluster growth driven by V point defects with an Ostwald ripening mechanism.

### *5.4.3.3 Antimony (Sb)*

PLM on Ge implanted with Sb has been first investigated by Thareja et al. [279,280]. They implanted Sb at 10 keV, $5 \times 10^{15}$ cm$^2$ at room temperature and performed PLM with pulses at 532-nm wavelength and energy densities in the range 0.15–0.7 J/cm$^2$. The ion implantation induced the formation of surface honeycomb structures. After PLM, the sheet resistance decreases by increasing the laser energy density, while the junction depth increases, but only at energy densities above 0.4 J/cm$^2$, the sample recrystallizes without defects and the surface smoothness is almost restored. A remarkable maximum carrier concentration estimated by SRP above $2 \times 10^{20}$ cm$^{-3}$ is obtained.

Bruno et al. [281], by performing PLM with 12-ns, 532-nm laser pulses both on Ge implanted with Sb and on Ge epitaxially deposited on GeOI and doped in situ with Sb, confirmed the ability of PLM to recover the implantation damage including surface honeycomb structures, and obtained concentrations of substitutional Sb, estimated by c-RBS, higher than $1 \times 10^{21}$ cm$^{-3}$. Furthermore, they observed that, after PLM, Sb induces in Ge layers deposited on GeOI a significant lattice expansion, which then significantly reduces by performing post-PLM thermal annealing up to 488°C.

Two years later, Bruno et al. [282] further investigated PLM on Ge implanted with Sb, employing 28-ns, 308-nm laser pulses in vacuum, and showed that in particular conditions, Sb triggers the generation of an excess of vacancies in the form of $Sb_m$-$V_n$ complexes. Furthermore, they reported a maximum carrier concentration, estimated by SRP, of $1 \times 10^{21}$ cm$^{-3}$.

Borland et al. [272] still investigated PLM in Ge implanted with Sb, by using a 160-ns pulse at 308-nm wavelength. Thanks to the long pulse, they observed box-like profiles already with 1 pulse PLM process. Comparing sheet resistance measurements with SIMS Sb depth profiles, they estimated a maximum active Sb concentration of $8 \times 10^{20}$ cm$^{-3}$, i.e., in line with the results obtained by Bruno et al. [282]. However, it must be noted that neither Borland et al. [272] nor Bruno et al. [282] provided any detail of the actual resistivity-vs-concentration curve used for the resistivity-to-concentration calibration of the sheet resistance or SRP measurements. Given this uncertainty, and the fact that the resistivity curve for Ge doped with Sb is about a factor of two lower than the commonly used resistivity curve for n-type Ge, which has been extracted instead for As and P donors [283], we have not considered the data reported by Borland et al. [272] and Bruno et al. [282] as a benchmark of the maximum active Sb concentration achieved in Ge by PLM (cf. Fig. 5.15).

Sgarbossa et al. [284] recently developed a new method for incorporating Sb in Ge using PLM. They first observed that by annealing Ge above 600°C in an atmosphere rich in Sb atoms, a uniform self-limited Sb monolayer of $6 \times 10^{14}$ Sb/cm$^2$ forms over the Ge surface. Then, they successfully diffused the Sb monolayer within the Ge matrix thanks to PLM processes. Fully active Sb profiles were incorporated, with a maximum concentration of $1.2 \times 10^{20}$ cm$^{-3}$.

Carraro et al. [252] further developed the Sb surface deposition method in conjunction with PLM, by depositing with the sputtering technique thicker Sb layers on Ge in the range 1 to 8 nm, and then performed PLM with two different lasers, a Nd-Yag emitting 7-ns pulses at 355 nm and a KrF excimer laser emitting 22-ns pulses at 248 nm. Then, they performed an extensive chemical, electrical, and structural characterization. They obtained that the deposited Sb entirely diffuses within Ge during PLM at extremely high concentrations, above $5 \times 10^{21}$ cm$^{-3}$, without any sign of surface segregation nor out-diffusion. Fig. 5.21 (left panel) shows an example of such Sb profile after PLM, compared with the corresponding carrier depth profiles obtained using two different methods: the first one used Van der Pauw-Hall data in conjunction with SIMS and assuming the correct Sb resistivity curve and Hall scattering factor for Sb in Ge [283,285]; the second one by performing differential Hall measurements (and still considering the correct Hall scattering factor). As reported in Fig. 5.21 (left panel), the two methods gave results in agreement with each other. They obtained as a whole that (see Fig. 5.21, right panel) at concentrations lower than $1.5 \times 10^{20}$ cm$^{-3}$ Sb is fully active, whereas by increasing the concentration, Sb starts to deactivate reaching a saturation level of about $3 \times 10^{20}$ cm$^{-3}$, in line with the results by Thareja et al. [279,280], and similar to what reported previously for As (see Fig. 5.19), which, however, saturated at a significantly lower concentration. A further increase of the

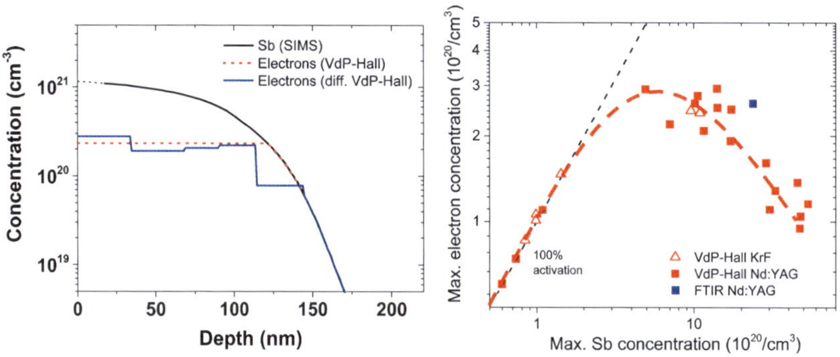


**FIG. 5.21** *Left panel*: Sb concentration depth profile extracted from the sample with 2-nm sputter deposition and treated with 8 laser pulses of Nd:YAG laser compared with the corresponding carrier depth profiles extracted by Van der Pauw-Hall and differential Van der Pauw-Hall. *Right panel*: Maximum electron concentration reported as a function of the maximum Sb concentration.

chemical concentration to higher values resulted in a greater deviation from the full activation line and eventually in a decrease of the carrier concentration.

It is worth reporting that the resistivity and mobility values extracted by Carraro et al. [252] are perfectly in agreement with the extrapolation of the existing values reported in the literature [283,285]. The obtained doped layers proved to be of excellent crystalline quality and did not show any sign of extended defects, allowing for record low resistivity of $1.4 \times 10^{-4}$ Ohm cm and record low plasma wavelength of 2.6 μm, which might provide a breakthrough for nanoelectronic and plasmonic applications based on Ge.

## 5.5 Conclusion

The need to fabricate extremely high-doped semiconductor thin layers is a crucial issue in the modern semiconductor nanoelectronics industry, which can be addressed by the use of nanosecond laser annealing in the melt regime. The most relevant results demonstrating the ability of this technique to achieve dopant activation well above the solid solubility have been presented in this chapter, together with a review of the distinctive physical phenomena occurring during the ultra-fast solidification of laser-induced liquid layers.

Due to the richness and relevance of numerous seminal works published in the late 1970s-the early 1980s in this domain, the results reported in this chapter have mainly been presented in the form of a historical review and perspective. In particular, the concepts of the above-equilibrium solubility and thermal stability have been discussed in general, with pulsed laser anneal in mind, applied to both Si and Ge, while for both materials, the main results were structured in terms of Group III elements (p-type dopants) and Group V elements (n-type dopants). Group VI elements for n-type doping of Si have also been discussed.

Several additional materials science issues related to laser kinetics in the melt regime were reviewed in this chapter, including melt temperature differences between amorphous and crystalline phases, melt front propagation, melt duration, recrystallization velocity, and solute trapping. Finally, the morphology, crystalline nature, and residual damage of Si, Ge, and SiGe materials submitted to nanosecond laser anneals in the melt regime have been discussed according to the material fabrication history as well as the laser anneal conditions.

## References


[1] A.G. Cullis, Transient annealing of semiconductors by laser, electron beam and radiant heating techniques, Reports Prog. Phys. 48 (1985) 1155–1233, https://doi.org/10.1088/0034-4885/48/8/002.

[2] D.H. Auston, J.A. Golovchenko, A.L. Simons, R.E. Slusher, P.R. Smith, C.M. Surko, T.N.C. Venkatesan, Dynamics of laser annealing, AIP Conf. Proc. 50 (1979) 11–26, https://doi.org/10.1063/1.31651.

[3] F. Cristiano, Ion implantation-induced extended defects: structural investigations and impact on Ultra-Shallow Junction properties, Habilitation à Diriger des Recherches, University Paul Sabatier Toulouse III, 2013. https://tel.archives-ouvertes.fr/tel-00919958/document.

[4] F. Spaepen, D. Turnbull, Kinetics of motion of crystal-melt interfaces, AIP Conf. Proc. 50 (1979) 73–83, https://doi.org/10.1063/1.31738.

[5] B.G. Bagley, H.S. Chen, A calculation of the thermodynamic first order amorphous semiconductor to metallic liquid transition temperature, AIP Conf. Proc. 50 (1979) 97–101, https://doi.org/10.1063/1.31740.

[6] E.P. Donovan, F. Spaepen, D. Turnbull, J.M. Poate, D.C. Jacobson, Calorimetric studies of crystallization and relaxation of amorphous Si and Ge prepared by ion implantation, J. Appl. Phys. 57 (1985) 1795–1804, https://doi.org/10.1063/1.334406.

[7] E.P. Donovan, F. Spaepen, D. Turnbull, J.M. Poate, D.C. Jacobson, Heat of crystallization and melting point of amorphous silicon, Appl. Phys. Lett. 42 (1983) 698–700, https://doi.org/10.1063/1.94077.

[8] E.P. Donovan, F. Spaepen, J.M. Poate, D.C. Jacobson, Homogeneous and interfacial heat releases in amorphous silicon, Appl. Phys. Lett. 55 (1989) 1516–1518, https://doi.org/10.1063/1.101593.

[9] R. Hultgren, P.D. Desai, D.T. Hawkins, M. Gleiser, K.K. Kelley, D.D. Wagman, Selected Values of the Thermodynamic Properties of the Elements, American Society for Metals, Metals Park, OH, 1973.

[10] M. Homa, N. Sobczak, Measurements of temperature and heat of phase transformation of pure silicon by using differential scanning calorimetry, J. Therm. Anal. Calorim. 138 (2019) 4215–4221, https://doi.org/10.1007/s10973-019-08716-5.

[11] M.O. Thompson, G.J. Galvin, J.W. Mayer, P.S. Peercy, J.M. Poate, D.C. Jacobson, A.G. Cullis, N.G. Chew, Melting temperature and explosive crystallization of amorphous silicon during pulsed laser irradiation, Phys. Rev. Lett. 52 (1984) 2360–2363, https://doi.org/10.1103/PhysRevLett.52.2360.

[12] P. Baeri, G. Foti, J.M. Poate, A.G. Cullis, Phase transitions in amorphous Si produced by rapid heating, Phys. Rev. Lett. 45 (1980) 2036–2039, https://doi.org/10.1103/PhysRevLett.45.2036.

[13] S.A. Kokorowski, G.L. Olson, J.A. Roth, L.D. Hess, Investigation of the melting temperature of amorphous silicon, Phys. Rev. Lett. 48 (1982) 498–501, https://doi.org/10.1103/PhysRevLett.48.498.

[14] S. Roorda, W.C. Sinke, J.M. Poate, D.C. Jacobson, S. Dierker, B.S. Dennis, D.J. Eaglesham, F. Spaepen, P. Fuoss, Structural relaxation and defect annihilation in pure amorphous silicon, Phys. Rev. B 44 (1991) 3702–3725, https://doi.org/10.1103/PhysRevB.44.3702.

[15] C. Spinella, S. Lombardo, F. Priolo, Crystal grain nucleation in amorphous silicon, J. Appl. Phys. 84 (1998) 5383–5414, https://doi.org/10.1063/1.368873.

[16] M.G. Grimaldi, P. Baeri, M.A. Malvezzi, Melting temperature of unrelaxed amorphous silicon, Phys. Rev. B 44 (1991) 1546–1553, https://doi.org/10.1103/PhysRevB.44.1546.

[17] L. Pelaz, L.A. Marqués, J. Barbolla, Ion-beam-induced amorphization and recrystallization in silicon, J. Appl. Phys. 96 (2004) 5947–5976, https://doi.org/10.1063/1.1808484.

[18] Ioffe. http://www.ioffe.ru/SVA/NSM/Semicond/SiGe/index.html.

[19] R.W. Olesinski, G.J. Abbaschian, The Ge-Si (germanium-silicon) system, Bull. Alloy Phase Diagr. 5 (1984) 180–183, https://doi.org/10.1007/BF02868957.

[20] M. Weizman, N.H. Nickel, I. Sieber, B. Yan, Laser-induced self-organization in silicon-germanium thin films, J. Appl. Phys. 103 (2008), https://doi.org/10.1063/1.2919772, 093536.

[21] T. Tabata, J. Aubin, K. Huet, F. Mazzamuto, Y. Mori, A. La Magna, L.M. Rubin, P. Kopalidis, H.C. Tsai, D. Roh, R. Reece, Super activation of highly surface segregated dopants in high Ge content SiGe obtained by melt UV laser annealing, in: V. Haublein, H. Ryssel (Eds.), Proceedings of International Conference on Ion Implantation Technology, IEEE., 2018, pp. 353–356, https://doi.org/10.1109/IIT.2018.8807912.

[22] P. Baeri, S.U. Campisano, G. Foti, E. Rimini, A melting model for pulsing-laser annealing of implanted semiconductors, J. Appl. Phys. 50 (1979) 788–797, https://doi.org/10.1063/1.326046.

[23] C.M. Surko, A.L. Simons, D.H. Auston, J.A. Golovchenko, R.E. Slusher, T.N.C. Venkatesan, Calculation of the dynamics of surface melting during laser annealing, Appl. Phys. Lett. 34 (1979) 635–637, https://doi.org/10.1063/1.90619.

[24] M.O. Thompson, G.J. Galvin, J.W. Mayer, P.S. Peercy, R.B. Hammond, Melt dynamics of silicon-on-sapphire during pulsed laser annealing, Appl. Phys. Lett. 42 (1983) 445–447,-https://doi.org/10.1063/1.93965.

[25] W.L. Brown, Transient laser-induced processes in semiconductors, in: C.W. White, P.S. Peercy (Eds.), Laser and Electron Beam Processing of Materials, Academic Press, 1980, pp. 20–36, https://doi.org/10.1016/b978-0-12-746850-1.50008-6.

[26] D.H. Auston, C.M. Surko, T.N.C. Venkatesan, R.E. Slusher, J.A. Golovchenko, Time-resolved reflectivity of ion-implanted silicon during laser annealing, Appl. Phys. Lett. 33 (1978) 437–440, https://doi.org/10.1063/1.90369.

[27] D.H. Auston, J.A. Golovchenko, P.R. Smith, C.M. Surko, T.N.C. Venkatesan, Cw argon laser annealing of ion-implanted silicon, Appl. Phys. Lett. 33 (1978) 539–541, https://doi.org/10.1063/1.90429.

[28] D.H. Auston, J.A. Golovchenko, A.L. Simons, C.M. Surko, T.N.C. Venkatesan, Dynamics of Q-switched laser annealing, Appl. Phys. Lett. 34 (1979) 777–779, https://doi.org/10.1063/1.90670.

[29] J. Boneberg, P. Leiderer, On the interpretation of time-resolved surface reflectivity measurements during the laser annealing of Si thin films, Phys. Status Solidi 166 (1998) 643–650, https://doi.org/10.1002/(SICI)1521-396X(199804)166:2<643::AID-PSSA643>3.0.CO;2-L.

[30] F.C. Voogt, R. Ishihara, F.D. Tichelaar, Melting and crystallization behavior of low-pressure chemical-vapor-deposition amorphous Si films during excimer-laser annealing, J. Appl. Phys. 95 (2004) 2873–2879, https://doi.org/10.1063/1.1642286.

[31] C.C. Kuo, Solidification velocity in liquid silicon during excimer laser crystallization, Appl. Phys. A Mater. Sci. Process. 95 (2009) 573–578, https://doi.org/10.1007/s00339-008-4953-9.

[32] G.J. Galvin, M.O. Thompson, J.W. Mayer, R.B. Hammond, N. Paulter, P.S. Peercy, Measurement of the velocity of the crystal-liquid interface in pulsed laser annealing of Si, Phys. Rev. Lett. 48 (1982) 33–36, https://doi.org/10.1103/PhysRevLett.48.33.

[33] G.J. Galvin, M.O. Thompson, J.W. Mayer, P.S. Peercy, R.B. Hammond, N. Paulter, Time-resolved conductance and reflectance measurements of silicon during pulsed-laser annealing, Phys. Rev. B 27 (1983) 1079–1087, https://doi.org/10.1103/PhysRevB.27.1079.

[34] M.O. Thompson, J.W. Mayer, A.G. Cullis, H.C. Webber, N.G. Chew, J.M. Poate, D.C. Jacobson, Silicon melt, regrowth, and amorphization velocities during pulsed laser irradiation, Phys. Rev. Lett. 50 (1983) 896–899, https://doi.org/10.1103/PhysRevLett.50.896.

[35] O. Bostanjoglo, R.P. Tornow, W. Tornow, Nanosecond-exposure electron microscopy of laser-induced phase transformations, Ultramicroscopy 21 (1987) 367–372, https://doi.org/10.1016/0304-3991(87)90034-9.

[36] L. Nikolova, T. Lagrange, B.W. Reed, M.J. Stern, N.D. Browning, G.H. Campbell, J.C. Kieffer, B.J. Siwick, F. Rosei, Nanocrystallization of amorphous germanium films observed with nanosecond temporal resolution, Appl. Phys. Lett. 97 (2010) 203102, https://doi.org/10.1063/1.3518069.

[37] L. Nikolova, T. Lagrange, M.J. Stern, J.M. MacLeod, B.W. Reed, H. Ibrahim, G.H. Campbell, F. Rosei, B.J. Siwick, Complex crystallization dynamics in amorphous germanium observed with dynamic transmission electron microscopy, Phys. Rev. B: Condens. Matter Mater. Phys. 87 (2013), https://doi.org/10.1103/PhysRevB.87.064105, 064105.

[38] L. Nikolova, M.J. Stern, J.M. Macleod, B.W. Reed, H. Ibrahim, G.H. Campbell, F. Rosei, T. Lagrange, B.J. Siwick, In situ investigation of explosive crystallization in a-Ge: experimental determination of the interface response function using dynamic transmission electron microscopy, J. Appl. Phys. 116 (2014), https://doi.org/10.1063/1.4894397, 093512.

[39] M.K. Santala, S. Raoux, G.H. Campbell, Kinetics of liquid-mediated crystallization of amorphous Ge from multi-frame dynamic transmission electron microscopy, Appl. Phys. Lett. 107 (2015) 252106, https://doi.org/10.1063/1.4938751.

[40] H.J. Leamy, J.C. Bean, J.M. Poate, G.K. Celler, Nonequilibrium incorporation of impurities during rapid solidification, J. Cryst. Growth 48 (1980) 379–382, https://doi.org/10.1016/0022-0248(80)90032-9.

[41] A.G. Cullis, H.C. Webber, J.M. Poate, A.L. Simons, Segregation and increased dopant solubility in Pt-implanted and laser-annealed Si layers, Appl. Phys. Lett. 36 (1980) 320–322,-https://doi.org/10.1063/1.91478.

[42] P. Baeri, J.M. Poate, S.U. Campisano, G. Foti, E. Rimini, A.G. Cullis, Dependence of trapping and segregation of indium in silicon on the velocity of the liquid-solid interface, Appl. Phys. Lett. 37 (1980) 912–914, https://doi.org/10.1063/1.91856.

[43] M. Aziz, Crystal growth and solute trapping, Mater. Res. Soc. Symp. Proc. 23 (1984) 369–374, https://doi.org/10.1557/proc-23-369.

[44] M.J. Aziz, T. Kaplan, Continuous growth model for interface motion during alloy solidification, Acta Metall. 36 (1988) 2335–2347, https://doi.org/10.1016/0001-6160(88)90333-1.

[45] L.M. Goldman, M.J. Aziz, Aperiodic stepwise growth model for the velocity and orientation dependence of solute trapping, J. Mater. Res. 2 (1987) 524–527.

[46] R. Reitano, P.M. Smith, M.J. Aziz, Solute trapping of group III, IV, and V elements in silicon by an aperiodic stepwise growth mechanism, J. Appl. Phys. 76 (1994) 1518–1529, https://doi.org/10.1063/1.357728.

[47] R. Reitano, P.M. Smith, M.J. Aziz, Trends in solute segregation behavior during silicon solidification, MRS Proc. 321 (1993) 479, https://doi.org/10.1557/PROC-321-479.

[48] S. Lombardo, P.M. Smith, M.J. Uttormark, D.P. Brunco, K. Kramer, M.O. Thompson, Demonstration of laser-assisted epitaxial deposition of Ge xSi1-x alloys on single-crystal Si, Appl. Phys. Lett. 58 (1991) 1768–1770, https://doi.org/10.1063/1.105085.

[49] S. Lombardo, K. Kramer, M.O. Thompson, D.R. Smith, Pulsed laser assisted epitaxy of $Ge_xSi_{1-x}$ alloys on Si <100>, Appl. Phys. Lett. 59 (1991) 3455–3457, https://doi.org/10.1063/1.105676.

[50] M. Berti, G. Mazzi, L. Calcagnile, A.V. Drigo, P.G. Merli, A. Migliori, Composition and structure of Si-Ge layers produced by ion implantation and laser melting, J. Mater. Res. 6 (1991) 2120–2126.

[51] F. Repplinger, E. Fogarassy, A. Grob, J.J. Grob, D. Muller, B. Prévot, J.P. Stoquert, S. De Unamuno, Preparation of Si1-xGex thin crystalline films by pulsed excimer laser annealing of heavily Ge implanted Si, Thin Solid Films 241 (1994) 155–158, https://doi.org/10.1016/0040-6090(94)90417-0.

[52] D.P. Brunco, M.O. Thompson, D.E. Hoglund, M.J. Aziz, H.J. Gossmann, Germanium partitioning in silicon during rapid solidification, J. Appl. Phys. 78 (1995) 1575–1582, https://doi.org/10.1063/1.360251.

[53] L.J. Willis, D.C. Emmony, Laser damage in germanium, Opt. Laser Technol. 7 (1975) 222–228, https://doi.org/10.1016/0030-3992(75)90043-2.

[54] W.G. Hawkins, D.K. Biegelsen, Origin of lamellae in radiatively melted silicon films, Appl. Phys. Lett. 42 (1983) 358–360, https://doi.org/10.1063/1.93939.

[55] J.F. Young, J.E. Sipe, H.M. van Driel, Laser-induced periodic surface structure. III. Fluence regimes, the role of feedback, and details of the induced topography in germanium, Phys. Rev. B 30 (1984) 2001–2015.

[56] G.E. Jellison, D.H. Lowndes, D.N. Mashburn, R.F. Wood, Time-resolved reflectivity measurements on silicon and germanium using a pulsed excimer KrF laser heating beam, Phys. Rev. B 34 (1986) 2407–2415, https://doi.org/10.1103/PhysRevB.34.2407.

[57] T. Surek, B. Chalmers, The direction of growth of the surface of a crystal in contact with its melt, J. Cryst. Growth 29 (1975) 1–11, https://doi.org/10.1016/0022-0248(75)90041-X.

[58] K.-H. Heinig, M. Voelskow, J. Matthäi, A. Zetzsche, C. Treutler, Local melting by nucleation on the surface of single crystalline silicon, in: K. Hennig (Ed.), Proceedings of the International Conference on Energy Pulse Modification of Semiconductors Related Materials, 25–28 September 1984, ZfK Rossendorf Dresden, Germany, 1984, p. 280.

[59] K.-H. Heinig, Effect of local melting on semiconductor surfaces, in: K. Hennig (Ed.), Proceedings of the International Conference on Energy Pulse Modification of Semiconductors Related Materials, 25–28 September 1984, ZfK Rossendorf Dresden, Germany, 1984, p. 265.

[60] M.P. Smith, R.A. McMahon, M. Voelskow, W. Skorupa, J. Stoemenos, G. Ferro, Process control and melt depth homogenization for SiC-on-Si structures during flash lamp annealing by carbon implantation, J. Appl. Phys. 100 (2006), https://doi.org/10.1063/1.2359684, 094909.

[61] D. Skarlatos, V. Ioannou-Sougleridis, M. Barozzi, G. Pepponi, D. Velessiotis, M.C. Skoulikidou, N.Z. Vouroutzis, K. Papagelis, P. Dimitrakis, C. Thomidis, B. Colombeau, Phosphorous diffusion in $N_2^+$-implanted germanium during flash lamp annealing: influence of nitrogen on Ge substrate damage and capping layer engineering, ECS J. Solid State Sci. Technol. 6 (2017) P418–P428, https://doi.org/10.1149/2.0201707jss.

[62] L. Dagault, S. Kerdilès, P. Acosta Alba, J.M. Hartmann, J.P. Barnes, P. Gergaud, E. Scheid, F. Cristiano, Investigation of recrystallization and stress relaxation in nanosecond laser annealed Si1 − xGex/Si epilayers, Appl. Surf. Sci. 527 (2020) 146752, https://doi.org/10.1016/j.apsusc.2020.146752.

[63] Q.S. Mei, K. Lu, Melting and superheating of crystalline solids: from bulk to nanocrystals, Prog. Mater. Sci. 52 (2007) 1175–1262, https://doi.org/10.1016/j.pmatsci.2007.01.001.

[64] H.T.T. Nguyen, V. Van Hoang, L.N.T. Minh, Melting of crystalline silicon thin films, Comput. Mater. Sci. 89 (2014) 97–101, https://doi.org/10.1016/j.commatsci.2014.03.039.

[65] P.L. Liu, R. Yen, N. Bloembergen, R.T. Hodgson, Picosecond laser-induced melting and resolidification morphology on Si, Appl. Phys. Lett. 34 (1979) 864–866, https://doi.org/10.1063/1.90703.

[66] R. Tsu, R.T. Hodgson, T.Y. Tan, J.E. Baglin, Order-disorder transition in single-crystal silicon induced by pulsed UV laser irradiation, Phys. Rev. Lett. 42 (1979) 1356–1358, https://doi.org/10.1103/PhysRevLett.42.1356.

[67] A.G. Cullis, H.C. Webber, N.G. Chew, J.M. Poate, P. Baeri, Transitions to defective crystal and the amorphous state induced in elemental Si by laser quenching, Phys. Rev. Lett. 49 (1982) 219–222, https://doi.org/10.1103/PhysRevLett.49.219.

[68] G. Foti, S.U. Campisano, P. Baeri, E. Rimini, W.F. Tseng, Laser annealing of self-ion damaged silicon, Appl. Phys. Lett. 35 (1979) 701–703, https://doi.org/10.1063/1.91260.

[69] R.A. McMahon, H. Ahmed, A.G. Cullis, Comparative structural and electrical characterization of scanning-electron- and pulsed-laser-annealed silicon, Appl. Phys. Lett. 37 (1980) 1016–1018, https://doi.org/10.1063/1.91724.

[70] D.K. Sadana, M. Strathman, J. Washburn, G.R. Booker, Transmission electron microscopy and Rutherford backscattering studies of single and double discrete buried damage layers in P + implanted Si on subsequent laser annealing, J. Appl. Phys. 52 (1981) 744–747, https://doi.org/10.1063/1.328756.

[71] S. Earles, M. Law, R. Brindos, K. Jones, S. Talwar, S. Corcoran, Nonmelt laser annealing of 5-KeV and 1-KeV boron-implanted silicon, IEEE Trans. Electron Dev. 49 (2002) 1118–1123, https://doi.org/10.1109/TED.2002.1013265.

[72] Y. Qiu, F. Cristiano, K. Huet, F. Mazzamuto, G. Fisicaro, A. La Magna, M. Quillec, N. Cherkashin, H. Wang, S. Duguay, D. Blavette, Extended defects formation in nanosecond laser-annealed ion implanted silicon, Nano Lett. 14 (2014) 1769–1775, https://doi.org/10.1021/nl4042438.

[73] F. Cristiano, M. Shayesteh, R. Duffy, K. Huet, F. Mazzamuto, Y. Qiu, M. Quillec, H.H. Henrichsen, P.F. Nielsen, D.H. Petersen, A. La Magna, G. Caruso, S. Boninelli, Defect evolution and dopant activation in laser annealed Si and Ge, Mater. Sci. Semicond. Process. 42 (2016) 188–195, https://doi.org/10.1016/j.mssp.2015.09.011.

[74] L.A. Marqués, M. Aboy, I. Santos, P. López, F. Cristiano, A. La Magna, K. Huet, T. Tabata, L. Pelaz, Ultrafast generation of unconventional {001} loops in Si, Phys. Rev. Lett. 119 (2017) 205503, https://doi.org/10.1103/PhysRevLett.119.205503.

[75] L.A. Marqués, M. Aboy, I. Santos, P. López, F. Cristiano, A. La Magna, K. Huet, T. Tabata, L. Pelaz, On the anomalous generation of {001} loops during laser annealing of ion-implanted silicon, Nucl. Instrum. Methods Phys. Res. Sect. B Beam Interact. Mater. Atoms 458 (2019) 179–183, https://doi.org/10.1016/j.nimb.2018.09.030.

[76] K.S. Jones, H. Banisaukas, J. Glassberg, E. Andideh, C. Jasper, A. Hoover, A. Agarwal, M. Rendon, Transient enhanced diffusion after laser thermal processing of ion implanted silicon, Appl. Phys. Lett. 75 (1999) 3659–3661, https://doi.org/10.1063/1.125420.

[77] H. Banisaukas, K.S. Jones, S. Falk, D.F. Downey, Defect reduction in laser thermal processing, Mater. Res. Soc. Symp. Proc. 610 (2000) 1–7, https://doi.org/10.1557/proc-610-b10.3.
[78] E. Kuryliw, K.S. Jones, D. Sing, M.J. Rendon, S. Talwar, Effect of laser thermal processing on defect evolution in silicon, Mater. Res. Soc. Symp. Proc. 717 (2002) 39–44, https://doi.org/10.1557/proc-717-c1.9.
[79] E.A. Kuryliw, Analyzing the thermal annealing behavior of laser thermal processed silicon, Ph.D. Thesis, University of Florida, 2003.
[80] P. Baeri, G. Foti, J.M. Poate, A.G. Cullis, Phase transitions in amorphous Si produced by pulsed laser or electron irradiation, in: J.F. Gibbons, L.D. Hess, T.W. Sigmon (Eds.), Laser and Electron-Beam Solid Interactions and Materials Processing, Elsevier, North Holland, 1981, pp. 39–44.
[81] H.C. Webber, A.G. Cullis, N.G. Chew, Computer simulation of high speed melting of amorphous silicon, Appl. Phys. Lett. 43 (1983) 669–671, https://doi.org/10.1063/1.94440.
[82] W. Sinke, F.W. Saris, Evidence for a self-propagating melt in amorphous silicon upon pulsed-laser irradiation, Phys. Rev. Lett. 53 (1984) 2121–2124, https://doi.org/10.1103/PhysRevLett.53.2121.
[83] J.J.P. Bruines, R.P.M. Van Hal, H.M.J. Boots, A. Polman, F.W. Saris, Time-resolved reflectivity measurements during explosive crystallization of amorphous silicon, Appl. Phys. Lett. 49 (1986) 1160–1162, https://doi.org/10.1063/1.97453.
[84] D.H. Lowndes, G.E. Jellison, S.J. Pennycook, S.P. Withrow, D.N. Mashburn, Direct measurements of the velocity and thickness of "explosively" propagating buried molten layers in amorphous silicon, Appl. Phys. Lett. 48 (1986) 1389–1391, https://doi.org/10.1063/1.96918.
[85] G. Götz, Explosive crystallization processes in silicon, Appl. Phys. A 40 (1986) 29–36.
[86] J.S. Im, H.J. Kim, M.O. Thompson, Phase transformation mechanisms involved in excimer laser crystallization of amorphous silicon films, Appl. Phys. Lett. 63 (1993) 1969–1971,-https://doi.org/10.1063/1.110617.
[87] J.H. Yoon, Rapid melting dynamics of amorphous Si films, Met. Mater. Int. 5 (1999) 525–532, https://doi.org/10.1007/BF03026299.
[88] J.J.P. Bruines, R.P.M. Van Hal, H.M.J. Boots, W. Sinke, F.W. Saris, Direct observation of resolidification from the surface upon pulsed-laser melting of amorphous silicon, Appl. Phys. Lett. 48 (1986) 1252–1254, https://doi.org/10.1063/1.96995.
[89] S.F. Lombardo, S. Boninelli, F. Cristiano, I. Deretzis, M.G. Grimaldi, K. Huet, E. Napolitani, A. La Magna, Phase field model of the nanoscale evolution during the explosive crystallization phenomenon, J. Appl. Phys. 123 (2018) 105105, https://doi.org/10.1063/1.5008362.
[90] Q. Yu, Computer simulation and experimental investigation of crystal growth in group IV semiconductor alloys, Ph.D. Thesis, Cornell University, 1995.
[91] E.J. Albenze, M.O. Thompson, P. Clancy, Molecular dynamics study of explosive crystallization of SiGe and boron-doped SiGe alloys, Ind. Eng. Chem. Res. 45 (2006) 5628–5639,-https://doi.org/10.1021/ie051361w.
[92] Q. Yu, M.O. Thompson, P. Clancy, Solidification kinetics in SiGe alloys, Phys. Rev. B 53 (1996) 8386–8397, https://doi.org/10.1103/PhysRevB.53.8386.
[93] S. Sedky, M. Gromova, T. Van Der Donck, J.P. Celis, A. Witvrouw, Characterization of KrF excimer laser annealed PECVD Si xGe 1-x for MEMS post-processing, Sens. Actuators A: Phys. 127 (2006) 316–323, https://doi.org/10.1016/j.sna.2006.01.035.
[94] G. Gore, On a peculiar phenomenon in the electro-deposition of antimony, Philos. Mag. Ser. 4 9 (1855) 73–74, https://doi.org/10.1080/14786445508641829.

[95] M. Kikuchi, A. Matsuda, T. Kurosu, A. Mineo, K.J. Callanan, 'Shock-crystallization' on the basis of a domino-type model, Solid State Commun. 14 (1974) 731–734, https://doi.org/10.1016/0038-1098(74)90874-6.

[96] A. Mineo, A. Matsuda, T. Kurosu, M. Kikuchi, Velocity of propagation in the shock-crystallization of sputtered amorphous germanium, Solid State Commun. 13 (1973) 329–331, https://doi.org/10.1016/0038-1098(73)90603-0.

[97] T. Takamori, R. Messier, R. Roy, New noncrystalline germanium which crystallizes "explosively" at room temperature, Appl. Phys. Lett. 20 (1972) 201–203, https://doi.org/10.1063/1.1654108.

[98] H.J. Zeiger, J.C.C. Fan, B.J. Palm, R.P. Gale, R.L. Chapman, Solid-phase crystallization produced by laser scanning of amorphous Ge films: the role of latent heat in crystallization-front dynamics, in: C.W. White, P.S. Peercy (Eds.), Laser and Electron Beam Processing of Materials, Academic Press, 1980, pp. 234–240, https://doi.org/10.1016/B978-0-12-746850-1.50036-0.

[99] G.H. Gilmer, H.J. Leamy, An analysis of the explosive crystallization of amorphous layers, in: C.W. White, P.S. Peercy (Eds.), Laser and Electron Beam Processing of Materials, Academic Press, 1980, pp. 227–233, https://doi.org/10.1016/b978-0-12-746850-1.50035-9.

[100] R.B. Gold, J.F. Gibbons, T.J. Magee, J. Peng, R. Ormond, V.R. Deline, C.A. Evans, Laser-induced explosive radial crystallization of deposited Ge and Si thin films, in: C.W. White, P. S. Peercy (Eds.), Laser and Electron Beam Processing of Materials, Academic Press, 1980, pp. 221–226, https://doi.org/10.1016/B978-0-12-746850-1.50034-7.

[101] H.J. Leamy, W.L. Brown, G.K. Celler, G. Foti, G.H. Gilmer, J.C.C. Fan, Explosive crystallization of amorphous germanium, Appl. Phys. Lett. 38 (1981) 137–139, https://doi.org/10.1063/1.92277.

[102] J.R. Abelson, T.W. Sigmon, K.B. Kim, K.H. Weiner, Epitaxial GexSi1-x/Si (100) structures produced by pulsed laser mixing of evaporated Ge on Si (100) substrates, Appl. Phys. Lett. 52 (1988) 230–232, https://doi.org/10.1063/1.99528.

[103] N. Frangis, J. Van Landuyt, R. Larciprete, S. Martelli, E. Borsella, S. Chiussi, J. Castro, B. León, High resolution electron microscopy and x-ray photoelectron spectroscopy studies of heteroepitaxial SixGe(1-x) alloys produced through laser induced processing, Appl. Phys. Lett. 72 (1998) 2877–2879, https://doi.org/10.1063/1.121487.

[104] T. Kociniewski, F. Fossard, J. Boulmer, D. Bouchier, Synthesis of strained SiGe on Si(100) by pulsed laser induced epitaxy, Thin Solid Films 518 (2010) 2542–2545, https://doi.org/10.1016/j.tsf.2009.09.154.

[105] C.Y. Ong, K.L. Pey, K.K. Ong, D.X.M. Tan, X.C. Wang, H.Y. Zheng, C.M. Ng, L. Chan, A low-cost method of forming epitaxy SiGe on Si substrate by laser annealing, Appl. Phys. Lett. 94 (2009), https://doi.org/10.1063/1.3086881, 082104.

[106] L. Vincent, F. Fossard, T. Kociniewski, L. Largeau, N. Cherkashin, M.J. Hÿtch, D. Debarre, T. Sauvage, A. Claverie, J. Boulmer, D. Bouchier, Nanoscale concentration and strain distribution in pseudomorphic films Si 1-x Ge x /Si processed by pulsed laser induced epitaxy, Appl. Surf. Sci. 258 (2012) 9208–9212, https://doi.org/10.1016/j.apsusc.2011.07.074.

[107] G.V. Luong, S. Wirths, S. Stefanov, B. Holländer, J. Schubert, J.C. Conde, T. Stoica, U. Breuer, S. Chiussi, M. Goryll, D. Buca, S. Mantl, Study of dopant activation in biaxially compressively strained SiGe layers using excimer laser annealing, J. Appl. Phys. 113 (2013) 204902, https://doi.org/10.1063/1.4807001.

[108] L. Dagault, P. Acosta-Alba, S. Kerdilès, J.P. Barnes, J.M. Hartmann, P. Gergaud, T.T. Nguyen, A. Grenier, A.M. Papon, N. Bernier, V. Delaye, J. Aubin, F. Cristiano, Impact of UV nanosecond laser annealing on composition and strain of Undoped Si 0.8 Ge 0.2 epitaxial

layers, ECS J. Solid State Sci. Technol. 8 (2019) P202–P208, https://doi.org/10.1149/2.0191903jss.

[109] C.Y. Ong, K.L. Pey, X. Li, X.C. Wang, C.M. Ng, L. Chan, Laser annealing induced high Ge concentration epitaxial SiGe layer in Si1-x Gex virtual substrate, Appl. Phys. Lett. 93 (2008), https://doi.org/10.1063/1.2962991, 041112.

[110] D.C. Paine, D.J. Howard, N.G. Stoffel, J.A. Horton, The growth of strained $Si_{1-x}Ge_x$ alloys on <001> silicon using solid phase epitaxy, J. Mater. Res. 5 (1990) 1023–1031, https://doi.org/10.1557/JMR.1990.1023.

[111] F. Cristiano, A. Nejim, Y. Suprun-Belevich, A. Claverie, P.L.F. Hemment, Formation of extended defects and strain relaxation in ion beam synthesised SiGe alloys, Nucl. Instrum. Methods Phys. Res. Sect. B Beam Interact. Mater. Atoms 147 (1999) 35–42, https://doi.org/10.1016/S0168-583X(98)00589-8.

[112] J.M. Hartmann, A. Abbadie, S. Favier, Critical thickness for plastic relaxation of SiGe on Si (001) revisited, J. Appl. Phys. 110 (2011), https://doi.org/10.1063/1.3656989, 083529.

[113] G.L. Olson, J.A. Roth, Kinetics of solid phase crystallization in amorphous silicon, Mat. Sci. Rep. 3 (1988) 1–78.

[114] J. Narayan, O.W. Holland, Formation of metastable supersaturated solid solutions in ion implanted silicon during solid phase crystallization, Appl. Phys. Lett. 41 (1982) 239–242,-https://doi.org/10.1063/1.93480.

[115] J.S. Williams, Solid phase epitaxial regrowth phenomena in silicon, Nucl. Instrum. Methods Phys. Res. 209–210 (1983) 219–228, https://doi.org/10.1016/0167-5087(83)90803-7.

[116] C.W. White, S.R. Wilson, B.R. Appleton, F.W. Young, Supersaturated substitutional alloys formed by ion implantation and pulsed laser annealing of group-III and group-V dopants in silicon, J. Appl. Phys. 51 (1980) 738–749, https://doi.org/10.1063/1.327334.

[117] R. Duffy, T. Dao, Y. Tamminga, K. Van Der Tak, F. Roozeboom, E. Augendre, Groups III and v impurity solubilities in silicon due to laser, flash, and solid-phase-epitaxial-regrowth anneals, Appl. Phys. Lett. 89 (2006), https://doi.org/10.1063/1.2337081, 071915.

[118] F.A. Trumbore, Solid solubilities of impurity elements in germanium and silicon, Bell Syst. Tech. J. 39 (1960) 205–233, https://doi.org/10.1002/j.1538-7305.1960.tb03928.x.

[119] R. Duffy, M. Aboy, V.C. Venezia, L. Pelaz, S. Severi, B.J. Pawlak, P. Eyben, T. Janssens, W. Vandervorst, J. Loo, J. Roozeboom, Boron pocket and channel deactivation in nMOS transistors with SPER junctions, IEEE Trans. Electron. Dev. 53 (2006) 71–77.

[120] S.H. Jain, P.B. Griffin, J.D. Plummer, S. McCoy, J. Gelpey, T. Selinger, D.F. Downey, Low resistance, low-leakage ultrashallow p(+)-junction formation using millisecond flash anneals, IEEE Trans. Electron. Dev. 52 (2005) 1610–1615.

[121] R. Duffy, A. Heringa, V.C. Venezia, J. Loo, M.A. Verheijen, M.J.P. Hopstaken, K. van der Tak, M. de Potter, J.C. Hooker, P. Meunier-Beillard, R. Delhougne, Quantitative prediction of junction leakage in bulk-technology CMOS devices, Solid State Electron. 54 (2010) 243–251.

[122] C. Claeys, E. Simoen, A. Poyai, A. Czerwinski, Electrical quality assessment of epitaxial wafers based on p-n junction diagnostics, J. Electrochem. Soc. 146 (1999) 3429.

[123] D.K. Schröder, Carrier lifetimes in silicon, IEEE Trans. Electron. Dev. 44 (1997) 160–170.

[124] G. Mannino, S. Solmi, V. Privitera, M. Bersani, Electrical activation of B in the presence of boron-interstitials clusters, Appl. Phys. Lett. 79 (2001) 3764, https://doi.org/10.1063/1.1423775.

[125] L. Pelaz, M. Jaraiz, G.H. Gilmer, H.-J. Gossmann, C.S. Rafferty, D.J. Eaglesham, J.M. Poate, B diffusion and clustering in ion implanted Si: the role of B cluster precursors, Appl. Phys. Lett. 70 (1997) 2285, https://doi.org/10.1063/1.118839.

[126] L. Pelaz, G.H. Gilmer, H.-J. Gossmann, C.S. Rafferty, M. Jaraiz, J. Barbolla, B cluster formation and dissolution in Si: a scenario based on atomistic modeling, Appl. Phys. Lett. 74 (1999) 3657, https://doi.org/10.1063/1.123213.

[127] P. Pichler, current understanding and modeling of boron-interstitial clusters, Mat. Res. Soc. Symp. Proc. 717 (2002) 103–114.

[128] M. Aboy, L. Pelaz, L.A. Marques, P. López, J. Barbolla, R. Duffy, Atomistic analysis of the evolution of boron activation during annealing in crystalline and preamorphized silicon, J. Appl. Phys. 97 (2005) 103520, https://doi.org/10.1063/1.1904159.

[129] B.J. Pawlak, R. Surdeanu, B. Colombeau, A.J. Smith, N.E.B. Cowern, R. Lindsay, W. Vandervorst, B. Brijs, O. Richard, F. Cristiano, Evidence on the mechanism of boron deactivation in Ge-preamorphized ultrashallow junctions, Appl. Phys. Lett. 84 (2004) 2055, https://doi.org/10.1063/1.1682697.

[130] J.A. Sharp, N.E.B. Cowern, R.P. Webb, K.J. Kirkby, D. Giubertoni, S. Gennaro, M. Bersani, M.A. Foad, F. Cristiano, P.F. Fazzini, Deactivation of ultrashallow boron implants in preamorphized silicon after nonmelt laser annealing with multiple scans, Appl. Phys. Lett. 89 (2006) 192105, https://doi.org/10.1063/1.2385215.

[131] E. Schroer, M. Uematsu, Simulation of clustering and pile-up during post-implantation annealing of phosphorus in silicon, Jpn. J. Appl. Phys. 38 (1999) 7–11, https://doi.org/10.1143/JJAP.38.7.

[132] Y. Takamura, S.H. Jain, P.B. Griffin, J.D. Plummer, Thermal stability of dopants in laser annealed silicon, J. Appl. Phys. 92 (2002) 230–234, https://doi.org/10.1063/1.1481975.

[133] Y. Takamura, P.B. Griffin, J.D. Plummer, Physical processes associated with the deactivation of dopants in laser annealed silicon, J. Appl. Phys. 92 (2002) 235, https://doi.org/10.1063/1.1481974.

[134] M. Uematsu, Simulation of high-concentration phosphorus diffusion in silicon taking into account phosphorus clustering and pile-up, Jpn. J. Appl. Phys. 38 (1999) 6188–6192,-https://doi.org/10.1143/JJAP.38.6188.

[135] R. Pinacho, M. Jaraiz, P. Castrillo, I. Martin-Bragado, J.E. Rubio, J. Barbolla, Modeling arsenic deactivation through arsenic-vacancy clusters using an atomistic kinetic Monte Carlo approach, Appl. Phys. Lett. 86 (2005) 252103, https://doi.org/10.1063/1.1948533.

[136] R.O. Schwenker, R.S. Pan, R.F. Lever, Arsenic clustering in silicon, J. Appl. Phys. 42 (1971) 3195, https://doi.org/10.1063/1.1660706.

[137] S.A. Harrison, T.F. Edgar, G.S. Hwang, Interstitial-mediated arsenic clustering in ultrashallow junction formation, Elec. Sol. State. Lett. 9 (2006) G354, https://doi.org/10.1149/1.2359084.

[138] S.A. Harrison, T.F. Edgar, G.S. Hwang, Structure, stability, and diffusion of arsenic-silicon interstitial pairs, Appl. Phys. Lett. 87 (2006) 231905, https://doi.org/10.1063/1.2130398.

[139] C.K. Ong, H.S. Tan, E.H. Sin, Calculations of melting threshold energies of crystalline and amorphous materials due to pulsed-laser irradiation, Mater. Sci. Eng. 79 (1986) 79–85,-https://doi.org/10.1016/0025-5416(86)90389-7.

[140] R.T. Young, C.W. White, G.J. Clark, J. Narayan, W.H. Christie, M. Murakami, P.W. King, S. D. Kramer, Laser annealing of boron-implanted silicon, Appl. Phys. Lett. 32 (1978) 139–141,-https://doi.org/10.1063/1.89959.

[141] J. Narayan, R.T. Young, C.W. White, A comparative study of laser and thermal annealing of boron-implanted silicon, J. Appl. Phys. 49 (1978) 3912–3917, https://doi.org/10.1063/1.325398.

[142] J. Narayan, R.T. Young, R.F. Wood, W.H. Christie, P-n junction formation in boron-deposited silicon by laser-induced diffusion, Appl. Phys. Lett. 33 (1978) 338–340, https://doi.org/10.1063/1.90368.

[143] B.C. Larson, C.W. White, B.R. Appleton, Unidirectional contraction in boron-implanted laser-annealed silicon, Appl. Phys. Lett. 32 (1978) 801–803, https://doi.org/10.1063/1.89936.

[144] B.C. Larson, J.F. Barhorst, X-ray study of lattice strain in boron implanted laser annealed silicon, J. Appl. Phys. 51 (1980) 3181–3185, https://doi.org/10.1063/1.328069.

[145] H. Engstrom, J.B. Bates, Raman scattering from boron-implanted laser-annealed silicon, J. Appl. Phys. 50 (1979) 2921–2925, https://doi.org/10.1063/1.326212.

[146] H. Engstrom, Infrared reflectivity and transmissivity of boron-implanted, laser-annealed silicon, J. Appl. Phys. 51 (1980) 5245–5249, https://doi.org/10.1063/1.327476.

[147] G. Masetti, M. Severi, S. Solmi, Modeling of carrier mobility against carrier concentration in arsenic-, phosphorus-, and boron-doped silicon, IEEE Trans. Electron. Dev. 30 (1983) 764–769, https://doi.org/10.1109/T-ED.1983.21207.

[148] J. Narayan, O.W. Holland, W.H. Christie, J.J. Wortman, Rapid thermal and pulsed laser annealing of boron fluoride-implanted silicon, J. Appl. Phys. 57 (1985) 2709–2716,-https://doi.org/10.1063/1.335411.

[149] S. Solmi, E. Landi, F. Baruffaldi, High-concentration boron diffusion in silicon: simulation of the precipitation phenomena, J. Appl. Phys. 68 (1990) 3250–3258, https://doi.org/10.1063/1.346376.

[150] K.S. Jones, E. Kuryliw, R. Murto, M. Rendon, S. Talwar, Boron diffusion upon annealing of laser thermal processed silicon, in: Proceedings of International Conference on Ion Implantation Technology, 2000, pp. 111–114, https://doi.org/10.1109/.2000.924103.

[151] S. Whelan, V. Privitera, M. Italia, G. Mannino, C. Bongiorno, C. Spinella, G. Fortunato, L. Mariucci, M. Stanizzi, A. Mittiga, Redistribution and electrical activation of ultralow energy implanted boron in silicon following laser annealing, J. Vac. Sci. Technol. B Microelectron. Nanom. Struct. 20 (2002) 644, https://doi.org/10.1116/1.1459725.

[152] C.H. Poon, B.J. Cho, Y.F. Lu, M. Bhat, A. See, Multiple-pulse laser annealing of preamorphized silicon for ultrashallow boron junction formation, J. Vac. Sci. Technol. B Microelectron. Nanom. Struct. 21 (2003) 706, https://doi.org/10.1116/1.1547747.

[153] S.M. Jung, C.J. Park, J.H. Kim, M.W. Shin, Selective annealing of $Al_2O_3$/silicon interface by using an $Nd3^+$: YAG laser with a wavelength of 532 nm, Mater. Sci. Semicond. Process. 109 (2020) 104956, https://doi.org/10.1016/j.mssp.2020.104956.

[154] V. Gonda, A. Burtsev, T.L.M. Scholtes, L.K. Nanver, Near-ideal implanted shallow-junction diode formation by excimer laser annealing, in: 2005 13th International Conference on Advanced Thermal Processing of Semiconductors, IEEE, 2005, pp. 93–99, https://doi.org/10.1109/RTP.2005.1613688.

[155] P.H.D. Lu, D. Lin, X. Wang, A. Lennon, S. Wenham, Laser doping through anodic aluminium oxide silicon solar cell, Sol. Energy Mater. Sol. Cells 145 (2016) 349–357, https://doi.org/10.1016/j.solmat.2015.10.043.

[156] O. Nast, S. Brehme, S. Pritchard, A.G. Aberle, S.R. Wenham, Aluminium-induced crystallisation of silicon on glass for thin-film solar cells, Sol. Energy Mater. Sol. Cells 65 (2001) 385–392, https://doi.org/10.1016/S0927-0248(00)00117-3.

[157] Z. Liu, X. Hao, F. Qi, A. Ho-Baillie, M.A. Green, Epitaxial growth of single-crystalline silicon–germanium on silicon by aluminium-assisted crystallization, Scr. Mater. 71 (2014) 25–28, https://doi.org/10.1016/j.scriptamat.2013.09.026.

[158] Z. Liu, X. Hao, J. Huang, A. Ho-Baillie, M.A. Green, Laser-induced aluminium-assisted crystallization of Ge-rich SixGe1-x epitaxy on Si, Thin Solid Films 679 (2019) 55–57, https://doi.org/10.1016/j.tsf.2019.04.005.

[159] T. Tabata, J. Aubin, K. Huet, F. Mazzamuto, A. La Magna, L.M. Rubin, P. Kopalidis, H.-C. Tsai, D. Roh, R. Reece, Impact of solidification velocity on activation of Ga, In, and Al segregated in high Ge content SiGe by UV melt laser anneal, in: Electron Devices Technology and Manufacturing Conference (EDTM), 2019, p. 130, https://doi.org/10.1109/EDTM.2019.8731304.

[160] M.Y. Tsai, B.G. Streetman, V.R. Deline, C.A. Evans, Gallium distribution and electrical activation in Ga+-implanted Si, J. Electron. Mater. 8 (1979) 111–126, https://doi.org/10.1007/BF02663267.

[161] B.M. Arora, J.M. Castillo, M.B. Kurup, R.P. Sharma, Thermal annealing and electrical activation of high dose gallium implanted silicon, J. Electron. Mater. 10 (1981) 845–862, https://doi.org/10.1007/BF02661003.

[162] R. Elliman, G. Carter, Epitaxial recrystallisation of gallium implanted (100) silicon, Nucl. Instrum. Methods Phys. Res. 209–210 (1983) 663–669, https://doi.org/10.1016/0167-5087(83)90864-5.

[163] M. Tamura, S. Shukuri, M. Moniwa, M. Default, Focused ion beam gallium implantation into silicon, Appl. Phys. A Solids Surfaces 39 (1986) 183–190, https://doi.org/10.1007/BF00620733.

[164] S.Y. Shiryaev, A. Nylandsted Larsen, N. Safronov, Rapid thermal annealing of indium-implanted silicon single crystals, J. Appl. Phys. 65 (1989) 4220–4224, https://doi.org/10.1063/1.343304.

[165] S. Kurnaev, E. Uggerhøj, Laser annealing of indium-implanted silicon, Radiat. Eff. 40 (1979) 91–94, https://doi.org/10.1080/00337577908234498.

[166] R.F. Reihl, G.A. Smith, W. Katz, E.F. Koch, Migration of implanted indium in silicon as a function of thermal annealing, Appl. Phys. Lett. 42 (1983) 575–577, https://doi.org/10.1063/1.94006.

[167] G.J. Kemerink, F. Pleiter, Indium-vacancy interaction in laser-annealed silicon, Phys. Lett. A 121 (1987) 367–370, https://doi.org/10.1016/0375-9601(87)90308-2.

[168] M.H. Clark, K.S. Jones, Strain compensation in boron-indium coimplanted laser thermal processed silicon, J. Appl. Phys. 97 (2005), https://doi.org/10.1063/1.1891282, 093525.

[169] G. Battaglin, G.D. Mea, A.V. Drigo, G. Foti, G.G. Bentini, M. Servidori, Two-stage laser annealing of lattice disorder in phosphorus implanted silicon, Phys. Status Solidi 49 (1978) 347–352, https://doi.org/10.1002/pssa.2210490144.

[170] R.T. Young, J. Narayan, Laser annealing of diffusion-induced imperfections in silicon, Appl. Phys. Lett. 33 (1978) 14–15, https://doi.org/10.1063/1.90164.

[171] K. Affolter, W. Lüthy, M. Vvon Allmen, Properties of laser-assisted doping in silicon, Appl. Phys. Lett. 33 (1978) 185–187, https://doi.org/10.1063/1.90302.

[172] M. Miyao, H. Tamura, K. Ohyu, T. Tokuyama, Annealing behavior of phosphorus implanted silicon irradiated by several lasers of different wavelength, AIP Conf. Proc. 50 (1979) 325, https://doi.org/10.1063/1.31677.

[173] H. Tamura, M. Miyao, T. Tokuyama, Laser-annealing behavior of a phosphorus-implanted silicon substrate covered with a $SiO_2$ film, J. Appl. Phys. 50 (1979) 3783–3784, https://doi.org/10.1063/1.326296.

[174] M. Miyao, K. Itoh, M. Tamura, H. Tamura, T. Tokuyama, Furnace annealing behavior of phosphorus implanted, laser annealed silicon, J. Appl. Phys. 51 (1980) 4139–4144, https://doi.org/10.1063/1.328233.

[175] H. Nakashima, Y. Shiraki, M. Miyao, Photoluminescence study of laser annealing in phosphorus-implanted and unimplanted silicon, J. Appl. Phys. 50 (1979) 5966–5969, https://doi.org/10.1063/1.326699.

[176] T. Inada, T. Sugiyama, N. Okano, Y. Ishikawa, Pulsed electron-beam annealing of phosphorus-implanted silicon, Electron. Lett. 16 (1980) 54–55, https://doi.org/10.1049/el:19800041.

[177] M. Finetti, P. Negrini, S. Solmi, D. Nobili, Electrical properties and stability of supersaturated phosphorus-doped silicon layers, J. Electrochem. Soc. 128 (1981) 1313–1317, https://doi.org/10.1149/1.2127626.

[178] D. Nobili, A. Armigliato, M. Finetti, S. Solmi, Precipitation as the phenomenon responsible for the electrically inactive phosphorus in silicon, J. Appl. Phys. 53 (1982) 1484–1491, https://doi.org/10.1063/1.330646.

[179] J. Grun, R.P. Fischer, M. Peckerar, C.L. Felix, B.C. Covington, W.J. DeSisto, D.W. Donnelly, A. Ting, C.K. Manka, Athermal annealing of phosphorus-ion-implanted silicon, Appl. Phys. Lett. 77 (2000) 1997–1999, https://doi.org/10.1063/1.1312259.

[180] N. Loubet, T. Adam, M. Raymond, Q. Liu, K. Cheng, R. Sreenivasan, A. Reznicek, P. Khare, W. Kleemeier, V. Paruchuri, B. Doris, R. Sampson, Ultra-low resistivity in-situ phosphorus doped Si and SiC epitaxy for source/drain formation in advanced 20 nm n-type field effect transistor devices, Thin Solid Films 520 (2012) 3149–3154, https://doi.org/10.1016/j.tsf.2011.10.106.

[181] J. Krynicki, J. Suski, S. Ugniewski, R. Grötzschel, R. Klabes, U. Kreissig, J. Rüdiger, Laser annealing of arsenic implanted silicon, Phys. Lett. A 61 (1977) 181–182, https://doi.org/10.1016/0375-9601(77)90286-9.

[182] P. Baeri, S.U. Campisano, G. Foti, E. Rimini, Arsenic diffusion in silicon melted by high-power nanosecond laser pulsing, Appl. Phys. Lett. 33 (1978) 137–140, https://doi.org/10.1063/1.90283.

[183] T.N.C. Venkatesan, J.A. Golovchenko, J.M. Poate, P. Cowan, G.K. Celler, Dose dependence in the laser annealing of arsenic-implanted silicon, Appl. Phys. Lett. 33 (1978) 429–431,-https://doi.org/10.1063/1.90411.

[184] M. Takai, P.H. Tsien, S.C. Tsou, D. Röschenthaler, M. Ramin, H. Ryssel, I. Ruge, CW $CO_2$-laser annealing of arsenic implanted silicon, Appl. Phys. 22 (1980) 129–136, https://doi.org/10.1007/BF00885995.

[185] M. Delfino, CO laser annealing of arsenic-implanted silicon, J. Appl. Phys. 53 (1982) 3923–3925, https://doi.org/10.1063/1.331102.

[186] P.H. Tsien, J. Götzlich, H. Ryssel, I. Ruge, $CO_2$ laser annealing characteristics of high-dose boron- and arsenic-implanted silicon, J. Appl. Phys. 53 (1982) 663–668, https://doi.org/10.1063/1.329974.

[187] W.K. Chu, Stability study of laser irradiation of silicon diffused with arsenic, Appl. Phys. Lett. 36 (1980) 273–275, https://doi.org/10.1063/1.91450.

[188] J.S. Williams, W.L. Brown, G.K. Celler, H.J. Leamy, J.M. Poate, G.A. Rozgonyi, T.T. Sheng, Characterization of pulsed Nd:YAG laser-annealed, arsenic-ion-implanted silicon, J. Appl. Phys. 52 (1981) 1038–1049, https://doi.org/10.1063/1.328801.

[189] A. Erbil, W. Weber, G.S. Cargill, R.F. Boehme, Lattice distortions for arsenic in single-crystal silicon, Phys. Rev. B 34 (1986) 1392–1394, https://doi.org/10.1103/PhysRevB.34.1392.

[190] S. Solmi, M. Severi, R. Angelucci, L. Baldi, R. Bilenchi, Electrical properties of thermally and laser annealed polycrystalline silicon films heavily doped with arsenic and phosphorus, J. Electrochem. Soc. 129 (1982) 1811–1818, https://doi.org/10.1149/1.2124299.

[191] P.M. Rousseau, P.B. Griffin, J.D. Plummer, Electrical deactivation of arsenic as a source of point defects, Appl. Phys. Lett. 65 (1994) 578–580, https://doi.org/10.1063/1.112301.

[192] P.M. Rousseau, P.B. Griffin, W.T. Fang, J.D. Plummer, Arsenic deactivation enhanced diffusion: a time, temperature, and concentration study, J. Appl. Phys. 84 (1998) 3593–3601, https://doi.org/10.1063/1.368593.

[193] J.S. Williams, K.T. Short, Metastable doping behavior in antimony-implanted (100) silicon, J. Appl. Phys. 53 (1982) 8663–8667, https://doi.org/10.1063/1.330463.

[194] C.W. White, W.H. Christie, B.R. Appleton, S.R. Wilson, P.P. Pronko, C.W. Magee, Redistribution of dopants in ion-implanted silicon by pulsed-laser annealing, Appl. Phys. Lett. 33 (1978) 662–664, https://doi.org/10.1063/1.90456.

[195] J.S. Williams, W.L. Brown, H.J. Leamy, J.M. Poate, J.W. Rodgers, D. Rousseau, G.A. Rozgonyi, J.A. Shelnutt, T.T. Sheng, Solid-phase epitaxy of implanted silicon by cw Ar ion laser irradiation, Appl. Phys. Lett. 33 (1978) 542–544, https://doi.org/10.1063/1.90430.

[196] C.W. White, P.P. Pronko, S.R. Wilson, B.R. Appleton, J. Narayan, R.T. Young, Effects of pulsed ruby-laser annealing on As and Sb implanted silicon, J. Appl. Phys. 50 (1979) 3261–3273, https://doi.org/10.1063/1.326366.

[197] J. Narayan, Interface instability and cell formation in ion-implanted and laser-annealed silicon, J. Appl. Phys. 52 (1981) 1289–1293, https://doi.org/10.1063/1.329753.

[198] E. Fogarassy, R. Stuck, J.J. Grob, P. Siffert, Silicon solar cells realized by laser induced diffusion of vacuum-deposited dopants, J. Appl. Phys. 52 (1981) 1076–1082, https://doi.org/10.1063/1.328806.

[199] R. Stuck, E. Fogarassy, J.J. Grob, P. Siffert, Solubility limit of impurities in silicon after laser induced melting, Appl. Phys. 23 (1980) 15–19, https://doi.org/10.1007/BF00899564.

[200] A.P. Pogany, T. Preuss, K.T. Short, H.K. Wagenfeld, J.S. Williams, Metastable solid solutions of antimony in (100) silicon, Nucl. Instrum. Methods Phys. Res. 209–210 (1983) 731–736, https://doi.org/10.1016/0167-5087(83)90875-X.

[201] S.J. Pennycook, J. Narayan, O.W. Holland, Formation of partially coherent antimony precipitates in ion implanted thermally annealed silicon, J. Appl. Phys. 54 (1983) 6875–6878, https://doi.org/10.1063/1.331992.

[202] S.J. Pennycook, J. Narayan, O.W. Holland, Point defect trapping in solid-phase epitaxially grown silicon-antimony alloys, J. Appl. Phys. 55 (1984) 837–840, https://doi.org/10.1063/1.333179.

[203] J.L. Allain, A. Bourret, J.R. Regnard, A. Armigliato, X-ray absorption study of the atomic environment in Sb+ and Sb+/B+ implanted silicon, Appl. Phys. Lett. 61 (1992) 264–266, https://doi.org/10.1063/1.107964.

[204] A. Desmur, B. Bourguignon, J. Boulmer, J.B. Ozenne, J.P. Budin, D. Débarre, A. Aliouchouche, Pulsed laser etching of silicon: dopant profile modification and dopant desorption induced by surface melting, J. Appl. Phys. 76 (1994) 3081–3087, https://doi.org/10.1063/1.357490.

[205] R. Murto, K. Jones, M. Rendon, S. Talwar, Activation and deactivation studies of laser thermal annealed boron, arsenic, phosphorus, and antimony ultra-shallow abrupt junctions, in: 2000 International Conference on Ion Implantation Technology Proceedings. Ion Implantation Technology - 2000 (Cat. No. 00EX432), IEEE, 2000, pp. 155–158, https://doi.org/10.1109/.2000.924113.

[206] Y. Takamura, A.F. Marshall, A. Mehta, J. Arthur, P.B. Griffin, J.D. Plummer, J.R. Patel, Diffuse x-ray scattering and transmission electron microscopy study of defects in antimony-implanted silicon, J. Appl. Phys. 95 (2004) 3968–3976, https://doi.org/10.1063/1.1666975.

[207] T. Tabata, P.E. Raynal, K. Huet, J.L. Everaert, Segregation and activation of Sb implanted in Si by UV nanosecond-laser-anneal-induced non-equilibrium solidification, J. Appl. Phys. 127 (2020) 135701, https://doi.org/10.1063/5.0005176.

[208] R. Schaub, G. Pensl, M. Schulz, C. Holm, Donor states in tellurium-doped silicon, Appl. Phys. A Solids Surfaces 34 (1984) 215–222, https://doi.org/10.1007/BF00616575.

[209] M.A. Sheehy, B.R. Tull, C.M. Friend, E. Mazur, Chalcogen doping of silicon via intense femtosecond-laser irradiation, Mater. Sci. Eng. B Solid-State Mater. Adv. Technol. 137 (2007) 289–294, https://doi.org/10.1016/j.mseb.2006.10.002.

[210] G. Foti, S.U. Campisano, E. Rimini, G. Vitali, Lattice location of Te in laser-annealed Te-implanted silicon, J. Appl. Phys. 49 (1978) 2569, https://doi.org/10.1063/1.325068.

[211] G.J. Kemerink, D.O. Boerma, H. De Waard, L. Niesen, Tellurium in silicon. Part 1. Channeling and Rutherford backscattering studies of tellurium implanted silicon, Radiat. Eff. 69 (1983) 83–99, https://doi.org/10.1080/00337578308221727.

[212] G.J. Kemerink, H. De Waard, L. Niesen, D.O. Boerma, Tellurium and iodine in silicon. Part 2. Hall effect and resistivity measurements on ion implanted silicon, Radiat. Eff. 69 (1983) 101–112, https://doi.org/10.1080/00337578308221728.

[213] G.J. Kemerink, H. de Waard, L. Niesen, D.O. Boerma, Mössbauer spectroscopy of laser annealed tellurium implanted silicon (II).129I, Hyperfine Interact. 14 (1983) 53–88, https://doi.org/10.1007/BF02098295.

[214] M. Wang, Y. Berencen, E. Garcia-Hemme, S. Prucnal, R. Hubner, Y. Yuan, C. Xu, L. Rebohle, R. Bottger, R. Heller, H. Schneider, W. Skorupa, M. Helm, S. Zhou, Extended infrared photoresponse in Te-hyperdoped Si at room temperature, Phys. Rev. Appl. 10 (2018), https://doi.org/10.1103/PhysRevApplied.10.024054, 024054.

[215] M. Wang, A. Debernardi, Y. Berencen, R. Heller, C. Xu, Y. Yuan, Y. Xie, R. Bottger, L. Rebohle, W. Skorupa, M. Helm, S. Prucnal, S. Zhou, Breaking the doping limit in silicon by deep impurities, Phys. Rev. Appl. 11 (2019), https://doi.org/10.1103/PhysRevApplied.11.054039, 054039.

[216] C. Claeys, E. Simoen, Germanium-Based Technologies from Materials to Devices, Elsevier BV, 2007.

[217] R. Pillarisetty, Academic and industry research progress in germanium nanodevices, Nature 479 (2011) 324–328, https://doi.org/10.1038/nature10678.

[218] J. Michel, J. Liu, L.C. Kimerling, High-performance Ge-on-Si photodetectors, Nat. Photon. 4 (2010) 527–534, https://doi.org/10.1038/nphoton.2010.157.

[219] J. Liu, X. Sun, R.E. Camacho-Aguilera, L.C. Kimerling, J. Michel, Ge-on-Si laser operating at room temperature, Opt. Lett. 35 (2010) 679–681, https://doi.org/10.1364/OL.35.000679.

[220] S.-L. Cheng, J. Lu, G. Shambat, H.-Y. Yu, K. Saraswat, J. Vuckovic, Y. Nishi, Room temperature 16 μm electroluminescence from Ge light emitting diode on Si substrate, Opt. Express 17 (2009) 10019, https://doi.org/10.1364/OE.17.010019.

[221] R. Soref, Mid-infrared photonics in silicon and germanium, Nat. Photonics 4 (2010) 495–497, https://doi.org/10.1038/nphoton.2010.171.

[222] J. Frigerio, A. Ballabio, G. Isella, E. Sakat, G. Pellegrini, P. Biagioni, M. Bollani, E. Napolitani, C. Manganelli, M. Virgilio, A. Grupp, M.P. Fischer, D. Brida, K. Gallacher, D.J. Paul, L. Baldassarre, P. Calvani, V. Giliberti, A. Nucara, M. Ortolani, Tunability of the dielectric function of heavily doped germanium thin films for mid-infrared plasmonics, Phys. Rev. B 94 (2016), https://doi.org/10.1103/PhysRevB.94.085202, 085202.

[223] A. Jiménez, S.M. Carturan, R. Milazzo, A. Datas, D. de Salvador, C. del Cañizo, E. Napolitani, N-type doping of Ge by P spin on dopant and pulsed laser melting, Semicond. Sci. Technol. 35 (2020), https://doi.org/10.1088/1361-6641/ab7bed, 065002.

[224] S. Dushenko, M. Koike, Y. Ando, T. Shinjo, M. Myronov, M. Shiraishi, Experimental demonstration of room-temperature spin transport in n-type germanium epilayers, Phys. Rev. Lett. 114 (2015) 196602, https://doi.org/10.1103/PhysRevLett.114.196602.

[225] H. Watzinger, J. Kukučka, L. Vukušić, F. Gao, T. Wang, F. Schäffler, J.-J. Zhang, G. Katsaros, A germanium hole spin qubit, Nat. Commun. 9 (2018) 3902, https://doi.org/10.1038/s41467-018-06418-4.

[226] A. Chroneos, H. Bracht, Diffusion of n-type dopants in germanium, Appl. Phys. Rev. 1 (2014), https://doi.org/10.1063/1.4838215, 011301.

[227] S. Mirabella, D. De Salvador, E. Napolitani, E. Bruno, F. Priolo, Mechanisms of boron diffusion in silicon and germanium, J. Appl. Phys. 113 (2013), https://doi.org/10.1063/1.4763353, 031101.

[228] E. Simoen, J. Mitard, G. Hellings, G. Eneman, B. De Jaeger, L. Witters, B. Vincent, R. Loo, A. Delabie, S. Sioncke, M. Caymax, C. Claeys, Challenges and opportunities in advanced Ge pMOSFETs, Mater. Sci. Semicond. Process. 15 (2012) 588–600, https://doi.org/10.1016/j.mssp.2012.04.017.

[229] J. Vanhellemont, E. Simoen, Brother silicon, sister germanium, J. Electrochem. Soc. 154 (2007) H572, https://doi.org/10.1149/1.2732221.

[230] E. Napolitani, E. Bruno, G. Bisognin, M. Mastromatteo, D. De Salvador, G. Scapellato, S. Boninelli, F. Priolo, V. Privitera, A. Carnera, Role of ion mass on damage accumulation during ion implantation in Ge, Phys. status solidi 211 (2014) 118–121, https://doi.org/10.1002/pssa.201300324.

[231] E. Napolitani, G. Impellizzeri, Ion Implantation Defects and Shallow Junctions in Si and Ge, in: Defects in Semiconductors, Elsevier-Academic Press, 2015, pp. 93–122, https://doi.org/10.1016/bs.semsem.2015.01.001.

[232] I.H. Wilson, The effects of self-ion bombardment (30–500 keV) on the surface topography of single-crystal germanium, J. Appl. Phys. 53 (1982) 1698, https://doi.org/10.1063/1.331636.

[233] S. Fischler, Correlation between maximum solid solubility and distribution coefficient for impurities in Ge and Si, J. Appl. Phys. 33 (1962) 1615, https://doi.org/10.1063/1.1728792.

[234] R.W. Olesinski, G.J. Abbaschian, The B-Ge (boron-germanium) system, Bull. Alloy Phase Diagr. 5 (1984) 476–478, https://doi.org/10.1007/BF02872899.

[235] R.W. Olesinski, G.J. Abbaschian, The Ga-Ge (gallium-germanium) system, Bull. Alloy Phase Diagr. 6 (1985) 258–262, https://doi.org/10.1007/BF02880411.

[236] R.W. Olesinski, G.J. Abbaschian, The As-Ge (arsenic-germanium) system, Bull. Alloy Phase Diagr. 6 (1985) 250–254, https://doi.org/10.1007/BF02880409.

[237] R.W. Olesinski, G.J. Abbaschian, The Ge-Sb (germanium-antimony) system, Bull. Alloy Phase Diagr. 7 (1986) 219–222, https://doi.org/10.1007/BF02868990.

[238] R.W. Olesinski, N. Kanani, G.J. Abbaschian, The Ge-In (germanium-indium) system, Bull. Alloy Phase Diagr. 6 (1985) 536–539, https://doi.org/10.1007/BF02887153.

[239] R.W. Olesinski, N. Kanani, G.J. Abbaschian, The Ge-P (germanium-phosphorus) system, Bull. Alloy Phase Diagr. 6 (1985) 262–266, https://doi.org/10.1007/BF02880412.

[240] G. Impellizzeri, S. Mirabella, E. Bruno, A.M. Piro, M.G. Grimaldi, B activation and clustering in ion-implanted Ge, J. Appl. Phys. 105 (2009), https://doi.org/10.1063/1.3091289, 063533.

[241] G. Impellizzeri, S. Mirabella, A. Irrera, M.G. Grimaldi, E. Napolitani, Ga-implantation in Ge: electrical activation and clustering, J. Appl. Phys. 106 (2009), https://doi.org/10.1063/1.3159031, 013518.

[242] G. Impellizzeri, E. Napolitani, S. Boninelli, V. Privitera, T. Clarysse, W. Vandervorst, F. Priolo, Aluminium implantation in germanium: uphill diffusion, electrical activation, and trapping, Appl. Phys. Express 5 (2012), https://doi.org/10.1143/APEX.5.021301, 021301.

[243] M. Posselt, B. Schmidt, W. Anwand, R. Grotzschel, V. Heera, A. Mucklich, C. Wundisch, W. Skorupa, H. Hortenbach, S. Gennaro, M. Bersani, D. Giubertoni, A. Moller, H. Bracht, P implantation into preamorphized germanium and subsequent annealing: solid phase epitaxial regrowth, P diffusion, and activation, J. Vac. Sci. Technol. B Microelectron. Nanom. Struct. 26 (2008) 430, https://doi.org/10.1116/1.2805249.

[244] S. Prucnal, F. Liu, M. Voelskow, L. Vines, L. Rebohle, D. Lang, Y. Berencén, S. Andric, R. Boettger, M. Helm, S. Zhou, W. Skorupa, Ultra-doped n-type germanium thin films for sensing in the mid-infrared, Sci. Rep. 6 (2016) 27643, https://doi.org/10.1038/srep27643.

[245] S. Prucnal, V. Heera, R. Hübner, M. Wang, G.P. Mazur, M.J. Grzybowski, X. Qin, Y. Yuan, M. Voelskow, W. Skorupa, L. Rebohle, M. Helm, M. Sawicki, S. Zhou, Superconductivity in single-crystalline aluminum- and gallium-hyperdoped germanium, Phys. Rev. Mater. 3 (2019) 1–10, https://doi.org/10.1103/PhysRevMaterials.3.054802.

[246] G. Impellizzeri, E. Napolitani, S. Boninelli, G. Fisicaro, M. Cuscuna, R. Milazzo, A. La Magna, G. Fortunato, F. Priolo, V. Privitera, B-doping in Ge by excimer laser annealing, J. Appl. Phys. 113 (2013) 113505, https://doi.org/10.1063/1.4795268.

[247] V. Mazzocchi, C. Sabatier, M. Py, K. Huet, C. Boniface, J. Barnes, L. Hutin, V. Delayer, D. Morel, M. Vinet, C. Le Royer, J. Venturini, K. Yckache, Boron and phosphorus dopant activation in germanium using laser annealing with and without preamorphization implant, in: 2009 17th International Conference on Advanced Thermal Processing of Semiconductors, IEEE, 2009, pp. 1–5, https://doi.org/10.1109/RTP.2009.5373447.

[248] R. Milazzo, E. Napolitani, G. Impellizzeri, G. Fisicaro, S. Boninelli, M. Cuscunà, D. De Salvador, M. Mastromatteo, M. Italia, A. La Magna, G. Fortunato, F. Priolo, V. Privitera, A. Carnera, N-type doping of Ge by As implantation and excimer laser annealing, J. Appl. Phys. 115 (2014), https://doi.org/10.1063/1.4863779, 053501.

[249] R. Milazzo, M. Linser, G. Impellizzeri, D. Scarpa, M. Giarola, A. Sanson, G. Mariotto, A. Andrighetto, A. Carnera, E. Napolitani, p-type doping of Ge by Al ion implantation and pulsed laser melting, Appl. Surf. Sci. 509 (2020), https://doi.org/10.1016/j.apsusc.2019.145230, 145230.

[250] R. Milazzo, C. Carraro, J. Frigerio, A. Ballabio, G. Impellizzeri, D. Scarpa, A. Andrighetto, G. Isella, E. Napolitani, Ex-situ doping of epitaxially grown Ge on Si by ion-implantation and pulsed laser melting, Appl. Surf. Sci. 509 (2020) 145277, https://doi.org/10.1016/j.apsusc.2020.145277.

[251] S. Huang, F. Lu, W. Huang, C. Huang, C.W. Liu, Specific contact resistivity of phosphorus-doped Ge on Si by in-situ chemical vapor deposition doping and laser annealing, IEEE Electron Dev. Lett. 36 (2015) 1114–1117.

[252] C. Carraro, R. Milazzo, F. Sgarbossa, D. Fontana, G. Maggioni, W. Raniero, D. Scarpa, L. Baldassarre, M. Ortolani, A. Andrighetto, D.R. Napoli, D. De Salvador, E. Napolitani, N-type heavy doping with ultralow resistivity in Ge by Sb deposition and pulsed laser melting, Appl. Surf. Sci. 509 (2020) 145229, https://doi.org/10.1016/j.apsusc.2019.145229.

[253] S. Koffel, R.J. Kaiser, A.J. Bauer, B. Amon, P. Pichler, J. Lorenz, L. Frey, P. Scheiblin, V. Mazzocchi, J.-P. Barnes, A. Claverie, Experiments and simulation of the diffusion and activation of the n-type dopants P, As, and Sb implanted into germanium, Microelectron. Eng. 88 (2011) 458–461, https://doi.org/10.1016/j.mee.2010.09.023.

[254] E. Napolitani, G. Bisognin, E. Bruno, M. Mastromatteo, G. Scapellato, S. Boninelli, D. De Salvador, S. Mirabella, C. Spinella, A. Carnera, F. Priolo, Transient enhanced diffusion of B mediated by self-interstitials in preamorphized Ge, Appl. Phys. Lett. 96 (2010) 201906, https://doi.org/10.1063/1.3429084.

[255] S.F. Lombardo, S. Boninelli, F. Cristiano, G. Fisicaro, G. Fortunato, M.G. Grimaldi, G. Impellizzeri, M. Italia, A. Marino, R. Milazzo, E. Napolitani, V. Privitera, A. La Magna,

Laser annealing in Si and Ge: anomalous physical aspects and modeling approaches, Mater. Sci. Semicond. Process. 62 (2017) 80–91, https://doi.org/10.1016/j.mssp.2016.10.047.

[256] G. Fisicaro, K. Huet, R. Negru, M. Hackenberg, P. Pichler, N. Taleb, A. La Magna, Anomalous impurity segregation and local bonding fluctuation in l-Si, Phys. Rev. Lett. 110 (2013) 117801, https://doi.org/10.1103/PhysRevLett.110.117801.

[257] L.-L. Wang, H. Yu, M. Schaekers, J.-L. Everaert, A. Franquet, B. Douhard, L. Date, J. del Agua Borniquel, K. Hollar, F.A. Khaja, W. Aderhold, A.J. Mayur, J.Y. Lee, H. van Meer, D. Mocuta, N. Horiguchi, N. Collaert, K. De Meyer, Y.-L. Jiang, Comprehensive study of Ga activation in Si, SiGe and Ge with $5 \times 10^{-10}$ $\Omega{\cdot}cm^2$ contact resistivity achieved on Ga doped Ge using nanosecond laser activation, in: 2017 IEEE International Electron Devices Meeting (IEDM), IEEE, 2017, pp. 22.4.1–22.4.4, https://doi.org/10.1109/IEDM.2017.8268441.

[258] G. Impellizzeri, E. Napolitani, R. Milazzo, S. Boninelli, M. Cuscunà, G. Fisicaro, A. La Magna, G. Fortunato, F. Priolo, V. Privitera, Role of oxygen on the electrical activation of B in Ge by excimer laser annealing, Phys. Status Solidi 211 (2014) 122–125, https://doi.org/10.1002/pssa.201300308.

[259] R. Milazzo, G. Impellizzeri, M. Cuscunà, D. De Salvador, M. Mastromatteo, A. La Magna, G. Fortunato, F. Priolo, V. Privitera, A. Carnera, E. Napolitani, Oxygen behavior in germanium during melting laser thermal annealing, Mater. Sci. Semicond. Process. 42 (2016) 196–199, https://doi.org/10.1016/j.mssp.2015.07.066.

[260] R. Milazzo, G. Impellizzeri, D. Piccinotti, A. La Magna, G. Fortunato, D. De Salvador, A. Carnera, A. Portavoce, D. Mangelinck, V. Privitera, E. Napolitani, Impurity and defect interactions during laser thermal annealing in Ge, J. Appl. Phys. 119 (2016), https://doi.org/10.1063/1.4940737, 045702.

[261] R. Milazzo, M. Linser, D. Scarpa, A. Carnera, A. Andrighetto, E. Napolitani, Indiffusion of oxygen in germanium induced by pulsed laser melting, Mater. Sci. Semicond. Process. 88 (2018) 93–96, https://doi.org/10.1016/j.mssp.2018.07.036.

[262] J. Huang, N. Wu, Q. Zhang, C. Zhu, A.A.O. Tay, G. Chen, M. Hong, Germanium n[sup +]/p junction formation by laser thermal process, Appl. Phys. Lett. 87 (2005) 173507, https://doi.org/10.1063/1.2115078.

[263] D.J. McCulloch, S.D. Brotherton, Surface roughness effects in laser crystallized polycrystalline silicon, Appl. Phys. Lett. 66 (1995) 2060–2062, https://doi.org/10.1063/1.113902.

[264] S. Heo, S. Baek, D. Lee, M. Hasan, H. Jung, J. Lee, H. Hwang, Sub-15 nm n+p-germanium shallow junction formed by $PH_3$ plasma doping and excimer laser annealing, Electrochem. Solid-State Lett. 9 (2006) G136, https://doi.org/10.1149/1.2172470.

[265] P. Tsouroutas, D. Tsoukalas, A. Florakis, I. Zergioti, A. Serafetinides, N. Cherkashin, B. Marty, A. Claverie, Laser annealing for n+/p junction formation in germanium, Mater. Sci. Semicond. Process. 9 (2006) 644–649, https://doi.org/10.1016/j.mssp.2006.08.013.

[266] P. Tsouroutas, D. Tsoukalas, I. Zergioti, N. Cherkashin, A. Claverie, Diffusion and activation of phosphorus in germanium, Mater. Sci. Semicond. Process. 11 (2008) 372–377, https://doi.org/10.1016/j.mssp.2008.09.005.

[267] P. Tsouroutas, D. Tsoukalas, I. Zergioti, N. Cherkashin, A. Claverie, Modeling and experiments on diffusion and activation of phosphorus in germanium, J. Appl. Phys. 105 (2009) 094910, https://doi.org/10.1063/1.3117485.

[268] M. Shayesteh, D. O'Connell, F. Gity, P. Murphy-Armando, R. Yu, K. Huet, I. Toque-Tresonne, F. Cristiano, S. Boninelli, H.H. Henrichsen, P.F. Nielsen, D.H. Petersen, R. Duffy, Optimized laser thermal annealing on germanium for high dopant activation and low leakage current, IEEE Trans. Electron. Dev. 61 (2014) 4047–4055, https://doi.org/10.1109/TED.2014.2364957.

[269] K. Huet, F. Mazzamuto, T. Tabata, I. Toqué-Tresonne, Y. Mori, Doping of semiconductor devices by laser thermal annealing, Mater. Sci. Semicond. Process. 62 (2017) 92–102.

[270] B. Filippone, C. Donaldson, M. Shayesteh, D. O'Connell, K. Huet, I. Toqué-Tresonne, F. Crupi, R. Duffy, Electrical characterization of the influence of the annealing energy density on carrier lifetimes in germanium, ECS J. Solid State Sci. Technol. 5 (2016) P3013–P3017.

[271] S. Boninelli, R. Milazzo, R. Carles, F. Houdellier, R. Duffy, K. Huet, A. La Magna, E. Napolitani, F. Cristiano, Nanoscale measurements of phosphorous-induced lattice expansion in nanosecond laser annealed germanium, APL Mater. 6 (2018), 058504.

[272] J. Borland, J. Herman, S. Novak, H. Onoda, Y. Nakashima, K. Huet, W. Johnson, A. Joshi, P, Sb and Sn ion implantation with laser melt-LPC (liquid phase crystallization) for high activation n+ ultra shallow junction in Ge epilayer and surface strain-Ge formation for mobility enhancement, in: 2015 15th International Workshop on Junction Technology (IWJT), IEEE, 2015, pp. 15–18.

[273] J. Frigerio, A. Ballabio, K. Gallacher, V. Gilberti, L. Baldassarre, R. Millar, R. Milazzo, L. Maiolo, A. Minotti, F. Bottegoni, P. Biagioni, D.J. Paul, M. Ortolani, A. Pecora, E. Napolitani, G. Isella, Optical properties of highly n-doped germanium obtained by in-situ doping and laser annealing, J. Phys. D. Appl. Phys. 50 (46) (2017), 465103.

[274] D. Pastor, H.H. Gandhi, C.P. Monmeyran, A.J. Akey, R. Milazzo, Y. Cai, E. Napolitani, R.M. Gwilliam, I.F. Crowe, J. Michel, L.C. Kimerling, A. Agarwal, E. Mazur, M.J. Aziz, High level active n + doping of strained germanium through co-implantation and nanosecond pulsed laser melting, J. Appl. Phys. 123 (2018) 165101.

[275] T. Kalliovaara, J. Slotte, I. Makkonen, J. Kujala, F. Tuomisto, R. Milazzo, G. Impellizzeri, G. Fortunato, E. Napolitani, Electrical compensation via vacancy–donor complexes in arsenic-implanted and laser-annealed germanium, Appl. Phys. Lett. 109 (2016) 182107, https://doi.org/10.1063/1.4966947.

[276] R. Milazzo, G. Impellizzeri, D. Piccinotti, D. De Salvador, A. Portavoce, A. La Magna, G. Fortunato, D. Mangelinck, V. Privitera, A. Carnera, E. Napolitani, Low temperature deactivation of Ge heavily n-type doped by ion implantation and laser thermal annealing, Appl. Phys. Lett. 110 (2017), https://doi.org/10.1063/1.4973461, 011905.

[277] A. La Magna, V. Privitera, G. Fortunato, M. Cuscunà, B.G. Svensson, E. Monakhov, K. Kuitunen, J. Slotte, F. Tuomisto, Vacancy generation in liquid phase epitaxy of Si, Phys. Rev. B 75 (23) (2007) 235201, https://doi.org/10.1103/PhysRevB.75.235201.

[278] G. Mannino, V. Privitera, A. La Magna, E. Rimini, E. Napolitani, G. Fortunato, L. Mariucci, Depth distribution of B implanted in Si after excimer laser irradiation, Appl. Phys. Lett. 86 (5) (2005), https://doi.org/10.1063/1.1856696, 051909.

[279] G. Thareja, J. Liang, S. Chopra, B. Adams, N. Patil, S.-L. Cheng, A. Nainani, E. Tasyurek, Y. Kim, S. Moffatt, R. Brennan, J. McVittie, T. Kamins, K. Saraswat, Y. Nishi, High performance germanium n-MOSFET with antimony dopant activation beyond $1 \times 10^{20} cm^{-3}$, in: 2010 International Electron Devices Meeting, IEEE, 2010, pp. 10.5.1–10.5.4, https://doi.org/10.1109/IEDM.2010.5703336.

[280] G. Thareja, S. Chopra, B. Adams, Y. Kim, S. Moffatt, K. Saraswat, Y. Nishi, High n-type antimony dopant activation in germanium using laser annealing for n+/p junction diode, IEEE Trans. Electron. Dev. 32 (2011) 838–840.

[281] E. Bruno, G.G. Scapellato, G. Bisognin, E. Carria, L. Romano, A. Carnera, F. Priolo, High-level incorporation of antimony in germanium by laser annealing, J. Appl. Phys. 108 (2010) 124902, https://doi.org/10.1063/1.3520671.

[282] E. Bruno, G. Scapellato, A. La Magna, M. Cuscunà, E. Napolitani, S. Boninelli, F. Priolo, G. Fortunato, V. Privitera, Anomalous transport of Sb in laser irradiated Ge, Appl. Phys. Lett. 101 (2012) 172110, https://doi.org/10.1063/1.4764069.

[283] D.B. Cuttriss, Relation between surface concentration and average conductivity in diffused layers in germanium, Bell Syst. Tech. J. 40 (1961) 509–521, https://doi.org/10.1002/j.1538-7305.1961.tb01627.x.

[284] F. Sgarbossa, G. Maggioni, G.A. Rizzi, S.M. Carturan, E. Napolitani, W. Raniero, C. Carraro, F. Bondino, I. Píš, D. De Salvador, Self-limiting Sb monolayer as a diffusion source for Ge doping, Appl. Surf. Sci. 496 (2019) 143713, https://doi.org/10.1016/j.apsusc.2019.143713.

[285] C. Xu, C.L. Senaratne, P. Sims, J. Kouvetakis, J. Menéndez, Ultralow resistivity Ge:Sb heterostructures on Si using hydride epitaxy of deuterated stibine and trigermane, ACS Appl. Mater. Interfaces 8 (2016) 23810–23819, https://doi.org/10.1021/acsami.6b06161.

Chapter 6

# Continuum modeling and TCAD simulations of laser-related phenomena in CMOS applications

Salvatore Francesco Lombardo, Ioannis Deretzis, Alberto Sciuto, and Antonino La Magna
*Consiglio Nazionale delle Ricerche (National Research Council), Istituto per la Microelettronica e Microsistemi CNR-IMM, Catania, Italy*

## 6.1 Introduction: Complexity and multiple time and space scales in the simulation of irradiation processing

Predicting the modification of materials and structures induced by laser irradiation is a key element for qualifying this process in nanoelectronic applications and device manufacturing. Probably, the limited integration of reliable and complete laser annealing (LA) simulators (also in terms of material calibration) in commercial technology computer-aided design (TCAD) tools, commonly used in the technological development of devices, is still the major bottleneck that hinders the application of LA in micro- and nanoelectronics.

An apparent contradictory characteristic of the current claim for the LA use is that the appeal of this process grows with the critical issue of its control. Indeed, the interest for LA features (e.g., focused heating/melting, see the discussions in Chapters 1, 4, and 5 and Ref. [1] for a presentation of LA applications in current and future device technologies), as an alternative to conventional thermal processes, increases with the complexity of the device structures to be processed: it is easy to figure out that the irradiation effects in complex two-dimensional (2D)/three-dimensional (3D) distributions of different materials are difficult to predict/control. In addition, the eventual successful process integration needs a complete retraining also in the case of limited changes of the design in the development of the same device class (i.e., in the case of new device generations).

Laser Annealing Processes in Semiconductor Technology https://doi.org/10.1016/B978-0-12-820255-5.00002-7

For example, a query for the localized heating, which LA could guarantee, is invoked when the advanced device design needs some components on top of prefabricated layers which could be damaged if the heating extends in the full wafer (see Chapter 4). Obviously, in this case, the propagation of the electromagnetic (e.m.) field, its absorption, and all the phenomena activated by the structure – e.m. field interaction depend critically on the usually complex geometry and material design of the devices.

Apart from the critical dependence of the LA process feature on the device design, the paradigm "interest vs critical control" emerges in other usually invoked applications of this particular manufacturing technique. Indeed, another important application class for LA is the activation of (often localized and ultrafast) phase transitions. This class includes the reference LA application in semiconductors' doping, that is, the high dopant activation above the solubility level after melting and crystal regrowth [2–4], but also more challenging and critical applications like the (explosive) crystallization of amorphous nanostructures or the silicide formation when processing a metal layer deposited on semiconducting substrates.

In addition to the difficult prediction of phases and chemical component evolution during LA, ultrafast phase transitions are also correlated to the peculiar behavior of the physical observables of interest, which should be also modeled and controlled. Among them we cite: the alloy redistribution [5–8], the electrical activation of ion-implanted impurities [9–11], the defect evolution [12–14], and the observation of nonconventional extended defects [1, 15].

In any case, LA induces a far from equilibrium kinetics in the irradiated materials and the benefits of this nonequilibrium phenomenon (i.e., the potential process performance well beyond conventional annealing) are counter balanced by the difficult process control, especially when the process design is purely based on experimental design of experiments (DoE) procedures.

As discussed in Chapter 3, the accurate modeling of many of these far from equilibrium phenomena require atomistic approaches such as molecular dynamics or kinetic Monte-Carlo. However, the complete TCAD simulation of the LA process is hard to achieve with atomistic techniques only, since structures with multispace scales react to the irradiation (e.g., accurate simulations of nanostructures' melting are barely obtained without computing the temperature field evolution in the close-to-millimeter scale of wafer thickness). Moreover, for many cases, also the full process timescale (including the quenching stage and the pulse to pulse time in a multipulse mode) is difficult to model with atomistic simulations. As a consequence, continuum modeling, eventually integrated with more accurate atomistic simulation results, is a key approach for the predictive understanding and control of the LA process.

This chapter will focus on the current status of continuum modeling of LA as a device technological design able to predict the ultrarapid evolution of average fields related to the microstructural and atomistic evolution of the processed systems. We mainly discuss the LA modeling of structures and materials of interest for nanoelectronics. Anyhow, the modeling approaches are often transferable to other technological applications of LA.

The chapter is structured as follows: in Section 6.2, we discuss briefly (see Chapter 2 for a detailed presentation) the laser-matter interaction considering optical excitations of electrons in solids activated by photons with ~eV energies and the inner interaction leading to the energy transfer from the electron system to the phonon bath (i.e., the thermalization); Section 6.3 focuses on the simulation of the nonmelting laser processing with a proper focus on geometry and space-scale dependence of the results, discussing also the failure of the Fourier law in nanoscaled structures and the relative possible corrections in the continuum modeling of the thermal field evolution; in Section 6.4, the modeling of LA processes inducing the crystal-liquid-crystal phase transitions is discussed, reviewing mainly the computational prediction on nonequilibrium impurity kinetics and activation; in Section 6.5, the critical issues for the modeling of difficult phenomena (anomalous or explosive) in terms of process control are presented; and finally, Section 6.6 reports conclusive remarks and perspectives related to future challenges.

## 6.2 Computation of the energy transfer between laser and matter

The fundamentals of the interaction between laser and matter have been presented in Chapter 2. Here, for the sake of consistency, we will briefly resume the main concepts with the specific scope to insert the thermalization approximation in the proper context. Indeed, the current state-of-the-art TCAD models for LA assume this approximation, which in turn allows for affordable numerical simulations of the process also in complex device structures. In all the subsequent sections, we will present simulation examples where the full thermalization of the transferred energy between laser light and irradiated matter is imposed. As a consequence, we would like to direct the reader to the correct use of current TCAD tools.

### 6.2.1 Optical excitations and relaxation phenomena

The modification of solid materials by laser radiation involves a complex chain of processes [16, 17]. The characteristic of the excitation is strongly dependent on the pulse duration. With the use of ultrashort laser pulses, nonequilibrium energy distributions with large excess population in the excited states can be produced. The processes which play an important role in laser-solid interaction on the ultrafast timescale open new routes for modifying the structure and the morphology of materials and offer interesting perspectives in laser material processing. A detailed computational theory for the description of optical excitations and the subsequent phonon-induced relaxation dynamics of nonequilibrium electrons (see the schematic in Fig. 6.1) is presented in the work of [18].

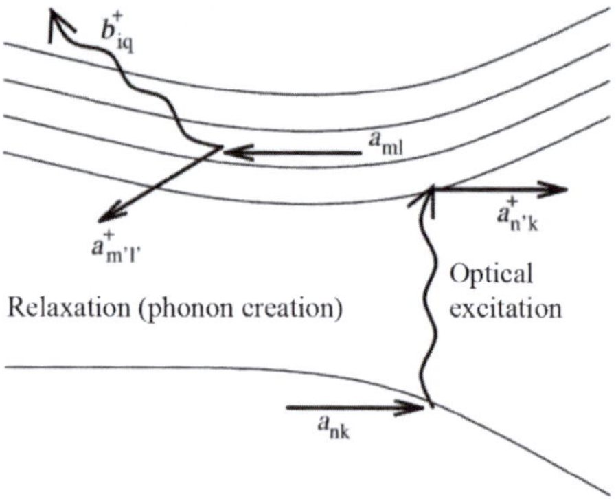


**FIG. 6.1** Optical excitation and relaxation phenomena scheme. *(From N. Buecking, M. Scheffler, P. Kratzer, A. Knorr, Theory of optical excitation and relaxation phenomena at semiconductor surfaces: linking density functional and density matrix theory, Appl. Phys. A 88 (3) (2007) 505–518, https://doi.org/10.1007/s00339-007-4043-4.)*

Irradiation with ultrashort laser pulses can probe the dynamics of electrons in a solid or near its surface, for timescales of tens of femtoseconds. On such short timescales, coherent evolution of excited states of the systems can be observed in some cases, as long as only few distinct channels for energy dissipation are available.

After the coherent timescale, relaxation processes destroy phase coherence and lead to the formation of quasiequilibrium electron distribution functions. While the coherent and subsequent relaxation dynamics of the system calls for a description in terms of density matrix theory (DMT), the still rather well-defined excitation energies allow for a description in terms of a quasiparticle band structure. Determination of the latter first requires a self-consistent electronic structure calculation as starting point for calculations of the quasiparticle band structure by more elaborate methods.

While it is possible in principle to describe the dynamics of excited states by a single-step theory like ab initio theories, for example, by time-dependent density functional theory, such approach is computationally cumbersome and currently still has to rely on a number of approximations. As an alternative, we can use the two-step approach, where the relevant degrees of freedom of the dynamics are described within DMT, while the information about the microscopic processes (entering as parameters into the DMT equations) are provided by DFT. The dynamical equations for the ultrafast response of the solid derived from DMT are valid for a general multiband system with arbitrary phonon modes. This rather general setting allows a flexible modeling of the dynamics, either by retaining only a small number of bands and fixing the few remaining parameters empirically, or by working with a rather elaborate description of the electronic quasiparticle band structure and the phonon modes as delivered by ab initio methods. DFT is capable of providing a realistic description of the surface atomic structure. However, for the excited states, we can use methods of many-particle perturbation theory (i.e., the GW approximation for single quasiparticle states and solutions of the Bethe-Salpeter equation for two-particle states [19]). These methods can also be incorporated in the DMT equations via the electron-electron interaction of optically excited states [20].

### 6.2.2 Ultrashort laser pulse (~*ps* timescale): Electrons and phonons kinetics

The full quantum approach presented in the previous section is necessary for describing phenomena occurring in fs pulses, but it is not applicable for laser pulses of significantly larger duration. Anyhow, a trade-off between pulse duration and accuracy of the numerical method leads to reliable solutions, with successive approximations, for longer duration of the irradiation.

The transition from a quantum description to a semiclassical one is justified by the following qualitative considerations. At first, a quick transient coherent polarization is associated with the primary electronic excitation. This polarization is destroyed by a dephasing process occurring at a timescale of 10 fs. Next, the occupation of these states is modified by nonthermal carrier-carrier interactions on a timescale of about 100 fs. The hot electron state cools down by emission of phonons. Electron cooling and phonon relaxation occur in $<1$ ps. Subsequently, thermal cooling occurs on timescales $>1$ ps.

Semiclassical methods based on the particle-like Monte-Carlo technique permit sufficiently accurate simulations of the charge carrier kinetics for pulses within the $10^{-12}$ to $10^{-10}$ duration range. These methods are calibrated according to the known band structure and electron-phonon coupling parameters. Time-dependent Maxwell equations can be self-consistently solved in a closed form with the carrier semiclassical evolution, which also considers electron-hole generation. An example of the results obtained with this method is shown in Fig. 6.2 for a continuum wave and pulsed irradiation cases.

The local average energy $E_{ave}(t)$ of the overall electrons in the bands tends to follow the electric field oscillation but it deviates from the pure ballistic energy $E_{rms}$ due to the effect of the different scattering events leading also to a different estimate of the self-consistent optical coefficients (see, e.g., the absorption coefficient at the excitation wavelength $\alpha(E_{rms})$ and $\alpha(E_{ave}(t))$) estimated in the two cases and reported in Fig. 6.2. In the case of the pulsed irradiation, this provokes an asymmetric time distribution of the electron energy with respect to the electromagnetic one.

The rate of energy transfer from the electron system to the phonons bath can be evaluated with this stochastic simulation model. Anyhow, we notice that for pulses in the ps timescale (cf. Fig. 6.2, right panel), a consistent part of the total energy is still stored in the electron system where the equivalent average temperature $T_e$ (electron temperature) could be orders of magnitude larger with

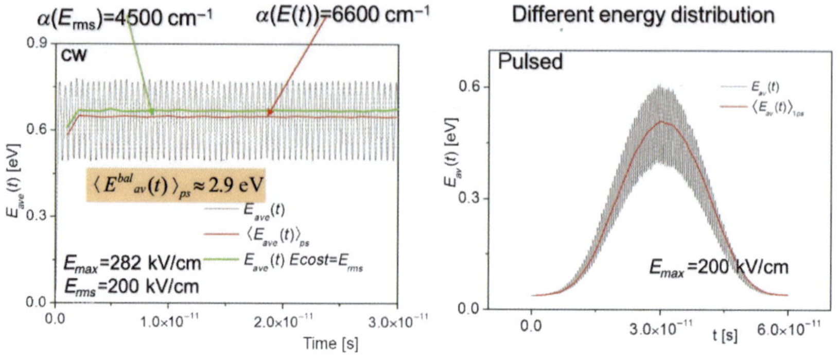


**FIG. 6.2** Average electron energy $E_{ave}(t)$ for a CW (*left panel*) and pulsed (*right panel*) excitation in the THz range. The maximum (max) value of the oscillating electric field and its root mean square values (rms) are reported. The average pure ballistic estimate of the electron energy is $\{E^{bal}\} \sim$ 2.9 eV. *(A. La Magna (original figures).)*

respect to the lattice one ($kT_e \sim 1$ eV and maximum electron energy of the order of few eVs).

A reliably calibrated continuum model for pulses in this regime of duration uses two separate temperature fields, that is, $T_e$ and the standard "lattice" one $T$ [21, 22]. Solutions obtained with this two-temperature theory indicate that, for long timescales ~1 ns and for smooth variations of the laser power density, the electron temperature $T_e$ converges toward the lattice temperature, that is, $T_e \sim T$. Therefore, they can also be used to assess the validity of the thermalization approximation where the delayed mediation by the electron system is neglected and the whole absorbed laser energy is used as a temperature source.

### 6.2.3 Thermalization approximation and self-consistent electromagnetic calculations for the heat sources in laser annealing

In laser annealing, using pulsed laser configurations, the spatial distribution of the electromagnetic power can be considered uniform over areas $\approx$cm$^2$. The coherence length $Lc$ of the e.m. field is in general much smaller than the pulse dimension $Lc \ll 1$ cm but much larger than the feature size of the irradiated structures. As a consequence, the heat source can be approximated as the one generated by a generic incident coherent electromagnetic plane wave with fixed angular frequency $\omega$ in a given polarization state. The laser pulse is usually specified in terms of the total fluence (energy density [ED]) $E_{laser}$ in [J cm$^{-2}$] units and the normalized power density $P_{norm}(t)$

$$\int_0^{\Delta t_{pulse}} P_{norm}(t)dt = 1 \tag{6.1}$$

where $P_{norm}(t) = 0$ when $t < 0$ or $t > \Delta t$.

For pulse duration longer than ~1 ns and photon energy of ~1 eV, $\Delta t_{pulse}$ is usually much larger than the inverse of the frequency

$$\nu^{-1} = (\omega/2\pi)^{-1} = (c/\lambda)^{-1} \tag{6.2}$$

where $\omega$ is the angular frequency and $c$ is the light speed. These considerations make ineffective any fully time-dependent solution of the Maxwell equations for the absorbed energy calculations. An "adiabatic-like" formulation of the heating problem is sufficient, assuming that the change of the optical constants during the annealing is slow with respect to the system response to the electromagnetic excitation. In this approximation, with the additional assumption that the electromagnetic excitation is efficiently transferred from the electrons to the photons we can determine the heat sources using a stationary (or time-harmonic) evaluation of the resistive heat averaging the "ultrafast" timescales of the oscillating electromagnetic field

$$S_{laser}(t,r) = \frac{\varepsilon''}{2\rho}\left|\boldsymbol{E}^{th}(r)\right|^2 \cdot P_{norm}(t) \tag{6.3}$$

where $\varepsilon''$ is the imaginary part of the complex dielectric constant $\varepsilon = \varepsilon' + j\varepsilon''$ of the material and $E^{th}$ is the time-harmonic electric field $\boldsymbol{E} = \boldsymbol{E}^{th} \times \exp(-j\omega t + \phi)$ that can be evaluated by the following time-independent form (i.e., the time-harmonic Maxwell curl-curl problem)

$$\nabla \times (\mu^{-1} \nabla \times \boldsymbol{E}^{th}) - j\omega(\sigma - j\omega'')\boldsymbol{E}^{th} = j\omega \boldsymbol{J}, \text{ in } \Omega \tag{6.4}$$

and

$$n \times \boldsymbol{E}^{th} = n \times \boldsymbol{E}_0^{th}, \text{ on } \delta\Omega_e \tag{6.5}$$

$$\boldsymbol{n} \times (\mu^{-1} \nabla \times \boldsymbol{E}^{th}) + j\omega \boldsymbol{n} \times (\lambda(\boldsymbol{n} \times \boldsymbol{E}^{th})) = j\omega \boldsymbol{J}_f + j\omega \boldsymbol{n} \times \boldsymbol{H}_0, \text{ on } \delta\Omega_h \tag{6.6}$$

where

$$\delta\Omega = \delta\Omega_e \cup \delta\Omega_h \tag{6.7}$$

and

$$\delta\Omega_e \cap \delta\Omega_h = 0 \tag{6.8}$$

where $\Omega$ is an open, bounded domain with boundary $\Gamma = \delta\Omega$, $\delta\Omega_e$, and $\delta\Omega_h$ are, respectively, the total boundary and the boundaries where a condition on the electric or magnetic field is applied, $\boldsymbol{J}$ is the internal source and $\boldsymbol{n}$ is the outward normal vector to the boundary. Simplified equation can be solved for fixed polarization cases. For example in the case of a transverse electric (TE) mode, we can modify Eq. (6.5) with the relations:

$$\boldsymbol{E}^{th} = E_z^{th} \boldsymbol{z} \tag{6.9}$$

$$\frac{\delta E_z^{\ th}}{\delta z} = 0 \tag{6.10}$$

Obtaining the Helmotz equation:

$$-\nabla(\mu^{-1}\nabla E_z^{th}) - j\omega(\sigma - j\omega\varepsilon)E_z^{th} = 0 \tag{6.11}$$

and

$$E_z^{th} = E_0^{th}, \quad \text{on} \quad \delta\Omega_e \tag{6.12}$$

$$-\boldsymbol{n}\cdot\nabla E_z^{th} - jkE_z^{th} = j\omega\boldsymbol{n}\times\boldsymbol{H}_0|_z \quad \text{on} \quad \delta\Omega_h \tag{6.13}$$

### 6.2.4 Calibration issues

Since the optical parameters depend significantly on temperature and phase, Eqs. (6.3), (6.11) have to be solved self-consistently with the phase and temperature evolution equations to evaluate the time and space-dependent heat sources, as can be seen in Fig. 6.3.

In this context, a good calibration of such parameters is mandatory, and experimental data are not always reliable (especially for higher temperatures), since temperature-dependent ellipsometry is not available at LTA temperatures. More information on a reliable calibration strategy can be found in [24], where details on the implementation for complex alloy structures are discussed.

Briefly, calibration parameters are the complex relative permittivity (or complex dielectric constant) at the irradiation wavelength of the topmost absorbing material layer and the species diffusion properties in the liquid phase

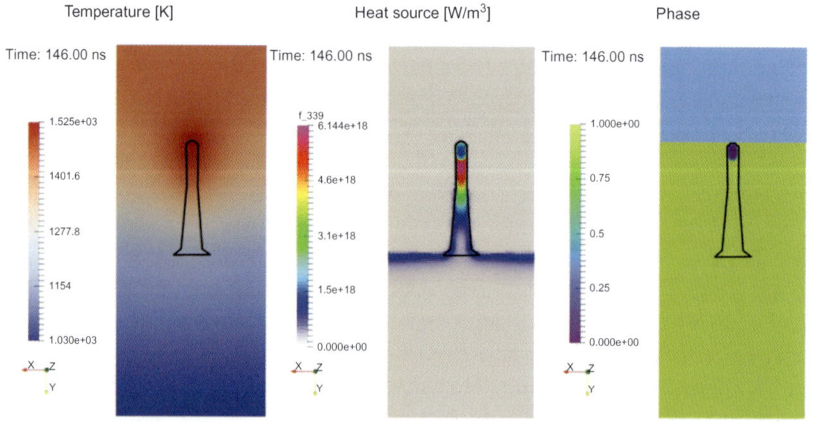


**FIG. 6.3** From left to right, snapshots of temperature, heat source, and phase field for a FinFET with a pitch of 135 nm, irradiated at the fluence of 0.6 J cm$^{-2}$ and 90 degrees of polarization (transverse magnetic) at $t$ = 146 ns. In the phase-field section, *green* represents the solid domains, while *violet* is the liquid part. The *black line* is the shape of the FinFET structure [23]. *(From S.F. Lombardo, G. Fisicaro, I. Deretzis, A. La Magna, B. Curvers, B. Lespinasse, K. Huet, Theoretical study of the laser annealing process in FinFET structures, Appl. Surf. Sci. 467–468 (2019) 666–672, https://doi.org/10.1016/j.apsusc.2018.10.155.)*

(diffusivity and segregation). The first enables reproducing the experimental melt depth relationship to the laser ED, while the latter enables reproducing the diffused species profile shape.

More specifically, given an experimental dataset, the calibration sequence used in this work is as follows. The initial simulation parameter set is defined based on the available literature data, temperature-dependent ellipsometry measurements for the solid-phase optical properties, and linear interpolation for alloy fraction dependence when not available in the literature (especially for thermal properties). Second, the simulated ED is adjusted to match the experimental melt depth for each condition. Third, the species diffusion parameters are adjusted to reproduce the chemical profiles. Finally, the optical refractive index of the liquid phase is adjusted so that the simulated ED corresponds to the experimental one at each melt depth.

## 6.3 Laser annealing TCAD simulation: Nonmelting processes

The prediction of heating for nonmelting laser processes can be obtained in the diffusive limit by means of a self-consistent solution of the time-dependent Fourier law:

$$C(T)\frac{\partial T}{\partial t} = \nabla \cdot [k_{bulk}(T)\nabla T] + S_T(t) \tag{6.14}$$

where $C(T)$ is the thermal capacitance, $k_{bulk}(T)$ is the bulk thermal conductivity, and $S_T(t)$ is the time-dependent internal heat source, which implicitly depends on the $T$ field as discussed in Section 6.2.

In Ref. [25], Peraud et al. demonstrate that Eq. (6.14) is the correct zero-order approximation for the phononic-mediated energy transport in the diffusive regime: that is, small Knudsen number $\langle Kn \rangle$. We notice that $\langle Kn \rangle$ is a dimensionless number defined as the ratio between the average mean free path of the phonons and the representative length of the system in study. Moreover, in this limit [25] the usual relationships:

1. Dirichelet: $T = T_0$
2. Neunmann: $\nabla T = 0$
3. Continuity $- k^1\nabla T_1|_\Gamma = -k^2\nabla T_2|_\Gamma$ and $T_1|_\Gamma = T_2|_\Gamma = T$

define the correct boundary conditions (BCs), respectively, for: (1) the thermostat at $T_0$ temperature; (2) the diffusely specular wall; (3) the interface $\Gamma$ between two materials 1 and 2 with thermal conductivities $k_1$ and $k_2$.

In the following sections, we discuss some key features for the TCAD simulations of laser processes in the thermalization approximation for the laser-matter interaction. We will here consider the nonmelting processes only, while the cases where transient phase transitions could occur during the process will be the topic of the sections. In particular, we will focus on: the patterning effect in the energy absorption and heating kinetics, the prediction of ultrafast solid-phase evolution, and the role of nanoscale constrains in the thermal transport.

### 6.3.1 Predictions of the heating process in electronic device structures

The role of the geometric constrains in the solution of the Maxwell equation makes the laser-induced heating strongly dependent on the structure geometry and material distribution in the geometry itself. As a consequence, for example, the transfer of the process design from a blanket to a structured sample is unfeasible. The application of the simulation is extremely important also for a preliminary estimation of the temperature trend. In this section, we will discuss this issue, by comparing the simulation predictions in different geometries, while also changing the simulation dimensionality.

#### *6.3.1.1 1D simulation example*

An extended discussion of the submelting thermal simulations in blanket samples is reported in [26]. In this analysis, the simulation box is composed of a stack of three layers. The top layer is 200 nm thick, and it is composed of air, the thickness of the middle zone is 100 nm, the bottom layer is 10 μm thick and it is composed of crystalline Si (*c*-Si). A large number of tests have been discussed, including all the materials in the calibration database presented in the Appendix of Ref. [26]. As an example here, we just present the case of a Ge film over a Si substrate in order to show the typical trend.

The initial state of the simulation reproduces a planar wafer at room temperature ($T_0 = 300$ K). A zero flux (Neumann) boundary condition is imposed on the top (air) boundary. The incident wave is coming from the top of the air with a descending direction.

In Fig. 6.4, we report the maximum temperature in the system (obtained for this case at the surface location) as a function of time for an irradiation process with an ED of 0.5 J cm$^{-2}$ and wave length $\lambda = 308$ nm, obtained using COMSOL™ (blue line and triangles) and LIAB (red line and square) [23]. The plot shows also the laser pulse considered as a magenta line and circles. Without lack of generality, we will often consider this pulse shape and frequency in the solutions discussed in the following: the change of the irradiation parameters surely affects quantitatively the results but the overall scenario is usually confirmed.

The interaction of the laser with the thermal response of the system can be divided in multiple zones:

1. From 0 to 36 ns, the source is zero, so the temperature remains unchanged and equal to $T_0$, as imposed at the beginning of the simulation.
2. From 36 to 120 ns, there is the ignition of laser and consequently, the heating up of the sample. The slope depends on the material and the fluence.

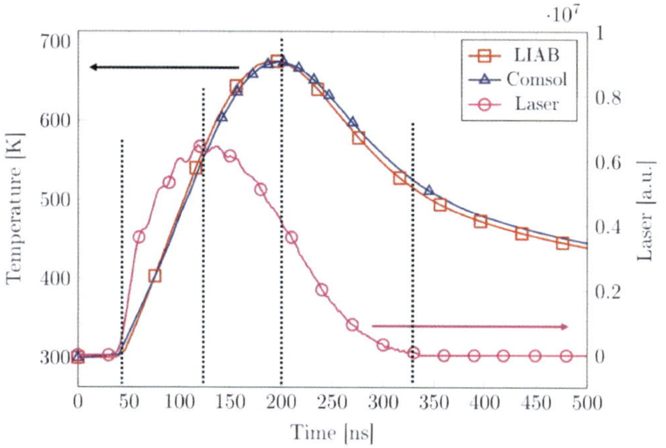


**FIG. 6.4** Comparison between LIAB (*red line and squares*) and COMSOL (*blue line and triangles*) simulated maximum temperature for a 1D structure made of germanium/silicon stack for an LA process with a fluence of 0.5 J cm$^{-2}$ (nonmelting case). *(From S.F. Lombardo, Coupled Kinetic and Electromagnetic Approaches for the Simulation of Complex Processes (Ph.D. thesis), 2018.)*

3. From 130 to 210 ns, despite the reduction of the laser power density, the temperature continues to rise.
4. From 210 to 320 ns, the system begins the thermal quenching, since the laser is now too weak to sustain the high temperature against heat diffusion.
5. From 320 to 500 ns, the laser reaches again the zero value. Consequently the temperature continues to decrease (the external heat source is zero). The overall thermal inertia of the system changes the slope of this state.
6. At larger times (not shown in the figure), the system continues to reduce the temperature, and reaches again 300 K after a couple of microseconds.

The difference between the simulated curves of maximum temperature is very small in spite of the fact that the two simulations run on different codes (different mesh and different finite element method implementation).

By modifying the material and/or the irradiation conditions (pulse shape, fluence, and wavelength), the trend of the thermal field is similar, apart from the particular interval of time of the different zones and the maximum temperature obtained [26]. The role of the temperature dependence of the optical properties (which strongly affects the absorption of electromagnetic waves) on the transmitted energy for one-dimensional (1D) solutions is discussed in Chapter 2. Anyhow, these effects are accurately evaluated in coupled numerical solutions of the Maxwell equations and the heat diffusion.

### *6.3.1.2 Patterning effects: FinFET arrays*

In order to discuss the role of subwavelength patterning in the process design, we will refer to an interesting application case, that is, the study of the pulsed irradiation for the implementation of vertical integration in innovative nanoelectronic device design. Indeed, laser annealing provides localized heating sources that should in principle allow for the processing of the upper structures in a staking sequence of nanotransistors without promoting degradation of the lower ones. Anyhow, due to the peculiarity of the LA process already discussed in the previous section, the process design is critical and the process window difficult to determine. In this section, we discuss the heating dependence on the system geometry design for a fin field-effect transistor (FinFET)-like structure.

Considering the elongated FinFET structure, it is possible to study a simplified 2D system using appropriate boundary conditions. In the simulation box, the substrate is made of silicon with a total depth of 10 μm (not totally shown in Fig. 6.5). The Fin structure is also made of silicon and is surrounded by silicon dioxide (in green). The upper part of silicon (the top-Fin) is surrounded by a cap layer of silicon dioxide of ~2 nm. This layer reproduces the presence of the native oxide, which is simulated assuming the same properties of the other $SiO_2$ layers. The system is completed with a top layer of 200 nm of air. For what concerns the temperature, the whole system at the beginning of the simulations is initialized with the value of $T_0 = 300$ K. A condition of symmetry between the left and right boundaries is imposed. This condition is consistent with the periodic space distribution of the FinFETs' array in the wafer. The incident wave is coming from the top of the air and is directed along the $y$-axis in a descending direction. This kind of study shows not only the effect of the laser energy densities (like the 1D case already discussed), but also the mutual interdependence between geometric effects (which are, therefore, specific to the structure under examination) and optical polarization effects. We notice that light polarization

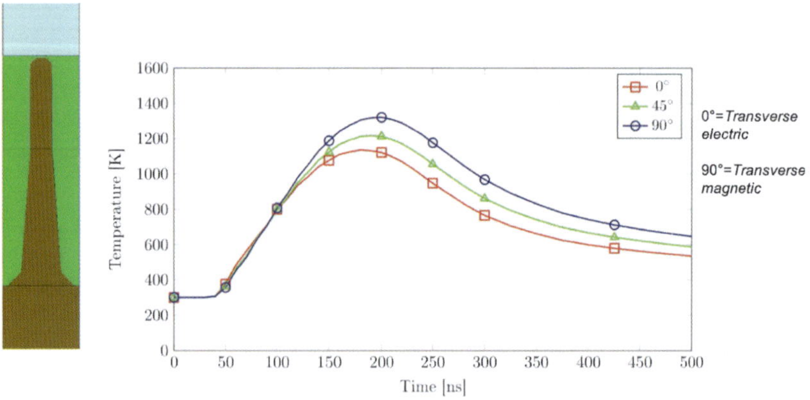


**FIG. 6.5** Maximum temperature in the top-Fin region as dependent of the laser polarization. Pitch value of 35 nm and 0.6 J cm$^{-2}$ of laser fluence. *(From S.F. Lombardo, Coupled Kinetic and Electromagnetic Approaches for the Simulation of Complex Processes (Ph.D. thesis), 2018.)*

can be also selected as a process parameter in the industrial laser annealing tools.

In Fig. 6.5, the maximum temperature for a FinFET with pitch of 35 nm as a function of the time is shown for three different polarization values. In the 2D simulation, the time dependence of $T$ is qualitatively similar to the one reported in Fig. 6.4 for the blanket case, although the peak temperatures are significantly larger with respect to the blanket case ($T_{peak} \sim 800$ K for a simulated process with the same fluence of 0.6 J $cm^{-2}$ in a pure Si substrate). However, the remarkable feature is the behavior with different polarization angle with respect to the Fin elongation direction. Indeed, a 0 degree angle (transverse electric TE polarization) tends to heat up less than 90 degrees (transverse magnetic TM polarization), while a polarization angle of 45 degrees shows an intermediate heating.

In Fig. 6.6, we show the same temperature curves as the previous image but for two different periodicities, the one previously discussed on the right and a wider one (a pitch of 135 nm instead of 35 nm) on the left. By changing the geometry and fixing the fluence in the 135 nm case, melting occurs, while for the 35 nm pitch melting never occurs. Note that melting is evidenced by the plateau of the red curve relative to the TE polarization (see also the following section). The spreading of the curves, when the polarization changes, is generally significantly larger in the 135 nm case (the difference in the peak temperature is almost 1000 K). Finally, the most important (qualitative) difference is the inverse polarization effect. In the 135 nm pitch, the polarization which allows obtaining the maximum (minimum) temperature is 90 degrees one (0 degree one) with the 45 degrees polarization lying between the TE

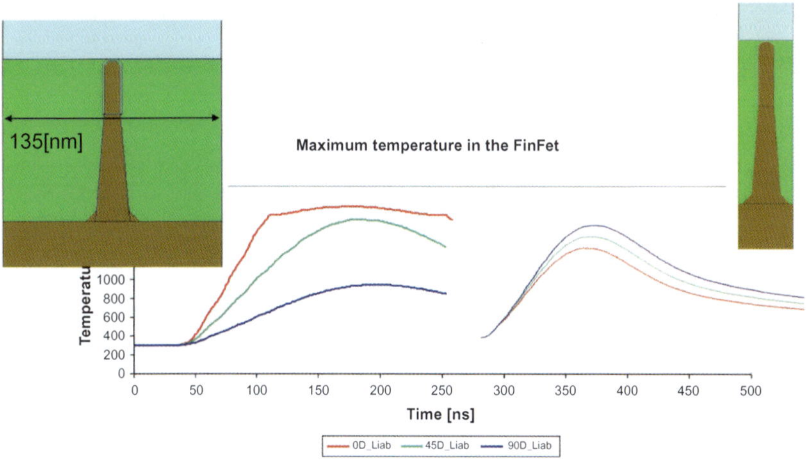


**FIG. 6.6** Maximum temperature in the top-Fin region as dependent of the laser polarization. Pitch values of 135 and 35 nm and 0.6 J $cm^{-2}$ of laser fluence. *(From S.F. Lombardo, Coupled Kinetic and Electromagnetic Approaches for the Simulation of Complex Processes (Ph.D. thesis), 2018.)*

and TM cases. Different geometries and structure patterns change the way electromagnetic waves are absorbed and consequently the interaction with the sample itself. This is a straightforward justification of the necessity to use accurate and efficient tools for the laser process design, which is a problem that cannot be delegated solely to the experience of the process engineer. We point out that additional complexity is added by phase transitions (see the following sections), which cause severe nonlinear effects and often some counter intuitive trends.

### 6.3.2 Diffusion, reactions, and alloy mixing in the solid phase

Ultrafast heating/cooling cycles induced by laser irradiation especially for the case of pulsed laser with pulses in the tens-hundreds ns range, as in the case discussed in the previous section, lead to the usual assumption that such laser annealing processes are "diffusion-less" in the nonmelting regime. Actually, an important reorganization of defective regions, that is, a fast and important kinetic evolution, could occur also in these timescales at very high temperature as discussed in Chapter 3. Moreover, in general, mass transport could occur for systems showing strong intermixing in the solid phase.

A relevant example also for the application in microelectronics is the Ni-Si system. Indeed, a fast interdiffusion of the two elements can be activated also by a low thermal budget, which favors the formation of Ni-silicide compounds at the timescale of few tens of nanoseconds [27, 28] in the case of pulsed laser processes.

The kinetics of silicide formation can be modeled by coupling the heating simulations discussed in the previous sections with a diffusion-reaction model for the prediction of the formation kinetics of Ni-Si compounds (see, e.g., Refs. [28, 29]). This evolution is characterized by the stages dominated by different Ni-Si phases, whereas the Si-rich ones usually need larger fluence to become stabilized with respect to the Ni-rich one.

An example of the model solutions is reported in Fig. 6.7, where the density of the $Ni_2Si$ compound forming at the Ni-Si interface is shown as a function of the time for increasing laser fluence, for laser processes with a pulse of 28 ns ($\lambda = 308$nm). The fast depletion for fluences $\geq 0.4$ J $cm^{-2}$ is related to the formation of the NiSi compound in the same regions.

Melting can also be considered in the study of this complex evolution, and its occurrence, after the significant transformation of the materials in the solid phase during the fast heating further enriches the evolution scenario (see, e.g., Ref. [30]).

### 6.3.3 Energy transport mediated by phonons in the nanoscale: Failure of Fourier law and corrections

The simulation method discussed so far is based on the Fourier law (Eq. 6.14) and it cannot be extended in this form when the structure feature size is further

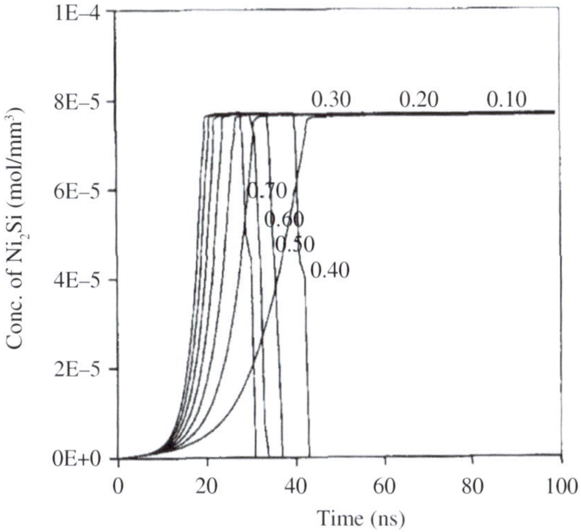


**FIG. 6.7** $NiSi_2$ at initial Ni-Si interface for irradiation processes from 0.1 to 0.7 J $cm^{-2}$ (initial thickness 500 nm). *(From R. Černý, V. Cháb, P. Přikryl, Numerical simulation of the formation of Ni silicides induced by pulsed lasers, Comput. Mater. Sci. 4 (3) (1995) 269–281, https://doi.org/10.1016/0927-0256(95)00047-T.)*

reduced. Indeed, the Fourier law fails, both for the time-dependent and stationary cases, when the critical dimension of the investigated structures is of the order of the average phonon mean free path (i.e., when $\langle Kn \rangle \sim 1$). In this condition, Monte-Carlo evaluation of the phononic Boltzmann transport equation (BTE) and molecular dynamics (MD) simulations demonstrate that the transport of the average thermal energy is apparently ruled by effective coefficients (i.e., apparent thermal conductivity $k_{app}$), which depend on the nanostructure geometry. As an example in Fig. 6.8, the apparent conductivity evaluated by Cartoixà et al. [31] by means of molecular dynamics simulation is shown for two Si nanowires (NWs) of different size: the qualitative trend is a decrease of $k_{app}$ when the critical dimensions shrink.

A possible approach for the TCAD of a laser annealing process in systems with very small structures is the preliminary investigation based on MD [31] or BTE [32] of the analyzed structure and the use of the estimated dependence of $k_{app}(T)$ in Eq. (6.14). However, this procedure is computationally expensive and practically unfeasible for general applications.

An alternative method has been formalized in Ref. [33] and extended in Ref. [34] to the LA simulation. This method is consistent with the general result [25] that the temperature field is ruled by Eq. (6.14) with the bulk value of the conductivity $k_{bulk}(T)$, while corrections due to the finite phonon mean free path have to be considered only for the boundary conditions. In particular, using jump-lick BCs, exact solutions in the diffusive $\langle Kn \rangle \ll 1$ and ballistic $\langle Kn \rangle \gg 1$

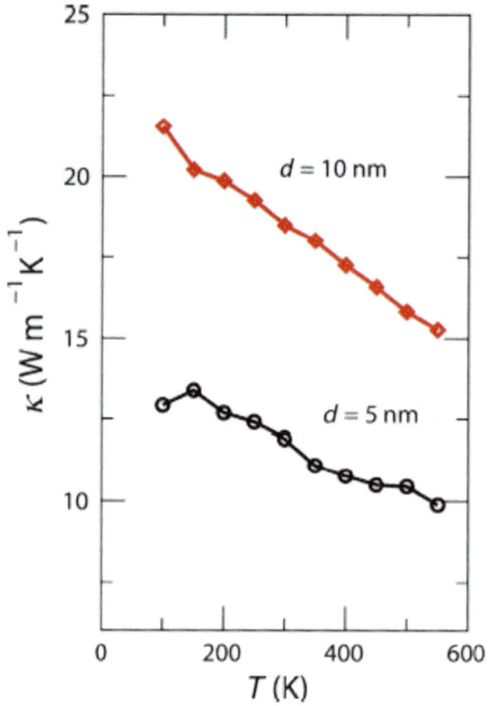


**FIG. 6.8** Apparent thermal conductivity as calculated by MD in two Si nanowires with diameter of 10 and 5 nm. *(From X. Cartoixà, L. Colombo, R. Rurali, Thermal rectification by design in telescopic Si nanowires, Nano Lett. 15 (12) (2015) 8255–8259, https://doi.org/10.1021/acs.nanolett.5b03781.)*

limit can be obtained, while the solutions deviate only by a few percent with respect to the BTE solution in the intermediate $\langle Kn \rangle \sim 1$ region [33]. In this formalism, Dirichelet boundary conditions for a system in contact with a thermostat at a $T_0$ is substituted by

$$\hat{n} \cdot \vec{F}_Q = k_{bulk} \lambda^{-1} (T - T_0) \tag{6.15}$$

where $\vec{F}_Q = \vec{F}_Q^{+} - \vec{F}_Q^{-}$ is the combined phonon flux coming from the left and right size of the junction between the nanosystem and the thermostat, $k_{bulk}$ is the bulk thermal conductivity of the material, and $\lambda$ is the average phonon scattering length, which can be related to known and measurable material properties as

$$\lambda(T) = 4k_{bulk}(T)/C(T)v_s(T) \tag{6.16}$$

where $v_s$ is the sound speed in the material. Moreover, the material boundary is ruled by the jump condition

$$-k_{bulk}^{1}\nabla T_1=\widehat{\Pi}(T_1|_{\Gamma}^{+}\hat{n}_1+T_2|_{\Gamma}^{-}\hat{n}_2) \tag{6.17}$$

for the temperatures $T_1$ and $T_2$ at the interface $\Gamma$ between two regions made of different materials. In Eq. (6.17), $\hat{n}_1=-\hat{n}_2$ are the local outward normal unit vectors to the two sides of the boundary and $\widehat{\Pi}=(k_{bulk}^{1}\lambda_1^{-1}+k_{bulk}^{2}\lambda_2^{-1})|_{\Gamma}$ is the local phonon scattering functional. Moreover, the energy conservation condition at the boundary has to be imposed as [25]

$$-k_{bulk}^{1}\nabla T_1|_{\Gamma}=-k_{bulk}^{2}\nabla T_2|_{\Gamma} \tag{6.18}$$

Eq. (6.18) acts as a closure of the model at the interface location. We note that an external boundary simulating a virtual interface with the same material is still ruled by the Neunmann relation $\nabla T=0$.

The reliability of the approach can be demonstrated by calculating the apparent conductivity $k_{app}$, which can be directly compared with MD estimates or experimental data [34]. According to Ref. [33], $k_{app}$ can be calculated by

$$k_{app}(T)=k_{bulk}(T)\frac{\langle T_{standard}-T_0\rangle}{\langle T_{corrected}-T_0\rangle} \tag{6.19}$$

where $T_{correct}$ and $T_{standard}$ are the corrected and diffusive temperature fields evaluated numerically, the symbol $T_0$ is the thermostat temperature, while $\langle\rangle$ indicates the average of the field expression over the space region occupied by the structures.

In Fig. 6.9A, the apparent conductivity $k_{app}(T)$, evaluated in a broad temperature range for Si NWs of different sizes and in the cases of the two types of thermal contacts is compared with the $k_{bulk}(T)$ of bulk Si. $k_{app}(T)$ is close to the bulk value for meso-structures in the μm range, while there is an important average reduction of $k_{app}(T)$ and a strong geometric effect for sizes of the order of ~100 nm and below. Consistently, we note that the quantitative value $k_{app}(T)$ significantly depends on the contact realization procedure at the nanoscale, while this dependence is less important at the mesoscale.

Comparisons between estimates of $k_{app}(T)$ and experimental data are shown in Fig. 6.9B, where experimental data (empty symbols) are taken by Ref. [35] and the numerical evaluations of $k_{app}(T)$ are plotted for single (solid lines) and double (dashed lines) thermal contacts. The agreement between calculations and measurements is noteworthy.

The jump corrections can be implemented in the TCAD code using a proper integration method, which allows for the computation of discontinuous solutions for the thermal field (for the details see Ref. [34]). An example of laser annealing TCAD simulation considering the discussed phononic correction is shown in Fig. 6.10. The system (see the computational mesh and the laser source distribution in Fig. 6.10A and B) is composed by a Ge telescopic NW (10 nm + 15 nm diameters) immersed in air and deposited over a Si substrate. In this geometry, a thermal contact forms between the substrate and the larger

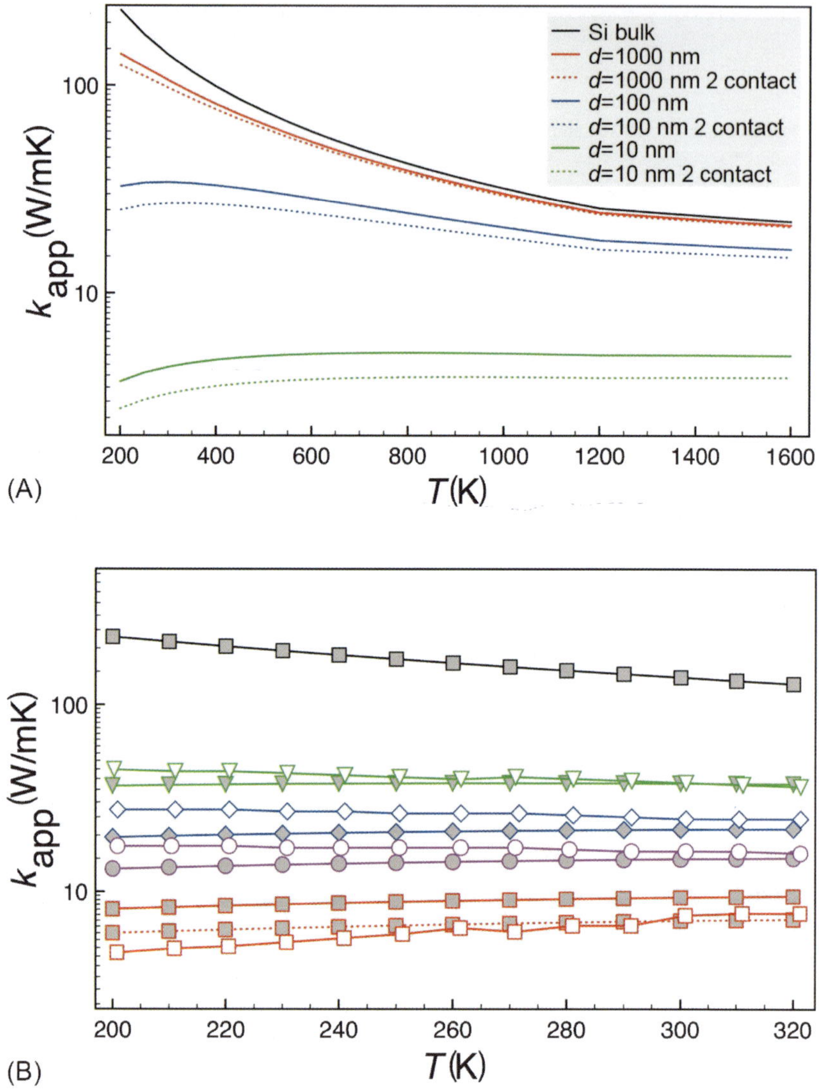


**FIG. 6.9** (A) Evaluation of the apparent thermal conductivity for Si NWs with one and two thermal contacts and different sizes: 10 nm (*green lines*), 100 nm (*blue lines*), and 1000 nm (*red lines*). Bulk Si thermal conductivity is shown as a *black line*. (B) Comparison of simulated (*filled gray markers*, *plain line* for one contact, *dotted* for two contacts) and observed (*empty markers*) thermal conductivity for Si NWs with different sizes: 22 nm (*red lines*), 37 nm (*purple lines*), 56 nm (*blue lines*), and 115 nm (*green lines*). Bulk Si thermal conductivity is shown as a *black line*. *(From A. Sciuto, I. Deretzis, G. Fisicaro, S.F. Lombardo, M.G. Grimaldi, K. Huet, B. Curvers, B. Lespinasse, A. Verstraete, A. La Magna, Phononic transport and simulations of annealing processes in nanometric complex structures, Phys. Rev. Mater. (2020) 056007, https://doi.org/10.1103/PhysRevMaterials.4.056007.)*

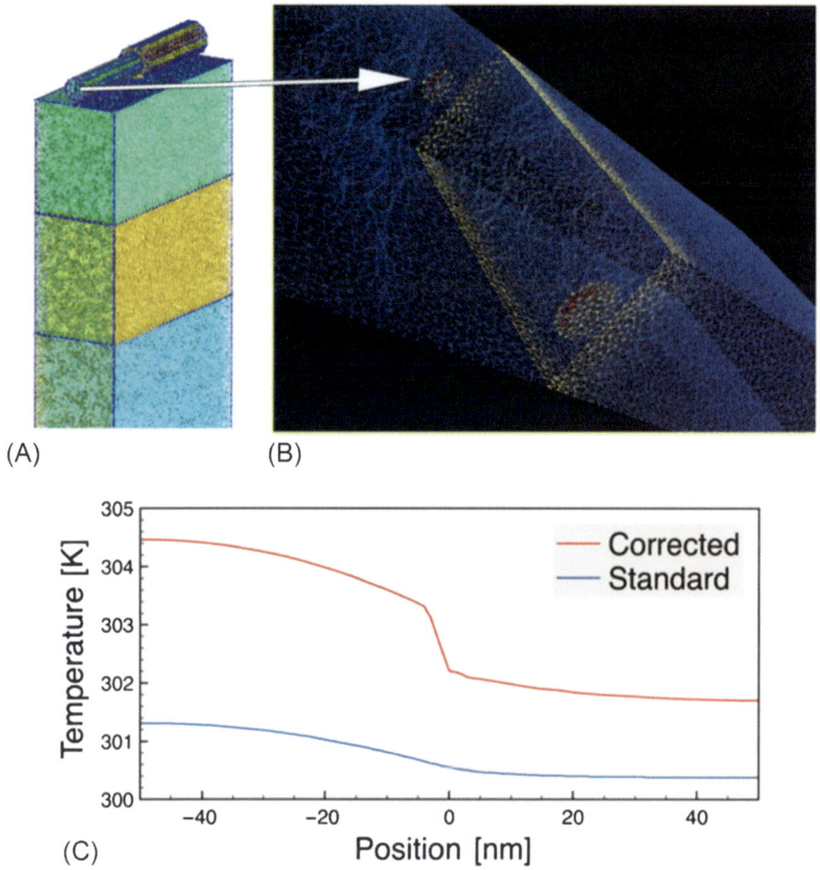


**FIG. 6.10** (A) CAD model and mesh of a Ge telescopic NW on an Si substrate: the colors distinguish the different elements used to compose the CAD geometry. (B) Heat source distribution mapped in the structure mesh used for FEM calculations. The *arrow* indicates equivalent regions in the panels (A) and (B). (C) Comparison of thermal profiles obtained after 5 ns of simulated LTA using standard and corrected BCs. *(From A. Sciuto, I. Deretzis, G. Fisicaro, S.F. Lombardo, M.G. Grimaldi, K. Huet, B. Curvers, B. Lespinasse, A. Verstraete, A. La Magna, Phononic transport and simulations of annealing processes in nanometric complex structures. Phys. Rev. Mater. (2020) 056007. https://doi.org/10.1103/PhysRevMaterials.4.056007.)*

portion of the telescopic NW. The used value of the speed of sound in Ge is $v_s = 5400\ \mathrm{m\ s^{-1}}$ [36], while the interface with the air is ruled by an effective value of $\lambda_{air} = 3$ nm. The calibration of the thermal parameters for Ge taken from Ref. [37]. The laser parameters are: wave length 308 nm, pulse width ~160 ns and fluence 0.4 J cm$^{-2}$.

The solutions (corrected and standard) of the thermal field distribution along the NW axis obtained after 5 ns of simulated annealing are shown in Fig. 6.10C. A completely different heat distribution is calculated for the corrected solution.

Apart from low apparent conductivity effects already discussed, an abrupt change of the temperature at the junction of the telescopic NW can be observed. This jump is obtained only including corrections for the boundary conditions and, as a consequence, considering a finite scattering length for the phonons at the material interfaces.

## 6.4 Laser annealing TCAD simulation: Melting processes

Transient and localized melting adds further potential functionalities to the laser process activating phenomena, which are too slow or not occurring in the solid phase. Obviously, the control of a laser process in the melting regime is far more difficult and the TCAD approaches should consider the nucleation and the evolution of the molten front in the modeling of all the cited phenomena. In this section, we will discuss the simulations for the case of two-phase boundaries (i.e., solid-liquid or crystal-liquid) with a special focus on the doping of elemental semiconductors, for example, Si or Ge semiconductors and an evolving impurity density profile.

### 6.4.1 Liquid-phase transition fundamentals: Free energy barrier, nucleation, and evolution

Melting/solidification is a first-order phase transition due to the energetic unbalance between the free energies of the solid/liquid phase of the material at a given temperature value $T_M$. A proper generic continuum formulation of the free energy $F(\phi, T)$ behavior can be obtained within the Landau-Ginsberg formalism, where $F(\phi, T)$ is a double minimum function of an order parameter $\phi$, where the minimum at $\phi = 0$ represents the liquid phase and the one at $\phi = 1$ represents the solid one. A possible expression of $F(\phi, T)$ which is at the basis of the phase-field model [3] for the melting simulation is [2]

$$-F(\phi,T) = f(\phi) + g(\phi)U(T - T_M) \tag{6.20}$$

where

$$f(\phi) = -\phi^2(1-\phi)^2 \tag{6.21}$$

$$g(\phi) = 8\left(\frac{1}{3}\phi^3 - \frac{1}{2}\phi^4 + \frac{1}{5}\phi^5\right) \tag{6.22}$$

and $U(T - T_M)$ a drift term, which is 0 for $T = T_M$ and $U > 0 (U < 0)$ for $T > T_M (T < T_M)$.

The behavior of this free energy function is shown in Fig. 6.11. The two minima separated by a free energy barrier are degenerate at $T = T_M$ (green curve), while the absolute minimum is obtained at the proper $\phi$ representative of the two phases $T < T_M$ (blue line) and $T > T_M$ (red line). The free energy barrier embeds the nucleation mechanism of the melting/solidification initial event:

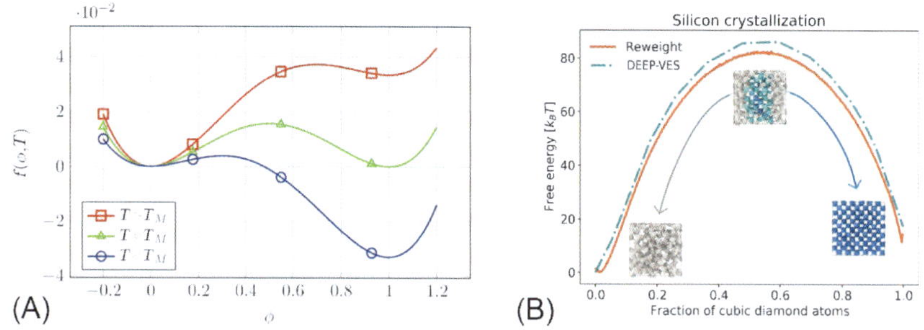


**FIG. 6.11** (A) The free energy density $F(\phi T)$ of Eq. (6.20), displayed as a function for three values of temperature near the melting temperature $T_M$: for $T < T_M$, $T = T_M$, and $T > T_M$. (B) Liquid/solid free energy barrier calculated by machine learning MD in Si. *((A) From S.F. Lombardo, Coupled Kinetic and Electromagnetic Approaches for the Simulation of Complex Processes (Ph.D. thesis), 2018; (B) from L. Bonati, Y.-Y. Zhang, M. Parrinello, Proc. Natl. Acad. Sci. 116 (36) (2019) 17641, https://doi.org/10.1073/pnas.1907975116.)*

for example, a model based on Eq. (6.20) will maintain the solid value of $\phi$ even at $T > T_M$ until the first liquid nucleus forms. The nucleation mechanism and related barrier is very difficult to investigate with atomistic simulations. Only recently the nucleation has been evaluated by means of a machine learning analysis of molecular dynamics (MD) trajectories for Si [38] (see, e.g., Fig. 6.11B), confirming the reliability of the Landau-Ginsberg approach.

Note that, consistently to the nucleation theory character of the phase-field-based formalism, the Landau-Ginsberg approach can also consider the relative stability of the nucleated phase with respect to the nucleus size: for example, the curvature effects in the kinetics of the evolving phase (for a detailed discussion, see the seminal paper [39]).

From the previous discussion, it emerges that in transient and strongly nonequilibrium conditions the liquid/solid interface is not necessarily equivalent to one defined by the $T = T_M$ condition, while, at equilibrium, the $T_M$ level of the thermal field coincides with the melting front. As a consequence, a fully reliable theory should also include the dependence of the interface speed on $T$. Fig. 6.12 shows a Fulcher-Vogel [37]-type expression fitted on experimental data for the amorphous and crystal Si. The curves are approximately linear close to $T = T_M$ with a maximum in the undercooling region ($T < T_M$) and monotonically increase in absolute value for $T > T_M$.

### 6.4.2 Phase-field and enthalpy formalisms: Limits and merits

The phase-field formalism introduces the features of melting/solidification phenomena discussed in the previous sections in evolution equations for $\phi$ and $T$, which can be properly integrated in TCAD tools. The derivation is reported in several papers (see, e.g., Ref. [40] and references therein for a complete

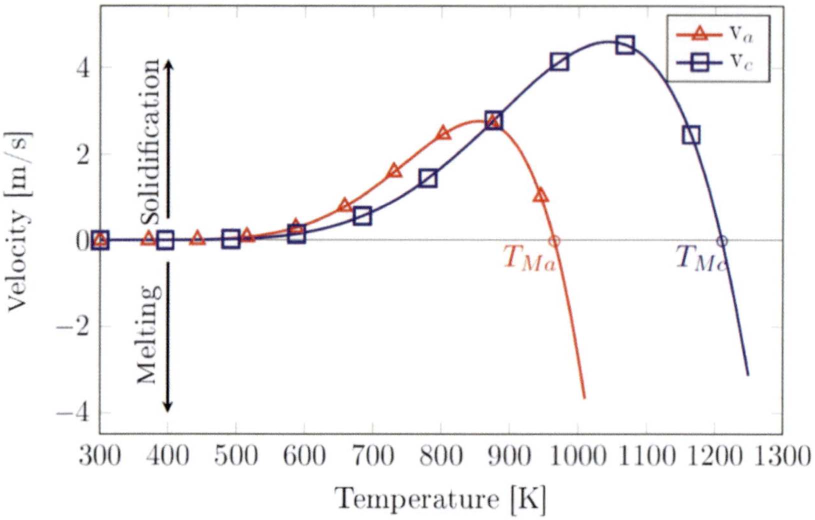


**FIG. 6.12** Interface speed for amorphous (*red line and triangles*) and crystalline (*blue line and squares*) phases. The velocity reaches a 0 value in correspondence of the melting temperature of each phase. *(From S.F. Lombardo, Coupled Kinetic and Electromagnetic Approaches for the Simulation of Complex Processes (Ph.D. thesis), 2018.)*

discussion regarding laser annealing phase-field modeling). Here, we report the coupled PDEs:

$$\tau\frac{\partial\phi}{\partial t} = W^2\nabla^2\phi - \frac{\partial f}{\partial\phi} - U\frac{\partial g}{\partial\phi} \tag{6.23}$$

$$C(T)\frac{\partial T}{\partial t} = \nabla[k_{bulk}(\phi,T)\nabla T] + 30L\phi^2(1-\phi)^2\frac{\partial\phi}{\partial t} + S \tag{6.24}$$

where Eq. (6.23) is a diffuse interface equation, where $W$ rules the thickness of the solid/liquid front, while the temperature equation (6.24) includes here a latent heat term $L$ is appositely expressed in a way that the correct amount of latent heat is consumed (released) in the melting (solidification).

The phase-field equations allow for an accurate continuum modeling of the melting kinetics once a reliable nucleation model is integrated in the formalism (i.e., a method which is able to introduce the stochastic formation of the early nucleus of the other phase [37] in the deterministic PDE-based approach).

The numerical integration of nonlinear Eqs. (6.23), (6.24) could need intensive CPU use especially for large and complex 2D and 3D systems. As a consequence, approximated methods are often applied [41] in the class of the so-called "enthalpy" method, which is reliable for flat liquid/solid interfaces and close to equilibrium evolution. The "enthalpy" method is based on a heat equation formally written as energy conservation law in terms of the (volumetric) enthalpy $H$:

$$\partial_t H + \nabla \cdot \vec{Q} = S \tag{6.25}$$

where $H$ is the enthalpy per unit of volume, $\vec{Q}$ is the heat flux, and $S$ is the heat source. This law allows to find the heat and phase changes, since in the enthalpy formulation the latent heat is also included. However, this equation has to be coupled with an appropriate equation for relating $H$ and $T$. A simpler equivalent derivation [42] uses a modified $C_p$ function in the Fourier law that takes into account also the heat absorbed (released) by the phase changes. The jump discontinuity of the modified $C_p$ function in the transition zone is essential to know the position of the interface itself. This enthalpy model formulation incorporates the latent heat change in the $C_p$ function of $T$, using a convenient Gaussian approximation of the corresponding Dirac $\delta$ function. As a consequence, the $C_p(T)$ function can be written as

$$\tilde{C}_p = C_p(T) + \frac{1}{\Delta T\sqrt{\pi}} e^{-\left(\frac{T-T_M}{\Delta T}\right)^2} L_f \tag{6.26}$$

where $T_M$ is the melting point, $L_f$ is the latent heat of fusion, and $\Delta T$ is a numerical parameter that sets the width of the Gaussian (the phase transition region). The last one is the only parameter that the user can adjust to correctly calculate the phase change but also to have a numerical convergence. The "enthalpy" method considers the melting as a relatively straightforward modification of the conventional heat equation. Therefore, it allows for an easy implementation and a less demanding numerical integration. Anyhow, the limitations with respect to the phase-field model should be considered in order to choose the correct numerical approach:

- The curvature role is not included, therefore, the enthalpy model solutions can deviate significantly from the phase-field ones in 2D and 3D cases especially in the presence of a constrained materials' interface.
- The $T = T_M$ level sets the solid/liquid front. This could cause numerical instability when $T_M$ depends on other variables (e.g., the alloy fraction, see the following section).
- The studied problem has to belong to quasiequilibrium ones: that is, the temperature at the solid-liquid front must stay close to $T_M$. This is not the case of, e.g., nonequilibrium phenomena as the explosive crystallization (see again the following section).

### 6.4.3 Simulation examples of melting processes in 1D, 2D, and 3D systems

The melt depth (i.e., the maximum extension of the molten film) in blanket samples can be measured as a function of the irradiation condition (see, e.g.,

Chapter 5). As a consequence, comparisons between the measured and predicted values of this quantity are a reliable approach to validate the simulations (model + parameter calibration). Fig. 6.13 shows the prediction between measured and phase-field simulated melt depth in a pure Si substrate as a function of the fluence for a laser process with the pulse shown in Fig. 6.4 ($\lambda = 308$ nm). The agreement between simulation and experiment is excellent, thus demonstrating also the mature parameter calibration for Si. In this case, as widely expected, phase-field and enthalpy models give equivalent results.

As already anticipated, the complex behavior of the laser process (namely its dependence on the geometric constrains in patterned samples) is amplified in the case of melting processes. An example of this nonlinear trend is shown in Fig. 6.14, where the simulated total melt area is shown for the FinFET structure described in the previous section. A strongly nonlinear dependence of the melt area is obtained simply changing the periodicity of the FinFET array and the peak of these curves as a function of the pitch occurs at different pitch values when the fluence changes.

Due to the complex interplay between electromagnetic energy absorption, heating, and melting phenomena, the process design with TCAD cannot, in general, be properly addressed with some simplification of the structure in study. For example, approximated simulations of 2D cuts, which are usually considered in the study of other thermal processes, are not very useful for the laser annealing process if the structure is intrinsically 3D (due to, e.g., the presence of a tungsten gate depicted as green region in the example of Fig. 6.15A). In order to demonstrate the relevance of this issue, in Fig. 6.15B the maximum temperature simulated in the 3D FinFET geometry (solid lines) and in one possible 2D (dashed lines) cut of the same system are compared. Nonmelting and

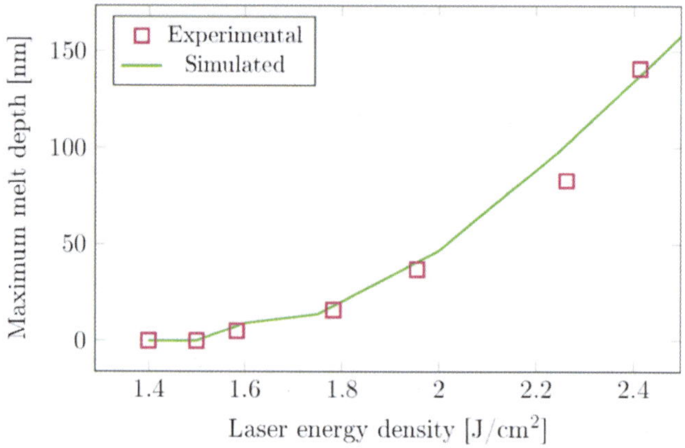


**FIG. 6.13** A comparison between LIAB [23] simulated (*green line*) and experimental (*magenta squares*) maximum melt depth. The central part of the 1D structure is made of silicon. *(From S.F. Lombardo, Coupled Kinetic and Electromagnetic Approaches for the Simulation of Complex Processes (Ph.D. thesis), 2018.)*

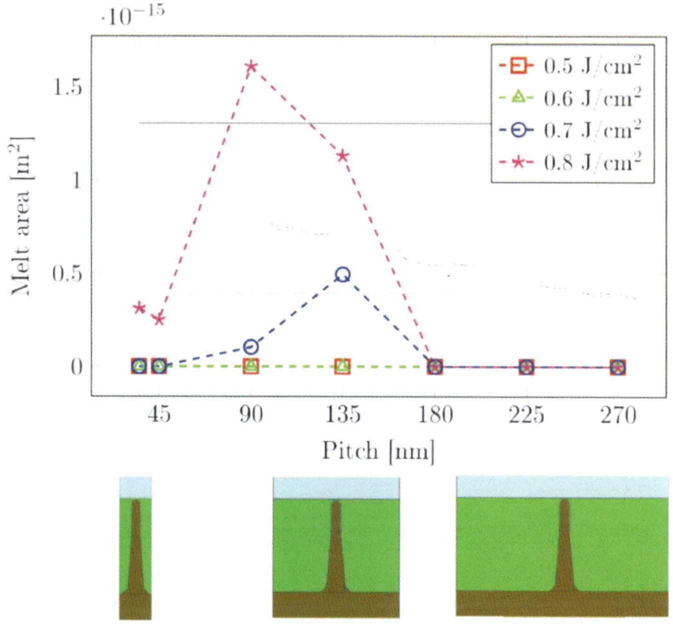


**FIG. 6.14** Global maximum melt area for different laser energies, for the 45-polarization case. Total top-Fin and bottom-Fin areas are indicated by *horizontal lines*. *(From S.F. Lombardo, Coupled Kinetic and Electromagnetic Approaches for the Simulation of Complex Processes (Ph.D. thesis), 2018.)*

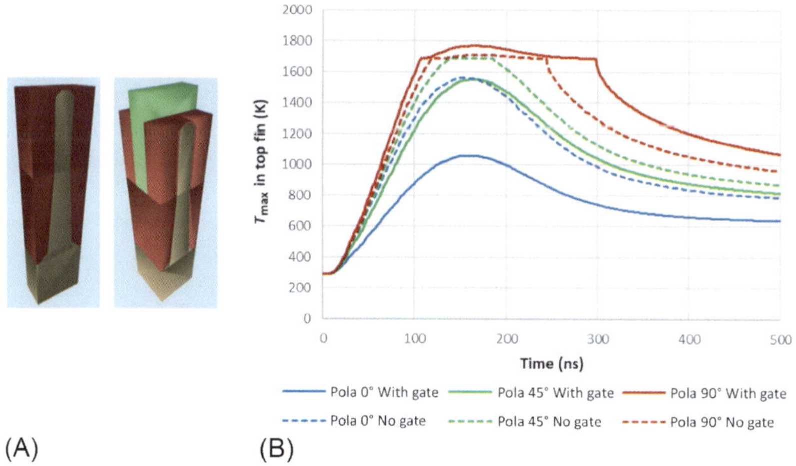


**FIG. 6.15** (A) Two- and three-dimensional models of a FinFET structure. Three-dimensional models allow the possibility to consider also a spatially confined top gate. (B) Maximum temperature of the top-Fin as a function of time for a 3D laser annealing simulation for different polarization values and the presence/absence of a top gate. *(From S.F. Lombardo, G. Fisicaro, I. Deretzis, A. La Magna, B. Curvers, B. Lespinasse, K. Huet, Theoretical study of the laser annealing process in FinFET structures, Appl. Surf. Sci. 467–468 (2019) 666–672, https://doi.org/10.1016/j.apsusc.2018.10.155.)*

melting (see the plateau of the temperature curves) processes are simulated as a function of the polarization for both cases with equivalent qualitative trend. However, the quantitative values are strongly different, which indicates the necessity of the fully 3D study.

### 6.4.4 Impurity evolution in melting processes: Diffusion, segregation, and solute trapping

Transient melting obtained with pulsed laser processes has a further advantage as a doping process, since highly active (above the maximum solid solubility level) impurity profiles can be potentially obtained due to nonequilibrium phenomena. Therefore, the prediction of the impurity evolution in a melting laser annealing process is a key feature for the TCAD simulations.

Contrarily to the nonmelting case, important mass transport in the nanosecond scale is ubiquitous in the melting stage, which is due to the high diffusivity $\sim 10^{-4}\,\text{cm}^2\,\text{s}^{-1}$ (strong intermixing) in liquid phase of all the impurities, including the doping ones (see Chapter 5 for the experimental results). As a consequence, impurity chemical profiles, which are essentially unchanged during the ultrafast heating, are strongly affected by the phase transition.

Two main effects should be modeled to achieve sufficiently reliable predictions of the impurity redistribution, the bulk diffusion and the segregation behavior at the solid-liquid interface. The latter phenomenon is related to the density jump at the solid-liquid interface and in quasiequilibrium conditions are ruled by the impurity equilibrium segregation coefficient

$$k_0 = C_{solid}/C_{liquid} \simeq Constant < 1 \tag{6.27}$$

Phase-field and enthalpy models suitably map the molten region (e.g., $\phi = 0$ in the phase field and $T > T_M$ in the enthalpy case), therefore, they can properly address both diffusion and equilibrium segregation. We note that the kinetic effect of the density jump (Eq. 6.27), especially for low values of $k_0$, strongly affects the resulting final impurity profile with an accumulation peak (segregation peak) for the impurity density, which forms during the solidification stage in the shrinking molten region [3].

Actually, modeling is also complicated by an additional issue: the (nonequilibrium) segregation coefficient depends on the local speed of the liquid-solid front and tends to 1 for very high speed [43]. This effect, known as solute trapping, is very useful for the doping applications but it complicates further the numerical approach due to the necessary inclusion of the front speed in the dopant diffusion equation, and it also makes less feasible the use of enthalpy models. Anyhow, a global formulation of the phase and density fields in a generalized phase-field approach [2, 3] allows a proper formulation of the solute trapping effect within the coupled phase-field equations for the phase and density variables. No speed front evaluation is needed and the model is suitable for integration in any dimension.

The integration of the dopant evolution equations in the laser annealing TCAD models exhibits a capacity of predictability for the dopant profiles in blanket and structured samples. Examples of agreement between experimental and simulated profiles, when these approaches are applied, are reported in Fig. 6.16 for a blanket case (initial implanted phosporus P 1D profile) and Fig. 6.17 for a patterned case (initial implanted as 2D profile).

## 6.5 Complex phenomena and advanced simulations

This final technical section deals with some phenomena, observed in irradiation experiments, whose related continuum models have been only recently and somehow partially integrated in TCAD tools for laser annealing processes. All these kinetic features have some relevance for applications and should be definitively assessed in mature TCAD, allowing for accurate predictions of the LA effects in any condition imposed by the technological specifications. Indeed, we notice that the advanced modeling, subject of the following sections, is necessary in the case of: melting of alloy materials, formation of ultrashallow junctions, partial melting of implanted regions, and the presence of amorphous

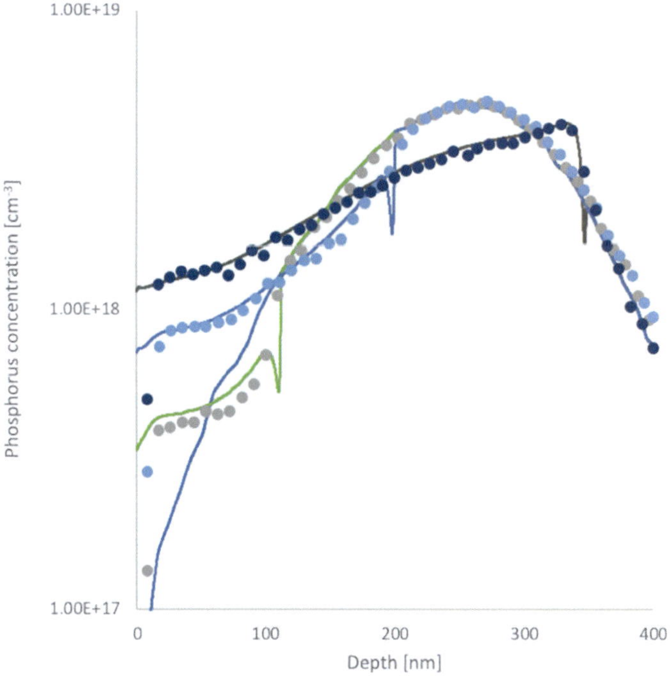


**FIG. 6.16** A comparison between LIAB [23] simulated continuum lines and SIMS profiles (point) of implanted P in Si (P 200 keV $10^{14}$ cm$^{-2}$, *gray points*) after laser annealing processes at different fluences (~160 ns pulse and $\lambda = 308$ nm). *(Original figure.)*

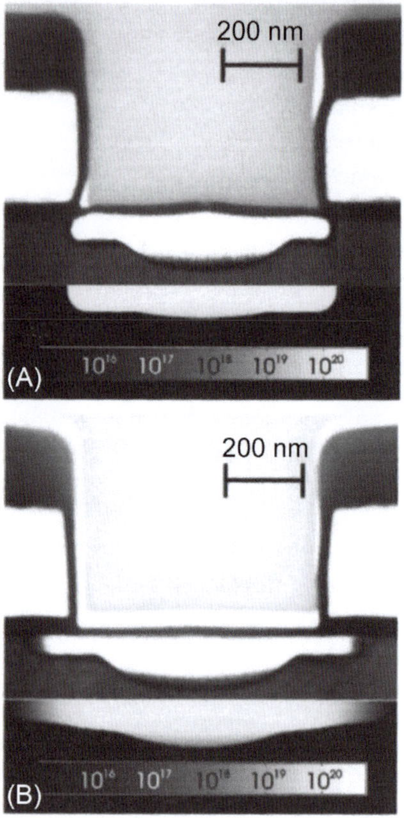


**FIG. 6.17** Comparison between a transmission electron microscopy of dopant profile delineated using the selective etching technique and the simulated dopant distribution. The sample was annealed using one pulse at fluence 0.887 J cm$^{-2}$ (A) and 1 J cm$^{-2}$ (B) (~20 ns pulse and $\lambda$ = 308 nm). *(From A. La Magna, P. Alippi, V. Privitera, G. Fortunato, Role of light scattering in excimer laser annealing of Si, Appl. Phys. Lett. 86 (16) (2005) 161905, https://doi.org/10.1063/1.1906318.)*

pockets. All these requirements can be commonly encountered in the fabrication flow of devices.

### 6.5.1 Alloy problem in melting laser annealing simulations

SiGe is nowadays a material of choice for a large class of devices and it plays a particular role in the vertical integration concept (see, e.g., Chapter 4), where melting LA is invoked as a low thermal budget process. The difficult simulation of the melting processes of $Si_{1-X}Ge_X$ alloy is mainly due to the dependence of the material properties on the local alloy fraction $X$; which, as the impurity profile, evolves during the transient melting ruled by a diffusion/segregation equation. This dependence is particularly critical for numerical simulations in the case of melting temperature $T_M(X)$, since it largely impacts the dynamics of the solid/liquid interface and increases the nonlinear behavior of the diffusion/segregation model. Finally, the difficulties of the simulation approach are also related to the achievement of a reliable calibration of the thermal and optical parameters which, in addition to the temperature dependence, exploit also a dependence on $X$.

As already noted, the interplay between the segregation behavior of the $X$ field and the $X$-dependent melting point make unfeasible the application of the enthalpy approach, since the ultrafast change of $X$ at the solid/liquid interface artificially and instantly displaces the position of the interface itself (which is located at $T = T_M(X)$ level in the enthalpy model). In the phase-field model, the interface is ruled by the additional phase equation and the $T_M(X)$ variable, correctly affects the interface speed ($U$ term in Eq. 6.23), not its position, allowing for the suitable, although difficult, numerical implementation.

A reliable TCAD model of the LA process for SiGe has been recently [24] obtained. The phase-field-type approach and the relative calibration have been validated in blanket samples. In Fig. 6.18, simulated and experimental alloy fraction profiles and total melt depth are compared for an irradiated system composed by 30-nm-thick layer epitaxially grown SiGe on a Si substrate. The segregation effects are important and correctly simulated (see Fig. 6.18A) both for the cases where only the SiGe melts and for the case at 2.23 J $cm^{-2}$ fluence when the molten layer extends in the Si substrate.

This model has also been tested in complex device structures (see, e.g., Fig. 6.19 for an example of a TCAD simulation in SiGe-based nanodevice structure) also considering dopant atoms. In spite of the highly nonlinear character of the continuum model for the alloy, the numerical solutions are proved to be sufficiently stable at least in the 2D case after a careful setting of the FEM approach.

### 6.5.2 Anomalous redistribution of impurities in melting LA processes

Calibrated diffusion equations (Fick's law) and the segregation models presented in the previous section [3, 43] are in general sufficient to simulate impurity evolution during melting pulsed LA processes. Anyhow, in frequent cases, especially for ultraswallow profiles, the diffuse profiles show anomalies that require the refinement of the models. In particular, anomalous peaks (see, e.g., Fig. 6.20) close to the melt depth region are systematically observed in shallow profiles of implanted Si-$X$ and Ge-$X$ ($X$ = B, P, and As) alloys after LA [6, 8, 44, 45].

A particular model able to overcome this issue has been proposed in Ref. [5], which treats the anomalous redistribution of the impurities as an indirect effect

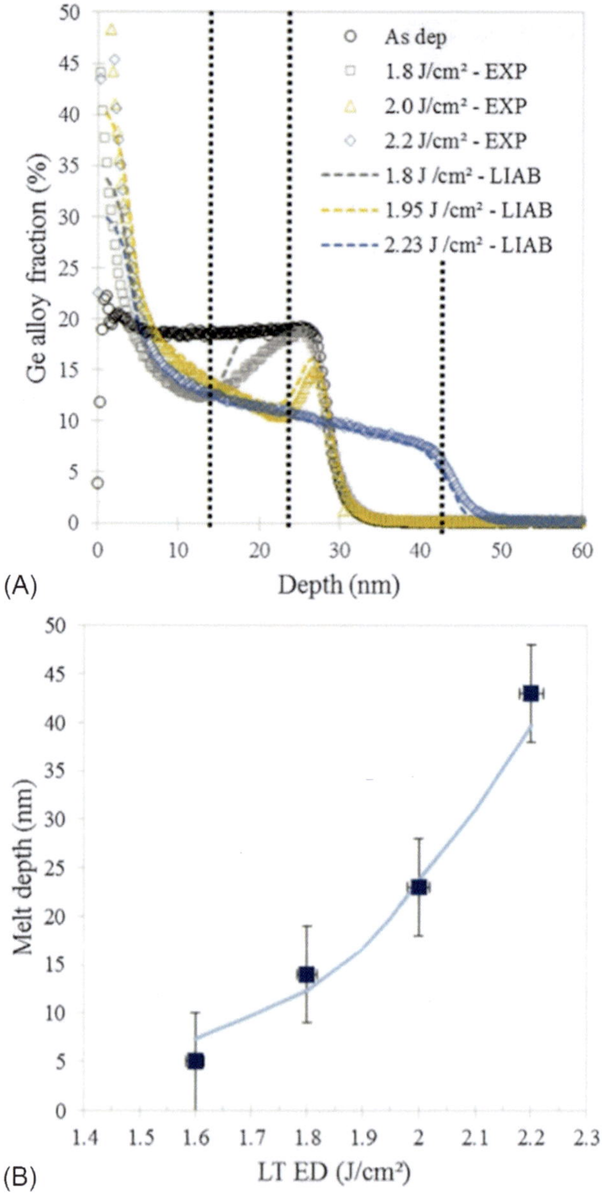


**FIG. 6.18** (A) Simulated (*dashed lines*) Si SIMS depth profiles for $Si_{0.8}Ge_{0.2}$ 30-nm-thick layer case superposed to experimental SIMS after laser annealing for fluences = 1.8, 2.0, and 2.2 J cm$^{-2}$. (B) Melting depth as a function of laser fluence extracted from Si SIMS profiles (*symbols*) and simulated after calibration of the $Si_{1-X}Ge_X$ optical properties (*solid line*). *(From K. Huet, J. Aubin, P.-E. Raynal, B. Curvers, A. Verstraete, B. Lespinasse, F. Mazzamuto, A. Sciuto, S.F. Lombardo, A. La Magna, P. Acosta-Alba, L. Dagault, C. Licitra, J.-M. Hartmann, S. Kerdilés, Pulsed laser annealing for advanced technology nodes: modeling and calibration, Appl. Surf. Sci. 505 (2020) 144470, https://doi.org/10.1016/j.apsusc.2019.144470.)*

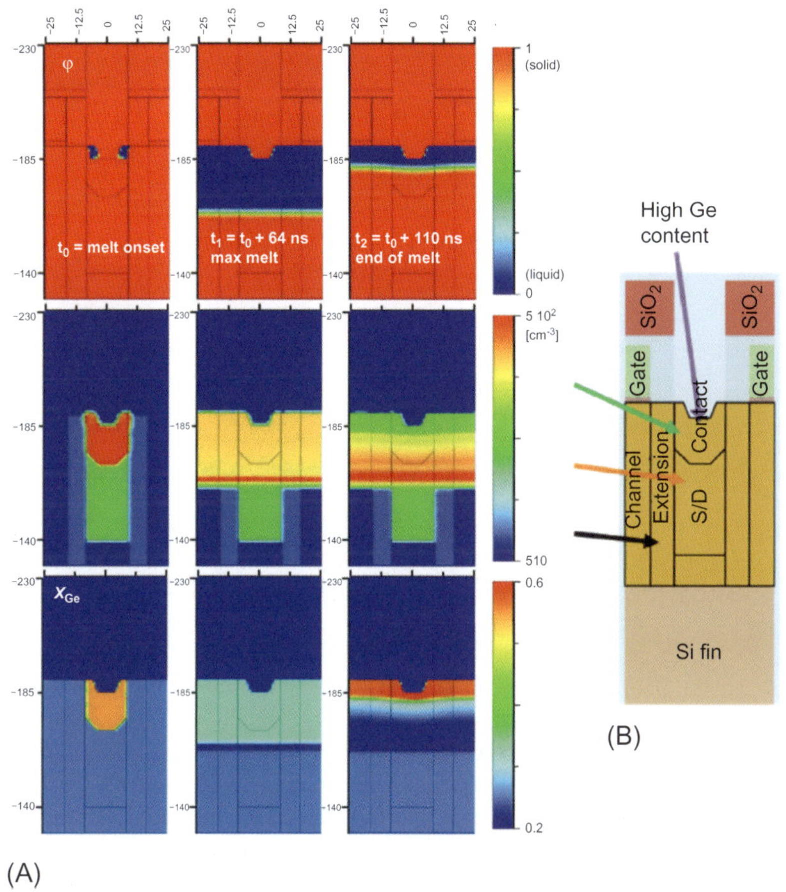


**FIG. 6.19** (A) Snapshots of phase (*top*, 1: solid, 0: liquid), B concentration (*center*) and Ge fraction (*bottom*) fields at time of surface melting onset (*left*), maximum melt (*middle*), and end of melting (*right*). (B) Schematic of the structure. *(From K. Huet, J. Aubin, P.-E. Raynal, B. Curvers, A. Verstraete, B. Lespinasse, F. Mazzamuto, A. Sciuto, S.F. Lombardo, A. La Magna, P. Acosta-Alba, L. Dagault, C. Licitra, J.-M. Hartmann, S. Kerdilès, Pulsed laser annealing for advanced technology nodes: modeling and calibration, Appl. Surf. Sci. 505 (2020) 144470, https://doi.org/10.1016/j.apsusc.2019.144470.)*

of their diffusion mechanism in the liquid semiconductor. The underlying assumption, partially supported also by ab initio studies [46–48], is the presence of two states with proper bonding configuration for the solute atoms dissolved in the liquid semiconductor. These states, characterized by strongly different mobilities, are in a dynamic equilibrium ruled by the local temperature. The balance of these two components of the solute density gives rise to a globally non-Fickian diffusion of impurities in *l*-Si(Ge). The diffusion model is formulated in terms of the following equations [5]:

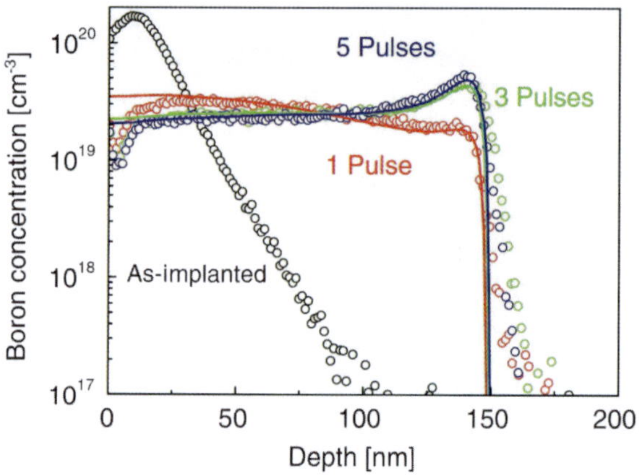


**FIG. 6.20** Experimental (*circles*) and simulated (*lines*) dopant redistribution in the melting phase in implanted Si after multipulse laser irradiation (2.6 J cm$^{-2}$). *(From G. Fisicaro, K. Huet, R. Negru, M. Hackenberg, P. Pichler, N. Taleb, A. La Magna, Anomalous impurity segregation and local bonding fluctuation in* l-Si*, Phys. Rev. Lett. 110 (11) (2013) 117801, https://doi.org/10.1103/PhysRevLett.110.117801.)*

$$\frac{\partial c}{\partial t} = \frac{\partial c^{HD}}{\partial t} + \frac{\partial c^{LD}}{\partial t} \tag{6.28}$$

$$\frac{\partial c^{HD}}{\partial t} = \nabla[D^{HD}\nabla c^{HD}] + k^{\tau}\left(c^{LD} - \overline{R}\cdot c^{HD}\right) \tag{6.29}$$

$$\frac{\partial c^{LD}}{\partial t} = \nabla[D^{LD}\nabla c^{LD}] - k^{\tau}\left(c^{LD} - \overline{R}\cdot c^{HD}\right) \tag{6.30}$$

where $D^{HD}$ and $D^{LD}$ are the impurity diffusivities in the higher and lower diffusivity state in liquid phase. $c^{HD}$ and $c^{LD}$ are the corresponding concentrations. $\overline{R}(T)$ is the average (equilibrium) ratio between low and high diffusivity states at constant $T$. Please note that for nonequilibrium processes the temperature in the liquid phase is not, in general, restricted to $T > T_m$. Impurity atoms in the lower diffusivity state dominate the migration in undercooled regions [48] and $\overline{R}(T) > 1$, while $\overline{R}(T) < 1$ for $T$ well above the melting point and atoms in the higher diffusivity state characterize the impurity kinetics. In the solid phase, standard diffusion occurs and $D_{sol} = D_{sol}^{HD} = D_{sol}^{LD}$ [49]. $k^{\tau}$ is a rate coefficient controlling the transition between the two states, which should be also related to the rapidity of the bonding order fluctuations in *l*-Si and *l*-Ge. Its value has been generally fixed as a constant [5, 6].

Dopant atoms in the lower diffusivity state are strongly favored in regions with $T < T_m$, while atoms in the higher diffusivity state characterize the impurity kinetics in the stable liquid regions. Temperature-dependent diffusivity in the liquid phase is an obvious consequence of the two-state model. However, the anomalous redistribution is an additional effect, being the impurity pile up mechanism inherently related to the local unbalance between states with high and low mobility.

Comparisons between the model's prediction of diffusion profiles and boron (B) density profiles measured by SIMS in implanted Si (B 3 keV energy, $5 \times 10^{14}$ $cm^{-2}$ dose, black circles in the figures) after single-pulse LA processes at different energy densities are shown in Fig. 6.20. The post-LA B profiles predicted by the diffusion model (solid lines in Fig. 6.20) are in satisfying agreement with the experimental ones.

### 6.5.3 Defect evolution and dopant activation in partial melting processes

Melting LA processes of implanted impurity profiles with extension of the melting front exceeding the defective implanted regions promote: a complete healing of the damage after the regrowth, a generally epitaxial regrowth mode, a high level of activation of the dopant profiles. Actually, some technological applications require an incomplete melting of the implanted layers (pockets). As a consequence, damage evolution and the related activation efficiency in regions not reached by the melting front are of certain interest.

The current continuum [11] and Monte-Carlo [12] modeling of the implanted defect evolution in solid regions during LA extends to the time and temperature scale of the LA process in TCAD models used for the standard annealing processes. The main modification, with respect to the standard formalism, is the dependence of the rate parameters on the local (space-dependent) temperature field [12]. Anyhow, in these LA models, as in the ones used for conventional annealing processes, the defect and defect-impurity aggregates evolve according to a (slow) Ostwald-type ripening mechanism mediated by the mobile species density field [11]. As a consequence, they do not account for the ultrarapid formation of extended defects evidenced experimentally (Chapter 5) and supported by molecular dynamics simulations (Chapter 3).

In spite of this strong deficiency, the continuum modeling of systems with defects-impurities reproduce with a certain accuracy the activation kinetics of dopant profiles during pulsed LA processes [11]. An example, the reliable prediction of the dopant activation when a suitable defect-impurity interaction model is implemented is shown in Fig. 6.21, where the simulated active profile of implanted P in Si after partial melting LA is compared with the active profile measured by means of spreading resistance probe (SRP).

### 6.5.4 Explosive crystallization

Annealing processes of amorphous materials are rather usual in device technology and in most of the cases the crystal phase is recovered at the end of the process. In the case of LA application, the process can easily deal with another

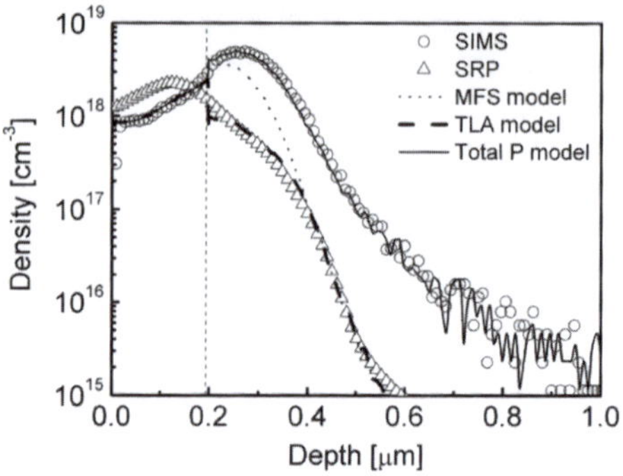


**FIG. 6.21** Experimental results from SRP measurements (*open triangles*) and the simulated (*thick dashes*) P active profile as a function of depth after single-pulse laser process (3.0 J cm$^{-2}$) of an implanted Si substrate (P 200 keV energy, $10^{14}$ cm$^{-2}$ dose). *Thin dashes*: the simulated P active profile with the simplified model. *Solid line*: the simulated total P redistribution. *Open circles* and the *vertical line* refer, respectively, to the SIMS analysis and the experimental melt depth. *(From G. Fisicaro, M. Italia, V. Privitera, G. Piccitto, K. Huet, J. Venturini, A. La Magna, Solid phase phosphorous activation in implanted silicon by excimer laser irradiation, J. Appl. Phys. 109 (11) (2011) 113513, https://doi.org/10.1063/1.3592262. MSF and TLA are models with different accuracy described in this paper.)*

complex ultrafast phenomenon known as explosive crystallization (see also Chapter 1). The explosive crystallization (EC) mechanism in covalent elemental semiconductors (Si and Ge) is known (see, e.g., Ref. [50] and references therein) as the result of a significant larger (negative) latent heat for the liquid to crystal phase transition with respect to the one (positive) for the amorphous to liquid transition. The explosive kinetics of the process has to be naturally recovered by the model since there is the unbalance of the thermodynamic parameters of the three concurring phases.

The quantitative modeling of the EC phenomenon should allow the computation of the contemporary kinetics of the crystal-liquid and liquid-amorphous front on the basis of the thermodynamic parameters of the substance in study. In Ref. [37], the authors propose a new method for the tracking of the interface. The proposed formalism is a natural extension of the original phase-field one and ensures an efficient computational way to solve a multiple-phase problem. The new formulation exploits three local minima of the potential ED at $-1, 0, 1$ (despite the two minima of the phase-field Wheeler's formulation [2] discussed in the previous section) that represent, respectively, the amorphous, liquid, and crystal phase. All the physical parameters (including the ones ruling the dopant evolution) in this approach must interpolate the calibrated values in the three phases. Here, we report an example of one of the parameters of the model that clearly depends on temperature and phase

$$K(\phi,T)=\theta(\phi)K_c(T)+[1-\theta(\phi)]K_a(T)+K_l(T)[1-|\phi|] \tag{6.31}$$

where $\theta(\phi)$ is the Heaviside step function, while the $c$, $a$, and $l$ subscripts indicate crystal, amorphous, and liquid phases, respectively.

An example of the solution of this three-phase model is reported in Fig. 6.22. The simulation refers to LA processes in samples with a ~35-nm amorphous layer of Ge ($\alpha$-Ge) laying on crystalline Ge ($c$-Ge). The amorphous layer is formed by phosphorus implantation at the energy of 15 keV up to the dose of $1 \times 10^{15}$ cm$^{-2}$.

The initial phase variable (see Fig. 6.22) matches the initial phase distribution: the top layer of 35 nm is in the amorphous phase, with value $\phi = -1$ of the phase variable, upon a crystalline bulk up to 50 μm (the substrate thickness) with a value of 1 of $\phi$. The initial temperature of the sample is set at the value of 300 K in the whole simulation box. The as-implanted P SIMS profile is also shown in Fig. 6.22 and is used for the initialization of the impurity density field in the LA simulations.

Fig. 6.22 shows snapshots of phase, temperature, and dopant fields obtained in the simulation of LA at a fluence of 1.2 J cm$^{-2}$ for different times. The red curves with asterisks represent the temperature (scale on the left side), the black curves with circle represent the phase (only three steps but no scale, the lower one is the amorphous, the liquid is in the middle, and the upper one is the crystal phase); the blue curves with the rhombus represent the phosphorous distribution and the scale is on the right side. The upper right part of the figures shows the melting temperature of the amorphous phase (magenta line with squares) and the crystalline phase (green line with triangles). In Fig. 6.22A, the blue curve represents the as-implanted P SIMS profile.

Fig. 6.22A shows the simulation results after $t = 155$ ns before the occurrence of the EC nucleation event, where it is possible to see the initial melting of the amorphous film on the surface. There is the liquid phase on the left side ($\phi = 0$), the amorphous one in the middle ($\phi = -1$), and the crystalline in the bulk ($\phi = 1$). The temperature exhibits the typical diffuse decay in the bulk and a plateau (at temperatures slightly larger than the melting point of $\alpha$-Ge) close to the surface due to both the latent heat absorption and the presence of the liquid phase with its high thermal conductivity.

Fig. 6.22B shows the simulation results after $t = 156$ns during the EC event. In this situation, the liquid phase is in an undercooled and unstable state, which then solidifies in the crystalline phase. The solidification releases the latent heat that heats up locally the sample (consider that the solidification of the crystalline phase takes place at a higher temperature than the amorphous one). The heat released tends to melt the amorphous part and, therefore, the liquid/amorphous front moves to the bulk and is followed by the crystalline/amorphous front. From left to right, the black curve (the phase) starts with value 1, that is, the crystal phase, then it drops to 0, which is the cushion of the liquid phase (this layer is about 15 nm thick), and below the liquid layer the value $-1$ indicates that the remaining part is in the amorphous phase at this time step. The last part of the phase function is the original crystalline bulk ($\phi = 1$).

The red curve represents the temperature of the specimen in this stage and it shows (from left to right): a significant higher value with respect to the melting point of $\alpha$-Ge due to the solidification on the surface (i.e., the heat

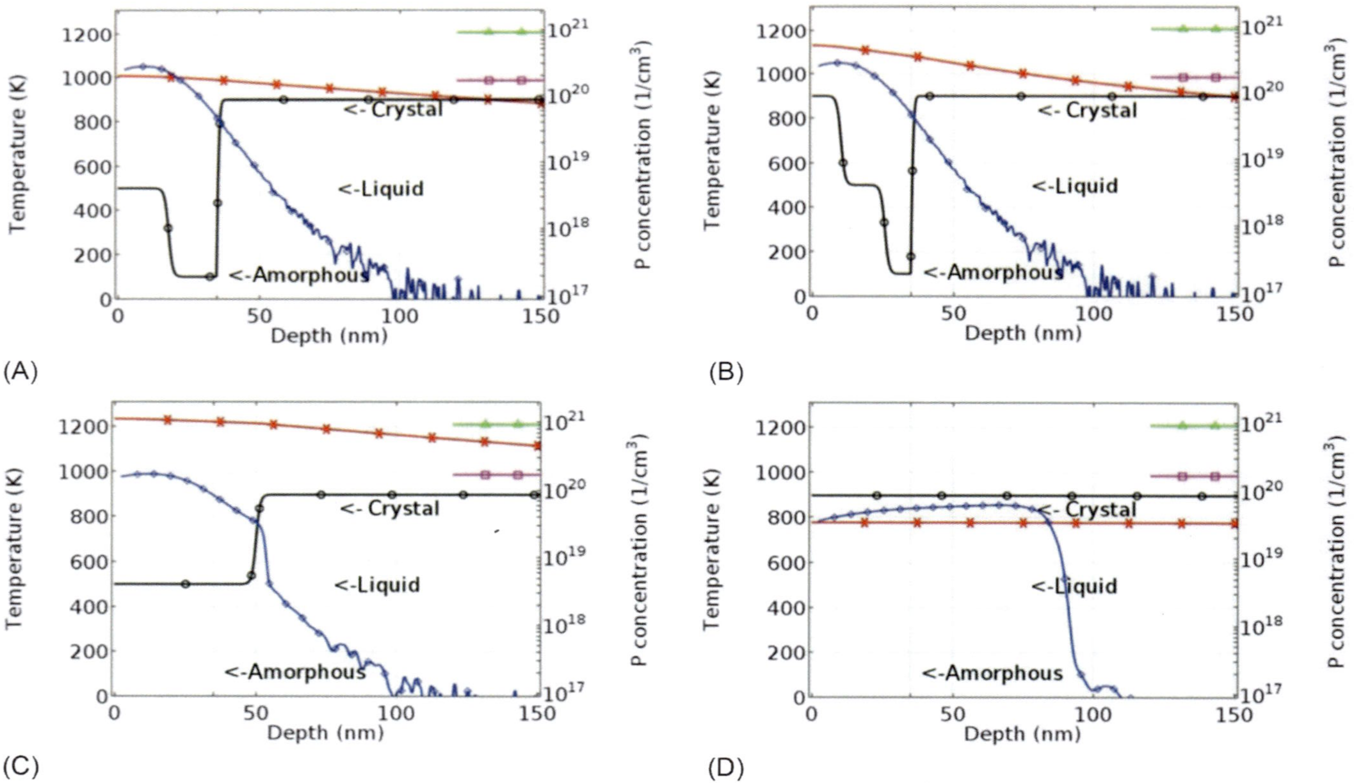


**FIG. 6.22** Snapshots of the simulation at different times for the 1.2 J cm$^{-2}$ fluence case. The scale on the left is for the temperature values (*red solid line and asterisks*), the one on the right for the dopant density (*blue solid line and rhombus*), and a text description is reported for the phase (*black solid line*), representing the three main values of the phase-field function $\phi$: $-1$, 0, and 1 corresponding to the amorphous, liquid, and crystalline phases, respectively. The melting temperatures of the amorphous (*magenta segment and squares*) and crystal phase (*green segment and triangles*) are also indicated. (A) Snapshot of the simulation at $t = 155$ ns (before the EC event). (B) Snapshot of the simulation at $t = 156$ ns (EC event). (C) Snapshot of the simulation at $t = 210$ ns (secondary melting). (D) Snapshot of the simulation at $t = 500$ ns (final state) [37]. *(From S.F. Lombardo, S. Boninelli, F. Cristiano, I. Deretzis, M.G. Grimaldi, K. Huet, E. Napolitani, A. La Magna, Phase field model of the nanoscale evolution during the explosive crystallization phenomenon, J. Appl. Phys. 123 (10) (2018) 105105, https://doi.org/10.1063/1.5008362.)*

is released); a descending value in the liquid layer; a lower value close to the liquid/amorphous interface (where heat is absorbed) and beyond the amorphous/crystal interface due to the typical diffusion decay. The primary front activated by the EC consumes the amorphous film, which is replaced by a nanocrystalline one at the end of the EC phenomenon after a very short-time interval ($EC_{time} \sim 1$ ns).

The simulated evolution at $t = 210$ is reproduced in Fig. 6.22C. A secondary (two-phase) front starts to evolve as soon as the (poly)crystal surface reaches the $c$-Ge melting point and, for the particular irradiation conditions considered here, it reaches a melt depth beyond the original A/C interface.

Depending on the fluence values considered, different scenarios are possible, which we do not report here, but which can be found in full in Ref. [37]. In this case, after the EC there is the remelting of the newly formed crystalline layer, well beyond the initial amorphous crystalline interface.

Fig. 6.22D shows the simulation at $t = 500$ ns when the evolution of the phase and dopant fields is practically quenched. Here, the sample is fully crystalline ($\phi = 1$). The temperature, the red curve in the figure, shows a rather uniform value in the region of interest and it continues to decrease with time recovering the 300 K value after few tens of microseconds.

## 6.6 Conclusion and future perspectives

We hope that, after the reading of this document presenting the current scenario of TCAD models dedicated to pulsed laser annealing processes, two general "takeaway" concepts emerge: (1) the use of TCAD is mandatory for the integration of LA in the nanodevice manufacturing flow, as the usual DoE protocol is doomed to fail and (2) the accuracy of the model predictions can reach that obtained for standard annealing processes after an appropriate research and development effort is ensured. We believe that the maturity level of the continuum modeling and related calibration is very high in some specific regimes (elemental semiconductors, nonamorphized preprocessing, full melting) and TCAD can be routinely used in these conditions for the process design of complex device structures also with fully 3D geometry.

Some phenomena whose emergence is essentially exclusive of the pulsed laser annealing need to receive a definitive assessment (see Section 6.5), which in consideration to the relative advancement status could impact: the material-dependent parameter calibration, the numerical stability of the model in any device geometry (including the 3D ones), the improvement of the formalism, and the new development of the formalism. These issues could be arguments of dedicated computational research in the near future.

Finally, an interesting future perspective which stays between the topics of Chapter 3 and those of the present one is the development of fully atomistic models of LA (e.g., molecular dynamics or kinetic lattice Monte-Carlo). Actually, regions which can be simulated at an atomic resolution cannot exceed the hundred nanometer size, while the Maxwell equations and temperature evolution in TCAD models of LA require large simulations boxes for the proper simulation of the complete process. An interesting method could be a hybrid approach where atomic description is achieved in a subportion of the simulated structure. In this respect, a challenging coupled MD-FEM or KLMC-FEM approach could be developed, similarly to the methods recently applied to a different class of evolution problems [51].

## References


[1] F. Cristiano, M. Shayesteh, R. Duffy, K. Huet, F. Mazzamuto, Y. Qiu, M. Quillec, H.H. Henrichsen, P.F. Nielsen, D.H. Petersen, A.L. Magna, G. Caruso, S. Boninelli, Defect evolution and dopant activation in laser annealed Si and Ge, Mater. Sci. Semicond. Process. 42 (Part 2) (2016) 188–195, https://doi.org/10.1016/j.mssp.2015.09.011.

[2] A.A. Wheeler, W.J. Boettinger, G.B. McFadden, Phase-field model for isothermal phase transitions in binary alloys, Phys. Rev. A 45 (10) (1992) 7424–7439, https://doi.org/10.1103/PhysRevA.45.7424.

[3] A. La Magna, P. Alippi, V. Privitera, G. Fortunato, M. Camalleri, B. Svensson, A phase-field approach to the simulation of the excimer laser annealing process in Si, J. Appl. Phys. 95 (9) (2004) 4806–4814, https://doi.org/10.1063/1.1690861.

[4] A. La Magna, P. Alippi, V. Privitera, G. Fortunato, Role of light scattering in excimer laser annealing of Si, Appl. Phys. Lett. 86 (16) (2005) 161905, https://doi.org/10.1063/1.1906318.

[5] G. Fisicaro, K. Huet, R. Negru, M. Hackenberg, P. Pichler, N. Taleb, A. La Magna, Anomalous impurity segregation and local bonding fluctuation in *l*-Si, Phys. Rev. Lett. 110 (11) (2013) 117801, https://doi.org/10.1103/PhysRevLett.110.117801.

[6] G. Impellizzeri, E. Napolitani, S. Boninelli, G. Fisicaro, M. Cuscunà, R. Milazzo, A.L. Magna, G. Fortunato, F. Priolo, V. Privitera, B-doping in Ge by excimer laser annealing, J. Appl. Phys. 113 (11) (2013) 113505, https://doi.org/10.1063/1.4795268.

[7] E. Bruno, G.G. Scapellato, A. La Magna, M. Cuscunà, E. Napolitani, S. Boninelli, F. Priolo, G. Fortunato, V. Privitera, Anomalous transport of Sb in laser irradiated Ge, Appl. Phys. Lett. 101 (17) (2012) 172110, https://doi.org/10.1063/1.4764069.

[8] R. Milazzo, E. Napolitani, G. Impellizzeri, G. Fisicaro, S. Boninelli, M. Cuscunà, D. De Salvador, M. Mastromatteo, M. Italia, A. La Magna, G. Fortunato, F. Priolo, V. Privitera, A. Carnera, N-type doping of Ge by As implantation and excimer laser annealing, J. Appl. Phys. 115 (5) (2014) 053501, https://doi.org/10.1063/1.4863779.

[9] K. Huet, G. Fisicaro, J. Venturini, H. Besaucèle, A. La Magna, Defect kinetics and dopant activation in submicrosecond laser thermal processes, Appl. Phys. Lett. 95 (23) (2009) 231901, https://doi.org/10.1063/1.3268472.

[10] G. Fisicaro, A.L. Magna, G. Piccitto, V. Privitera, Laser annealing of SiGe and Ge based devices, Microelectron. Eng. 88 (4) (2011) 488–491, https://doi.org/10.1016/j.mee.2010.09.014.

[11] G. Fisicaro, L. Pelaz, M. Aboy, P. Lopez, M. Italia, K. Huet, F. Cristiano, Z. Essa, Q. Yang, E. Bedel-Pereira, M. Quillec, A.L. Magna, Kinetic Monte Carlo simulations of boron activation in implanted Si under laser thermal annealing, Appl. Phys. Express 7 (2) (2014) 021301.

[12] G. Fisicaro, L. Pelaz, P. Lopez, A. La Magna, Kinetic Monte Carlo simulations for transient thermal fields: computational methodology and application to the submicrosecond laser processes in implanted silicon, Phys. Rev. E 86 (3) (2012) 036705, https://doi.org/10.1103/PhysRevE.86.036705.

[13] G. Fisicaro, A. La Magna, Modeling of laser annealing, J. Comput. Electron. 13 (1) (2014) 70–94, https://doi.org/10.1007/s10825-013-0545-9.

[14] R. Milazzo, G. Impellizzeri, D. Piccinotti, A. La Magna, G. Fortunato, D. De Salvador, A. Carnera, A. Portavoce, D. Mangelinck, V. Privitera, E. Napolitani, Impurity and defect interactions during laser thermal annealing in Ge, J. Appl. Phys. 119 (4) (2016) 045702, https://doi.org/10.1063/1.4940737.

[15] Y. Qiu, F. Cristiano, K. Huet, F. Mazzamuto, G. Fisicaro, A. La Magna, M. Quillec, N. Cherkashin, H. Wang, S. Duguay, D. Blavette, Extended defects formation in nanosecond laser-annealed ion implanted silicon, Nano Lett. 14 (4) (2014) 1769–1775, https://doi.org/10.1021/nl4042438.

[16] D. von der Linde, K. Sokolowski-Tinten, J. Bialkowski, Laser-solid interaction in the femtosecond time regime, Appl. Surf. Sci. 109–110 (1997) 1–10, https://doi.org/10.1016/s0169-4332(96)00611-3.

[17] A. Hu, Y. Zhou, W.W. Duley, Femtosecond laser-induced nanowelding: fundamentals and applications, Open Surface Sci. J. 3 (1) (2010) 42–49, https://doi.org/10.2174/1876531901103010042.

[18] N. Buecking, M. Scheffler, P. Kratzer, A. Knorr, Theory of optical excitation and relaxation phenomena at semiconductor surfaces: linking density functional and density matrix theory, Appl. Phys. A 88 (3) (2007) 505–518, https://doi.org/10.1007/s00339-007-4043-4.

[19] G. Onida, L. Reining, A. Rubio, Electronic excitations: density-functional versus many-body Green's-function approaches, Rev. Mod. Phys. 74 (2) (2002) 601–659, https://doi.org/10.1103/RevModPhys.74.601.

[20] M. Reichelt, T. Meier, S.W. Koch, M. Rohlfing, Theory for the nonlinear optical response of semiconductor surfaces: application to the optical Stark effect and spectral oscillations of the Si(111)-(2 × 1) surface exciton, Phys. Rev. B 68 (4) (2003) 045330, https://doi.org/10.1103/PhysRevB.68.045330.

[21] S.I. Anisimov, B.L. Kapeliovich, T.L. Perel'man, Electron emission from metal surfaces exposed to ultrashort laser pulses, J. Exp. Theor. Phys. 39 (1974) 375.

[22] M.I. Kaganov, I.M. Lifshitz, L.V. Tanatarov, Relaxation between electrons and the crystalline lattice, J. Exp. Theor. Phys. 4 (1957) 173.

[23] S.F. Lombardo, G. Fisicaro, I. Deretzis, A. La Magna, B. Curvers, B. Lespinasse, K. Huet, Theoretical study of the laser annealing process in FinFET structures, Appl. Surf. Sci. 467–468 (2019) 666–672, https://doi.org/10.1016/j.apsusc.2018.10.155.

[24] K. Huet, J. Aubin, P.-E. Raynal, B. Curvers, A. Verstraete, B. Lespinasse, F. Mazzamuto, A. Sciuto, S.F. Lombardo, A. La Magna, P. Acosta-Alba, L. Dagault, C. Licitra, J.-M. Hartmann, S. Kerdilés, Pulsed laser annealing for advanced technology nodes: Modeling and calibration, Appl. Surf. Sci. 505 (2020) 144470, https://doi.org/10.1016/j.apsusc.2019.144470.

[25] J.-P.M. Péraud, N.G. Hadjiconstantinou, Extending the range of validity of Fourier's law into the kinetic transport regime via asymptotic solution of the phonon Boltzmann transport equation, Phys. Rev. B 93 (4) (2016) 045424, https://doi.org/10.1103/PhysRevB.93.045424.

[26] S.F. Lombardo, Coupled Kinetic and Electromagnetic Approaches for the Simulation of Complex Processes (Ph.D. thesis), 2018.

[27] A. Alberti, A. La Magna, M. Cuscunà, G. Fortunato, C. Spinella, V. Privitera, Nickel-affected silicon crystallization and silicidation on polyimide by multipulse excimer laser annealing, J. Appl. Phys. 108 (12) (2010) 123511, https://doi.org/10.1063/1.3531562.

[28] A. Alberti, A. La Magna, Role of the early stages of Ni-Si interaction on the structural properties of the reaction products, J. Appl. Phys. 114 (12) (2013) 121301, https://doi.org/10.1063/1.4818630.

[29] R. Černý, V. Cháb, P. Přikryl, Numerical simulation of the formation of Ni silicides induced by pulsed lasers, Comput. Mater. Sci. 4 (3) (1995) 269–281, https://doi.org/10.1016/0927-0256(95)00047-T.

[30] A. Alberti, A. La Magna, M. Cuscunà, G. Fortunato, V. Privitera, Simultaneous nickel silicidation and silicon crystallization induced by excimer laser annealing on plastic substrate, Appl. Phys. Lett. 96 (14) (2010) 142113, https://doi.org/10.1063/1.3377916.

[31] X. Cartoixà, L. Colombo, R. Rurali, Thermal rectification by design in telescopic Si nanowires, Nano Lett. 15 (12) (2015) 8255–8259, https://doi.org/10.1021/acs.nanolett.5b03781.

[32] J. Carrete, B. Vermeersch, A. Katre, A. van Roekeghem, T. Wang, G.K.H. Madsen, N. Mingo, almaBTE: a solver of the space−time dependent Boltzmann transport equation for phonons in structured materials, Comput. Phys. Commun. 220 (2017) 351–362, https://doi.org/10.1016/j.cpc.2017.06.023.

[33] J. Kaiser, T. Feng, J. Maassen, X. Wang, X. Ruan, M. Lundstrom, Thermal transport at the nanoscale: a Fourier's law vs. phonon Boltzmann equation study, J. Appl. Phys. 121 (4) (2017) 044302, https://doi.org/10.1063/1.4974872.

[34] A. Sciuto, I. Deretzis, G. Fisicaro, S.F. Lombardo, M.G. Grimaldi, K. Huet, B. Curvers, B. Lespinasse, A. Verstraete, A. La Magna, Phononic transport and simulations of annealing processes in nanometric complex structures, Phys. Rev. Mater. 4 (5) (2020) 056007, https://doi.org/10.1103/PhysRevMaterials.4.056007.

[35] J. Anaya, J. Jimenez, T. Rodriguez, Thermal transport in semiconductor nanowires, in: X. Peng (Ed.), Nanowires, IntechOpen, Rijeka, 2012, https://doi.org/10.5772/52588 (Chapter 11).

[36] G.V. Samsonov, Handbook of the Physicochemical Properties of the Elements, Springer, US, 1968, https://doi.org/10.1007/978-1-4684-6066-7.

[37] S.F. Lombardo, S. Boninelli, F. Cristiano, I. Deretzis, M.G. Grimaldi, K. Huet, E. Napolitani, A. La Magna, Phase field model of the nanoscale evolution during the explosive crystallization phenomenon, J. Appl. Phys. 123 (10) (2018) 105105, https://doi.org/10.1063/1.5008362.

[38] L. Bonati, Y.-Y. Zhang, M. Parrinello, Proc. Natl. Acad. Sci. 116 (36) (2019) 17641, https://doi.org/10.1073/pnas.1907975116.

[39] A. Karma, W.-J. Rappel, Quantitative phase-field modeling of dendritic growth in two and three dimensions, Phys. Rev. E 57 (4) (1998) 4323–4349, https://doi.org/10.1103/PhysRevE.57.4323.

[40] S.F. Lombardo, S. Boninelli, F. Cristiano, G. Fisicaro, G. Fortunato, M.G. Grimaldi, G. Impellizzeri, M. Italia, A. Marino, R. Milazzo, E. Napolitani, V. Privitera, A. La Magna, Laser annealing in Si and Ge: anomalous physical aspects and modeling approaches, Mater. Sci. Semicond. Process. 62 (2017) 80–91, https://doi.org/10.1016/j.mssp.2016.10.047.

[41] M. Hackenberg, K. Huet, R. Negru, J. Venturini, G. Fisicaro, A. La Magna, P. Pichler, Modeling boron profiles in silicon after pulsed excimer laser annealing, in: L. Pelaz, I. Santos, R. Duffy, F. Torregrosa, K. Bourdelle (Eds.), AIP Conf. Proc., vol. 1496, AIP, 2012, pp. 241–244, https://doi.org/10.1063/1.4766533.

[42] A.B. Tayler, Free and moving boundary problems, J. Fluid Mech. 158 (1985) 532–533, https://doi.org/10.1017/S0022112085212750.

[43] N.A. Ahmad, A.A. Wheeler, W.J. Boettinger, G.B. McFadden, Solute trapping and solute drag in a phase-field model of rapid solidification, Phys. Rev. E 58 (3) (1998) 3436–3450, https://doi.org/10.1103/PhysRevE.58.3436.

[44] E.V. Monakhov, B.G. Svensson, M.K. Linnarsson, A. La Magna, M. Italia, V. Privitera, G. Fortunato, M. Cuscunà, L. Mariucci, The effect of excimer laser pretreatment on diffusion and activation of boron implanted in silicon, Appl. Phys. Lett. 87 (19) (2005) 192109, https://doi.org/10.1063/1.2126144.

[45] K.K. Ong, K.L. Pey, P.S. Lee, A.T.S. Wee, X.C. Wang, Y.F. Chong, Dopant distribution in the recrystallization transient at the maximum melt depth induced by laser annealing, Appl. Phys. Lett. 89 (17) (2006) 172111, https://doi.org/10.1063/1.2364834.

[46] I. Štich, R. Car, M. Parrinello, Bonding and disorder in liquid silicon, Phys. Rev. Lett. 63 (20) (1989) 2240–2243, https://doi.org/10.1103/PhysRevLett.63.2240.

[47] J.T. Okada, P.H.-L. Sit, Y. Watanabe, Y.J. Wang, B. Barbiellini, T. Ishikawa, M. Itou, Y. Sakurai, A. Bansil, R. Ishikawa, M. Hamaishi, T. Masaki, P.-F. Paradis, K. Kimura, T. Ishikawa, S. Nanao, Persistence of covalent bonding in liquid silicon probed by inelastic X-ray scattering, Phys. Rev. Lett. 108 (6) (2012) 067402, https://doi.org/10.1103/PhysRevLett.108.067402.

[48] N. Jakse, A. Pasturel, Dynamics of liquid and undercooled silicon: an ab initio molecular dynamics study, Phys. Rev. B 79 (14) (2009) 144206, https://doi.org/10.1103/PhysRevB.79.144206.

[49] P.M. Fahey, P.B. Griffin, J.D. Plummer, Point defects and dopant diffusion in silicon, Rev. Mod. Phys. 61 (2) (1989) 289–384, https://doi.org/10.1103/RevModPhys.61.289.

[50] G. Götz, Explosive crystallization processes in silicon, Appl. Phys. A 40 (1) (1986) 29–36,-https://doi.org/10.1007/BF00616588.

[51] M. Cascio, D. Baroli, S. Bordas, I. Deretzis, G. Falci, A. Magliano, A. La Magna, Coupled molecular-dynamics and finite-element-method simulations for the kinetics of particles subjected to field-mediated forces, Phys. Rev. E 99 (6) (2019) 063307, https://doi.org/10.1103/PhysRevE.99.063307.

Chapter 7

# Laser engineering of carbon materials for optoelectronic applications

Frédéric Antoni and François Stock
*ICube, D-ESSP, Strasbourg, France*

## 7.1 Optoelectronic and display devices: State of the art

### 7.1.1 Current technologies

Nowadays, optoelectronic and display devices are deeply rooted in our all-day life. With the emerging and growing of new communication technologies, connected devices, Internet of Things, augmented reality (AR), virtual reality (VR), and smart objects, the use of optoelectronic devices achieves high marks. As the supply and demand always rise, research topics in this wide domain are constantly pushing back the limits of performances. With a very high number of various devices (flat panel displays, projecting devices, optical sensors, light sources, lasers, etc.), optoelectronic and display technologies cover broad research fields like fundamental physics, optics, material, electronic and photonic sciences, or even chemistry. In addition, the application range of those technologies is very important and concerns lot of domains like biomedical, telecommunication, computer science, solar cells, imaging sensors, smart screens, wearable devices, etc. With this ubiquity, current and emerging technologies always try to improve the performances of optoelectronic devices, in terms of efficiency, dimensions, resolution, quality, power consumption, and manufacturing costs.

Among these bushy technologies of display devices, three main categories can be defined considering the way to transport the information to the human eye. We distinguish here emitting devices, in which light is directly provided by an element of the device, from nonemitting devices wherein light comes from an external part of the component and from projecting systems that use indirect light interaction to convey information [1–3].

In the last years, cathode ray tubes (CRTs) and plasma display panels (PDPs) definitively disappeared from the market. Nowadays, flat panel display

Laser Annealing Processes in Semiconductor Technology https://doi.org/10.1016/B978-0-12-820255-5.00005-2

technologies can be divided into three main categories, all of them using light-emitting diode (LED) technologies with different approaches. In the liquid crystal display (LCD) family, LEDs are currently used to provide backlight. It is important to notice that the adjunction of quantum dot nanoparticles permits to improve the spectral range and therefore the image fidelity compared to a classical white backlight technology [4, 5]. The main market players put a lot on this association between LED and quantum dot technologies to develop low-cost, efficient screens based on mastered concepts. The other alternatives are based on the use of LEDs obtained by two entirely different technologies. In first, the OLED technology has allowed to develop screens with performances never reachable by conventional LCD technology. This is also the main reason why manufacturers of LCD screens seek to get as close as possible to the image quality of OLED screens that have nowadays from this point of view, not really competition. However, the OLED family of screens has great difficulty in establishing itself due to several intrinsic difficulties to this technology. The promise of being able to develop efficient displays at low cost faces few difficulties that currently force this technology to represent only a marginal share of the market [6–8]. Due to many challenges, such as the life span and the need to create innovative manufacturing processes in new factories, the selling price of screens using this technology remains very high, which does not promote its development. In addition, the image quality of the latest LCD screens is increasingly close to OLEDs for a lower production cost [9]. The use of inorganic LEDs in screens is not new, but due to the large size of these components, they were until very recently, mainly intended for advertising displays. The very high brightness of these components makes it possible to display a high-quality image even in direct sunlight. To allow the development of small screens based on these LEDs, we had to find a way to significantly reduce their size. In recent years, they have been found in the form of panels of around 1 $m^2$ which can be combined to form big size screens. They can be seen commonly in TV shows or used as billboards in football stadiums, for example. Some cinemas are also starting to equip themselves with this technology, offering the possibility of overcoming the constraints and limitations of projection. The pixels of this type of screen are composed of an association of subpixels of colored LEDs whose size is less than 1 mm. Until very recently, it was not possible to further reduce the size of these devices, but some manufacturers are now considering the possibility of placing televisions, computers, or telephone screens on the market within 2–3 years. The LEDs used would reach a size of less than 20 μm in these cases, which is remarkable. The use of inorganic components whose manufacturing technologies are perfectly mastered and their association with performances in terms of luminosity and colorimetry makes it possible to envisage a rapid development of this new approach presenting itself as a particularly attractive alternative in terms of performance and industrialization [10]. It is interesting to us that these three very different technologies converge towards very close performances. Each of them searching to get as close as possible to

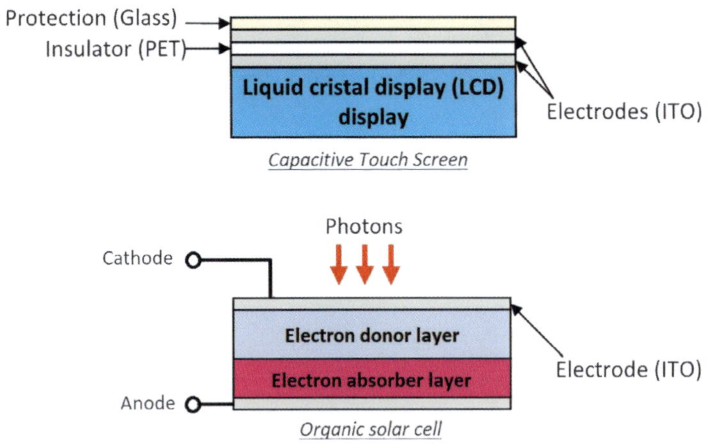


FIG. 7.1 Example of ITO applications as transparent electrodes in optoelectronic devices.

the main advantages of other families. It is very difficult to predict which of these technologies will prevail in a few years because at this stage, they all have advantages and drawbacks. If energy performance will be the key point in the future, micro-LED and OLED technologies will have a real advantage over LCD technology. The latter is still currently the most attractive in terms of manufacturing cost but for some time yet. To obtain performances similar to the two other aforementioned technologies, the manufacturing process becomes more and more complex by increasing the number of additional layers in the manufacturing process.

What all the screen devices have in common is the necessity to transport electric current through the surface of the screen without interfering with the emitted light. To achieve this aim, transparent conductive oxides are currently used as electrodes, whether in display devices such as liquid crystal touch screens, OLED, or in certain photovoltaic cells (Fig. 7.1). For many years, the most used transparent conductive oxide has been indium tin oxide (ITO). Its synthesis process is perfectly mastered and despite some limitations, in particular in terms of its flexibility, it still presents today the best cost-performance compromise on the market and for a few more years. It is, however, imperative to find an alternative in a few years because of the scarcity of indium, the constant increase in the demand for touch screens, and the development of solar cells based on organic components.

### 7.1.2 Future challenges in optoelectronic technologies

ITO offers very interesting performances owning high transparency in the visible range (between 80% and 90%) and a conductivity of around $10^4$ $(\Omega\ \text{cm})^{-1}$ [11–13]. Finding a reliable alternative to this transparent conductive oxide with suitable performances appears to be a key point for present and future generations of displays and optoelectronic devices. Various alternatives to ITO are studied since a few years, many of them trying to find out which metal oxide could reach similar performances in terms of resistivity and transmittance. The most studied metal oxide is ZnO. To obtain conductivity, it should be heavily doped with a metal. The most efficient associations are AZO and GZO (aluminum and gallium doped ZnO, respectively) [14–16]. Those oxides offer the immense advantage of not containing any rare metal while allowing obtaining performances similar to ITO in terms of transparency and conductivity. These are therefore particularly promising alternatives in the near future. Another approach studied in the last decade and supported by the raise of organic solar cells is based on polymeric transparent conductive thin films and specially PEDOT:PSS. This mixture of two ionomers (3,4-ethylenedioxythiophene and polystyrene sulfonate) combined with some optimizations presents a conductivity approaching 950 $(\Omega\ \mathrm{cm})^{-1}$, for high transparency [17, 18]. Nevertheless, this organic thin film remains fragile, mainly when it is exposed to UV rays, and also presents low temperature and hygrometric resistance, requesting then complementary materials and process steps to stabilize the film. Moreover, solutions based on carbon nanotubes [19, 20] or metallic nanowires [21, 22] added to electrodes start to appear, enhancing the film conductivity without lowering the transparency. This emerging technology shows also promising results. In all these alternative solutions to ITO, we also have to seriously consider graphene or graphene-like thin films.

## 7.2 Introduction to carbon in electronics

### 7.2.1 Graphene, graphene-like, and graphitic thin films

#### *7.2.1.1 Properties of thin graphitic layers*

Graphite is one of the most common crystalline forms of carbon. This material is only constituted by $sp^2$-hybridized carbon atoms. In graphite, atoms are organized in a succession of superposed two-dimension sheets (graphite plans) being separated by 3.4 Å from each other (Fig. 7.2). One graphite sheet is very compact and resistant due to a high and strong cohesion energy between atoms in the plan (around 7.4 eV). However, Van der Waals interactions between each of the graphite plans are very weak (typically 0.86 eV) and lead to a material showing anisotropic properties and low mechanical resistance. It is therefore very easy to exfoliate graphite. Typical conductivity reaches $3.5 \times 10^3$ $(\Omega\ \mathrm{cm})^{-1}$ but presents different values considering plan or volume conductivity due to anisotropic properties of this material. Graphite is nowadays used in various and wide application fields like in metallurgy (iron alloys and mold coatings), automobile industry (brake and gearbox linings), anticorrosive paintings, and battery industries as electrodes, etc. An interesting property of this material is its very high fusion temperature (3825°C). This particularity permits to use

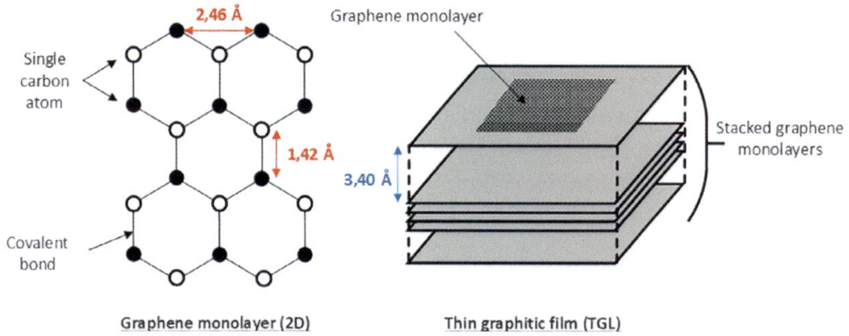


**FIG. 7.2** Graphene and thin graphitic layer structures.

graphite in applications needing high temperature resistance. Therefore, graphite is also used as an electrode for arc furnaces and electric motors.

If we consider a single graphite layer, we obtain a 2D material named graphene. Theorized in 1947 by P.R. Wallace and synthetized in 2004 by A. Geim, graphene presents a theoretical exceptional and record electrical mobility of above 200,000 $cm^2\ V^{-1}\ s^{-1}$ [23] auguring a very high electrical conductivity. This 2D structure is only made of carbon atoms organized in a honeycomb (hexagonal) pattern with an interatomic distance of 1.42 Å and a lattice parameter of 2.46 Å (Fig. 7.2). Showing very interesting optical properties and mainly a very low optical absorption of 2.3% for one monolayer [24], graphene is an excellent candidate for optoelectrical applications. While graphene monolayers are comparable to a metal (presenting a zero band gap), this material enters into the semiconductor field (band gap opening) as soon as the structure is made of at least two monolayers. Keeping advantageous electrical conductivity, a succession of graphene layers will form multilayer graphene (or few-layer graphene) and as a more commonly used term: thin graphitic layers (TGLs).

Properties of TGL could be very different considering the large type of materials classified under this domination. Indeed, transparency and conductivity are highly dependent on the structure and the thickness of the considered TGL as directly a function of the number of atomic layers. Moreover, doped graphene or TGL can also improve conductivity while inducing additional constraints in the process. Those properties are also related to the elaboration process used to synthetize the TGL. In the case of graphene and more specifically for single-layer graphene, the substrate has also a nonnegligible influence on the electrical properties of the 2D film.

### *7.2.1.2 Elaboration techniques*

To synthetize graphite thin layers, various techniques can be used. Due to the high fusion temperature of carbon, some techniques [25] consist of using a carbonized precursor (like polyimide) and then heat up the material to decompose noncarbon atom. After thermal treatment, it remains only the amorphous carbon thin films that can be turned into a graphitic structure by adding a high-temperature posttreatment (over 2800°C). This type of approach is highly indicated to obtain high conductive TGL or flexible graphite films (FGF) where transparency is not a criterion to be achieved. With this method, large surface films can be synthetized. Comparable processes based on organic chemical coatings are used to obtain graphene or graphite thin films [26].

Chemical vapor deposition is also on a major elaboration method used to synthetized TGL [27, 28]. Based on the decomposition of a carbonized gas mixture (like hydrogen-methane), this method permits the growth of high-quality carbon thin films over a large range of thicknesses. Process parameters are used to tune the obtained film physicochemical properties and especially the graphitic and/or diamond character of the layer. Physical deposition technics like physical vapor deposition (PVD) [29], sputtering [30], or pulsed laser deposition (PLD) [31] are also other ways to synthetize TGL.

In the case of more graphenic material (less than 10 graphite monolayers), another well-known process to elaborate graphene or TGL is based on mechanical delamination of pyrolytic graphite. This exfoliation leads to the extraction of a single or few graphite layers [32, 33]. Graphene synthetization has also been explored using aerosols, vaporization techniques [34, 35], or growing on metallic substrates [36]. All these techniques permit to obtain high-conductive graphene/TGL films. The transparency criterion is still achieved while considering graphene (mono-, bi-, or tri-layer) but starts to be considered by working with higher thicknesses of TGL.

However, in the case of high-quality graphene, all these techniques suffer in a lack of capability to form uniform and homogeneous sheets over wide areas. It is also very complicated to transfer graphene layers. Therefore, synthetizing highly homogeneous graphene (mono or multilayer) over large surfaces will be a technological lock for the forthcoming years and appear to be a key point for the use of this technology in optoelectronic devices.

### 7.2.1.3 Integration of TGL in microelectronic devices

Nowadays, TGLs are present in the electronic devices landscape being used in some specific applications. In fact, TGLs are well used in the energy storage field and especially in electrodes of lithium ion batteries. In most cases, TGLs intervene in the fabrication of the anodic part of the device and help to improve the performances of the battery [37–40]. Combined with other carbonized materials (porous graphite, carbon nanotubes, etc.) and other elements (metallic nanodots, oxides, etc.), TGLs contribute to better interaction between electrode and electrolyte, leading to obtain higher performances (charge and discharge behavior, capacity, and lifetime). TGLs are also used when flexibility is needed. Used as flexible electrodes, they can provide relatively high electric conductivity [41] over large surfaces.

The applications of graphene or graphene-like films are more deeply microelectronic oriented. As it has been established that graphene presents a very high electron mobility (much higher than for silicon), researchers try to integrate this material into transistor structures, with the objective to obtain devices owing to very high performances. Graphene starts to appear in different parts of transistors (gate bottom, source, drain, channel, and electrode contact) [42, 43]. The typical graphene-based field-effect transistor (GFET) embeds a graphene sheet between drain and source or even between emitter and collector inside graphene bipolar transistor (GBT) (Fig. 7.3). This sheet permits to enhance the transistor transconductance by offering very high channel charge carrier mobility permitting to reach very high conductivity and frequency performances [44, 45]. In addition, various transistor technologies (organic transistors, short channel devices, double gate structures, tunneling effect devices, etc.) start to integrate graphene into their structures [46–49]. It appears that graphene will be an unavoidable material in the future of transistor technology.

Graphene also starts to appear in photovoltaic devices. The high electrical performances combined with the very low optical absorption induced by only one carbon monolayer make it interesting to be used as electrodes in solar cells. Most of the studies focused on the integration of graphene [50–52] (sometimes combined with other technologies like semiconductor or metallic nanowires) as layers for high-efficiency electrical charge extraction (anode or cathode). Integrating this material permits to enhance the electrical performances of the cells without affecting the optical behavior of the device, being therefore a very interesting and promising material for future generations of photovoltaic devices.

Despite a very large integration of TGL in industrial applications, the case of graphene is not so easy. As presented in Section 7.2.1.2, it is hard to obtain high-quality graphene over large surfaces, narrowing therefore processes yield. This major drawback withdraws nowadays the industrialization of graphene-based technologies.

## 7.2.2 Diamond-like carbon thin films

### *7.2.2.1 Definition and properties of DLC*

Carbon materials exist in different forms depending on the atomic hybridizations formed by the atoms in the material structure. Three types of interatomic

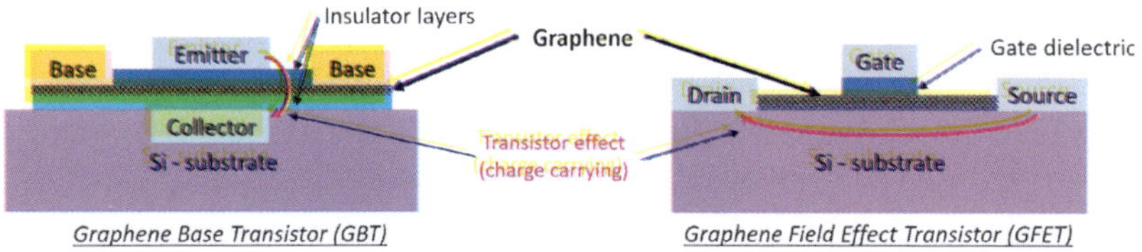


**FIG. 7.3** Example of two transistor structures integrating graphene [44].

bindings can be found. The $sp^1$ hybridization is present in carbyne-type organic chains such as gas like acetylene. The $sp^2$ binding is the stable form of carbon corresponding to a graphitic material (as presented in Section 7.2.1.1). Finally, carbon atoms can also bind in a $sp^3$ pattern, representing a three-dimensional structure of carbon. A pure $sp^3$ carbon material is diamond and presents properties like metastability (slowly graphitizing while heated over 300°C), a very high insulating character (due to a wide band gap of 5.5 eV), perfect optical transparency, and very high hardness. However, it exists carbon material presenting a mixture between those different types of interatomic bindings.

Diamond-like carbon (DLC) is a metastable variant of amorphous carbon (a-C) containing a nonnegligible $sp^3$ hybridization percentage in its structure, the rest being only $sp^2$ bindings. Depending on the elaboration technique, DLC can also present a relatively high concentration of hydrogen (leading to a-C:H) leading to a change in its properties. DLCs presenting a high amount of $sp^3$ are also known as tetrahedral amorphous carbons (ta-C or ta-C:H) and are located on the left edge and in the top corner of the carbon ternary phase diagram suggesting high proximity with diamond [53]. $Sp^2$, $sp^3$, and H percentages are highly influenced by the deposition techniques and by the related experimental parameters that will condition DLC properties.

In many cases, DLC shares many properties with diamond due to its strong kinship with this material. It presents a very high hardness (commonly over the 10 GPa range), and for very $sp^3$-rich DLC, it can reach 80 GPa in the case of hydrogen-free ta-C, approaching diamond (100 GPa). Deposited in thin films, DLC presents also a very low roughness (less than 1 nm) and is therefore a perfect material for mechanical purposes due to very low wear or friction resistances as well as strong hardness [53]. These advantageous properties permit its use as protective coatings for mechanical and industrial applications like drills, borers, sprockets, gears or motor parts, razor blades, and hard disk drive read heads. Moreover, the high biocompatibility and chemical inertness [54–57] of DLC combined with the mechanical resistance will place the DLC as an excellent material for biomedical applications. It is nowadays used as protective coatings for dental implants and hip or knee prosthesis. Another property of DLC linked with the diamond proximity is its perfect insulating character. Nevertheless, if DLC presents a very low $sp^3$ fraction and a high hydrogen concentration, the resistivity values can be relatively low, leading to a conductive material [58–61].

Another remarkable property of DLC is a high optical transparency in the visible range. The transmittance is highly related to the $sp^3$ percentage in the DLC meaning typically the more $sp^3$ bindings you have, the more transparency you get. This interesting optical property makes the DLC finely suitable for optical applications. Furthermore, DLC has a particularity compared to diamond in terms of transmittance in the ultraviolet range. Where diamond is fully transparent, DLC presents a relative high opacity in the UV and near-UV range (Fig. 7.4). Indeed, the absorption coefficient of DLC is increasing drastically

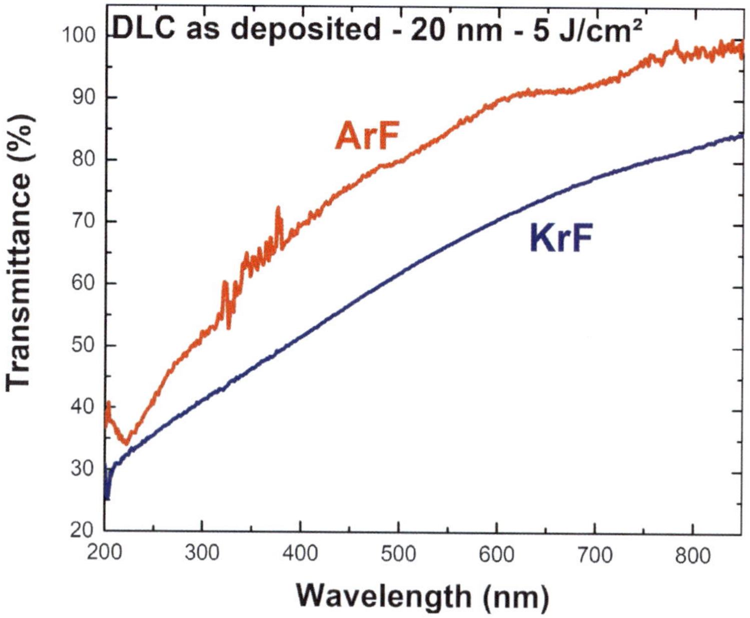


FIG. 7.4 Transparency of DLC layers (20 nm) elaborated by PLD at two different laser wavelengths.

soon as we enter the UV range and continue to increase while going in the deep UV. This particular property of DLC is very interesting in terms of UV laser treatments, permitting to interact preferably with the surface layer without affecting the whole film.

### *7.2.2.2 Elaboration techniques*

Different elaboration techniques permit to synthetize DLC. The properties of the resulting DLC are highly dependent on those elaboration techniques. The first main technique is chemical vapor deposition (CVD) [61, 62]. In this method, DLC is obtained by decomposition of carbonized gas precursors like methane ($CH_4$), ethylene ($C_2H_4$), acetylene ($C_2H_2$), or benzene ($C_6H_6$). CVD-grown DLC presents a high amount (up to 50%) of hydrogen in its structure, due to the presence of this element in the precursors (or present in the deposition chamber as carrier gas). Those C—H bindings will not necessarily influence the $sp^2/sp^3$ binding's ratio but lead to the elaboration of a-C:H or ta-C:H material. Nevertheless, this nonnegligible presence of hydrogen will mainly narrow the density of thin films, leading to move away from diamond properties. Typically, hydrogenated DLC thin films present a lower mechanical and wear resistance as well as lower hardness.

Another way to produce DLC thin layers is based on physical deposition processes. We can principally cite PVD like magnetron sputtering or vacuum arc [59, 63–65]. In these deposition methods, DLC properties are highly influenced by the deposition parameters, conditioning the energy of the atomic species reaching the substrate surface [66]. In fact, these techniques can lead to obtain various types of DLCs with a large spectrum of properties. DLC can vary from being nearby to diamond with a high amount of $sp^3$, high-hydrogenated DLC with a very low density or with relatively high graphitic bindings leading to a less adamant character. DLC also can be elaborated embedding nitrogen or metallic atoms leading to a material with a lower resistivity that can be a semimetal or a semiconductor [58]. In most cases, these DLCs present a lower density and lower optical transparency.

One of the well-known techniques to elaborated high-purity DLC thin films is PLD process [56, 66, 67]. Generally, carbon atoms are ripped out from a high-purity graphite target placed under ultrahigh vacuum. These deposition conditions lead to obtain thin films without hydrogen or impurity contaminations. Resulting DLC shows a very high density, very close to diamond. This kind of DLC (as type ta-C) also presents a high $sp^3$ percentage leading to a highly transparent and insulated material. It is also the type of DLC that presents the highest hardness and mechanical resistance properties. Being hydrogen-free, this material has the capability to endure laser treatments. Indeed, hydrogenated DLC layers will be easily destroyed by an annealing process due to the massive breaking of interatomic bindings. This effect leads to the liberation of hydrogen atoms causing then a progressive layer delamination. A same behavior happens when DLC contains oxygen. It is also important to notice that despite a hydrogen-free structure, PLD-DLC density and $sp^3/sp^2$ binding's ratio is highly dependent on the deposition parameters. This ratio is mainly dependent on the kinetic energy of the atoms (ions) present in the laser-generated plasma plume and reaching the substrate [68–70]. This energy is directly dependent on the laser fluence and conditions the properties of the obtain DLC thin films.

To summarize the DLC elaboration techniques and the related material physicochemical properties, it is important to precise that it exists various types of DLCs (Table 7.1). Despite some common points between also these materials, each DLC has its specificities depending on the thin-film composition and structure.

#### *7.2.2.3 Integration of DLC in optoelectronic devices*

DLC thin films are nowadays well implanted in industrial applications, especially as mechanical protective and/or biocompatible coatings. In optoelectronic devices, applications of DLC remain more sporadic and mainly kept in the state of laboratory research. Despite very interesting and promising properties in terms of optical behavior, DLC is not often integrated into electronic and optoelectronic components.

**TABLE 7.1** DLC physicochemical properties as a function of different elaboration processes [53].

| Material | Hydrogen concentration (%) | $sp^3$ binding's concentration (%) | Density ($g\ cm^{-3}$) | Hardness (GPa) | Elaboration technique |
|---|---|---|---|---|---|
| Diamond | 0 | 100 | 3.515 | 100 | – |
| ta-C | #0 | 80–88 | 3.1 | 80 | PLD |
| ta-C: H | 30 | 70 | 2.4 | 50 | CVD-PVD |
| a-C: H | 30–50 | 40–60 | 1.2–2.2 | 10–20 | |

One field in which DLC integration was already studied is solar cells. In fact, the use of DLC thin films as solar cells' top-protective coating was demonstrated, leveraging the advantageous mechanical properties of this material [71–74]. DLC is employed to provide cell encapsulation and protect it from environmental aggressions, preventing therefore cell aging and related performances and yield degradation. Moreover, DLC also provides an antireflexion behavior [75–77], being interesting to enhance photon collection in the solar cell. The use of DLC in photovoltaic devices shall be very promising by offering advantageous properties and helping to extend performances and service life of the cells. The integration of DLC in OLED was also studied [78, 79] and reports that nanometric DLC (with high resistivity) films located between the ITO top electrode and the structure offer an interesting electron injection layer, leading to higher device performances.

The integration of DLC was also studied in electronic devices like standard transistors. Showing high insulating properties, DLC was tested as gate dielectric substituting the silicon dioxide in a classical metal oxide semiconductor (MOS) structure [80]. This substitution appears to be an interesting alternative, also being economically advantageous due to process low cost (performed at low temperature). However, in this approach, one very important point is the substrate carbon contamination (especially in the channel region) which shall be as low as possible. This subsurface quality in the channel region conditions the quality and the performances of the obtained device. Therefore, having a drastic and precise control of the process parameters (considering this possible contamination) is a key point of DLC integration in microelectronic devices. Indeed, it also possible to combine this technology with graphene transistor (as shown in Ref. [81]). In GFET, the channel between drain and source regions has enhanced performances thanks to a thin graphene layer (Fig. 7.3). Gate insulator made of DLC is directly deposited on the graphene channel. This kind of transistor (DLC-GFET) belonging to the ambipolar transistor family shows promising performances for the future of carbon electronics. For a-C or a-C:H structure, low dielectric constants of these DLCs permit to integrate it into advanced storage electronic devices like nonvolatile memories combined with other technologies (like semiconductor nanocrystals or advanced MOS structures) [82, 83].

In conclusion on DLC, the broad spectrum of its optoelectrical properties suggests a very hopeful technology for the future of optoelectronic and photovoltaic devices. Deposition methods and process parameters will permit to elaborate on various types of DLCs designed for many applications, providing low cost and high efficiency.

## 7.3 Laser engineering of carbon materials

### 7.3.1 Pulsed laser deposition of carbon

#### *7.3.1.1 Principle and experimental setup*

PLD or pulsed laser ablation (PLA) is a well-known thin-film elaboration technique. Registered as a physical deposition technique and initiated in 1965 by H·W Smith and A.F Turner, PLD was firstly considered by its own contrivers as a nonpredictive technology hold back by the high number of unknown parameters influencing on the generated plasma plume and thereby on the deposited thin-film properties [84, 85]. It is finally D. Dijkkamp et al., in 1987 who presents PLD as interesting to synthetize high-quality superconductive oxide ($YBa_2Cu_3O_7$) [86]. The progress in laser and vacuum technologies quickly supported PLD development and leads to establish this deposition technique as a very promising technology. Widely used nowadays in laboratories, PLD offers the possibility to deposit a very large panel of materials like metallic and semiconductor films, superconductive materials, ceramic layers, oxides, and nanocrystalline structures. Industrial PLD firstly suffers from a lack of possibilities to obtain homogeneous films over large areas. However, over the last decade, industrial applications of PLD start to appear and grow, leading to the development of wide-area processes contributing to install PLD as a promising industrial thin-film deposition method.

The principle of PLD is based on the laser-matter interaction that happens when a high-energy laser beam hits a target. Species contained in the solid (or liquid) target are ripped out and generate a plasma plume. Ions present in the plasma will then be condensed on a substrate surface. Repetition of laser shots leads to an accumulation of matter onto the substrate surface driving to the formation of a thin film. This very simple principle bestows to PLD a very high flexibility over a large panel of deposited materials.

A typical PLD setup is mainly composed of a high-energy laser and a deposition chamber. Various types of pulsed lasers (from femtosecond to nanosecond pulse duration) can be used regarding the species to vaporize but in a large majority, PLD facilities are equipped with excimer gas lasers. This kind of source delivers very-high-energy beams reaching a few hundreds of MW. The laser beam is focused onto a target placed in the ablation chamber with generally a 45° incidence angle. With this setup, the impact surface is highly reduced and able to reach a large panel of on-target fluences (power densities). The target is set in rotation in order to avoid any crater formation on the target surface that could be caused by repeated laser shots on the same area (with a rotation speed adapted to the laser repetition rate). The substrate holder is set parallel to the target at a fixed distance. Most of PLD chambers are equipped with a heating system able to set the substrate at a high temperature during the deposition process. PLD can be performed under ultrahigh vacuum leading to high-purity thin films or structures but also under controlled neutral gas atmosphere permitting to slow down the velocity of the species contained in the plasma plume. While plasma ions are deliberately reacting with ambient gas, reactive PLD offers the possibility to deposit oxides or nitrides layers, for example.

In the case of DLC elaboration (Fig. 7.5), a high-purity graphite target (nuclear grade) is used as a carbon source. In Fig. 7.5A, Lambda Physik COMPEX 201 excimer gas laser is considered, which is able to work at two different wavelengths depending on the gas mixture used. An ArF gas mixture leads to a 193 nm ray with 20-ns pulses, while a KrF mixture gives a 248-nm beam for 25-ns pulses. With an on-target impact surface of around 1 $mm^2$, fluences between 2 and 15 $J/cm^2$ can be reached in KrF mode, while in ArF mode, fluences between 2 and 7 $J/cm^2$ are obtained. Lower fluences in ArF are explained by the reduced efficiency of the laser at this wavelength (only 400 mJ in ArF while being around 600 mJ in KrF), by the higher optical element losses, and by beam energy losses caused by the interaction between the 193-nm laser beam and the oxygen present in the air leading to ozone formation. All depositions are performed under a high vacuum at room temperature to ensure efficient DLC formation (which is subject to early graphitization by working at higher temperatures). The distance between target and substrate is fixed at 5 cm to ensure a homogeneous thickness of the thin film on the substrate surface. It is important to notice that if large amount of samples are required, the target-substrate distance has to be increased in order to provide a more homogeneous coverage by the plasma plume. Any kind of substrate could be used for DLC deposition while the process is performed at room temperature. Depositions are made on transparent substrates like quartz or glass (low cost) permitting optical measurements. Oxidized silicon wafers are also interesting substrates in order to perform electrical measurements when a transparent character is not requested.

### *7.3.1.2 DLC properties regarding laser parameters*

In the case of DLC, it is mainly the laser parameters that will influence the obtained thin-film properties. Where target-substrate distance and geometrical parameters will mostly affect the homogeneity of the DLC-deposited layer,

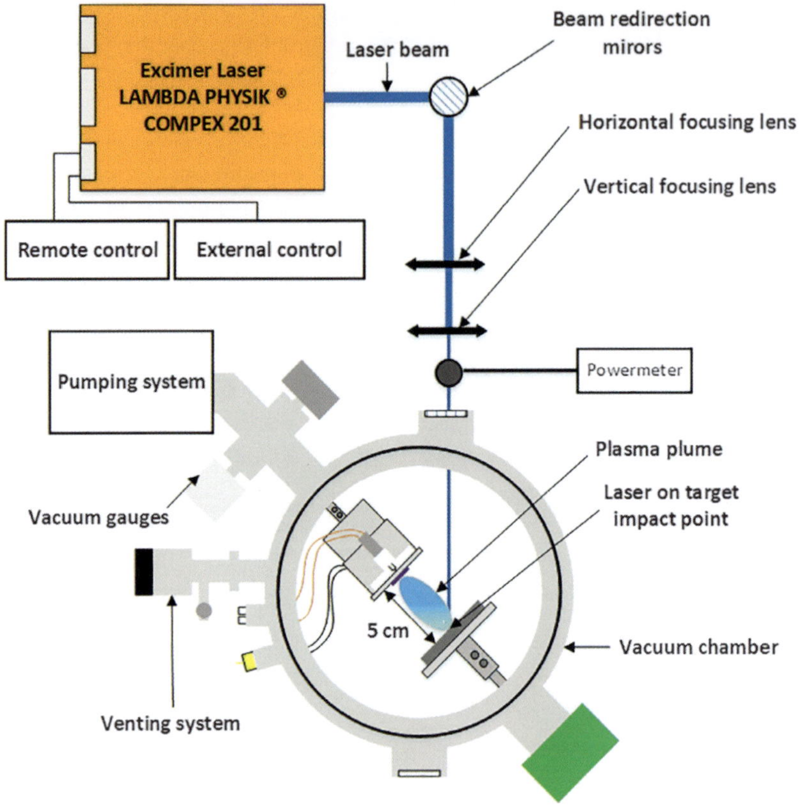


**FIG. 7.5** Schematic view of the PLD setup used to produce DLC thin films.

laser parameters like power and wavelength will impact more deeply its physicochemical properties. A very strong influence of laser fluence and wavelength on the adamant character of the DLC has been demonstrated in Ref. [87]. Here the main results are outlined.

DLC depositions were performed at 193 and 248 nm wavelengths for various fluences from 2.5 up to 7 J/cm$^2$ in ArF mode and up to 10 J/cm$^2$ in KrF mode. DLC (20 nm thick) has been deposited mainly on oxidized (63 nm) silicon, a suitable substrate to perform density measurement using nuclear reaction analysis (NRA). Very high densities for ArF-mode DLC are obtained, varying from 2.77 g cm$^{-3}$ at 2.5 J/cm$^2$ up to 3.46 g cm$^{-3}$ at 5 J/cm$^2$. This particular value at 5 J/cm$^2$ leads to obtain a density equal to 97% of that of diamond (3.52 g cm$^{-3}$) and appears to be the optimal fluence leading to the DLC with the most adamant character (Fig. 7.6). KrF-deposited DLC presents densities lower than the ArF ones, with values established between 1.84 g cm$^{-3}$

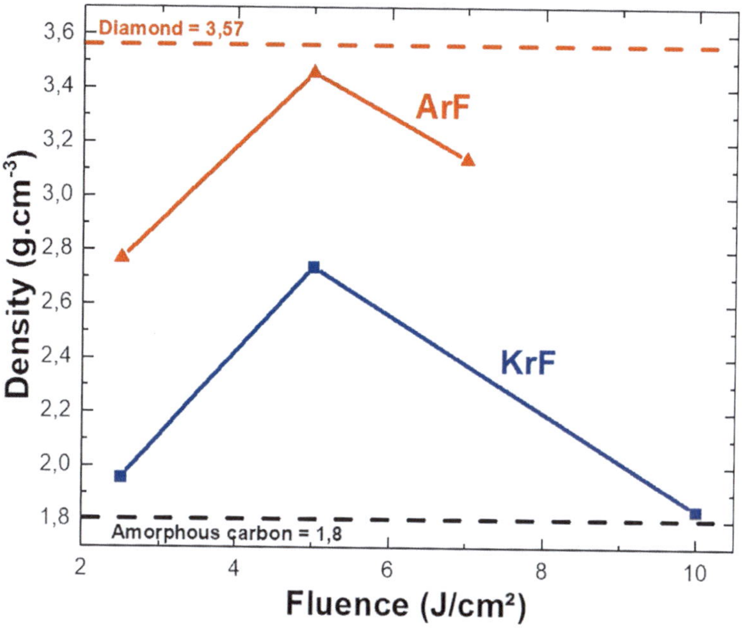


**FIG. 7.6** PLD-elaborated DLC density evolution regarding fluence and deposition wavelength.

(for depositions performed at 10 J/cm$^2$) and 2.74 g cm$^{-3}$ (at 5 J/cm$^2$). The same density optimum is reached for the fluence of 5 J/cm$^2$. In general, density measurements show that ArF-elaborated DLC has highest proximity with diamond material than KrF-mode DLC. The same DLC layers were analyzed by X-ray photoemission spectroscopy (XPS) to extract the relative percentages of sp$^2$ (graphite) and sp$^3$ (diamond) bindings. Oxygen contamination being very less (as proved by XPS full survey spectrum), analyses are performed without taking into account the interaction between carbon and oxygen. Therefore, we consider these DLC samples as pure and composed only of carbon atoms. These measurements show clear domination of sp$^3$ bindings in the case of ArF-mode DLC. Those DLC layers contain between 77.4% and 85.4% of diamond-bounded atoms, the highest sp$^3$ percentage being reached at 5 J/cm$^2$. For DLC deposited at 248 nm, we observe a lower amount of sp$^3$ bindings, varying between 25.6% and 60.9%. If depositions made at 2.5 and 10 J/cm$^2$ are clearly dominated by the sp$^2$ bindings (presenting then a much more graphitic character), DLC deposited at 5 J/cm$^2$ is still presenting the highest sp$^3$ percentage (60.9%). XPS measurements clearly show the same behavior for ArF and KrF layers in terms of sp$^3$ atomic binding's evolution (Fig. 7.7). We have also a very high correlation between density and interatomic bindings regarding wavelengths and deposition fluences. The reported 5 J/cm$^2$ fluence optimum

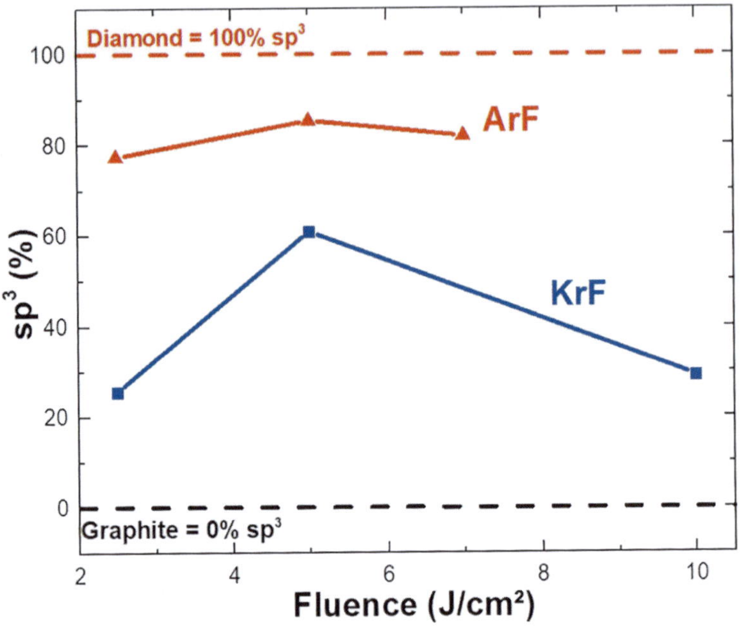


**FIG. 7.7** $sp^3$ relative concentration for PLD-elaborated DLC as a function of fluence and laser wavelength.

leads to preferably generate $sp^3$ hybridizations, related to an energy optimum of the atoms contained in the plasma plume. Overpassing this value, high kinetic energies of incoming atoms (a few hundreds of eV) on the DLC surface drive to break the existing $sp^3$ bindings (comparable to ionic bombardment effect or implantation of atoms). At lower fluences, kinetic energy is not enough to form a $sp^3$-hybridized structure while atoms preferably build $sp^2$ bindings. This behavior is much more observable in KrF mode when, at low energy, carbon atoms could be extracted from the target in $sp^2$-hybridized clusters ($C^{2+}$) while being only in atomic $C^+$ in ArF [68–70].

Considering the functional properties of the obtained DLC layers, a clear link between optical, electrical properties and proximity with diamond can be determined. First, in terms of transparency, ArF-deposited DLC always shows higher transparency than the KrF ones, approximately 10% higher over the complete visible range. There is a clear correlation between the percentage of $sp^3$ bindings and the resulting transparency, meaning that as soon as we approach a very high adamant DLC, the transparency is getting higher. Indeed, in both cases, DLC presents interesting transparency all over the visible range. As approaching the UV domain, the transparency of DLC is reducing, suggesting therefore a relatively high opacity of the DLC in the UV (Fig. 7.4). Regarding now the electrical properties of DLC, all obtained layers present a very high insulating character (well over the instrument measurement limitation), confirming also the kinship between DLC and diamond, making DLC a very interesting transparent and insulating substrate.

In addition to all the previously presented properties, it is important to notice that PLD-elaborated DLC is hydrogen-free. This particularity combined with the opacity of this material in the UV range permits to consider UV laser annealing (Fig. 7.8). The UV opacity of DLC able to limit the laser absorption at its surface and the hydrogen-free structure helps to avoid any destructive layer delamination.

## 7.3.2 Laser surface annealing of DLC thin films

### *7.3.2.1 Objectives of DLC surface annealing*

As presented in previous parts of this chapter, DLC is mainly transparent and insulated. If we plan to use DLC as transparent conductive electrodes, an additional step is needed to bring conductivity to this material. As already presented, DLC is (in most of the cases) dominated by $sp^3$ bindings. Therefore, on a typical DLC, we need to break existing $sp^3$ bindings and "convert" them to conductive graphite. This transformation has not to happen over the entire layer but must be enough to result in a homogeneous conductive structure set on the top of the DLC. To keep the DLC high transparency, it should be better to restrict graphitization to the first atomic layers, keeping unchanged the underneath DLC.

To induce this transformation, a sufficient quantity of energy has to be applied on the DLC to break existing surface diamond bindings (Fig. 7.8), taking advantage of the DLC high opacity in the UV range. By using the same high-power excimer UV laser (248 nm), it is possible to spread the available energy in order to force $sp^3$ bindings to rebind into $sp^2$ homogeneously over a large area. The coherent laser beam goes through a beam homogenizer (nine horizontal elements followed by seven vertical elements and a collimation lens) permitting to obtain a $1 \times 1\ cm^2$ homogeneous beam (Fig. 7.9). A redirection mirror is used to irradiate the sample surface vertically. A focusing lens completes the optical bench and allows to control the size of the beam on the surface of the sample. This sample is attached to a *XYZ* software-controlled table. The *Z*-axis displacement permits to control the beam size going from less than a few $mm^2$ to greater than 1 $cm^2$. *XY*-axis controls the position of the sample while driving the placement under the laser beam. The software is designed to automatically set the sample position as well as trigger the laser beam. With this setup, we are able to work from small surfaces by keeping the sample at a fix position up to very large areas by using *XY*-sweeping. To ensure high homogeneity while working in sweeping mode, an overlap between two contiguous beam positions (in *X* and in *Y* direction) is 90% typically. The facility is also completed with a chamber in which the laser treatment is performed under primary vacuum or under controlled neutral gas ambience.

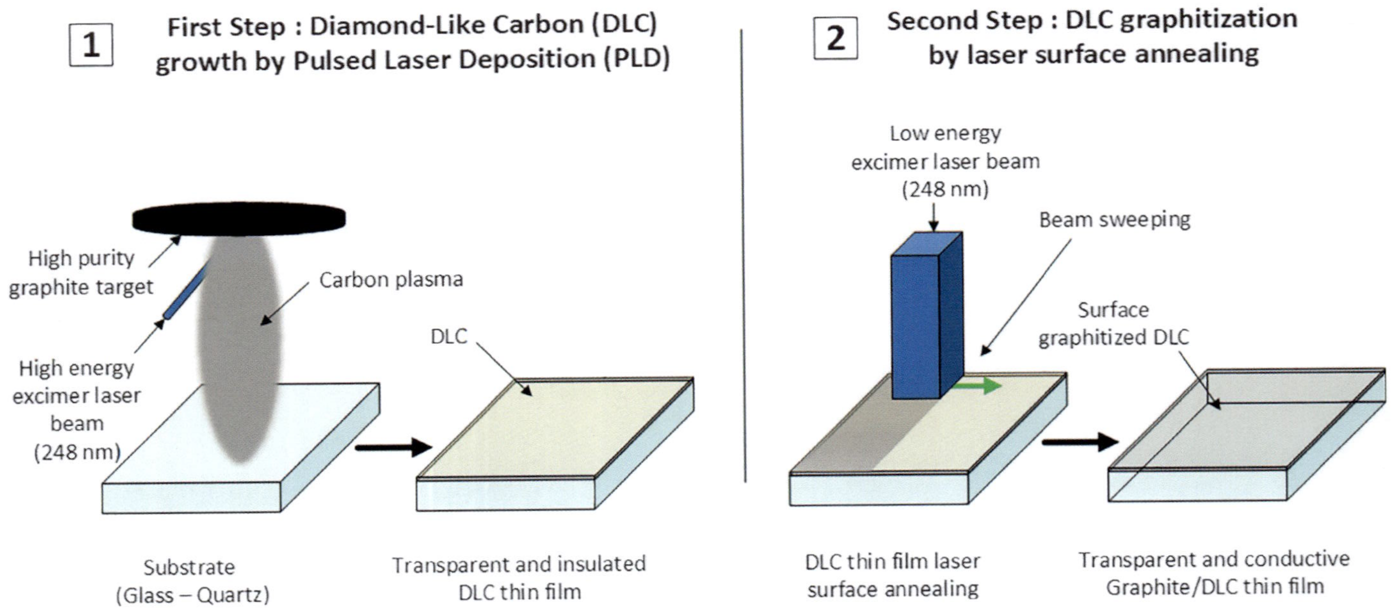


**FIG. 7.8** Description of the two-step laser process used to produce pure-carbon transparent electrodes.

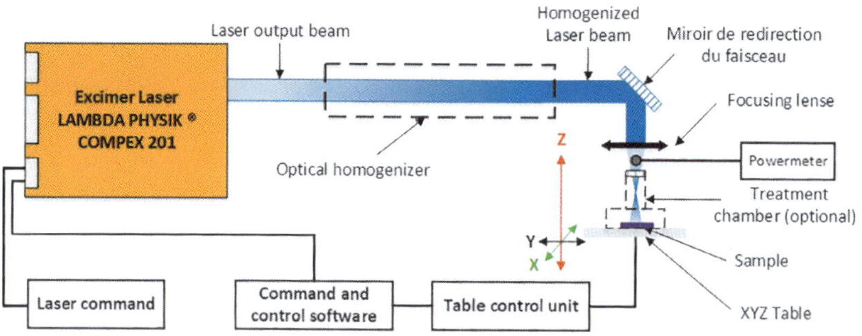


**FIG. 7.9** Schematic view of the UV laser annealing setup used to graphitize the DLC thin-film surface.

It is important to notice that the laser treatment must bring enough energy on the surface without overpassing the threshold leading to ablation or vaporization of the DLC [88–90]. Therefore, energy densities used for treatments are deeply lower than those used in the case of PLD. We are typically located under 0.3 $J/cm^2$ to avoid DLC destruction or delamination that will lead to a full transparency loss. As demonstrated [88–91], working over this fluence threshold, the whole DLC layer will be destroyed. There is also a low fluence threshold below which no significant change or conductivity property is noticeable. Both thresholds are dependent on the properties of the initial DLC thin films.

### *7.3.2.2 Effects of laser characteristics on the TGL + DLC thin-film structure*

Laser annealing of DLC layers has in all cases to be performed under vacuum or neutral gas atmosphere. It has been observed in preliminary studies that by annealing under air, carbon atoms react with oxygen forming carbonized gases (like CO and $CO_2$) leading to a progressive layer degradation resulting in complete delamination. This effect corresponds to the apparition of large fully graphitized areas (as confirmed by Raman spectroscopy) that extend progressively to the whole surface and finally lead to a full transparency loss. This observed behavior is similar to DLC produced in KrF mode and in ArF mode. Complementary experiments have concluded that ArF-produced DLCs are not suitable for UV laser annealing. Indeed, when annealing is performed on ArF-mode DLC, the fluence threshold corresponding to destruction is very low (under 0.2 $J/cm^2$) while the fluence threshold requested to obtain conductivity is greater than the fluence value. In this case, it is not possible to find suitable conditions making it possible to obtain a conductivity on ArF-DLC. A slight increase of $sp^2$ bindings is effectively observed, but not enough to form a conductive layer.

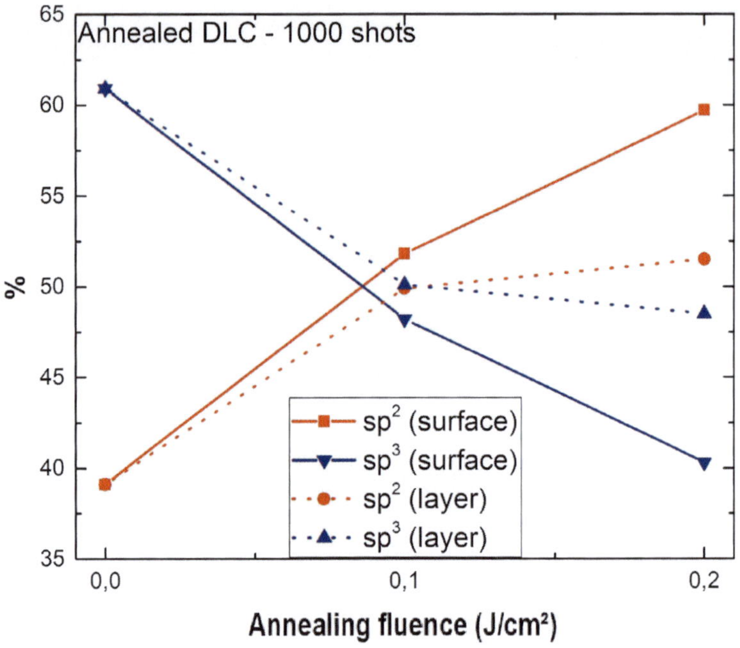


**FIG. 7.10** Evolution of atomic bindings in KrF-DLC after UV laser annealing at different fluences for 1000 shots.

When laser annealing is applied on KrF-elaborated DLC (at 5 J/cm$^2$ and for a 20 nm thickness), a significant increase of sp$^2$ bindings is observed. Before annealing, this DLC presents 39.1% of sp$^2$ bindings, homogeneously distributed in the entire layer. After annealing at 0.2 J/cm$^2$ for 1000 shots, an increase of 14% of sp$^2$ concentration is measured by XPS (Fig. 7.10). For the same annealing fluence but for 10,000 shots, an increase of 21.5% is obtained (Fig. 7.11). Angular-resolved XPS (AR-XPS) permits to give evidence that the increase of sp$^2$ bindings is much more significant when measurements are performed at a grazing incidence of 60°, corresponding to a 4.8-nm-layer depth analysis. A similar increase is also measured when the analysis is done at a relative normal incidence of 20° corresponding to a 9.1-nm-layer depth analysis but much lower than for the grazing analysis. We can therefore conclude that the presence of a few atomic layers made of pure sp$^2$ bindings is integrated into both angular measurements and enhances the measured mean values for both analyzed thicknesses. Finally, we also observe that the relative sp$^2$ binding's increase is highly correlated to the fluence and/or number of shots (while keeping under the layer destruction threshold).

Thanks to AR-XPS measurements combined with structure modeling and considering a fully graphitized layer (100% of sp$^2$) obtained on the top of the DLC layer, we can deduct its thickness by knowing the initial untreated

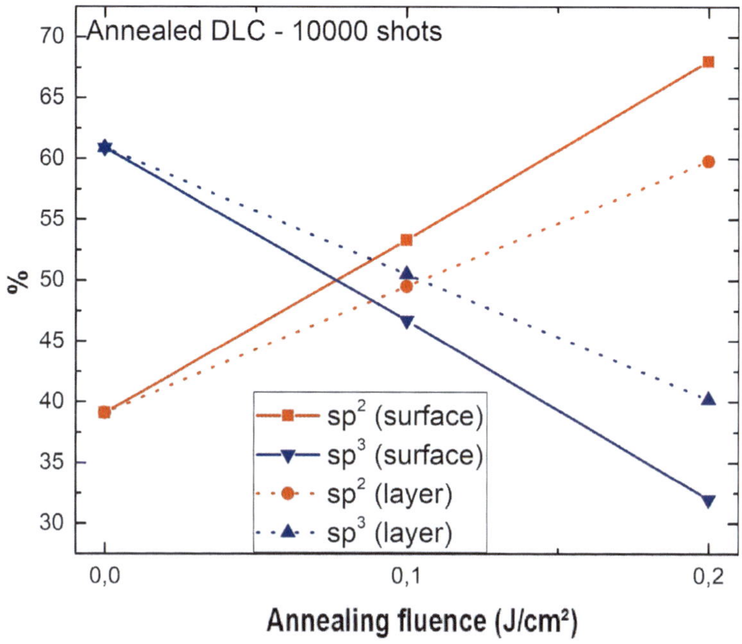


FIG. 7.11 Evolution of atomic bindings in KrF-DLC after UV laser annealing at different fluences for 10,000 shots.

DLC homogeneous atomic binding's proportions [88, 89]. Indeed, for a laser annealing at 0.1 J/cm$^2$ (1000 or 10,000 shots), a 1.35-nm graphitic layer is obtained. It is important to notice here that when annealing is performed at 0.2 J/cm$^2$ and 10,000 shots (limit of the efficient annealing parameter range, especially the number of shots), a higher graphitic thickness is determined, artificially boosted by the presence of some sporadic graphitic defaults on the surface. Then, for the optimal parameter set (0.2 J/cm$^2$ and 1000 shots), a 1.74-nm fully graphitized layer is obtained which is equivalent to 6 graphene monolayers.

In addition, Raman spectroscopy also confirms the presence of a graphitic material at the surface after annealing. In fact, when a typical DLC amorphous carbon spectrum is observed directly after deposition (large and mingled D and G bands with no 2D peak), after annealing, D and G bands (associated with a graphitic material) are present and clearly dissociated (Fig. 7.12), despite the presence of an only weak 2D peak. It is also important to precise that annealing parameters only sparsely affect Raman spectra, meaning that, as soon as a graphitic material is obtained, the same spectrum is measured. All these measurements permit to give evidence of the formation of a structure combining a TGL set on a pure DLC layer. The thickness of the obtained TGL can be adjusted (by choosing adapted annealing parameters) to obtain the best performances.

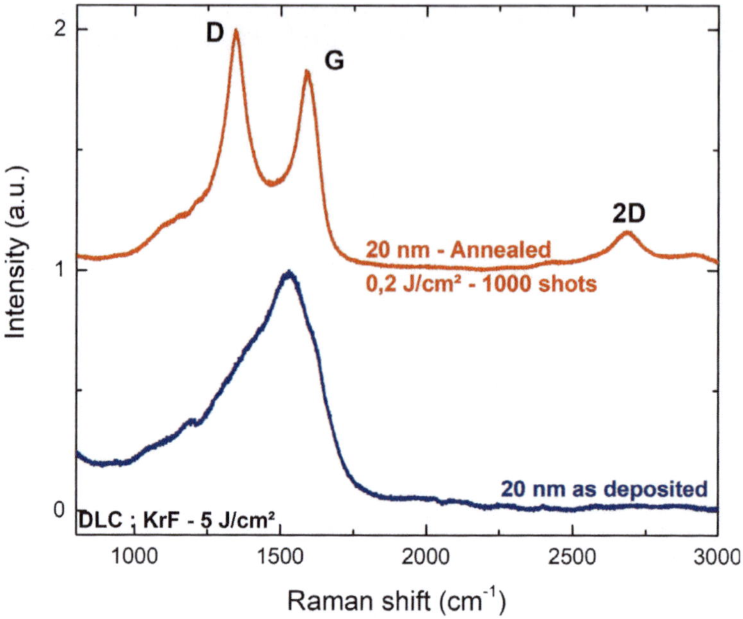


**FIG. 7.12** Raman spectra of KrF-mode DLC before and after UV laser annealing.

### *7.3.2.3 Electrical and optical properties of annealed DLC*

To use laser-annealed DLC as efficient transparent electrodes, the most important properties to evaluate are conductivity and transparency. Evaluating, controlling, and optimizing these properties are the key points to provide the best electrodes. As previously presented in Section 7.3.2.2 and despite better transparency, ArF-mode DLCs do not allow to obtain conductivity. Therefore, we only focus on conductivity and optical properties of DLC layers deposited in KrF mode.

It has been demonstrated that, when the annealing fluence is set to values between previously explained low and high thresholds and associated with an adapted number of shots, conductivity is obtained on KrF-mode DLC layers. While working at low fluence (0.1 $J/cm^2$), a high number of shots (10,000) are required to obtain conductivity which remains very low ($2.13 \times 10^{-3}$ $(\Omega\ cm)^{-1}$). For 0.2 $J/cm^2$ and 10,000 shots, a record conductivity is measured at $5 \times 10^4$ $(\Omega\ cm)^{-1}$. Nevertheless, with this parameter set, some sporadic graphitic defaults are observed on the DLC after annealing, narrowing therefore transmittance. The optimal parameters correspond to 0.2 $J/cm^2$ and 1000 shots, which allows to reach a $2.91 \times 10^3$ $(\Omega\ cm)^{-1}$ conductivity (Fig. 7.13). All these values are measured considering the entire DLC thickness (20 nm in most cases). However, we demonstrated in the previous section that the layer responsible for conductivity is set at the surface and measures only a few nanometers. The previously given conductivity is to consider worst-case conductivity.

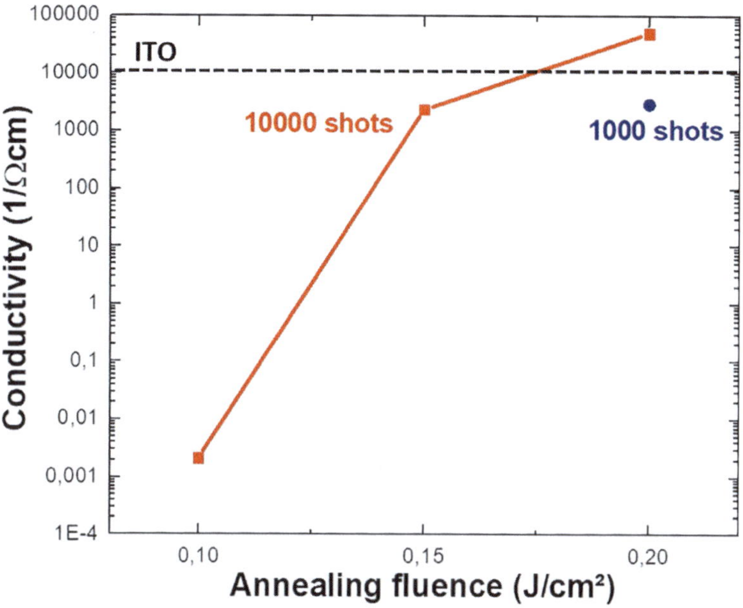


**FIG. 7.13** Conductivity of laser annealed KrF-mode DLC as a function of annealing fluence and number of shots.

Knowing the real thickness of the conductive layer, we can determine a more realistic value of conductivity. For the best parameter set (0.2 $J/cm^2$ and 1000 shots), a fully graphitized thickness of 1.74 nm corresponds finally to a $3.35 \times 10^4$ $(\Omega\ cm)^{-1}$ conductivity.

In terms of transparency, only a very slight influence of the annealing parameters is noticed. Indeed, after annealing (and especially on 20 nm of DLC on glass substrates), we observe high transparency around 85% all over the visible range (Fig. 7.14). On glass substrates, we observe an increase of the transmittance after laser annealing (Figs. 7.14 and 7.15). This transparency is the highest at 800 nm (90% typically) and reduces as shifting to the near UV range (78% at 400 nm). For different laser fluences, almost the same values of transmittances are obtained demonstrating therefore the very small influence of annealing parameters.

It is important to notice here that laser annealing on DLC has been performed on layers deposited on various substrates ($SiO_2$/Si, glass, and quartz). Nevertheless, conductivity and transparency values are in the same range for all kinds of substrates, even if a small adaptation of the annealing fluence can be necessary to obtain the best performances.

### *7.3.2.4 Performances ratings*

We have demonstrated that the combination of PLD-deposited DLC and laser annealing of those layers permits to obtain a transparent and conductive structure.

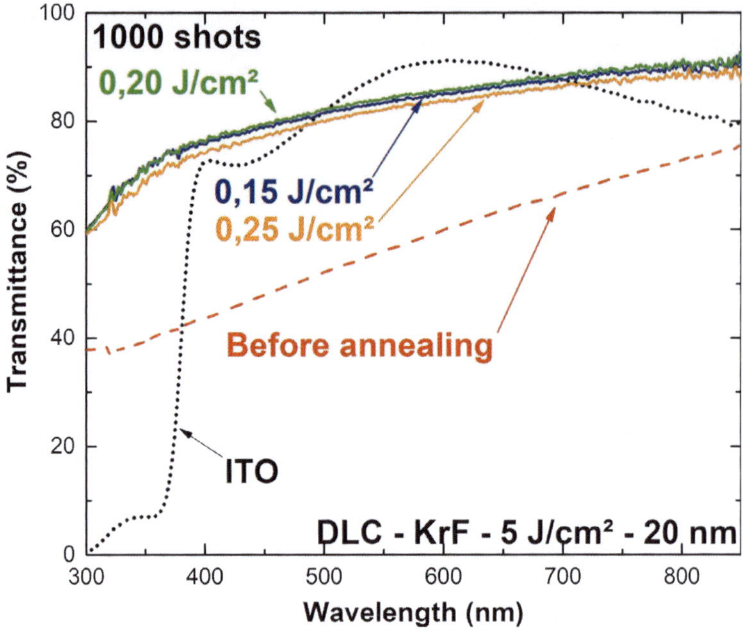


FIG. 7.14 Transmittance of DLC layers deposited on glass before and after annealing for 1000 shots at various fluences. *(ITO reference is extracted from T. Liu, X. Zhang, J. Zhang, W. Wang, L. Feng, L. Wu, W. Li, G. Zeng, B. Li, Int. J. Photoenergy (2013) 2013.)*

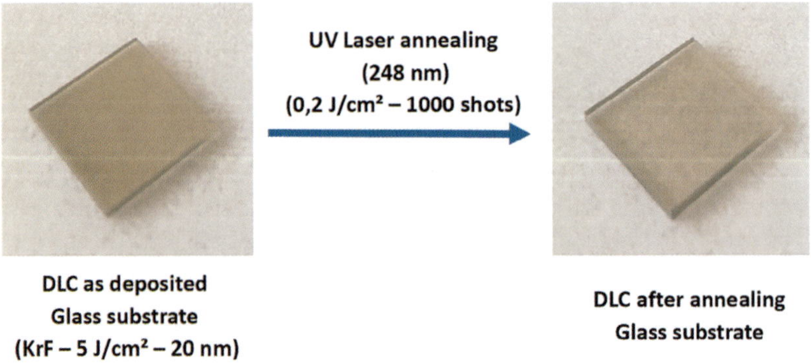


FIG. 7.15 Example of a KrF-mode DLC layer on glass before and after laser annealing at 0.2 J/cm$^2$ for 1000 shots.

To evaluate the potential of this technology, it is important to compare the produced electrodes to ITO. First, if we focus on conductivity, we observe that the obtained performances are in the same range as ITO (typically $10^4$ $(\Omega\ \text{cm})^{-1}$) [11–13]. Conductivity in the same decade can be obtained, some being slightly less and some being more (considering worst-case conductivity measured over the entire layer). With sheet resistances globally located between 170 Ω□ and 1 k Ω□, we also approach ITO values (conventionally being around 100 Ω□ for commercial ITO). Considering the transparency, DLC electrodes are in the same range as ITO. Transmittance behavior appears to be slightly different (Fig. 7.14). ITO has a maximum of 90% transmittance around 600 nm, when the maximum of treated DLC is at 800 nm (being also 90% when deposited on glass substrate). Typically, on the whole visible range, comparable mean values are obtained. We can conclude from these raw performance ratings that treated DLC layers show performances approaching those of commercial ITO.

Functional performances of treated DLC layers have also being rated by integrating samples in electrical circuits. Indeed, DLC electrodes can be used as transparent resistors and placed in an electric circuit with a LED. For example, a treated DLC sample showing a 4.2 kΩ resistance has been used in a circuit supplied with voltage values up to 30 VDC. It has been observed that, with a 30 VDC supply, a 6.8 mA current is driven by the DLC layer, corresponding to around 200 mW power dissipation in the TGL/DLC. The experiment was repeated and performed for longer than 6 h without any performance degradation or DLC deterioration, revealing the stability of the electrode and its ability of efficiently dissipating power. In addition, the same behaviors are obtained on various samples demonstrating therefore the reproducibility of our process.

To conclude on TGL/DLC electrode performances, we have demonstrated that this original two-step process, only based on laser technologies and carbon materials, permits to obtain functional electrodes with performances comparable to ITO.

## 7.4 Conclusion and perspectives

DLC thin films and TGL present some very interesting properties for optoelectrical devices. With high optical transparency and duality being an insulating or conductive material, a wide range of applications is open to the integration of carbon technology. Having precise control of the process and the related parameters is essential to ensure the best DLC or TGL properties required by various applications. This is the case when UV layer annealing is performed on PLD-elaborated DLC. This double-step process (Fig. 7.8) permits to start from a high-purity DLC layer (which is transparent and insulating) to a surface graphitization providing electrical conductivity. A transparent conductive electrode is obtained with performances comparable to other transparent conductive materials (Figs. 7.13–7.15). DLC-TGL is a pure carbon alternative to the use of rare earth elements like indium. With performance levels like these, other applications to this full carbon technology shall be interested in. This technology can also be able to project full transparent electronic devices.

Moreover, being only based on laser processes, this approach appears to be a serious alternative to more standard technologies. Laser processing offers many advantages like generating low process costs (compared to heavier deposition and treatment technologies), being scalable over large surfaces and being applicable on various substrates. In addition, PLD and laser posttreatments performed at room temperature ensure high compatibility with all the standard microelectronic, optoelectronic, and photovoltaic processes.

All these points demonstrate that laser engineering and integration of carbon materials in optoelectronics are very high hopeful technologies for the future. Elaborating electrical and optoelectrical components based on carbon materials and using laser processes shall be an essential key point for the future, being economic, ecologic, and viable for an endless time.

## References


[1] J.-H. Lee, D.N. Liu, S.-T. Wu, Introduction to Flat Panel Displays, Vol. 20 of Wiley Series in Display Technology, John Wiley & Sons, 2008.

[2] L.-B. Desroches, M. Ganeshalingam, Technol. Forecast. Soc. Change 90 (Part B) (2015) 562–574.

[3] J. Ma, Advanced MEMS-based technologies and displays, Displays 37 (2015) 2–10.

[4] Z. Yang, M. Gao, W. Wu, X. Yang, X.W. Sun, J. Zhang, H.-C. Wang, R.-S. Liu, C.-Y. Han, H. Yang, W. Li, Mater. Today 24 (2019) 69.

[5] X. Yanga, Z.-H. Zhangb, T. Dingc, N. Wangc, G. Chena, C. Dangc, H.V. Demirc, X.W. Sund, Nano Energy 46 (2018) 229–233.

[6] M. Eritt, C. May, K. Leo, M. Toerker, C. Radehaus, Thin Solid Films 518 (2010) 3042–3045.

[7] T. Dobbertin, E. Becker, T. Benstem, G. Ginev, D. Heithecker, H.-H. Johannes, D. Metzdorf, H. Neuner, R. Parashkov, W. Kowalsky, Thin Solid Films 442 (2010) 132–139.

[8] D. Bradley, Mater. Today 28 (2019) 6.

[9] R.H. Chen, Liquid Crystal Displays: Fundamental Physics and Technology, first ed., Wiley, 2011.

[10] T. Wu, C.W. Sher, Y. Lin, C.F. Lee, S. Liang, Y. Lu, S.W. Huang Chen, W. Guo, H.C. Kuo, Z. Chen, Appl. Sci. 8 (9) (2018) 1557.

[11] Z. Chen, W. Li, R. Li, Y. Zhang, G. Xu, H. Cheng, Langmuir 29 (45) (2013) 13836–13842.

[12] H. Kim, C.M. Gilmore, J. Appl. Phys. 86 (1999) 6451.

[13] T. Liu, X. Zhang, J. Zhang, W. Wang, L. Feng, L. Wu, W. Li, G. Zeng, B. Li, Int. J. Photoenergy (2013) 2013.

[14] U. Ozgur, Y.I. Alivov, C. Liu, A. Teke, M.A. Reshchikov, S. Dogan, V. Avrutin, S.-J. Cho, H. Morkoc, J. Appl. Phys. 98 (2005), 041301.

[15] G. Wisz, I. Virt, P. Sagan, P. Potera, R. Yavorskyi, Nanoscale Res. Lett. (2017) 12–253.

[16] J. Kim, G.V. Naik, A.V. Gavrilenko, K. Dondapati, V.I. Gravrilenko, S.M. Prokes, O.J. Glembocki, V.M. Shalaev, A. Boltasseva, Phys. Rev. X 3 (2013), 041037.

[17] Z. Zhu, C. Liu, J. Xu, Q. Jiang, H. Shi, E. Liu, Electron. Mater. Lett. 12 (1) (2016) 54–58.

[18] J. Saghaei, A. Fallahzadeh, M.H. Yousefi, Org. Electron. 19 (2015) 70–75.

[19] Z. Wu, Z. Chen, X. Du, J.M. Logan, J. Sippel, M. Nikolou, K. Kamaras, J.R. Reynolds, D.B. Tanner, A.F. Hebard, A.G. Rinzler, Science 305 (5688) (2004) 1273–1276.

[20] P.R. Bandaru, J. Nanosci. Nanotechnol. 7 (2007) 1–29.

[21] J. Liu, L. Guo, X. Liu, S. Li, X. Liu, X. Shen, S. Chang, K. Cheng, Z.-L. Du, Appl. Energy Mater. 1 (2018) 783–789.

[22] H. Zhai, Y. Li, X. Wang, R. Wang, L. Shi, R. Wang, J. Sun, Nano Res. 11 (2018) 1956–1966.

[23] S. Cho, M.S. Fuhrer, Phys. Rev. B 77 (8) (2008) 1–4, https://doi.org/10.1103/PhysRevB.77.081402.
[24] R.R. Nair, P. Blake, A.N. Grigorenko, K.S. Novoselov, T.J. Booth, T. Sauber, N.M.R. Peres, A.K. Geim, Science 320 (5881) (2008) 1308.
[25] M. Murakami, A. Tatami, M. Tachibana, Carbon 145 (2019) 23–30.
[26] M. Kato, K. Uki, S. Harako, X. Zhao, Microelectron. Eng. 121 (2014) 96–99.
[27] A.N. Obraztsov, E.A. Obraztsova, A.V. Tyurnina, A.A. Zolotukhin, Carbon 45 (2007) 2017–2021.
[28] M. Marcinek, L.J. Hardwick, G.Z. Zukowska, R. Kostecki, Carbon 48 (2010) 1552–1557.
[29] A.A. Voznesenskaya, A.V. Zhdanov, L.V. Belyaev, Mater. Today Proc. 19 (2019) 2270–2273.
[30] W. Mróza, S. Burdyńskaa, A. Prokopiuka, M. Jedyńskia, B. Budnera, M.L. Korwin-Pawlowskib, Acta Phys. Pol. A 116 (2009) S-120.
[31] K.T. Chaudharya, R. Qindeelb, Saktiotoc, M.S. Hussaind, J. Alie, P.P. Yupapinf, Procedia Eng. 8 (2011) 423–427.
[32] K.S. Novoselov, A.K. Geim, S.V. Morozov, D. Jiang, Y. Zhang, S.V. Dubonos, I.V. Grigorieva, A.A. Firsov, Science 306 (5696) (2004) 666–669.
[33] J. Meyer, A.K. Geim, M.I. Katsnelson, K.S. Novoselov, T.J. Booth, S. Roth, Nature 446 (2007) 7131.
[34] E. Jabari, E. Toyserkani, Mater. Lett. 174 (2016) 40–43.
[35] E. Jabari, E. Toyserkani, Carbon 91 (2015) 321–329.
[36] J. Azpeitia, G. Otero-Irurueta, I. Palacio, J.I. Martinez, N. Ruiz del Arbol, G. Santoro, A. Gutierrez, L. Aballe, M. Foerster, M. Kalbac, V. Vales, F.J. Mompean, M. García-Hernandez, J.A. Martín-Gago, C. Munuera, M.F. Lopez, Carbon 119 (2017) 535–543.
[37] Z. Ma, Y. Zhuang, Y. Deng, X. Song, X. Zuo, X. Xiao, J. Nan, J. Power Sources 376 (2018) 91–99.
[38] J. Xu, X. Wang, N. Yuan, B. Hu, J. Ding, S. Ge, J. Power Sources 430 (2019) 74–79.
[39] D.C. Guo, C.S. Sun, W. Wu, Z.S. Wu, J. Gao, N. Su, J. Chen, Appl. Surf. Sci. 493 (2019) 177–184.
[40] X. Chang, B. Sun, Z. Xie, Z. Wang, J. Zheng, X. Li, J. Electroanal. Chem. 854 (2019) 113529.
[41] R. Song, Q. Wang, B. Mao, Z. Wang, D. Tang, B. Zhang, J. Zhang, C. Liu, D. He, Z. Wu, S. Mu, Carbon 130 (2018) 164–169.
[42] L. Liao, X. Duan, Mater. Sci. Eng. R 70 (2010) 354–370.
[43] M. Karbalaei, D. Dideban, H. Heidari, Results Phys. 15 (2019) 102806.
[44] S. Vaziri, G. Lupina, C. Henkel, A.D. Smith, M. Östling, J. Dabrowski, G. Lippert, W. Mehr, M.C. Lemme, Nano Lett. 13 (4) (2013) 1435–1439.
[45] A.A. Dubinov, V. Ya Aleshkin, S.V. Morozov, V. Ryzhii, T. Otsuji, Opto-Electron. Rev. 27 (4) (2019) 345–347.
[46] H. Shin, J. Oh, Y. Kim, J. Gon Son, C. Lee, K.Y. Shin, J. Ind. Eng. Chem. 81 (2020) 488–495.
[47] G.-W. Hsieh, Z.-R. Lin, C.-Y. Hung, S.-Y. Lin, C.-R. Yang, Org. Electron. 54 (2018) 27–33.
[48] M.F. Ben Aissa, H. Rezgui, F. Nasri, H. Belmabrouk, A. Guizani, Superlattice. Microst. 128 (2019) 265–273.
[49] B.-W. Hwang, H.-I. Yeom, D. Kim, C.-K. Kim, D. Lee, Y.-K. Choi, Solid State Electron. 141 (2018) 65–68.
[50] T. Mahmoudi, Y. Wang, Y.B. Hahn, Nano Energy (2018) 51–65.
[51] D.H. Shin, J.H. Kim, D.H. Jung, S.-H. Choi, J. Alloys Compd. 803 (2019) 958–963.
[52] Y.-K. Lin, Y.-T. Hong, J.-J. Shyue, C.-H. Hsueh, Superlattice. Microst. 126 (2019) 42–48.
[53] J. Robertson, Mater. Sci. Eng. R 37 (2002) 129–281.
[54] Z. Nibennaoune, D. George, F. Antoni, S. Ahzi, D. Ruch, J. Gracio, Y. Rémond, Diam. Relat. Mater. 22 (2012) 105–112.

[55] E. Salgueiredo, M. Vila, M.A. Silva, M.A. Lopes, J.D. Santos, F.M. Costa, R.F. Silva, P.S. Gomes, M.H. Fernandes, Diam. Relat. Mater. 17 (2008) 878–881.
[56] M. Jelinek, T. Kocourek, J. Remsa, J. Miksovsky, J. Zemek, K. Smetana, B. Dvorankova, T. Luxbacher, Appl. Phys. A 101 (2010) 579–583.
[57] M. Ikeyama, S. Miyagawa, Y. Miyagawa, Y. Hayakawa, T. Miyajima, Surf. Coat. Technol. 201 (2007) 8112–8115.
[58] A. Grill, Thin Solid Films 355-356 (1999) 189–193.
[59] T. Sonoda, S. Nakao, M. Ikeyama, Vacuum 84 (2010) 666–668.
[60] A.P. Mousinho, R.D. Mansano, Appl. Surf. Sci. 254 (2007) 189–192.
[61] A. Grill, B. Meyerson, in: K.E. Spear, J.P. Dismukes (Eds.), Synthetic Diamond: Emerging CVD Science and Technology, Wiley, New York, 1994, p. 91.
[62] A. Varade, K. Niranjan Reddy, D. Sasen, A. Krishna, M. Chellamalai, P.V. Shashikumar, Procedia Eng. 97 (2014) 1452–1456.
[63] O.D. Coskun, T. Zerrin, Diam. Relat. Mater. 56 (2015) 29–35.
[64] X. Shi, D. Flynn, B.K. Tay, S. Prawer, K.W. Nugent, S.R.P. Silva, Y. Lifshitz, W.I. Milne, Philos. Mag. B 76 (1997) 351–361.
[65] H.-J. Scheibe, B. Schultrich, Thin Solid Films 246 (1994) 92–102.
[66] A.A. Voevodin, M.S. Donley, Surf. Coat. Technol. 82 (1996) 199–213.
[67] G. Reise, B. Keiper, S. Weissmantel, U. Falke, Appl. Surf. Sci. 127–129 (1998) 500–506.
[68] A.A. Puretzky, D.B. Geohegan, G.E. Jellison Jr., M.M. McGibbon, Appl. Surf. Sci. 96-98 (1996) 859–865.
[69] J.-J. Gaumet, A. Wakisaka, Y. Shimizu, Y. Tamori, J. Chem. Soc. 89 (1993) 1667–1670.
[70] P.T. Murray, D.T. Peeler, Appl. Surf. Sci. 69 (1993) 225–230.
[71] H. Ito, K. Yamamoto, Kobelco Technol. Rev. (35) (2017).
[72] Q. Wei, A.K. Sharma, J. Sankar, J. Narayan, Compos. Part B 30 (1999) 675–684.
[73] L. Valentini, E. Braca, J.M. Kenny, Trans. Eng. Sci. 25 (1999) 55–69.
[74] X. Shi, D. Flynn, B.K. Tay, S. Prawer, K.W. Nugent, S.R.P. Silva, Y. Lifshitz, W.I. Milne, Philos. Mag. B 76 (1997) 351–361.
[75] J.F. Pern, K. Touryan, Z. Panosyan, A.A. Gippius, J.A. Kontsevoy, Conference Paper 2005, National Renewable Energy Laboratory, 2005.
[76] C.-R. Lin, H.-M. Chang, C.-K. Chang, Int. J. Photoenergy (2013) 612163.
[77] N.I. Klyui, V.G. Litovchenko, A.G. Rozhin, V.N. Dikusha, M. Kittler, W. Seifert, Sol. Energy Mater. Sol. Cells 72 (2002) 597–603.
[78] S.-S. Yap, T.-K. Yong, T.-Y. Tou, Thin Solid Films 517 (2009) 5311–5313.
[79] S.-S. Yap, R.-B. Yang, T.-K. Yong, W.-O. Siew, T.-Y. Tou, Diam. Relat. Mater. 18 (2009) 380–383.
[80] J. Szmidt, R.B. Beck, S. Mitura, S. Sokolowska, Diam. Relat. Mater. 3 (1994) 853–857.
[81] S. Takabayashi, S. Ogawa, Y. Takakuwa, H.-C. Kang, R. Takahashi, H. Fukidome, M. Suemitsu, T. Suemitsu, T. Otsuji, Diam. Relat. Mater. 22 (2012) 118–123.
[82] B. Ren, L. Wang, L. Wang, J. Huang, K. Tang, Y. Lou, D. Yuan, Z. Pan, Y. Xia, Vacuum 107 (2014) 1–5.
[83] S. Alotaibi, K.N. Manjunatha, S. Paul, Diam. Relat. Mater. 90 (2018) 172–180.
[84] F. Breech, L. Cross, Appl. Spectrosc. 16 (1962) 59.
[85] H.M. Smith, A.F. Turner, Appl. Opt. 4 (1965) 147–148.
[86] D. Dijkkamp, T. Venkatesan, et al., Appl. Phys. Lett. 51 (B) (1987) 619.
[87] F. Stock, F. Antoni, F. Le Normand, D. Muller, M. Abdesselam, N. Boubiche, I. Komissarov, Appl. Phys. A 123 (2017) 590.

[88] F. Stock, F. Antoni, L. Diebold, C. Chowde Gowda, S. Hajjar-Garreau, D. Aubel, N. Boubiche, F. Le Normand, D. Muller, Appl. Surf. Sci. 464 (2019) 562–566.
[89] F. Stock, F. Antoni, D. Aubel, S. Hajjar-Garreau, D. Muller, Appl. Surf. Sci. 505 (2020) 144505.
[90] T.V. Kononenko, V.V. Kononenko, S.M. Pimenov, E.V. Zavedeev, V.I. Konov, V. Romano, G. Dumitru, Diam. Relat. Mater. 14 (2005) 1368–1376.
[91] M. De Feudis, A.P. Caricato, M. Martino, E. Alemanno, P. Ossi, G. Maruccio, A.G. Monteduro and M. Corrado Conference: 4th Workshop-Plasmi, Sorgenti, Biofisica ed Applicazioni, 2014.

Chapter 8

# Optical hyperdoping

**Wenjie Yang, Shao Qi Lim, and Jim S. Williams**
*The Australian National University, Canberra, ACT, Australia*

## 8.1 Introduction

As discussed in the previous chapters, ion implantation followed by pulsed laser melting (PLM) is a well-studied process for achieving localized heavy doping in semiconductors, where the dopant concentration can exceed the equilibrium solid solubility limit by orders of magnitude. In addition to the useful electrical properties exhibited by semiconductors hyperdoped with p- and n-type dopants, novel material properties have also been observed over the recent years in single-crystalline semiconductors containing a nonequilibrium concentration of various *nondopant* impurities. In this chapter, we focus on the intentional introduction of impurities into semiconductors as a method to modify their electronic band structure and, therefore, optical properties, a process which has recently become known as "optical hyperdoping." The research effort in optical hyperdoping has almost exclusively focused on Si and most of this chapter is dedicated to reviewing the literature on optical hyperdoping in Si. However, there has been an emerging work on hyperdoped Ge which we summarize. In addition, a logical extension of the Ge work to dilute GeSn alloys, also formed by implantation and pulsed laser melting, has received attention recently and a brief overview of this body of research is provided.

This chapter is divided into five sections. Section 8.1 provides an overview of optical hyperdoping in Si as a method to fabricate an intermediate band semiconductor. In Section 8.2, the historical development of optical hyperdoping in Si, mainly involving chalcogens, is outlined and discussed. Sections 8.3 and 8.4 describe more recent progress in optical hyperdoping using transition metals. The development of optical hyperdoping in Ge is reviewed in Section 8.5, where early studies have investigated the possibility of modifying the bandgap of Ge through either the introduction of suitable deep-level impurities or by alloying with Sn. Finally, a summary of this chapter and future perspectives are presented in Section 8.6.

**Laser Annealing Processes in Semiconductor Technology https://doi.org/10.1016/B978-0-12-820255-5.00001-5**

### 8.1.1 Si as an intermediate band semiconductor

Optical hyperdoping of Si is motivated by the prospect of using Si in optoelectronic applications (e.g., sensing, photodetection, and photovoltaics) as a means to exploit the numerous technological advantages of Si as a semiconductor material. However, a key limitation in applying Si in such applications is its inherent transparency in the near-infrared (NIR) due to its bandgap ($E_g$ = 1.12 eV, causing absorption to diminish at $\lambda > 1100$ nm), which unfortunately coincides with the most useful spectral range in telecommunications where there is minimal attenuation. Thus, achieving efficient subbandgap absorption (i.e., $\lambda > 1100$ nm) in Si is crucial for utilizing Si for NIR optoelectronic applications.

The relevance of the intermediate band semiconductor to obtaining subbandgap absorption is demonstrated by considering the physical process of light absorption in semiconductors. Specifically, the absorption of photons in semiconductors can be understood as the transition of electrons from the valence band (VB) into the conduction band (CB), a process which converts optical/photon energy into electrical potential energy. This is illustrated in Fig. 8.1, in which a photon with energy $E_1$ ($E_1 > E_G$) is absorbed through the excitation of an electron from the VB into the CB (leaving behind a hole). Photons with energies less than the bandgap (e.g., with energies $E_2$ and $E_3$) do not facilitate such an excitation and thus cannot be absorbed efficiently.

In so-called intermediate band semiconductors, a third band, the intermediate band (IB), is situated between the CB and the VB. In this scenario, subbandgap photons can, in theory, be absorbed through a stepwise process.

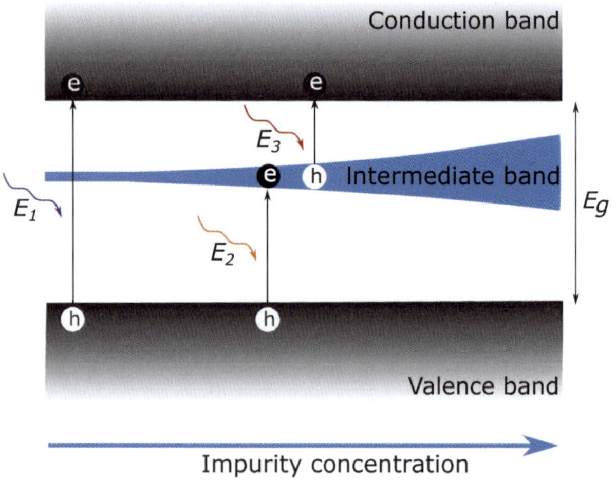


FIG. 8.1 Illustration of subbandgap absorption through an IB. Note that the absorption of light in indirect semiconductors such as Si also involves phonons, a process which is omitted here for simplicity.

As illustrated in Fig. 8.1, subbandgap photons with energies lower than $E_G$ but higher than $E_2$ or $E_3$ can then be subsequently absorbed by the excitation of electrons from the VB to the IB and from the IB to the CB, respectively. Ideally, the IB should not interfere with the absorption of above-bandgap photons, and should remain partially filled to allow the interband movement of electrons.

Numerous methods for implementing the intermediate band semiconductor concept have been explored, including quantum dots, mismatched alloys, and optical hyperdoping [1, 2]. Of those, the latter shall be the primary focus of this chapter. As will be discussed in the following section, optical hyperdoping has shown promise as a viable method for creating an IB because it allows deep-level impurities, which are required for forming an IB but are usually highly insoluble in Si, to be introduced at above-equilibrium concentrations. It should be noted that thin film deposition methods have been explored in lieu of ion implantation [3] and techniques such as flash lamp annealing (FLA) and rapid thermal annealing (RTA) [4, 5] have been shown to be alternatives to PLM, though the impurity concentration achieved in these cases is often significantly lower.

### 8.1.2 Intermediate band semiconductor by hyperdoping

Impurities with energy levels $E \gg 3k_BT$ ("deep" levels) away from the VB or CB are usually considered detrimental for semiconductor performance because such energy levels increase the probability of carrier recombination. On the other hand, these deep levels, even as discrete states (i.e., at/below equilibrium concentrations), also facilitate subbandgap absorption. However, strong subbandgap absorption requires a high density of intermediate states, which are expected to interact and form an IB due to Pauli's exclusion principle. It is interesting to note that such a delocalization is believed by some to suppress the nonradiative recombination rate, although such an effect has not been unequivocally demonstrated [6, 7].

As most deep-level impurities have very low solubilities in Si, the required impurity concentration is usually impossible to attain using equilibrium techniques such as thermal diffusion. Contrary to equilibrium techniques, hyperdoping by ion implantation and PLM gives rise to a nonequilibrium freezing-in of impurities into the single-crystalline matrix, which enables the incorporation of impurities at sufficient concentration levels required to form an IB. As shown in Fig. 8.2A, the distribution of impurities in hyperdoped layers is mostly determined by the diffusive speed of the impurity in the liquid phase and the resolidification rate of the crystalline matrix (usually in the order of a few meter per second) [8], while the thickness of the hyperdoped layer is approximately defined by the laser-induced melt depth (usually between a few tens to a few hundred nanometers). A typical impurity concentration profile, obtained from secondary ion mass spectrometry (SIMS) in this case, is shown in Fig. 8.2B using Au-hyperdoped Si as an example. Note that for this particular sample,

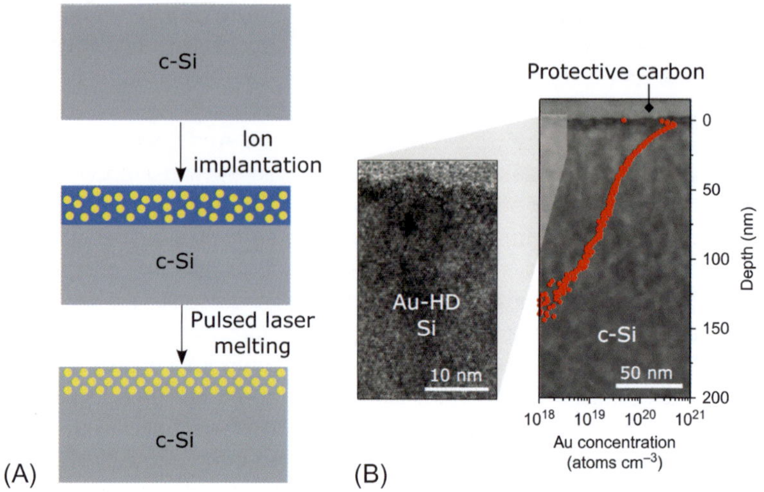


**FIG. 8.2** (A) Illustration of optical hyperdoping by ion implantation and PLM and (B) concentration profile (from SIMS) of Au in hyperdoped Si overlaid with an XTEM of the Au-hyperdoped layer. *((A) Modified from W. Yang, Characterising and Understanding Au-Hyperdoped Si for Sub-Band Gap Optical Absorption (Ph.D. thesis), The Australian National University, 2018 and (B) modified from J.P. Mailoa, A.J. Akey, C.B. Simmons, D. Hutchinson, J. Mathews, J.T. Sullivan, D. Recht, M.T. Winkler, J.S. Williams, J.M. Warrender, et al., Room-temperature sub-band gap optoelectronic response of hyperdoped silicon, Nat. Commun. 5 (1) (2014) 1–8.)*

cross-sectional transmission electron microscopy (XTEM) has indicated that the Si matrix remains single crystalline and that some Au atoms have segregated onto the surface.

## 8.2 Historic overview

In terms of experimental observations, enhanced NIR absorption by ion implantation followed by ns-PLM was first made on S-hyperdoped Si [11]. Prior to this, however, the concept of utilizing intermediate states, introduced by both intrinsic defects (e.g., point defects generated by self-implantation) and extrinsic defects (e.g., impurity species), has long been pursued for Si-based IR photodetection. The following section outlines the key results from these earlier efforts.

### 8.2.1 Defect-mediated all-Si NIR photodetectors

As mentioned in the previous section, intermediate states (or subbandgap states) can also arise from intrinsic crystalline defects in Si, such as those introduced by Si self-implantation. One of the earliest demonstrations of enhanced NIR absorption was observed in deuteron-irradiated Si by Fan and Ramdas [12] back in 1959, whose main results are shown in Fig. 8.3. Since then, numerous

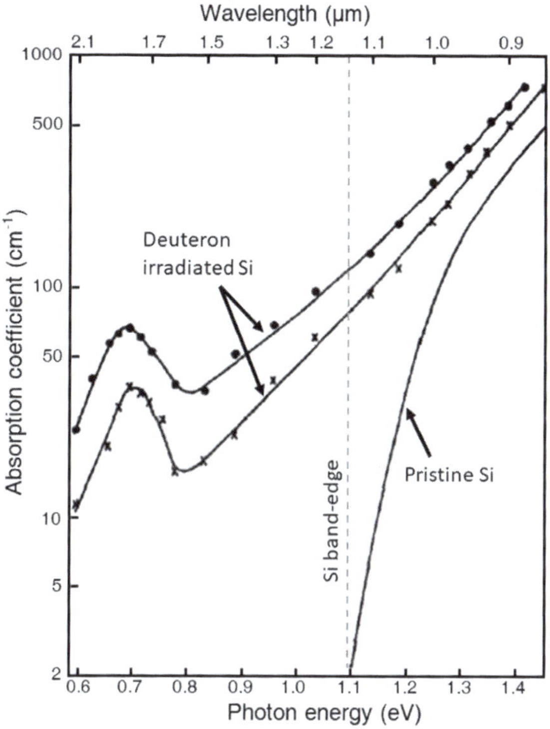


**FIG. 8.3** Absorption spectrum of high-resistivity Si before and after successive deuteron irradiations, showing enhanced absorption of NIR photons below the Si band-edge energy. *(Modified from H.Y. Fan, A.K. Ramdas, Infrared absorption and photoconductivity in irradiated silicon, J. Appl. Phys. 30 (8) (1959) 1127–1134.)*

research efforts have focused on the deliberate incorporation of crystalline defects (usually divacancies and interstitial clusters) into Si either by H, He, or Si self-implantation for improved subbandgap photon absorption [13].

It is easy to hypothesize that increasing the defect density (e.g., through increasing the implantation dose) can result in further enhancements in NIR absorption. However, this will inevitably result in deteriorated carrier transport properties and subsequently, detector performance. Thus, rigorous and often tedious detector optimization experiments are required to find the optimal balance between NIR light absorption and photocarrier collection. Consequently, it was not until more than 40 years later in 2003 when Knights et al. [14] finally demonstrated with proton implantation into silicon-on-insulator material a promising all-Si integrated detector that is not only sensitive to NIR wavelengths between 1520 and 1620 nm (i.e., the telecommunication C- and L-bands), but also CMOS compatible. More specifically, a detector responsivity of 8 mA $W^{-1}$ at 1550 nm and reverse leakage currents as low as 3 nA were reported when operating at $-1$ V [14]. Other notable research groups involved in the ongoing optimization of implantation-induced defect-mediated (integrated) all-Si NIR detectors include Refs. [15–17]. Through further optimizing the implantation conditions, fabrication procedures, and device architecture, high-speed NIR detectors with responsivities up to 140 mA $W^{-1}$ and leakage currents as low as 0.2 nA were achieved (operating at $-10$ V), while simultaneously having a tunable range of resonant wavelengths between 1510 and 1600 nm [17].

Since the main distinction between defect- and impurity-induced subbandgap absorption is the nature of the subbandgap states, it will be interesting to see how these materials compare in terms of their performance as NIR detectors. However, impurity-hyperdoping only became a possibility after the advent of PLM for semiconductor processing in the late 1970s, much later than the initial work on defect-mediated Si detectors. Furthermore, complications such as insulator-to-metal transitions (IMT) [18] and/or cellular/filamentary breakdown [8, 19] caused by extremely high impurity concentrations present further challenges to material fabrication, as will be discussed later. Thus, much less device optimization experiments have been carried out on impurity-hyperdoped Si compared with the now well-established defect-mediated all-Si detectors. The upcoming sections further examine some of the main challenges faced in fabricating impurity-hyperdoped Si with chalcogens (Section 8.2.2) and transition metal (Section 8.3) impurities.

### 8.2.2 Chalcogen fs-hyperdoped Si

Research in optical hyperdoping was initiated by the discovery of near-unity absorption in Si, both in the visible range of the spectrum and in the NIR, after fs- and ns-laser irradiation in a $SF_6$ atmosphere [20–22], as shown in Fig. 8.4. Later, it was found that while the enhancement in the visible range was caused by spiked, light-trapping microstructures formed due to laser ablation, enhancement in the subbandgap range ($\lambda > 1100$ nm) was likely to be a result of S-related energy levels [23]. This was supported by post-PLM furnace annealing experiments, in which the enhanced absorption in the visible range remained largely stable while that in the NIR decreased drastically (see dashed lines in Fig. 8.4), an effect attributed to the thermal instability of S-related defect levels [20].

To decouple subbandgap effects from ablation-induced light-trapping effects, ion implantation, and ns-PLM were explored as means of producing optically flat hyperdoped Si. Several ensuing studies then showed that the incorporation of chalcogens (i.e., S, Se, and Te) in Si introduced intermediate states in the Si bandgap [11, 24, 25], consistent with predictions from density functional theory (DFT) [26] (Fig. 8.5A). As shown in Fig. 8.5B, all chalcogen-hyperdoped Si were shown to exhibit similarly high subbandgap absorption. A compilation of the optical and electrical properties of chalcogen-hyperdoped Si is presented elsewhere [27].

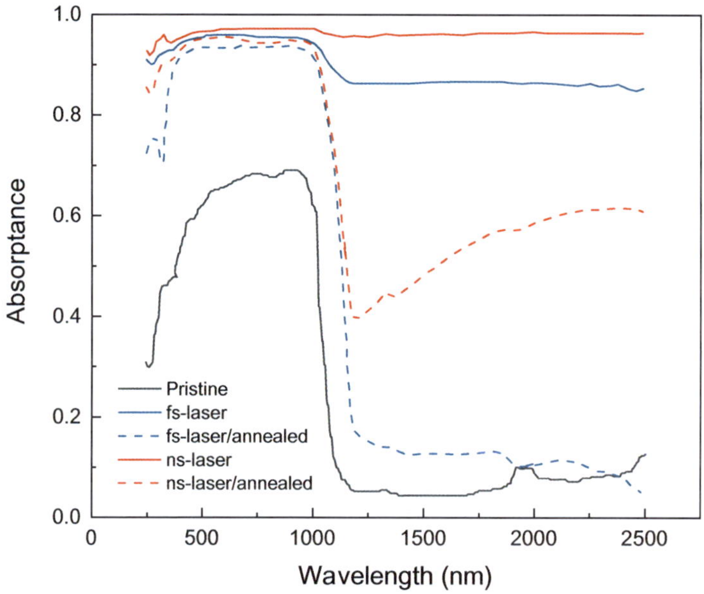


**FIG. 8.4** Absorption spectra for fs- and ns-laser processed samples in a $SF_6$ atmosphere. *Solid lines*: before subsequent furnace annealing and *dashed lines*: after subsequent furnace annealing at 875 K for 45 min in Ar. *(Modified from C. Wu, C.H. Crouch, L. Zhao, J.E. Carey, R. Younkin, J.A. Levinson, E. Mazur, R.M. Farrell, P. Gothoskar, A. Karger, Near-unity below-band-gap absorption by microstructured silicon, Appl. Phys. Lett. 78 (13) (2001) 1850–1852.)*

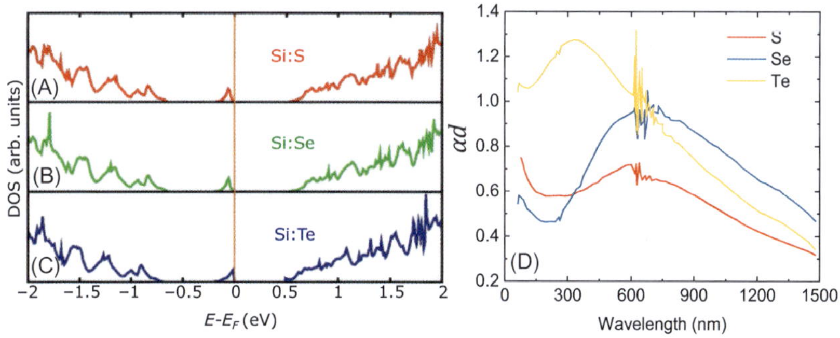


**FIG. 8.5** (A–C) Density of states diagrams for relaxed chalcogen (S, Se, and Te)-hyperdoped Si and (D) subbandgap optical absorptivity in S, Te, and Se-hyperdoped Si. *((A–C) Modified from K. Sánchez, I. Aguilera, P. Palacios, P. Wahnón, Assessment through first-principles calculations of an intermediate-band photovoltaic material based on Ti-implanted silicon: interstitial versus substitutional origin, Phys. Rev. B 79 (16) (2009) 165203; (D) modified from M. Tabbal, T. Kim, J.M. Warrender, M.J. Aziz, B.L. Cardozo, R.S. Goldman, Formation of single crystal sulfur supersaturated silicon based junctions by pulsed laser melting, J. Vac. Sci. Technol. B Microelectron. Nanometer Struct. Process. Meas. Phenom. 25 (6) (2007) 1847–1852.)*

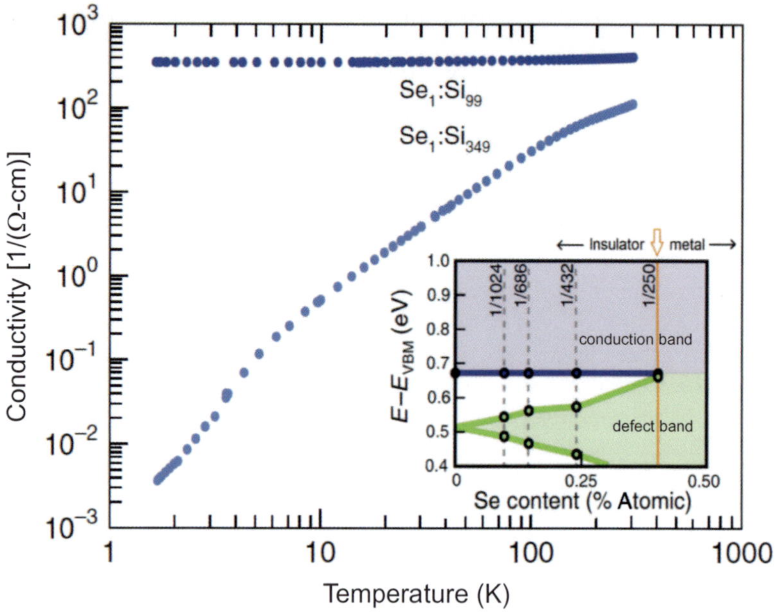


**FIG. 8.6** Temperature-dependent conductivity of Se-hyperdoped Si for $Se_1$:$Si_{349}$ and $Se_1$:$Si_{99}$. The conductivity of the sample with higher Se concentration appears to be insensitive to temperature, indicating a metal-like behavior. *Inset* shows the proposed IMT process. *(Modified from E. Ertekin, M.T. Winkler, D. Recht, A.J. Said, M.J. Aziz, T. Buonassisi, J.C. Grossman, Insulator-to-metal transition in selenium-hyperdoped silicon: observation and origin, Phys. Rev. Lett. 108 (2) (2012) 026401.)*

While initial optical absorption measurements on chalcogen-hyperdoped Si have shown promise, the high subbandgap absorption did not translate into correspondingly high photoresponse [27, 28]. Later, temperature-dependent conductivity measurements and first principle studies both suggested that the lack of photoresponse in S and Se-hyperdoped Si at room temperature can be attributed to an IMT that occurs at high impurity concentrations [18, 29], as illustrated in Fig. 8.6. As a result of the IB widening with increasing Se concentration, the IB overlaps with the CB at Se concentrations above ~0.35 at.% (see Fig. 8.6, inset). Thus, Se hyperdoping has effectively narrowed the Si bandgap rather than creating an IB that is separate from the VB and the CB, and the observed enhancement in subbandgap absorption is predominantly due to free carriers in the merged IB-CB and not from interband absorption.

In an attempt to avoid the undesirable effects of IMT, Sánchez et al. [26] proposed that the gap between the Fermi level and the CB edges could be further separated by codoping chalcogen-hyperdoped Si (in which the chalcogen behaves as an donor) with an acceptor-type impurity, and that the degree of compensation, which determines the exact location of the Fermi level, can be accurately controlled by the donor-to-acceptor ratio. This concept has been

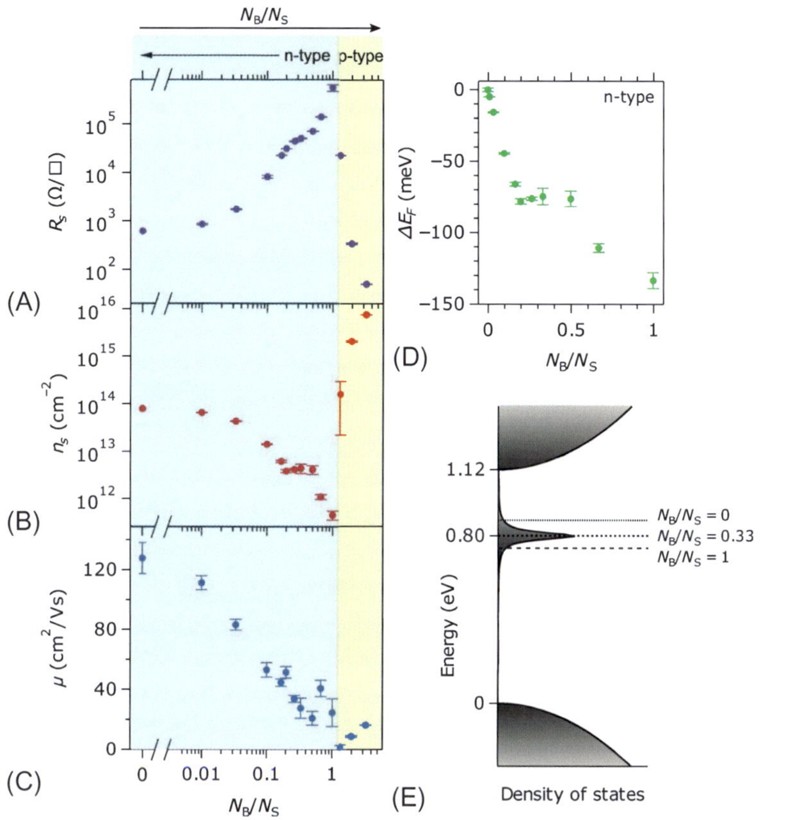


**FIG. 8.7** Electronic properties of S-hyperdoped Si codoped with B at various B-content $N_B$ to S-content $N_S$ ratios: (A) Sheet resistance $R_S$ determined from van der Pauw measurements; (B) sheet carrier concentration $n_S$ determined from Hall measurements; (C) carrier mobility $\mu$ calculated from (A) and (B); (D) shift in Fermi level $E_F$ calculated from (B); and (E) a schematic density of states diagram depicting the location of $E_F$ with respect to the IB [30].

later demonstrated experimentally [30]. As can be observed in Fig. 8.7A, the sheet resistivity of B codoped, S-hyperdoped Si samples increased as the B concentration increased, clearly demonstrating a compensation effect. Meanwhile, the sheet carrier density decreased as a result of the reduction in free carriers (Fig. 8.7B). It was demonstrated that codoping S-hyperdoped Si with B at a ratio of 1:1 increased the photoresponse by three orders of magnitude [30].

Although the idea of codoping achieved initial success, further studies and advances have been relatively limited to date. To this end, another more widely pursued method for achieving IB-induced subbandgap absorption in Si is to utilize alternative impurities that lie deeper still in the Si bandgap, such that the broadening of the IB does not overlap with the CB/VB edges. The following section discusses this body of work, most of which is focused on transition metal impurities.

## 8.3 Implantation and pulsed laser melting of transition metals in Si

Prior to the relatively recent developments in optical hyperdoping, studies on incorporating transition metals into Si were limited except for isolated studies on Au and Pt supersaturated Si in the 1980s [31, 32]. With this as a premise, Recht et al. [8] systematically explored the potential of hyperdoping Si with a range of transition metals, namely Au, Co, Cr, Cu, Fe, W, Zn, Pd, and Pt, using ion implantation followed by ns-PLM. Unique challenges arose from this study. First, unlike chalcogen impurities, transition metals were found to have very high diffusive velocities in molten Si and often have extremely low solid solubilities in Si. A direct consequence of this, as can be readily seen from the continuous growth model (Chapter 3), is that very high interface velocities are needed to trap a high concentration of rapidly diffusing impurities inside the regrown solid.

### 8.3.1 Titanium

Ti is the first transition metal studied for Si optical hyperdoping. The formation of an IB and subsequent subbandgap absorption in Ti-hyperdoped Si were first reported by Refs. [33–35]. In particular, high concentrations of Ti ($> 6 \times 10^{19}\ cm^{-3}$) were observed to be trapped within the PLM layer (as measured by SIMS). It was later suggested that these observations can be associated with the formation of an IB from deep levels introduced by *interstitial* Ti in Si ($E_C -$ 0.22 eV, $E_C - 0.30$ eV, and $E_V + 0.29$ eV in the dilute limit) [36–39]. A heterojunction solar cell and an NIR photodetector have since been demonstrated with such material, albeit with low efficiencies [40, 41].

Notwithstanding these advances, it is not unanimous in the literature whether a Ti-induced IB has truly been achieved in these materials. García-Hemme et al. [41, 42] and Mathews et al. [43] have independently shown that the supposedly "Ti-hyperdoped" Si layer was highly defective, as evidenced by TEM images showing stacking faults (specific to $\langle 111 \rangle$ Si) and filamentary breakdown, a phenomenon in which the impurities segregate into filament-like regions in hyperdoped Si. Representative TEM images showing extended defects within the resolidified layer are shown in Fig. 8.8. More recently, Akey et al. [44] have shown through three-dimensional (3D) atom probe tomography (APT) that for samples that have undergone filamentary breakdown, nearly all of the Ti atoms are confined to within filament structures. Such observations have led Mathews et al. [43] and Akey et al. [44] to conclude that the Ti concentration needed for IB formation cannot be achieved in crystalline regrowth regimes even in $\langle 111 \rangle$ Si. Note that although $\langle 111 \rangle$ Si is known to have a lower tolerance to amorphization compared with the $\langle 100 \rangle$ orientation, the $\langle 111 \rangle$ planes have the slowest crystallization speed compared with other orientations and are thus expected to trap the most solute [45]. Since successful hyperdoping

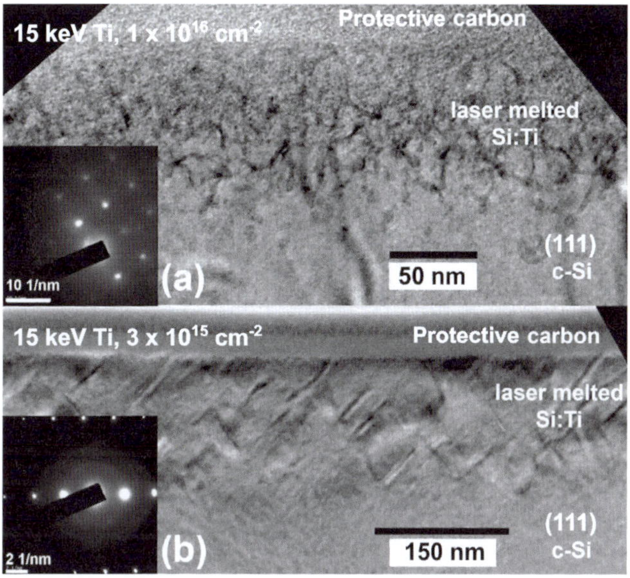


**FIG. 8.8** XTEM micrographs of the highly defective "Ti-hyperdoped" Si (111). Samples were implanted at 15 keV (A) $1 \times 10^{16}$ $cm^{-2}$ and; (B) $3 \times 10^{15}$ $cm^{-2}$, both receiving a PLM fluence of 0.75 J $cm^{-2}$. Stacking faults are apparent in (B), while a much more complicated network of extended defects is visible in (A). *(Modified from J. Mathews, A.J. Akey, D. Recht, G. Malladi, H. Efstathiadis, M.J. Aziz, J.M. Warrender, On the limits to Ti incorporation into Si using pulsed laser melting, Appl. Phys. Lett. 104 (11) (2014) 112102.)*

has not been conclusively demonstrated, these materials will be referred to as "Si-Ti" instead of "Ti-hyperdoped Si" from here on.

Similar conclusions have also been drawn by Liu et al. [46] when comparing Si-Ti samples fabricated by implantation followed by either PLM or FLA. Furthermore, based on conductivity mapping measurements, Liu et al. [15] concluded that the metallic-like electrical behavior observed in samples containing high Ti concentrations is most likely due to percolation conduction rather than the formation of a Ti-induced IB. In addition, atom location measurements by Liu et al. [46] (shown in Fig. 8.9) and Lim et al. [47] suggest that the majority of Ti atoms in the Si-Ti layer reside in random lattice positions (i.e., they are neither preferentially substitutional nor interstitial in the Si lattice). This is especially important since the formation of a Ti-induced IB requires high concentrations of *interstitial* Ti to be incorporated in Si [36, 39, 48]. Thus, given the highly defective nature of the Ti-rich filamentary layer, the subbandgap absorption and photoresponse observed previously most likely originated from (predominantly) defects rather than Ti-hyperdoping. This is consistent with optical absorption measurements performed on similarly defective and filament-rich Si-Au samples [9, 49], as will be discussed in the following section.

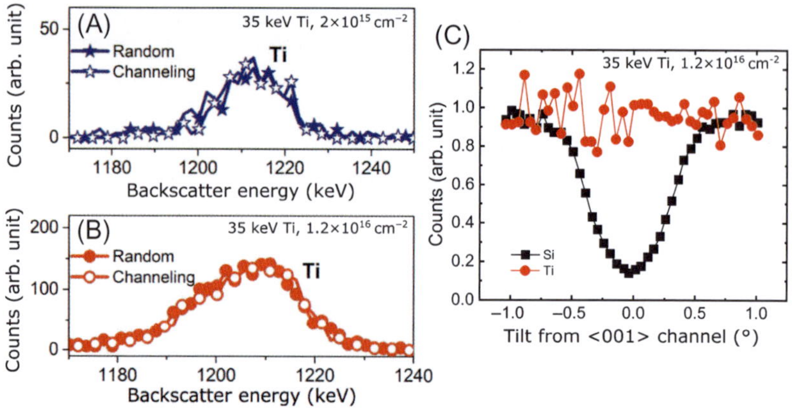


**FIG. 8.9** (A, B) Partial spectra of the 35 keV Ti-implanted and PLM regions measured from Rutherford backscattering spectrometry with channeling analysis. No channeling behavior was observed for Ti both at low ($2 \times 10^{15}$cm$^{-2}$) and high ($1.2 \times 10^{16}$cm$^{-2}$) Ti doses, indicating that the Ti atoms do not occupy substitutional lattice positions; (C) normalized angular scan profiles of Si and Ti from the $1.2 \times 10^{16}$cm$^{-2}$ dose sample showing that the Ti atoms are neither preferentially substitutional nor interstitial with respect to the Si lattice. *(Modified from F. Liu, M. Wang, Y. Berencén, S. Prucnal, M. Engler, R. Hübner, Y. Yuan, R. Heller, R. Böttger, L. Rebohle, et al., On the insulator-to-metal transition in titanium-implanted silicon, Sci. Rep. 8 (1) (2018) 1–8.)*

## 8.3.2 Gold

Several studies have demonstrated the ability of implantation and PLM at incorporating substantial quantities of substitutional Au in Si ($>6 \times 10^{19}$cm$^{-3}$) [8, 10, 31, 49]. In fact, the trapping of Au seems to require considerably less stringent hyperdoping conditions compared with other transition metals [8]. As a result, subbandgap photoresponse from Au-hyperdoped Si has already been demonstrated at room temperature with a relatively rudimentary device architecture [10]. As shown in Fig. 8.10, a photodetector based on Au-hyperdoped Si (Au dose = $10^{15}$cm$^{-2}$) exhibits an external quantum efficiency (EQE) of 0.01% under 1550 nm illumination, whereas the undoped reference sample exhibits no response. In a more recent study, a threefold improvement in the subbandgap absorption was achieved in samples with Au concentrations exceeding $10^{21}$cm$^{-3}$ (>1.4 at.%) [49]. It should be noted that Au-hyperdoped Si has also been made using Au-film evaporation followed by PLM [3], though the Au concentration attained is less than that achieved by ion implantation.

The enhanced subbandgap absorption in Au-hyperdoped Si was attributed to the formation of an IB from deep levels associated with substitutional Au in Si [48, 49]. More specifically, as depicted in Fig. 8.10C, it has been shown that the subbandgap absorption increased as a function of the substitutional Au dose and is largely independent of the implantation and PLM conditions. This is supported by DFT calculations [50] and deep-level transient spectroscopy (DLTS) measurements [48], both indicating that substitutional Au introduces mid-gap

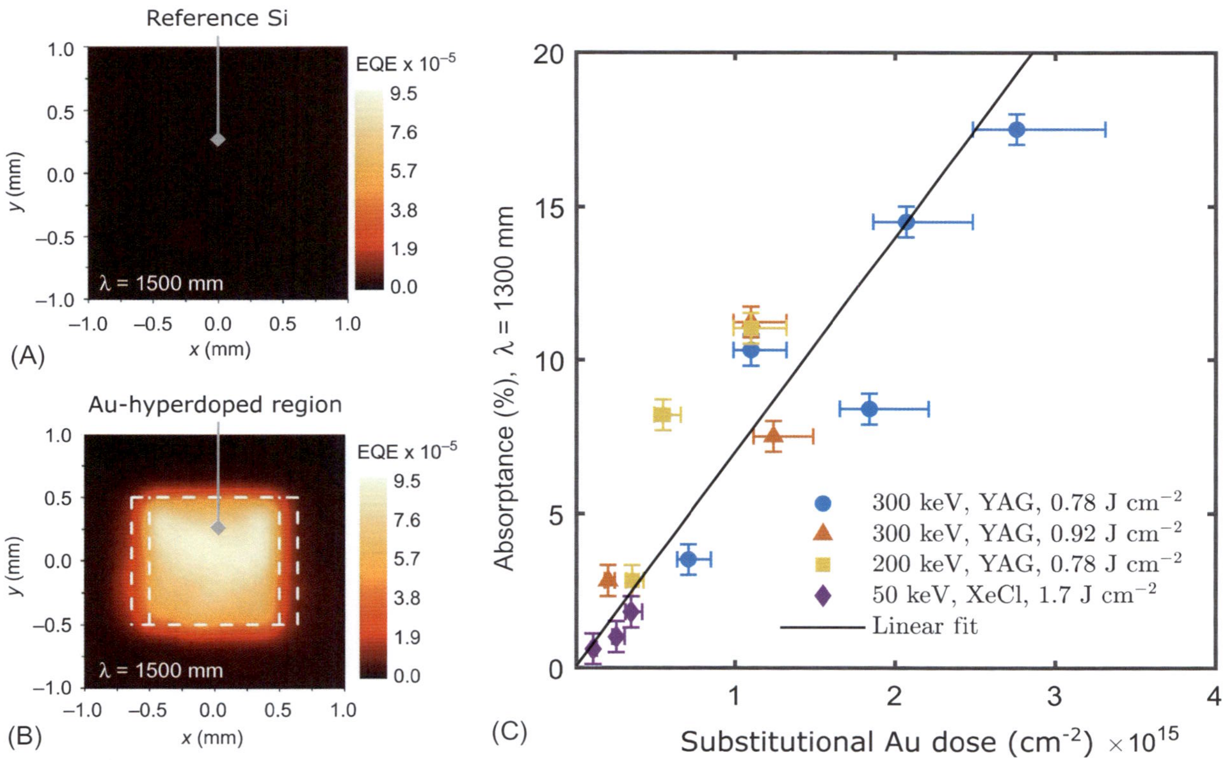


**FIG. 8.10** EQE map of pn-junction devices fabricated from (A) Au-hyperdoped Si; (B) B-doped Si as a reference; and (C) subbandgap absorption ($\lambda = 1300$ nm) in Au-hyperdoped Si as a function of substitutional Au dose. A linear fit is included to guide the eye. *((A and B) Modified from J.P. Mailoa, A.J. Akey, C.B. Simmons, D. Hutchinson, J. Mathews, J.T. Sullivan, D. Recht, M.T. Winkler, J.S. Williams, J.M. Warrender, et al., Room-temperature sub-band gap optoelectronic response of hyperdoped silicon, Nat. Commun. 5 (1) (2014) 1–8; (C) Modified from W. Yang, A.J. Akey, L.A. Smillie, J.P. Mailoa, B.C. Johnson, J.C. McCallum, D. Macdonald, T. Buonassisi, M.J. Aziz, J.S. Williams, Au-rich filamentary behavior and associated subband gap optical absorption in hyperdoped Si, Phys. Rev. Mater. 1 (2017) 074602.)*

states in Si and can give rise to subbandgap absorption in the NIR [50]. Interestingly, the majority of the Au mid-gap levels are associated with the partially filled $t_2$ dangling-bond states of Si in the presence of Au, and not the valance electronic states of the Au atoms themselves [50].

Although Au has proven to be a promising candidate for optical hyperdoping in Si, it has not yet been demonstrated whether the improved subbandgap absorption at higher Au concentrations than those in Mailoa et al. [10] can be fully translated into improvement in device efficiency. Several issues are pertinent here. For example, whereas lower implantation doses of Au result in near-perfect defect-free epitaxy following PLM as shown in Fig. 8.11A, it has been shown that the Au distribution becomes inhomogeneous, both in-plane and vertically, at the higher Au doses [49]. The latter is evident from TEM images shown in Fig. 8.11B and C. This behavior, widely referred to as "filamentary breakdown," is similar to that observed earlier in similarly prepared Si-Co samples [19], and more recently also in Si-Ti [44] and Si-Ag [47] samples (according to the APT measurements). In particular, filament-shaped, Au-rich regions were observed, in which the local Au concentration inside the filaments is estimated to be more than 3 at.%. It is important to emphasize that a key difference between filamentary and (conventional) cellular breakdown [45, 51, Chapter 5] is that the Au (and Co) atoms remain mostly randomly dissolved (i.e., not precipitated) in metal-rich filaments within the Si lattice, and can occupy substitutional lattice positions within the filaments. Furthermore, it was shown that the filamentary Si-Au samples exhibited further enhancements in NIR absorption compared with Au-hyperdoped samples that received a lower implantation dose (i.e., samples showing good crystal quality with no

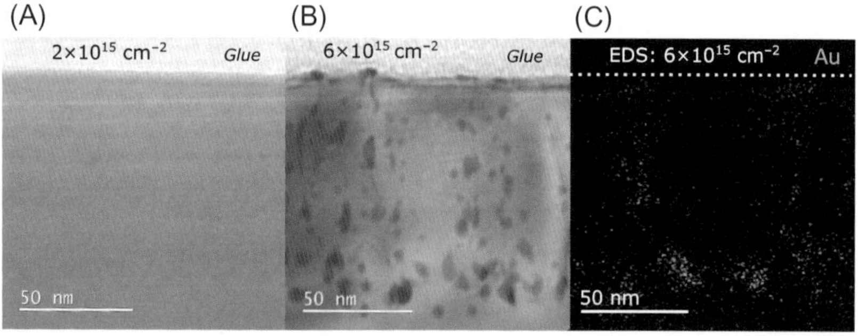


**FIG. 8.11** XTEM images of Au-hyperdoped Si (implanted at 300 keV and melted with Nd:YAG laser 0.78 J cm$^{-2}$) with a Au dose of (A) 2 × 10$^{15}$cm$^{-2}$; (B) 6 × 10$^{15}$cm$^{-2}$; and (C) energy-dispersive X-ray spectroscopy (EDS) map taken on the same sample as (B) but in a different region, in which signal from Au is shown in *green* and the *scale bar* is the same as that in (B). The *dashed line* indicates the top surface of the sample. *(Modified from W. Yang, A.J. Akey, L.A. Smillie, J.P. Mailoa, B.C. Johnson, J.C. McCallum, D. Macdonald, T. Buonassisi, M.J. Aziz, J.S. Williams, Au-rich filamentary behavior and associated subband gap optical absorption in hyperdoped Si, Phys. Rev. Mater. 1 (2017) 074602.)*

filamentary breakdown) [49]. Nevertheless, the effect of Au-rich filaments on the device performance requires further investigation, especially as this behavior could be a more general behavior in metal-rich filamentary Si, as will be discussed in the following section. In addition, due to the extremely nonequilibrium nature of the hyperdoping process, a range of Au and Au-related atomic configurations, such as the Au-Au dimer, the Au-V complex, and isolated vacancies are expected to be present in Au-hyperdoped Si (filamentary or not) at varying concentrations [52]. These thermally metastable atomic configurations are each associated with different energy levels in the Si bandgap and may affect the optical properties of the material [53]. For example, they can lower the expected level of subbandgap absorption from substitutional Au. These Au-related defects in the Au-hyperdoped layer are in addition to implantation and PLM-related defects (so-called process-induced defects), which are discussed in Section 8.4.

### 8.3.3 Other transition metal species

Compared with Ti and Au, studies on Si hyperdoped with other transition metals are much more limited. Enhanced NIR photoconductance has been reported in similarly prepared Si-V and Si-Co samples [38, 54], although efficient photodetector performance has not yet been demonstrated. On the other hand, implantation and PLM experiments have been carried out on Si-Ag samples in both ns [47] and fs [46] regimes. Table 8.1 provides a summary of the relevant literature on these different Si-metal materials fabricated from implantation followed by ns-PLM. Note that chalcogen-hyperdoped Si is also included for completeness, while prior works involving fs-PLM are not included due to additional complications that arise from microstructures introduced to the Si surface by fs-PLM.

First, from Table 8.1, it appears that not all transition metals are suitable for optical hyperdoping. As discussed earlier, atom location experiments have indicated that impurities such as Ti and Ag do not appear to have a preferential lattice configuration in Si after PLM [46, 47]. This is important because only the interstitial Ti and substitutional Ag/substitutional-interstitial Ag pairs are expected to introduce deep levels in the Si bandgap [39, 48, 55]. Furthermore, it is clear from Table 8.1 that cellular/filamentary breakdown occurs in most transition metal-hyperdoped Si at high impurity concentrations [8, 19, 42–44, 47, 49]. As a result, although it is possible to attain very high *average* concentrations of metal atoms ($>5 \times 10^{19} cm^{-2}$) in the recrystallized Si layer, APT suggests that effectively all of the metal atoms are confined within the filament structures [19, 44, 56]. It should be noted that, based mainly on APT studies, filamentary breakdown has been used to describe the defective filament-like structures observed recently following high-dose implantation and PLM, which could contain *dissolved* impurities. In contrast, much of the previous literature has referred to cellular breakdown [45, 57], which describes a near-equilibrium

**TABLE 8.1 Summary of impurities used for Si optical hyperdoping fabricated from implantation followed by ns-PLM.**

| Impurity | PLM conditions ($\lambda$, FWHM) | Comments and issues | Demonstrated detector | References |
|---|---|---|---|---|
| S | XeCl excimer (308 nm, 25 ns) 1.7–1.8 J cm$^{-2}$ | IMT; S is highly substitutional; Fermi level can be controlled by codoping | Yes | [18, 23, 28, 30, 58] |
| Se | XeCl excimer (308 nm, 25 ns) 1.7–2.3 J cm$^{-2}$ | IMT; Se is highly substitutional | No | [11, 24, 29] |
| Te | XeCl excimer (308 nm, 25–28 ns) 1.2–2.3 J cm$^{-2}$ | IMT; Te is highly substitutional | No | [11, 104, 105] |
| Ti | KrF excimer (248 nm, 16–20 ns) 1.0–1.8 J cm$^{-2}$ Nd:YAG (355 nm, 6 ns) 0.5–1.4 J cm$^{-2}$ | Strong Ti surface segregation[a]; Most Ti neither preferentially substitutional nor interstitial; Filamentary breakdown[a]; Percolation conduction; Ti confined inside filaments | Yes | [34, 35, 37, 41–44, 46, 47] |
| Au | Nd:YAG (355 nm, 5–6 ns) 0.7–1.0 J cm$^{-2}$ XeCl excimer (308 nm, 25 ns) 1.7 J cm$^{-2}$ | Less significant Au surface segregation[a]; Au is highly substitutional; Filamentary breakdown[a] (though Au is still highly substitutional); Evidence of vacancy trapping | Yes | [8–10, 49, 52, 53, 56 [60]] |

**TABLE 8.1** Summary of impurities used for Si optical hyperdoping fabricated from implantation followed by ns-PLM—cont'd

| Impurity | PLM conditions (λ, FWHM) | Comments and issues | Demonstrated detector | References |
|---|---|---|---|---|
| V | KrF excimer (248 nm) 1.0 J $cm^{-2}$ | Strong V surface segregation[a] suspected; No atom location data; Filamentary breakdown suspected[a]; Ti-like "metallic" conductivity | No | [54] |
| Co | Nd:YAG (512 nm, 15 ns) 0.7 J $cm^{-2}$ XeCl excimer (308 nm, 25 ns) 1.7 J $cm^{-2}$ | Strong Co surface segregation[a]; No atom location data; Filamentary breakdown[a]; Ti-like "metallic" conductivity; Co confined to within filaments | No | [8, 19, 38] |
| Ag | Nd:YAG (355 nm, 6 ns) 0.5–1.4 J $cm^{-2}$ | Strong Ag surface segregation[a]; Most Ag neither preferentially substitutional nor interstitial; Filamentary breakdown[a]; Ag confined to within filaments | No | [47] |

[a]*Cellular/filamentary breakdown occurs when the impurity concentration in the melt is sufficiently high (usually from increasing the impurity species implantation dose). At lower impurity concentrations, either impurity trapping (S, Se, Te, Au) or near-complete impurity surface segregation (Ti, V, Co, Ag) occurs.*

columnar defect structure with *precipitated* impurities (As, Bi, Ga, In, and Sb). Nevertheless, both regimes, filamentary and cellular, limit the concentration of impurity that can be uniformly trapped in good quality crystalline Si.

It is perhaps tempting to try circumvent the issue of filamentary/cellular breakdown by avoiding high metal concentrations (e.g., by having lower implantation doses, or by increasing the melt duration by using longer wavelength laser pulses). However, other than Au, most transition metals studied thus far have shown strong segregation toward the surface (Ti, Co, Ag, Zn, Cr, Cu, Fe, Mn, W, Mg, and Yb) at lower impurity concentrations (see references in Table 8.1 and [45, Chapter 5]). In cases where strong impurity surface segregation was reported, the soluble metal concentration below the surface is almost always insufficient for forming an IB ($\sim 5 \times 10^{19} cm^{-2}$) [8, 19, 44, 47]. In fact, Au appears to be the only metal that has been unequivocally shown to hyperdope Si to sufficiently high concentrations above $10^{20}$ $cm^{-3}$ (where an IB is expected), while simultaneously retaining good crystallinity following PLM. It is interesting to note that, even in the filamentary breakdown regime, Au remains largely substitutional both in regions *with* and without the filaments [49]. Unfortunately, it remains unclear why Au can effectively hyperdope Si, while other metals have been less successful, as the hyperdoping process involves an interplay between a large number of unknown parameters, such as metal diffusivities in the liquid and at the interface, the partition coefficient (related to ratio between the solubility of the impurity in liquid phase and solid phase), and intrinsic defects which can play a role on the diffusion mechanism of the impurity [8]. Thus, the only way to find other suitable metal impurities for hyperdoping in Si, at present, is through further experimentation.

Finally, due to the highly nonequilibrium nature of hyperdoping, it is important to note that most hyperdoped Si materials are thermally metastable, placing further constraints on subsequent device processing. As discussed in the prior sections, deterioration in the NIR absorption was observed in both S- and Au-hyperdoped Si following low- to moderate-temperature annealing (between 400 and 600°C), most likely due to S and Au atoms coming off substitutional lattice positions [53, 58].

## 8.4 Recent electrical and defect measurements

As mentioned in the previous section, Au is the only metal impurity that has consistently delivered promising results for Si-based optical hyperdoping to date and thus some of the more recent developments in photodetector optimization were focused on Au-hyperdoped Si. However, the current major setback for Au-hyperdoped Si NIR detectors is its very low photoresponse (around $10^{-4}$ A $W^{-1}$) and consequently low EQE (<0.01%) [10, 56, 59]. Indeed, it was pointed out by Lim et al. [56] who compared Au-hyperdoped

Si to two control samples that there is an apparent discrepancy between the optical absorption and photocurrent properties of Au-hyperdoped Si. These results are shown as Figs. 8.12 and 8.13, respectively. The corresponding sample fabrication conditions are summarized in Table 8.2. In particular, the process control sample was designed to isolate photo-electrical effects originating from implantation and PLM-induced defects (collectively known as the process-induced defects) from those induced by Au-hyperdoping.

In Fig. 8.12, it is clear that there is an enhanced absorption of below band-edge NIR photons (<1.1 eV) in the Au-hyperdoped Si sample, but not in the process control sample. On the other hand, in Fig. 8.13, the process control

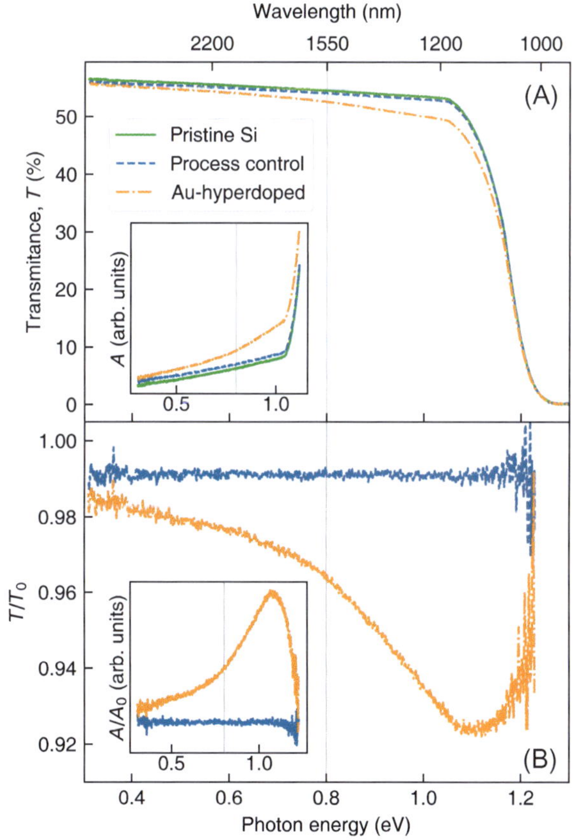


**FIG. 8.12** (A) Transmittance spectra ($T$) for the Au-hyperdoped Si, process control, and pristine Si samples; and (B) spectra in (A) normalized against the pristine Si reference spectra ($T_0$). *Insets* are the calculated absorptance ($A$) and normalized absorptance ($A/A_0$) spectra of the same samples. The *faint vertical line* indicates 1550 nm. *(Modified from S.Q. Lim, C.T.-K. Lew, P.K. Chow, J.M. Warrender, J.S. Williams, B.C. Johnson, Toward understanding and optimizing Au-hyperdoped Si infrared photodetectors, APL Mater. 8 (6) (2020) 061109.)*

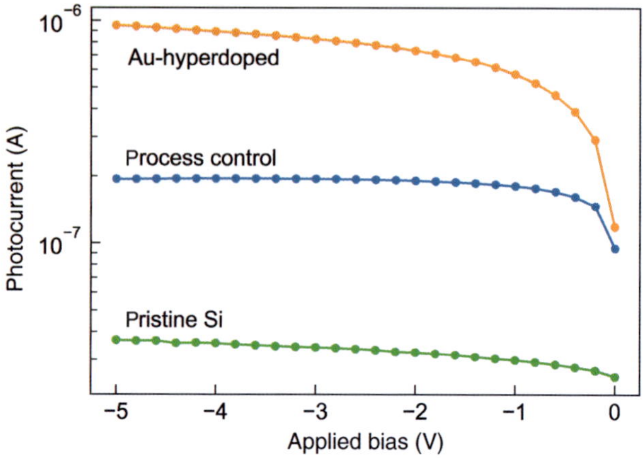


**FIG. 8.13** Photocurrent under 1550 nm laser illumination ($\Delta I$) as a function of reverse bias for photodetectors fabricated from Au-hyperdoped Si and two other control diodes. *(Modified from S.Q. Lim, C.T.-K. Lew, P.K. Chow, J.M. Warrender, J.S. Williams, B.C. Johnson, Toward understanding and optimizing Au-hyperdoped Si infrared photodetectors, APL Mater. 8 (6) (2020) 061109.)*

**TABLE 8.2 Implantation and PLM [Nd:YAG (308 nm, 6 ns)] conditions used by Lim et al. [56].**

| Sample | Implant conditions | PLM fluence |
|---|---|---|
| Au-hyperdoped Si | 300 keV Au $10^{15}$ cm$^{-2}$ | 0.9 J cm$^{-2}$ |
| Process control | 70 keV Si $10^{15}$ cm$^{-2}$ followed by 300 keV Au $10^{10}$ cm$^{-2}$ | 0.9 J cm$^{-2}$ |

(i.e., not hyperdoped) diode was observed to produce a measurable NIR photocurrent despite its insignificant optical absorption (note that the higher than expected photocurrent in the pristine Si reference diode most likely originates from surface states as discussed in Lim et al. [56]). The observed NIR photocurrent in the process control diode is unsurprising, since previous DLTS measurements have confirmed high densities of process-induced defects ($\sim 10^{14}$ cm$^{-3}$) within the device depletion layer of the same process control diode [56, 60] (also see Section 8.2.1).

However, in Fig. 8.13, it is surprising that the NIR photocurrent of Au-hyperdoped Si is only five times larger than the process control diode. Given the trends observed in the optical measurements and the difference between the Au impurity and process-induced defect densities ($\sim 10^{20}$ and $10^{14}$ cm$^{-3}$, respectively), the NIR photocurrent in the Au-hyperdoped Si sample should have been several orders of magnitude larger. Thus, there must be a large proportion of NIR photocarriers (generated by Au-hyperdoping) that are not being efficiently collected. Indeed, the poor carrier transport properties of Au-hyperdoped Si are not so surprising if one considers the short carrier lifetimes associated with having extremely high concentrations of Au impurities in Si (estimated to be around 10 ps). However, there also appears to be an obvious shortfall in the diode structure of such Au-hyperdoped Si materials. Since Au-hyperdoping results in heavy p-type doping [56], in all prior experiments involving the fabrication and characterization of Au-hyperdoped Si diodes, a vertical $p^+n$-junction was used (i.e., the Si substrate is n-type). As a result, the depletion region (electric field layer) is placed entirely within the n-type Si substrate, with little to no overlap with the (NIR absorbing) Au-hyperdoped layer. In principle, if the device structure can be tailored to strategically place the depletion layer within the Au-hyperdoped layer (e.g., through lateral junctions or a MOS structure), this can increase the photocarrier collection efficiency due to improved carrier transport properties within the depletion layer (e.g., larger drift velocities). Thus, it is very likely that optimizing the device architecture will result in improved NIR photocurrent and ultimately, improved detector performance.

## 8.5 Ge and GeSn

Compared with studies of optical hyperdoping in Si, as outlined in the previous sections, similar studies in Ge are only just emerging. The bandgap of Ge is 0.67 eV and hence, it efficiently absorbs in the near-infrared (NIR) up to around 1880 nm. In contrast to optical hyperdoping in Si, the aim of hyperdoping of Ge is to extend efficient absorption into the mid-infrared region, up to and beyond 3000 nm. The contribution below first gives a brief introduction to the recent pioneering work on optical hyperdoping in Ge by a team of our collaborators, where some encouraging progress has been made. Furthermore, the team has also extended the approach of ion implantation and PLM to produce dilute GeSn alloys that desirably alter the Ge bandgap such that it becomes a direct bandgap semiconductor, opening up interesting possibilities for optical devices, including lasers and photodetectors operating beyond 4000 nm. Although these recent alloy studies are not specifically optical hyperdoping since the solubility of Sn in Ge only needs to be exceeded by up to two orders of magnitude to access a range of direct bandgap materials, the method used to produce them, and analyses used to characterize them, are essentially the same as for optical hyperdoping. Therefore, the initial studies on the formation and properties of GeSn alloys by ion implantation and PLM are also briefly treated in this section.

### 8.5.1 Background

Although it has been known for several decades that the introduction of deep-level impurities into Ge (at concentrations near the equilibrium solubility limit) can extend its optical absorption beyond the near- to mid-infrared region, short-wavelength-infrared photodetectors produced by this method have exhibited low efficiency and only produced subbandgap response at low temperatures [61–65]. It has been only recently that optical hyperdoping has been attempted in Ge [66]. In this case, Ge was hyperdoped with Au using the same implantation and PLM processing approach as was used for Si. A peak Au concentration of around $3 \times 10^{19} cm^{-3}$ was achieved, which is three orders of magnitude above the solid solubility limit. It was found that this material exhibited sub-bandgap absorption of light at wavelengths up to 3000 nm and at room temperature. Furthermore, the absorption coefficient was found to be comparable with commercial photodetectors operating in the short-wavelength-infrared range, suggesting much promise for low-cost, room-temperature, silicon-compatible, and short-wavelength-infrared photodetection [66].

Another approach to enhancing the optical properties of Ge in the infrared range explored over the past two decades has been directed toward the fabrication of dilute GeSn alloys, with Sn concentrations up the around 15 at.%. At certain Sn concentrations, theoretical calculations indicate that this alloy is a direct bandgap semiconductor [67, 69–71], and recent estimates indicate that the alloy can exhibit direct bandgap behavior at Sn concentrations above about 6 at.% [67]. Fig. 8.14A shows a simple band structure diagram that illustrates the indirect-to-direct bandgap transition when a sufficient concentration of Sn is alloyed with Ge. On the left-hand panel, the normal indirect transition in Ge is from the lowest point of the CB (the $L$ valley) to the top of the VB. The direct transition is from the $\Gamma$ valley to the VB but this is higher than the indirect transition in pure Ge. When Sn is added, as shown in the right-hand panel, the bottom of the $\Gamma$ valley is lowered with respect to the $L$ valley, and hence a direct bandgap transition can occur at sufficiently high Sn concentrations. However, there are two important issues in arriving at a direct bandgap alloy. The first issue is the effect of strain on the direct transition and this is illustrated in Fig. 8.14B, where the Sn mole fraction is plotted against the biaxial strain [67]. The contour lines depict the energy difference in eV between the $L$ and the $\Gamma$ conduction band minima. Note that the color code merely indicates the direction of a red shift or blue shift for different combinations of Sn content and biaxial strain. Hence, negative values on the contour lines correspond to indirect bandgap, while positive values correspond to direct bandgap.

Whereas tensile strain can even result in a direct bandgap transition in pure Ge [72, 73], compressive strain will increase the Sn concentration required for a direct transition. Since compressive strain is usually a result of pseudomorphic growth of GeSn on both Ge and Si substrates, the larger lattice parameter of GeSn compared with that of these substrates can cause issues for fabricating

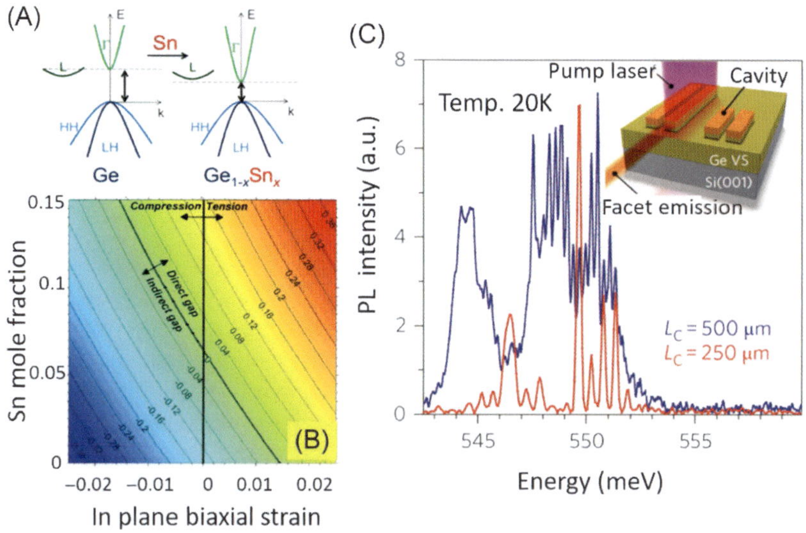


**FIG. 8.14** (A) Simplified band structure of indirect bandgap Ge (*left*) and direct bandgap GeSn alloy (*right*); (B) Sn mole fraction in GeSn versus in plane biaxial strain showing indirect and direct gap regions, with the contour lines depicting the energy difference (in eV) between the $L$ and the $\Gamma$ conduction band minima, and the colored regions indicating the direction of a blue and a red shift; and (C) power-dependent photoluminescence spectra at two different waveguide lengths $L_c$ taken at a power of 500 kW cm$^{-2}$, illustrating direct bandgap GeSn behavior; the inset shows the device structure where a pump laser excites the GeSn waveguide cavity and the light is emitted from one of the etched facets. *((A and B) Modified from S. Gupta, B. Magyari-Köpe, Y. Nishi, K.C. Saraswat, Achieving direct band gap in germanium through integration of Sn alloying and external strain, J. Appl. Phys. 113 (7) (2013) 073707; (C) modified from S. Wirths, R. Geiger, N. von den Driesch, G. Mussler, T. Stoica, S. Mantl, Z. Ikonic, M. Luysberg, S. Chiussi, J.-M. Hartmann, et al., Lasing in direct-bandgap GeSn alloy grown on Si, Nat. Photon. 9 (2) (2015) 88–92.)*

devices, as we discuss here. A second issue is that the equilibrium solubility of Sn in Ge is around 0.5 at.% at room temperature and hence nonequilibrium methods must be used to facilitate an alloy without Sn precipitation [74].

Prior to hyperdoping, GeSn alloys have been demonstrated using various nonequilibrium techniques, such as molecular beam epitaxy [75–77], solid-phase epitaxy [78], sputter deposition [79], and chemical vapor deposition (CVD) [80–82]. The first direct bandgap photoluminescence (PL) at room temperature from a GeSn alloy was reported by Mathews et al. [83] and light amplification and stimulated emission from the resulting materials has been demonstrated [84]. Furthermore, Wirths et al. [68] published the first lasing results from a GeSn alloy in 2015 as shown in Fig. 8.14C. The lasing experiment was carried out at 20 K with a CVD-grown, 560 nm thick partially relaxed GeSn layer on Ge, with a Sn concentration of around 12 at.%. Lasing, with a somewhat high threshold, was observed up to 90 K. Improved laser performance has been reported subsequently, such as by Margetis et al. [85], in which lasing was observed at higher temperatures than liquid nitrogen but not at room temperature. Both compressive strain and defects in the epitaxially grown alloy were considered the main limitations affecting laser performance. These results demonstrate that the GeSn alloy is progressing rapidly for integrated photonics applications but more work is needed to improve material quality. Here, we summarize the recent progress in fabrication of dilute GeSn alloys by ion implantation and PLM.

### 8.5.2 Ion beam synthesis of GeSn alloys

Ion implantation followed by PLM and subsequent solidification is a promising approach for producing GeSn alloys because the process is highly controllable, precursor free, and hence virtually free of contamination. This has been the approach of our collaborating team, where we have recently undertaken a pioneering study of GeSn alloys by ion implantation and PLM. A schematic diagram showing the concept is given in Fig. 8.15A and involves implantation with Sn ions that produce an amorphous GeSn alloy. PLM is then carried out in the nanosecond pulse length regime to rapidly melt into the underlying Ge substrate, followed by rapid epitaxial solidification. The earliest reports of using ion implantation followed by PLM to produce high Sn content GeSn alloys were not successful. For example, Bhatia et al. [86] showed that room temperature high-dose Sn implants into Ge, followed by PLM gave poor crystal quality and resulted in subsurface pores, whereas Gao et al. [87], although showing somewhat better material quality, reported that the material contained a columnar structure and the Sn concentration was too low for a direct bandgap transition. In fact, porosity induced in Ge by high-dose implantation is the reason for the low crystal quality in these early studies. Porosity caused by heavy ion

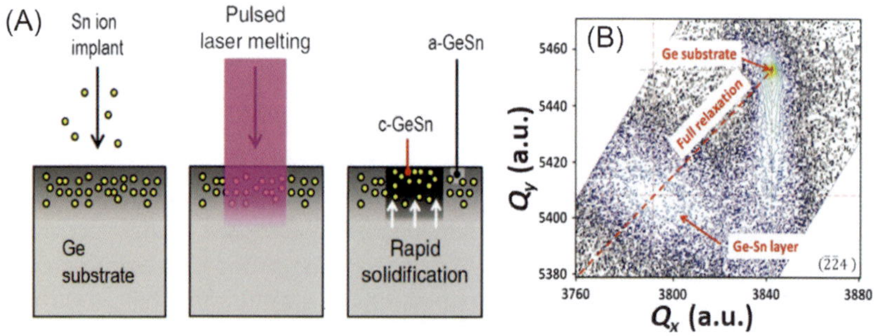


**FIG. 8.15** (A) Schematic illustrating the implantation and PLM process to produce dilute GeSn alloys and (B) XRD reciprocal space map for a GeSn film with 5.1 at.% Sn after PLM, showing that the layer is essentially fully relaxed. $Q_x$ and $Q_y$ are the reciprocal lattice vectors corresponding to $x$- and $y$-directions. *(Modified from T.T. Tran, Q. Hudspeth, Y. Liu, L.A. Smillie, B. Wang, R.A. Bruce, J. Mathews, J.M. Warrender, J.S. Williams, Ion beam synthesis and photoluminescence study of supersaturated fully-relaxed Ge-Sn alloys, Mater. Sci. Eng. B 262 (2020) 114702.)*

bombardment of Ge is well known and has been a subject of study for several decades [88–91]. It preferentially occurs during room temperature ion implantation and leads to surface craters and a subsurface columnar structure in ion-amorphized Ge. However, in most cases, undertaking the implantation into Ge at liquid nitrogen temperatures produces pore-free amorphous Ge [90, 91]. Using low-temperature implantation, Tran et al. [92] produced a reasonably good quality alloy up to about 6 at.% Sn following Sn ion implantation and PLM [92]. However, the layers were thin and compressively strained such that no direct bandgap was achieved. It was also found that above 6 at.% Sn, even during liquid nitrogen temperature Sn implantation, the amorphous-Ge turned porous [92], but porosity could be eliminated by capping of the Ge prior to implantation, enabling GeSn alloys up to 15 at.% Sn [93]. These initial studies produced alloys, which were around 100 nm thick and hence, although the alloy films were of high crystal quality, they were compressively strained (i.e., with a lattice parameter parallel to the surface that was the same as that of the underlying substrate), with lattice expansion normal to the surface [93].

In the subsequent studies, thicker GeSn alloys up to 300 nm thick were produced by using a higher implant energy (of 350 keV) into Ge followed by PLM, and such layers were observed to be near-fully relaxed, as illustrated by the X-ray reciprocal space map shown in Fig. 8.15B [94]. Such a map is for a 5.1 Sn at.% GeSn alloy film, showing the in-plane and out-of-plane strain. The full relaxation axis is shown on the map and corresponds to the line on which the in-plane and out-of-plane lattice constants ($a$ and $c$, respectively) have the same values. The X-ray mapping result, therefore, indicates that the GeSn layer is essentially fully relaxed since its peak is situated on the axis of full relaxation. When the lattice constants of the Ge and GeSn films were calculated from the X-ray data, they were consistent with existing data in the literature [95, 96] for fully relaxed films of the same composition. Note that it was necessary to introduce a bowing constant in a modified Vigard's law according to Beeler et al. [97] to obtain the dependence of lattice parameter on composition [94].

The quality of the relaxed GeSn films was analyzed by XTEM for the various compositions and showed threading-like defects throughout the films, as illustrated in Fig. 8.16A for a 4.5 Sn at.% film [94]. Besides occasional threading-type defects, the GeSn film is shown to be of high quality. Each threading defect is terminated on the surface with a small extrusion of the alloy out from the surface. The origin of these defects is not entirely clear. They are not equilibrium defects such as misfit and threading dislocations associated with alloys fully relaxed under near-equilibrium elevated temperature conditions [98, 99]. Tran et al. [94] suggested that ultrarapid solidification precludes the formation of equilibrium defects upon relaxation and proposed that the threading-like defects form via a stress-induced punching out of slip-like defects close to room temperature following solidification.

Finally, Tran et al. [94] measured the PL spectra from relaxed GeSn films as shown in Fig. 8.16B. The Ge substrate exhibits an intense indirect peak at

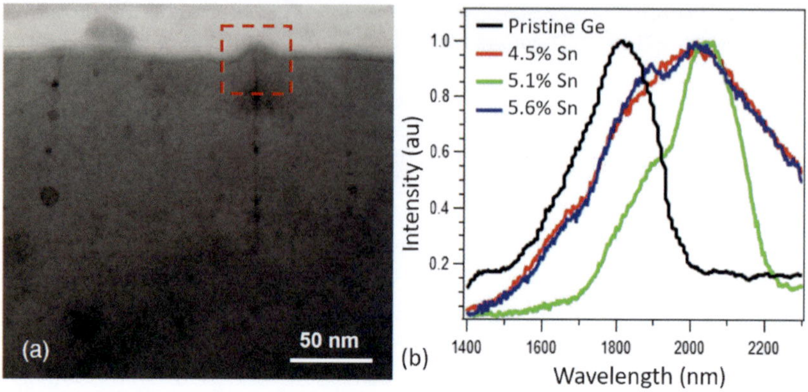


**FIG. 8.16** (A) XTEM bright field micrograph on ⟨100⟩ zone axis showing the morphology of a 4.5 Sn at.% GeSn film on Ge after PLM, where it is good single crystal with occasional threading-like defects extending out to the surface and surface humps (*red square*) extending along the defect lines and (B) PL spectra from Ge and various composition GeSn alloy layers. *(Modified from T.T. Tran, Q. Hudspeth, Y. Liu, L.A. Smillie, B. Wang, R.A. Bruce, J. Mathews, J.M. Warrender, J.S. Williams, Ion beam synthesis and photoluminescence study of supersaturated fully-relaxed Ge-Sn alloys, Mater. Sci. Eng. B 262 (2020) 114702.)*

1880 nm and a less intense direct peak at 1450 nm. All relaxed GeSn films show longer emission wavelength compared to Ge, as might be expected with the incorporation of Sn into the Ge lattice, noting that GeSn films typically exhibit PL spectra dominated by peaks resulting from the direct bandgap emissions [83, 100]. Using the relationship between composition and bandgap energy (emission wavelength) given by Gallagher et al. [101], as well as the values for bowing parameter by Jiang et al. [102], the peak positions and shoulders in Fig. 8.16B were explained by Tran et al. [94] in terms of the expected direct and indirect bandgaps of the various GeSn alloys. The fact that the spectra are skewed toward lower wavelengths was interpreted in terms of several competing effects: overlap with the indirect bandgap of pure Ge from the underlying substrate, splitting of the valance band due to residual strain in the material [103] (such as the shoulder at 1900 nm for the 5.1 at.% film), and nonuniform Sn concentration in the alloy. Overall, the PL results, although encouraging for ion implanted and PLM processed material, highlight the need to minimize or remove threading defects from strain-relaxed material, as well as producing more uniform Sn concentration, such as using multiple-energy Sn implants. The former issue of minimizing threading defects could be addressed by a second PLM treatment on the upper part of the relaxed film, as suggested by Tran et al. [94].

To conclude this section, it has been shown that, in the recent work of our collaborating team, ion implantation followed by PLM processing is an attractive alternative to CVD as a method for preparing good quality GeSn alloys grown on Ge substrates. Finally, although the initial attempts at optical hyperdoping of Ge with Au, and the formation of GeSn by ion implantation and PLM, are encouraging, the work is at an early stage and there remain many challenges to be overcome before efficient, commercially viable infrared photodetectors and lasers operating at room temperatures are realized.

## 8.6 Conclusion

In the last two decades, optical hyperdoping has emerged as an effective technique for achieving novel optoelectronic properties in Si and Ge, especially for enabling the detection of near- to mid-infrared light. To obtain the level of impurity supersaturation necessary for hyperdoping, very high concentrations of impurities (usually orders of magnitude beyond the equilibrium solubility) are needed, and this can be achieved by ion implantation followed by ns-PLM. In principle, a high concentration of impurities, which give rise to deep-level defect states, results in an IB within the Si or Ge bandgap. This configuration facilitates the absorption of subbandgap photons, thus extending the absorption of light into the near- to mid-infrared region, enabling infrared photodetection. Although implantation-related defects themselves have also been shown to give rise to subbandgap photoresponse, chalcogens and transition metals are the two main classes of impurities that have been explored for hyperdoping to provide a much more controllable and effective means of obtaining infrared photodetection. To this end, chalcogen-hyperdoped Si shows very high subbandgap absorption, yet an IMT at high impurity concentrations has limited its application for NIR photodetection.

On the other hand, hyperdoping Si with transition metals introduces states that lie deeper in the bandgap than chalcogens, and is thus less affected by the IMT. The challenge here has been to achieve hyperdoping (i.e., a sufficient level of supersaturation) that can give rise to efficient absorption in the infrared without compromising the ability to efficiently collect photo-induced carriers (i.e., required for an efficient photodetector). Au-hyperdoped Si has been shown to give rise to efficient NIR absorption and has led to the first hyperdoped Si NIR photodetector. However, current photodetectors fabricated from hyperdoped-Si have limited efficiency as a result of short carrier lifetimes and issues associated with carrier extraction in current device architectures. A further issue is the introduction of crystalline defects and inhomogeneous impurity-doped regions during PLM for ultra-high impurity concentrations. For Au in Si, filamentary breakdown limits the concentration of Au that can be uniformly trapped in the lattice. For other metals, it has been demonstrated, particularly for Ti, that hyperdoping is inhibited by two main effects: severe surface segregation of the impurity during PLM at lower concentrations, and filamentary/cellular breakdown with defective crystal quality at higher concentrations. As a result, to date, it is unlikely that the infrared photoresponse observed for Si hyperdoped with metals other than Au is a result of hyperdoping. Instead, it is suggested that inhomogeneous and defective material also produces small photoresponse in the subbandgap range. Indeed, it has been shown that the implantation and PLM processes themselves result in defects that are optically active and can contribute to absorption in the NIR and, to some extent, photodetection. At present, it would appear that Au hyperdoping is the best prospect for Si NIR photodetectors but issues with poor photocarrier collection need to be overcome with much improved device architectures.

Compared with Si, hyperdoped Ge remains a much less developed area of research, where only recently Ge has been hyperdoped in a similar fashion to Si. Au-hyperdoped Ge has been shown to exhibit much enhanced subbandgap absorption up to 3 μm with an absorption coefficient approaching that of current commercial photodetectors. This represents a promising alternative to achieving low-cost, Si-compatible photodetection in the near- to mid-infrared range. In addition, it was shown that GeSn alloys can modify the bandgap of Ge to enable an indirect-to-direct bandgap transition. To this end, the challenge is to greatly exceed the 0.5 at.% solubility of Sn in Ge and minimize or eliminate strain in alloys grown on Si or Ge. Such Sn-dilute GeSn alloys have been demonstrated successfully by ion implantation and PLM, as an attractive alternative to CVD growth. In particular, direct bandgap GeSn films of reasonable quality have opened up viable applications for photodetection out to 4 μm, and lasers operating in the NIR. These early results are encouraging and will stimulate further work in understanding and optimizing the properties of the material before practical devices can be demonstrated.

## References

[1] A. Luque, A. Martí, Increasing the efficiency of ideal solar cells by photon induced transitions at intermediate levels, Phys. Rev. Lett. 78 (26) (1997) 5014.

[2] A. Luque, A. Marti, The intermediate band solar cell: progress toward the realization of an attractive concept, Adv. Mater. 22 (2) (2010) 160–174.

[3] P.K. Chow, W. Yang, Q. Hudspeth, S.Q. Lim, J.S. Williams, J.M. Warrender, Observation of enhanced infrared absorption in silicon supersaturated with gold by pulsed laser melting of nanometer-thick gold films, J. Appl. Phys. 123 (13) (2018) 133101.

[4] S. Zhou, F. Liu, S. Prucnal, K. Gao, M. Khalid, C. Baehtz, M. Posselt, W. Skorupa, M. Helm, Hyperdoping silicon with selenium: solid vs. liquid phase epitaxy, Sci. Rep. 5 (2015) 8329.

[5] F. Liu, S. Prucnal, R. Hübner, Y. Yuan, W. Skorupa, M. Helm, S. Zhou, Suppressing the cellular breakdown in silicon supersaturated with titanium, J. Phys. D Appl. Phys. 49 (24) (2016) 245104.

[6] E. Antolín, A. Martí, J. Olea, D. Pastor, G. González-Díaz, I. Mártil, A. Luque, Lifetime recovery in ultrahighly titanium-doped silicon for the implementation of an intermediate band material, Appl. Phys. Lett. 94 (4) (2009) 042115.

[7] J.J. Krich, B.I. Halperin, A. Aspuru-Guzik, Nonradiative lifetimes in intermediate band photovoltaics—absence of lifetime recovery, J. Appl. Phys. 112 (1) (2012) 013707.

[8] D. Recht, M.J. Smith, S. Charnvanichborikarn, J.T. Sullivan, M.T. Winkler, J. Mathews, J. M. Warrender, T. Buonassisi, J.S. Williams, S. Gradečak, Supersaturating silicon with transition metals by ion implantation and pulsed laser melting, J. Appl. Phys. 114 (12) (2013) 124903.

[9] W. Yang, Characterising and Understanding Au-Hyperdoped Si for Sub-Band Gap Optical Absorption (Ph.D. thesis), The Australian National University, 2018.
[10] J.P. Mailoa, A.J. Akey, C.B. Simmons, D. Hutchinson, J. Mathews, J.T. Sullivan, D. Recht, M.T. Winkler, J.S. Williams, J.M. Warrender, Room-temperature sub-band gap optoelectronic response of hyperdoped silicon, Nat. Commun. 5 (1) (2014) 1–8.
[11] B.P. Bob, A. Kohno, S. Charnvanichborikarn, J.M. Warrender, I. Umezu, M. Tabbal, J.S. Williams, M.J. Aziz, Fabrication and subband gap optical properties of silicon supersaturated with chalcogens by ion implantation and pulsed laser melting, J. Appl. Phys. 107 (12) (2010) 123506.
[12] H.Y. Fan, A.K. Ramdas, Infrared absorption and photoconductivity in irradiated silicon, J. Appl. Phys. 30 (8) (1959) 1127–1134.
[13] M. Casalino, G. Coppola, M. Iodice, I. Rendina, L. Sirleto, Near-infrared sub-bandgap all-silicon photodetectors: state of the art and perspectives, Sensors 10 (12) (2010) 10571–10600.
[14] A.P. Knights, A.A. House, R. MacNaughton, F. Hopper, Optical Fiber Communication Conference, 2003, p. FJ3. https://www.osapublishing.org/abstract.cfm?uri=OFC-2003-FJ3.
[15] Y. Liu, C.W. Chow, W.Y. Cheung, H.K. Tsang, In-line channel power monitor based on helium ion implantation in silicon-on-insulator waveguides, IEEE Photon. Technol. Lett. 18 (17) (2006) 1882–1884.
[16] M.W. Geis, S.J. Spector, M.E. Grein, R.T. Schulein, J.U. Yoon, D.M. Lennon, S. Deneault, F. Gan, F.X. Kaertner, T.M. Lyszczarz, CMOS-compatible all-Si high-speed waveguide photodiodes with high responsivity in near-infrared communication band, IEEE Photon. Technol. Lett. 19 (3) (2007) 152–154.
[17] R. Shafiiha, D. Zheng, S. Liao, P. Dong, H. Liang, N.-N. Feng, B.J. Luff, D. Feng, G. Li, J. Cunningham, Silicon waveguide coupled resonator infrared detector, in: Optical Fiber Communication Conference, Optical Society of America, 2010, p. OMI8.
[18] M.T. Winkler, D. Recht, M.-J. Sher, A.J. Said, E. Mazur, M.J. Aziz, Insulator-to-metal transition in sulfur-doped silicon, Phys. Rev. Lett. 106 (17) (2011) 178701.
[19] A.J. Akey, D. Recht, J.S. Williams, M.J. Aziz, T. Buonassisi, Single-phase filamentary cellular breakdown via laser-induced solute segregation, Adv. Funct. Mater. 25 (29) (2015) 4642–4649.
[20] C. Wu, C.H. Crouch, L. Zhao, J.E. Carey, R. Younkin, J.A. Levinson, E. Mazur, R.M. Farrell, P. Gothoskar, A. Karger, Near-unity below-band-gap absorption by microstructured silicon, Appl. Phys. Lett. 78 (13) (2001) 1850–1852.
[21] C.H. Crouch, J.E. Carey, M. Shen, E. Mazur, F.Y. Genin, Infrared absorption by sulfur-doped silicon formed by femtosecond laser irradiation, Appl. Phys. A 79 (7) (2004) 1635–1641.
[22] C.H. Crouch, J.E. Carey, J.M. Warrender, M.J. Aziz, E. Mazur, F.Y. Génin, Comparison of structure and properties of femtosecond and nanosecond laser-structured silicon, Appl. Phys. Lett. 84 (11) (2004) 1850–1852.
[23] T.G. Kim, J.M. Warrender, M.J. Aziz, Strong sub-band-gap infrared absorption in silicon supersaturated with sulfur, Appl. Phys. Lett. 88 (24) (2006) 241902.
[24] I. Umezu, J.M. Warrender, S. Charnvanichborikarn, A. Kohno, J.S. Williams, M. Tabbal, D. G. Papazoglou, X.-C. Zhang, M.J. Aziz, Emergence of very broad infrared absorption band by hyperdoping of silicon with chalcogens, J. Appl. Phys. 113 (21) (2013) 213501.
[25] M. Tabbal, T. Kim, J.M. Warrender, M.J. Aziz, B.L. Cardozo, R.S. Goldman, Formation of single crystal sulfur supersaturated silicon based junctions by pulsed laser melting, J. Vac. Sci. Technol. B Microelectron. Nanometer Struct. Process. Meas. Phenom. 25 (6) (2007) 1847–1852.

[26] K. Sánchez, I. Aguilera, P. Palacios, P. Wahnón, Assessment through first-principles calculations of an intermediate-band photovoltaic material based on Ti-implanted silicon: interstitial versus substitutional origin, Phys. Rev. B 79 (16) (2009) 165203.
[27] J.M. Warrender, Laser hyperdoping silicon for enhanced infrared optoelectronic properties, Appl. Phys. Rev. 3 (3) (2016) 031104.
[28] A.J. Said, D. Recht, J.T. Sullivan, J.M. Warrender, T. Buonassisi, P.D. Persans, M.J. Aziz, Extended infrared photoresponse and gain in chalcogen-supersaturated silicon photodiodes, Appl. Phys. Lett. 99 (7) (2011) 073503.
[29] E. Ertekin, M.T. Winkler, D. Recht, A.J. Said, M.J. Aziz, T. Buonassisi, J.C. Grossman, Insulator-to-metal transition in selenium-hyperdoped silicon: observation and origin, Phys. Rev. Lett. 108 (2) (2012) 026401.
[30] C.B. Simmons, A.J. Akey, J.P. Mailoa, D. Recht, M.J. Aziz, T. Buonassisi, Enhancing the infrared photoresponse of silicon by controlling the Fermi level location within an impurity band, Adv. Funct. Mater. 24 (19) (2014) 2852–2858.
[31] F. Priolo, J.M. Poate, D.C. Jacobson, J.L. Batstone, J.S. Custer, M.O. Thompson, Trapping of Au in Si during pulsed laser irradiation: a comparison with ion beam induced segregation, Appl. Phys. Lett. 53 (25) (1988) 2486–2488.
[32] A.G. Cullis, H.C. Webber, J.M. Poate, A.L. Simons, Segregation and increased dopant solubility in Pt-implanted and laser-annealed Si layers, Appl. Phys. Lett. 36 (4) (1980) 320–322.
[33] J. Olea, M. Toledano-Luque, D. Pastor, G. González-Díaz, I. Mártil, Titanium doped silicon layers with very high concentration, J. Appl. Phys. 104 (1) (2008) 016105.
[34] J. Olea, G. González-Díaz, D. Pastor, I. Mártil, Electronic transport properties of Ti-impurity band in Si, J. Phys. D Appl. Phys. 42 (8) (2009) 085110.
[35] J. Olea, A. Del Prado, D. Pastor, I. Mártil, G. González-Díaz, Sub-bandgap absorption in Ti implanted Si over the Mott limit, J. Appl. Phys. 109 (11) (2011) 113541.
[36] D. Pastor, J. Olea, A. Muñoz-Martín, A. Climent-Font, I. Mártil, G. González-Díaz, Interstitial Ti for intermediate band formation in Ti-supersaturated silicon, J. Appl. Phys. 112 (11) (2012) 113514.
[37] D. Pastor, J. Olea, A. Del Prado, E. García-Hemme, R. García-Hernansanz, G. González-Díaz, Insulator to metallic transition due to intermediate band formation in Ti-implanted silicon, Sol. Energy Mater. Sol. Cells 104 (2012) 159–164.
[38] Y. Zhou, F. Liu, X. Song, The insulator-to-metal transition of Co hyperdoped crystalline silicon, J. Appl. Phys. 113 (10) (2013) 103702.
[39] V.P. Markevich, S. Leonard, A.R. Peaker, B. Hamilton, A.G. Marinopoulos, J. Coutinho, Titanium in silicon: lattice positions and electronic properties, Appl. Phys. Lett. 104 (15) (2014) 152105.
[40] S. Silvestre, A. Boronat, M. Colina, L. Castaner, J. Olea, D. Pastor, A. del Prado, I. Mártil, G. González-Díaz, A. Luque, Sub-bandgap external quantum efficiency in Ti implanted Si heterojunction with intrinsic thin layer cells, Jpn. J. Appl. Phys. 52 (12R) (2013) 122302.
[41] E. García-Hemme, R. García-Hernansanz, J. Olea, D. Pastor, A. Del Prado, I. Mártil, G. González-Díaz, Room-temperature operation of a titanium supersaturated silicon-based infrared photodetector, Appl. Phys. Lett. 104 (21) (2014) 211105.
[42] E. García-Hemme, R. García-Hernansanz, J. Olea, D. Pastor, A. Del Prado, I. Mártil, G. Gonzalez-Diaz, Sub-bandgap spectral photo-response analysis of Ti supersaturated Si, Appl. Phys. Lett. 101 (19) (2012) 192101.
[43] J. Mathews, A.J. Akey, D. Recht, G. Malladi, H. Efstathiadis, M.J. Aziz, J.M. Warrender, On the limits to Ti incorporation into Si using pulsed laser melting, Appl. Phys. Lett. 104 (11) (2014) 112102.

[44] A.J. Akey, J. Matthews, J.M. Warrender, Maximum Ti concentrations in Si quantified with atom probe tomography, J. Appl. Phys., 2020. Submitted for publication.

[45] J.M. Poate, Laser Annealing of Semiconductors, Elsevier, 2012.

[46] F. Liu, M. Wang, Y. Berencén, S. Prucnal, M. Engler, R. Hübner, Y. Yuan, R. Heller, R. Böttger, L. Rebohle, On the insulator-to-metal transition in titanium-implanted silicon, Sci. Rep. 8 (1) (2018) 1–8.

[47] S.Q. Lim, A.J. Akey, E. Napolitani, P.K. Chow, J.M. Warrender, J.S. Williams, Critical assessment of Ag and Ti for hyperdoping in Si, J. Appl. Phys., 2020. Submitted for publication.

[48] K. Graff, Metal Impurities in Silicon-Device Fabrication, vol. 24, Springer Science & Business Media, 2013.

[49] W. Yang, A.J. Akey, L.A. Smillie, J.P. Mailoa, B.C. Johnson, J.C. McCallum, D. Macdonald, T. Buonassisi, M.J. Aziz, J.S. Williams, Au-rich filamentary behavior and associated subband gap optical absorption in hyperdoped Si, Phys. Rev. Mater. 1 (2017) 074602.

[50] N. Ferdous, E. Ertekin, Atomic scale origins of sub-band gap optical absorption in gold-hyperdoped silicon, AIP Adv. 8 (5) (2018) 055014.

[51] C.W. White, Dopant incorporation during nonequilibrium solidification, J. Phys. Colloq. 44 (1983) C5–145-C5-155.

[52] W. Yang, N. Ferdous, P.J. Simpson, J.M. Gaudet, Q. Hudspeth, P.K. Chow, J.M. Warrender, A.J. Akey, M.J. Aziz, E. Ertekin, Evidence for vacancy trapping in Au-hyperdoped Si following pulsed laser melting, APL Mater. 7 (10) (2019) 101124.

[53] W. Yang, Q. Hudspeth, P.K. Chow, J.M. Warrender, N. Ferdous, E. Ertekin, G. Malladi, A.J. Akey, M.J. Aziz, J.S. Williams, Atomistic mechanisms for the thermal relaxation of Au-hyperdoped Si, Phys. Rev. Appl. 12 (2) (2019) 024015.

[54] E. García-Hemme, R. García-Hernansanz, J. Olea, D. Pastor, A. Del Prado, I. Mártil, G. González-Díaz, Far infrared photoconductivity in a silicon based material: vanadium supersaturated silicon, Appl. Phys. Lett. 103 (3) (2013) 032101.

[55] K. Sánchez, I. Aguilera, P. Palacios, P. Wahnón, Formation of a reliable intermediate band in Si heavily coimplanted with chalcogens (S, Se, Te) and group III elements (B, Al), Phys. Rev. B 82 (16) (2010) 165201.

[56] S.Q. Lim, C.T.-K. Lew, P.K. Chow, J.M. Warrender, J.S. Williams, B.C. Johnson, Toward understanding and optimizing Au-hyperdoped Si infrared photodetectors, APL Mater. 8 (6) (2020) 061109.

[57] J. Narayan, Development of morphological instability and formation of cells in silicon alloys during pulsed laser irradiation, J. Cryst. Growth 59 (3) (1982) 583–598.

[58] C.B. Simmons, A.J. Akey, J.J. Krich, J.T. Sullivan, D. Recht, M.J. Aziz, T. Buonassisi, Deactivation of metastable single-crystal silicon hyperdoped with sulfur, J. Appl. Phys. 114 (24) (2013) 243514.

[59] P.K. Chow, S.Q. Lim, J.S. Williams, J.M. Warrender, Sub-bandgap photoresponse and leakage current analysis in gold thin film-hyperdoped silicon photodiodes, 2020 (to be published).

[60] S.Q. Lim, C.T.-K. Lew, P.K. Chow, J.M. Warrender, J.S. Williams, B.C. Johnson, Process-induced defects in Au-hyperdoped Si photodiodes, J. Appl. Phys. 126 (22) (2019) 224502.

[61] E.H. Putley, Far infra-red photoconductivity, Phys. Status Solidi B 6 (3) (1964) 571–614.

[62] N. Sclar, Properties of doped silicon and germanium infrared detectors, Prog. Quantum Electron. 9 (3) (1984) 149–257.

[63] J. Zhu, H. Zhu, H. Xu, Z. Weng, H. Wu, Ge-based mid-infrared blocked-impurity-band photodetectors, Infrared Phys. Technol. 92 (2018) 13–17.

[64] J.W. Beeman, S. Goyal, L.A. Reichertz, E.E. Haller, Ion-implanted Ge: B far-infrared blocked-impurity-band detectors, Infrared Phys. Technol. 51 (1) (2007) 60–65.

[65] J. Bandaru, J.W. Beeman, E.E. Haller, S. Samperi, N.M. Haegel, Influer distribution on far infrared photoconductivity in Ge: Sb blocked impur Infrared Phys. Technol. 43 (6) (2002) 353–360.

[66] H.H. Gandhi, D. Pastor, T.T. Tran, S. Kalchmair, L.A. Smillie, J.P. Mailc Napolitani, M. Loncar, J.S. Williams, M.J. Aziz, E. Mazur, Au-hyperdoped room-temperature sub-band-gap optoelectronic response, Phys. Rev. Appl.

[67] S. Gupta, B. Magyari-Köpe, Y. Nishi, K.C. Saraswat, Achieving direct band ga through integration of Sn alloying and external strain, J. Appl. Phys. 113 (7) (

[68] S. Wirths, R. Geiger, N. von den Driesch, G. Mussler, T. Stoica, S. Mar M. Luysberg, S. Chiussi, J.-M. Hartmann, Lasing in direct-bandgap GeSn all Si, Nat. Photon. 9 (2) (2015) 88–92.

[69] D.W. Jenkins, J.D. Dow, Electronic properties of metastable $Ge_xSn_{1-x}$ alloys, Pl 36 (15) (1987) 7994.

[70] K.A. Mäder, A. Baldereschi, H. von Känel, Band structure and instability of $Ge_{1-x}S$ Solid State Commun. 69 (12) (1989) 1123–1126.

[71] W.-J. Yin, X.-G. Gong, S.-H. Wei, Origin of the unusually large band-gap bowing breakdown of the band-edge distribution rule in the $Sn_xGe_{1-x}$ alloys, Phys. Rev (16) (2008) 161203.

[72] J. Liu, X. Sun, D. Pan, X. Wang, L.C. Kimerling, T.L. Koch, J. Michel, Tensile-straine type Ge as a gain medium for monolithic laser integration on Si, Opt. Express 15 (18) (2 11272–11277.

[73] R.E. Camacho-Aguilera, Y. Cai, N. Patel, J.T. Bessette, M. Romagnoli, L.C. Kimerling, Michel, An electrically pumped germanium laser, Opt. Express 20 (10) (2012) 11316–11320.

[74] R.W. Olesinski, G.J. Abbaschian, The Ge-Sn (Germanium-Tin) system, Bull. Alloy Phase Diagr. 5 (3) (1984) 265–271.

[75] G. He, H.A. Atwater, Interband transitions in $Sn_xGe_{1-x}$ alloys, Phys. Rev. Lett. 79 (10) (1997) 1937.

[76] Y. Shimura, N. Tsutsui, O. Nakatsuka, A. Sakai, S. Zaima, Low temperature growth of $Ge_{1-x}Sn_x$ buffer layers for tensile-strained Ge layers, Thin Solid Films 518 (6) (2010) S2–S5.

[77] S.A. Ghetmiri, W. Du, J. Margetis, A. Mosleh, L. Cousar, B.R. Conley, L. Domulevicz, A. Nazzal, G. Sun, R.A. Soref, Direct-bandgap GeSn grown on silicon with 2230 nm photoluminescence, Appl. Phys. Lett. 105 (15) (2014) 151109.

[78] R.R. Lieten, J.W. Seo, S. Decoster, A. Vantomme, S. Peters, K.C. Bustillo, E.E. Haller, M. Menghini, J.-P. Locquet, Tensile strained GeSn on Si by solid phase epitaxy, Appl. Phys. Lett. 102 (5) (2013) 052106.

[79] S.I. Shah, J.E. Greene, L.L. Abels, Q. Yao, P.M. Raccah, Growth of single-crystal metastable $Ge_{1-x}Sn_x$ alloys on Ge (100) and GaAs (100) substrates, J. Cryst. Growth 83 (1) (1987) 3–10.

[80] J. Taraci, J. Tolle, J. Kouvetakis, M.R. McCartney, D.J. Smith, J. Menendez, M.A. Santana, Simple chemical routes to diamond-cubic germanium-tin alloys, Appl. Phys. Lett. 78 (23) (2001) 3607–3609.

[81] C.L. Senaratne, J.D. Gallagher, L. Jiang, T. Aoki, D.J. Smith, J. Menendez, J. Kouvetakis, $Ge_{1-y}Sn_y$ ($y = 0.01 - 0.10$) alloys on Ge-buffered Si: synthesis, microstructure, and optical properties, J. Appl. Phys. 116 (13) (2014) 133509.

[82] C.L. Senaratne, J.D. Gallagher, T. Aoki, J. Kouvetakis, J. Menendez, Advances in light emission from group-IV alloys via lattice engineering and n-type doping based on custom-designed chemistries, Chem. Mater. 26 (20) (2014) 6033–6041.

[83] J. Mathews, R.T. Beeler, J. Tolle, C. Xu, R. Roucka, J. Kouvetakis, J. Menendez, Direct-gap photoluminescence with tunable emission wavelength in $Ge_{1-y}Sn_y$ alloys on silicon, Appl. Phys. Lett. 97 (22) (2010) 221912.

[84] J. Mathews, Z. Li, Y. Zhao, J. Gallagher, I. Agha, J. Menendez, J. Kouvetakis, Toward GeSn lasers: light amplification and stimulated emission in GeSn waveguides at room temperature, ECS Trans. 75 (8) (2016) 163–176.

[85] J. Margetis, S. Al-Kabi, W. Du, W. Dou, Y. Zhou, T. Pham, P. Grant, S. Ghetmiri, A. Mosleh, B. Li, Si-based GeSn lasers with wavelength coverage of 2–3 μm and operating temperatures up to 180 K, ACs Photonics 5 (3) (2017) 827–833.

[86] A. Bhatia, W.M.H. Oo, G. Siegel, P.R. Stone, K.-M. Yu, M.A. Scarpulla, Synthesis of $Ge_{1-x}Sn_x$ alloy thin films using ion implantation and pulsed laser melting, J. Electron. Mater. 41 (5) (2012) 837–844.

[87] K. Gao, S. Prucnal, R. Huebner, C. Baehtz, I. Skorupa, Y. Wang, W. Skorupa, M. Helm, S. Zhou, $Ge_{1-x}Sn_x$ alloys synthesized by ion implantation and pulsed laser melting, Appl. Phys. Lett. 105 (4) (2014) 042107.

[88] B.R. Appleton, O.W. Holland, J. Narayan, O.E. Schow III, J.S. Williams, K.T. Short, E. Lawson, Characterization of damage in ion implanted Ge, Appl. Phys. Lett. 41 (8) (1982) 711–712.

[89] B. Stritzker, R.G. Elliman, J. Zou, Self-ion-induced swelling of germanium, Nucl. Inst. Methods Phys. Res. B 175 (2001) 193–196.

[90] L. Romano, G. Impellizzeri, M.V. Tomasello, F. Giannazzo, C. Spinella, M.G. Grimaldi, Nanostructuring in Ge by self-ion implantation, J. Appl. Phys. 107 (8) (2010) 084314.

[91] H.S. Alkhaldi, F. Kremer, T. Bierschenk, J.L. Hansen, A. Nylandsted-Larsen, J.S. Williams, M.C. Ridgway, Porosity as a function of stoichiometry and implantation temperature in Ge/$Si_{1-x}Ge_x$ alloys, J. Appl. Phys. 119 (9) (2016) 094303.

[92] T.T. Tran, D. Pastor, H.H. Gandhi, L.A. Smillie, A.J. Akey, M.J. Aziz, J.S. Williams, Synthesis of $Ge_{1-x}Sn_x$ alloys by ion implantation and pulsed laser melting: towards a group IV direct bandgap material, J. Appl. Phys. 119 (18) (2016) 183102.

[93] T.T. Tran, H.H. Gandhi, D. Pastor, M.J. Aziz, J.S. Williams, Ion-beam synthesis and thermal stability of highly tin-concentrated germanium-tin alloys, Mater. Sci. Semicond. Process. 62 (2017) 192–195.

[94] T.T. Tran, Q. Hudspeth, Y. Liu, L.A. Smillie, B. Wang, R.A. Bruce, J. Mathews, J.M. Warrender, J.S. Williams, Ion beam synthesis and photoluminescence study of supersaturated fully-relaxed Ge-Sn alloys, Mater. Sci. Eng. B 262 (2020) 114702.

[95] S. Wirths, D. Buca, S. Mantl, Si-Ge-Sn alloys: from growth to applications, Prog. Cryst. Growth Charact. Mater. 62 (1) (2016) 1–39.

[96] V.S. Vavilov, Handbook on the physical properties of Ge, Si, GaAs and InP by A Dargys and J Kundrotas, Phys. Usp. 39 (7) (1996) 757.

[97] R. Beeler, R. Roucka, A.V.G. Chizmeshya, J. Kouvetakis, J. Menendez, Nonlinear structure-composition relationships in the $Ge_{1-y}Sn_y$/Si (100) ($y < 0.15$) system, Phys. Rev. B 84 (3) (2011) 035204.

[98] E. Kasper, H.-J. Herzog, Elastic strain and misfit dislocation density in $Si_{0.92}Ge_{0.08}$ films on silicon substrates, Thin Solid Films 44 (3) (1977) 357–370.

[99] M.T. Currie, S.B. Samavedam, T.A. Langdo, C.W. Leitz, E.A. Fitzgerald, Controlling threading dislocation densities in Ge on Si using graded SiGe layers and chemical-mechanical polishing, Appl. Phys. Lett. 72 (14) (1998) 1718–1720.

[100] G. Grzybowski, R. Roucka, J. Mathews, L. Jiang, R.T. Beeler, J. Kouvetakis, J. Menendez, Direct versus indirect optical recombination in Ge films grown on Si substrates, Phys. Rev. B 84 (20) (2011) 205307.

[101] J.D. Gallagher, C.L. Senaratne, J. Kouvetakis, J. Menendez, Compositional dependence of the bowing parameter for the direct and indirect band gaps in $Ge_{1-y}Sn_y$ alloys, Appl. Phys. Lett. 105 (14) (2014) 142102.

[102] L. Jiang, J.D. Gallagher, C.L. Senaratne, T. Aoki, J. Mathews, J. Kouvetakis, J. Menendez, Compositional dependence of the direct and indirect band gaps in $Ge_{1-y}Sn_y$ alloys from room temperature photoluminescence: implications for the indirect to direct gap crossover in intrinsic and n-type materials, Semicond. Sci. Technol. 29 (11) (2014) 115028.
[103] D. Stange, S. Wirths, R. Geiger, C. Schulte-Braucks, B. Marzban, N. von den Driesch, G. Mussler, T. Zabel, T. Stoica, J.-M. Hartmann, Optically pumped GeSn microdisk lasers on Si, ACS Photonics 3 (7) (2016) 1279–1285.
[104] M. Wang, Y. Berencén, E. García-Hemme, S. Prucnal, R. Hübner, Y. Yuan, C. Xu, L. Rebohle, R. Böttger, R. Heller, Extended infrared photoresponse in Te-hyperdoped Si at room temperature, Phys. Rev. Appl. 10 (2) (2018) 024054.
[105] M. Wang, R. Hübner, C. Xu, Y. Xie, Y. Berencén, R. Heller, L. Rebohle, M. Helm, S. Prucnal, S. Zhou, Thermal stability of Te-hyperdoped Si: atomic-scale correlation of the structural, electrical, and optical properties, Phys. Rev. Mater. 3 (4) (2019) 044606.

Chapter 9

# Laser ultra-doped silicon: Superconductivity and applications

Francesca Chiodi[a], Richard Daubriac[b], and Sébastien Kerdilès[b]
[a]*Centre for Nanoscience and Nanotechnology, Université Paris Saclay, CNRS, Palaiseau, France*, [b]*CEA-LETI, Université Grenoble-Alpes, Grenoble, France*

## 9.1 Introduction

The first theoretical prediction of superconductivity in many valley-doped semiconductors was made in 1964 [1]. However, the first experimental proof of superconducting behavior in group IV covalent semiconductors was obtained much later, with highly boron-doped diamond. Indeed, Ekimov and his partners succeeded in 2004 in synthesizing such material at high pressure (8–9 GPa) and high temperatures (2500–2800 K), with a total boron content close to 2.8 at.% [2]. Electrical measurements revealed a type II superconductivity with a transition temperature around 4 K, surviving in a magnetic field up to 3.5 T. Already known for its remarkable properties such as excellent hardness and thermal conductivity, the best-in-class electrical insulator increased further its potential for applications, thanks to this discovery.

In the same work, Ekimov et al. also suggested that a similar behavior was to be expected in silicon and germanium under appropriate conditions, as they share the same crystallographic structure. A few months later, Boeri et al. [3] showed, by numerical and analytical calculations, that superconductivity in diamond originates from strong electron-phonon coupling, similarly to $MgB_2$. Using the same method, they demonstrated that the same superconducting behavior should be expected in hole-doped silicon and germanium by reaching 5–10 at.% active boron concentration. These levels of doping correspond to boron concentrations between $2.5 \times 10^{21}\ cm^{-3}$ and $5.0 \times 10^{21}\ cm^{-3}$, i.e., well above the solid solubility limit of boron in silicon.

To overcome the solubility limit, standard equilibrium doping techniques cannot be employed, as they limit both the activation level and junction depth, as was demonstrated during the past decades by Si microelectronic research.

Laser Annealing Processes in Semiconductor Technology. https://doi.org/10.1016/B978-0-12-820255-5.00009-X

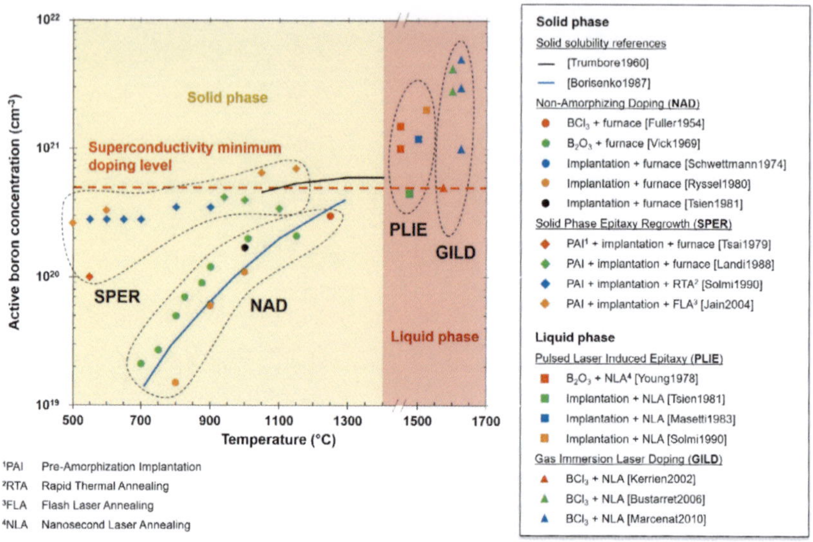


**FIG. 9.1** Active boron concentration in silicon as a function of the processing temperature using various techniques, and compared with the solid solubility limit [4–9,12,20,22,64–70].

Indeed, the miniaturization of MOSFET required sharper and sharper junctions with higher and higher active dopant concentrations in silicon, particularly at source-drain junctions. Fig. 9.1 illustrates the achieved *active* boron concentrations, $[B_{act.}]_{Si}$, in silicon using various techniques. The use of equilibrium processes restricts theoretically the concentration of boron atoms in substitutional position to around 0.8% ($4.0 \times 10^{20}$ cm$^{-3}$) [4] to 1.2% ($6.0 \times 10^{20}$ cm$^{-3}$) [5] in the 1300–1400°C range. In the solid-state regime, two approaches result in $[B_{act.}]_{Si}$ levels close to this solid solubility limit. They both require boron implantation. For solid-phase epitaxial regrowth (SPER), amorphizing implantation conditions are combined with a relatively long anneal at a limited temperature to recover a monocrystalline layer and limit boron diffusion [6, 7]. Note that for a given annealing temperature, non-amorphizing conditions coupled with a treatment in a furnace exhibit much lower active boron concentrations [8,9]. The second approach leverages subsecond anneals, from a few milliseconds using flash lamps for example, down to a fraction of a microsecond using pulsed lasers, at temperatures in the 1050–1300°C range [9,64]. In this approach, shorter and shorter anneals were implemented to limit boron diffusion and quench the silicon layer to maximize $[B_{act.}]_{Si}$. Thanks to this strategy, ultra-shallow junctions (~10 nm) with boron doping concentrations close to the solid solubility were obtained. However, to reach superconductivity requirements, i.e., doping levels above 1 at.%, these approaches were not sufficient. That is why research teams moved from these solid-state techniques to an alternative process leveraging liquid-phase epitaxial regrowth (LPER) based on ultraviolet nanosecond pulsed lasers.

UV light provided by excimer lasers such as XeCl (308 nm) or ArF (193 nm) is strongly absorbed by silicon (penetration depth $\sim$ 7 nm) so that most of the nonreflected intensity is converted into heat in the first nanometers. Thanks to a very short pulse duration (typically 10–500 ns), the deposited energy is confined in the near-surface region (quasiadiabatic process). With an appropriate choice of energy density, pulsed laser annealing easily allows to reach temperatures in the 1450–1700°C range, well beyond the melting point of Si (1414°C). This leads to a temporary and depth-controlled melt of the surface, and then, after pulsed laser energy dissipation, ultra-fast re-solidification.

Pulsed laser annealing in the melt regime gives access to boron contents in the liquid phase well beyond the limit observed in the solid phase, opening the way to higher levels of active boron after solidification. Once in the liquid Si phase, boron diffusivity is strongly enhanced, reaching $\sim 3.5 \times 10^{-4}$ cm$^2$/s [10], i.e., about seven orders of magnitude higher than in solid Si. During the melt sequence ($\sim$20–200 ns typically), boron atoms can spread in the whole melted region (typically $\sim$20–300 nm). Then, fast re-solidification acts like a quenching process that freezes boron atoms preferentially in substitutional sites, limiting the boron segregation into clusters often observed after standard annealing techniques.

In view of superconducting silicon, two techniques based on pulsed lasers are now used to obtain heavily boron-doped silicon (Fig. 9.2): gas immersion laser doping (GILD) and pulsed laser-induced epitaxy (PLIE). GILD consists in $BCl_3$ precursor gas chemisorption at the surface of the silicon wafer and laser annealing with melt conditions under ultra-high vacuum. The laser fluence and the number of pulses at the same location are tuned to reach the desired boron concentration down to a controlled depth. The first superconducting Si:B layer was fabricated using GILD in 2006 [12]. The first two parts of this chapter will be devoted to GILD, as it allows to simply obtain a large set of thickness and doping conditions to explore superconductivity in Si. The properties of superconducting Si:B layers and the fabrication of the first superconducting Si:B devices will be detailed. In the perspective of integrating superconducting silicon at a large scale, more conventional processes than GILD are required. PLIE, the other identified technique enabling the fabrication of superconducting Si:B layers, is developed in the second part of this chapter. PLIE, combining ion implantation and laser annealing at melt conditions, was identified as the best successor of GILD. Indeed, PLIE is more compatible with existing Si technologies and is employed with a wide choice of dopants, but the initial implanted dopant profile influences the final doped depth and concentration, and the total dopant dose is fixed. However, for industrial applications with the well-established dose parameters, PLIE is the natural choice.

In 2013, first studies on superconducting samples made with PLIE were done in high-vacuum conditions [13] but recent results showed the possibility of obtaining reproducibly superconducting heavily boron-doped silicon in industrially compatible conditions [14]. These recent developments will be discussed in the third and last part of this chapter.

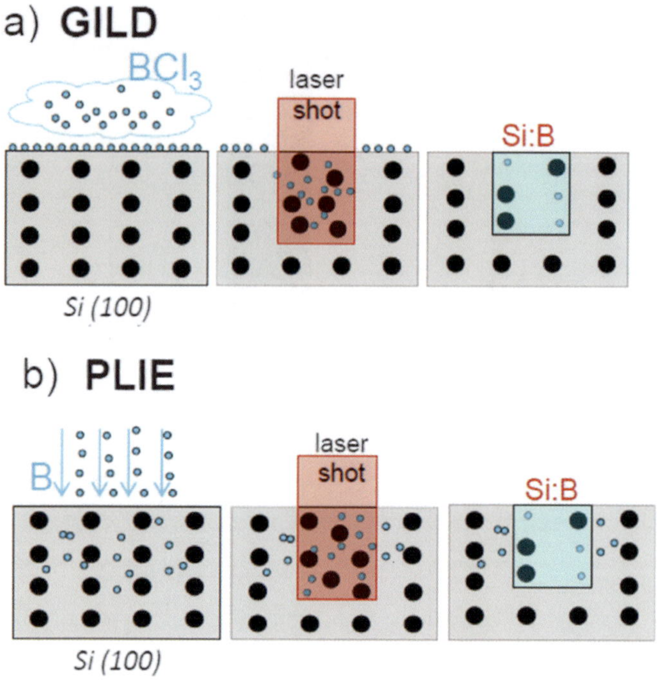


FIG. 9.2 (A) Gas immersion laser doping (GILD): chemisorption of the precursor gas $BCl_3$ onto the Si surface under ultra-high-vacuum conditions; laser melting of the substrate by an excimer laser pulse and boron diffusion throughout the liquid phase; fast epitaxy of ultra-doped Si:B crystal on the Si substrate. The presence of Cl induces the expulsion of SiCl dimers with a resulting surface etching of less than 0.056 nm [11]. (B) Pulsed laser-induced epitaxy (PLIE): implantation of B atoms; laser melting of the substrate by an excimer laser pulse and boron diffusion throughout the liquid phase; fast epitaxy of ultra-doped Si:B crystal on the Si substrate. Around the Si:B layer, the Si is implanted; crystal defects due to the implantation may also be present underneath the epitaxied layer.

## 9.2 Laser-doped superconducting silicon: Gas immersion laser doping

Superconducting silicon was first realized using GILD, and GILD samples currently represent the state of the art for superconducting Si in terms of superconducting critical temperature and control of the structural properties. Detailed below are the conditions, initially developed to fabricate ultra-doped, thin, Si:B layers for microelectronics, that have been since employed to fabricate GILD superconducting samples.

Si substrates are initially deoxidated with a HF etch to remove the native silica layer at the Si surface and the polluting organic compounds. GILD is then performed by first chemisorbing a precursor gas, $BCl_3$, onto the clean Si surface in a UHV chamber ($\sim 10^{-9}$ mbar, limiting the incorporation of unintentional impurities). The substrate is then melted by a pulsed excimer XeCl laser ($\lambda = 308$ nm, pulse duration 25 ns, repetition rate 2 Hz) of $E = 200$ mJ, providing an energy density up to 2.5 J/cm$^2$ over the 2 mm $\times$ 2 mm laser spot. The boron, rapidly diffusing into the liquid silicon, is incorporated in substitutional sites during the epitaxy of a Si:B crystal on the underlying silicon. Because of the short pulse duration and the high recrystallization speed, the boron does not have the time to attain its equilibrium distribution, and large boron concentrations, as high as 10 at.% ($5 \times 10^{21}$ cm$^{-3}$), could be obtained in thin layers (10–300 nm) without the formation of B aggregates, even in the samples with the stronger doping analyzed by atom probe tomography [15]. Both the thickness of the doped layer and the doping level can be independently tuned, the thickness (controlled within a few nm precision) being determined by the energy density at the sample level and the surface B dose ($1.2 \times 10^{14}$–$6 \times 10^{16}$ cm$^{-2}$) by the number of laser pulses shone at the same location.

The technique of gas immersion laser doping is described in more detail hereafter: we detail the laser-induced fusion and the dopant incorporation, relating the fabrication to the structural and electrical properties of Si:B layers.

### 9.2.1 Laser-induced fusion

The XeCl UV photons ($E = 4$ eV) are directly absorbed in Si, as their energy is larger than the Si direct gap ($E = 3.4$ eV). The hot gas of conducting electrons created relaxes its energy toward the phonon bath. This process is dominant as compared with the other relaxation paths, and has a characteristic time much shorter than the laser pulse duration. The laser energy can thus be considered as integrally and immediately converted into thermal energy (see Chapter 6 for a detailed discussion). The heat produced in the first surface nanometers diffuses then unidirectionally toward the substrate, the convection and radiation being negligible. This process is extremely rapid: during the 25-ns pulse duration, the heat diffuses over more than 1 μm. The temperature increase induces the melting of Si: a fusion front propagates, driven by the energy deposited on the surface. Once the energy dissipates, the fusion front inverses its direction and becomes a solidification front, proceeding at ~4 m/s for a pulse duration of 25 ns [16]. This velocity is low enough for a crystal to be epitaxied over the underneath Si without defects, but fast enough to introduce the boron atoms into the crystal lattice.

The depth of the melted zone is precisely controlled: above an energy threshold necessary to compensate the latent specific heat, the depth increases at first order linearly with the energy deposited on the sample. The melting threshold energy density was estimated for bulk Si and a 25-ns laser pulse at ~600 mJ/cm$^2$, with an increase in melted depth with the energy density of ~0.28 nm/(mJ/cm$^2$) [17]. However, such values depend on the laser temporal profile and the sample structure: e.g., the silica layer in SOI samples limits the heat diffusion, lowering the fusion threshold and increasing the melted depth.

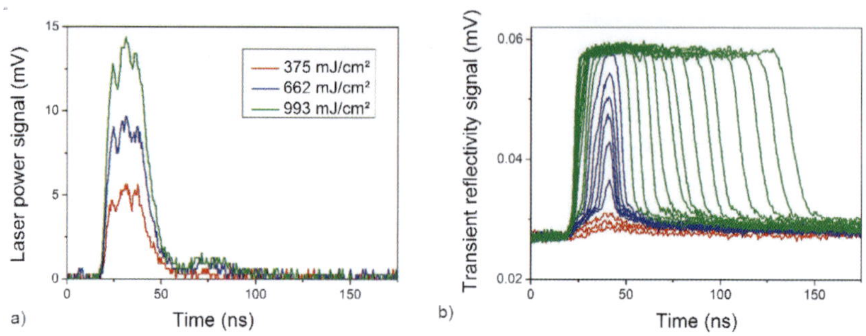


**FIG. 9.3** (A) Time profile of the XeCl laser pulse, (B) time-resolved reflectivity (TRR) of a 675-nm laser as a function of energy density: *(red)* in the 375–551 mJ/cm$^2$ energy density range, below the fusion threshold, the Si is heated, inducing a "bump" in the reflectivity; *(blue)* in the 650–882 mJ/cm$^2$ range, the fusion takes place, the melted depth increases with energy density; *(green)* in the 990–2051 mJ/cm$^2$ range, the melted layer is thicker than ~20 nm and the TTR signal saturates. However, it is possible to follow the evolution of the melted depth monitoring the fusion duration.

The melted depth is limited on one side by the available laser power, and on the other by the damage induced on the surface at large energy density, such as ablation and/or large roughness induced by the boiling of liquid Si.

The fusion can be followed in a real time by measuring the time-resolved reflectivity (TRR): indeed, the optical reflectivity varies between the metallic liquid Si and the semiconducting solid Si, a factor 2 for the reflection of a 675-nm red laser. In Fig. 9.3, the UV laser intensity and the corresponding TRR time profile are shown. At the lowest energy density, the reflectivity records a slight increase as the Si is heated but not melted. The appearance of the liquid phase is characterized by a sharp increase in the reflectivity. At low energy density, the liquid phase is preceded by ~10 ns heating, possibly inducing a partial desorption of the chemisorbed precursor gas, while at higher energy densities, the fusion is more rapidly engaged. At the onset of the melt, similarly to what has been evidenced for SiGe epilayers [18], some molten isolated islands may form first and then tend to form a continuous liquid layer with a thickness increasing with the laser energy density. The maximum reflectivity and the width of the TRR peak increase with the increase in the energy density and consequent melted depth. Above a depth of ~20 nm, the red laser is not reflected anymore on the melted layer bottom as it is completely absorbed in the metallic liquid Si. However, it is still possible to monitor the melted depth through the width of the fusion peak, i.e., the fusion duration, which increases linearly with the melted depth. Note that between the onset of the fusion ($d = 0$) and the signal saturation ($d = 20$ nm), the entire TRR curves conspicuously evolve, as do their widths above $d = 20$ nm, allowing to control the thickness down to a few nanometers precision. The TTR monitoring is in particular extremely important for GILD, as the optical properties of the Si:B layer evolve with the incorporated B concentration: when the B doping increases, the laser absorption is enhanced. To keep a constant melting depth, corresponding to a constant fusion duration, it is necessary to attenuate the laser energy during the doping, relying on the monitored TTR.

### 9.2.2 Surface homogeneity

Laser annealing is a threshold phenomenon: for small melted depths of a few tens of nanometers, the laser energy is extremely close to the threshold fusion energy, and small energy variations will induce large relative variations of the melted depth. Incoherent lasers, with longitudinal and lateral coherence smaller than 1 mm, are usually chosen to avoid interference effects (see, e.g., Ref. [19]) and thus insure homogeneous illumination. To improve the homogeneity of the laser beam, fly-eye homogenizers were furthermore employed to achieve variations in the energy density as small as ~1% over the 2 mm × 2 mm laser spot (Fig. 9.4).

### 9.2.3 Dopant incorporation

The dopants could be introduced directly from a vapor phase during the fusion, the number of pressure-dependent collisions on the surface determining the incorporated dose. However, a more controlled method was chosen by Ref. [16]: a chlorine-based precursor gas, $BCl_3$, was injected over the Si surface before the fusion, saturating the chemisorption sites. Since the number of such sites is fixed, and since the thermal-induced desorption is limited by the strong chemical bond (>1 eV) insured by the Cl compounds, the boron dose incorporated for each fusion-solidification cycle is constant.

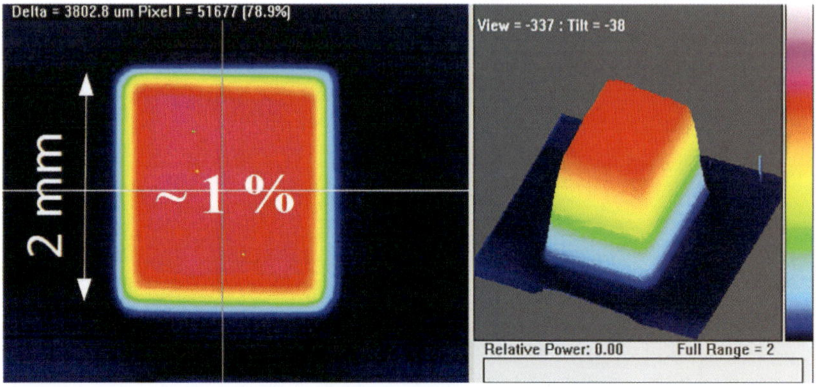


FIG. 9.4 Density of energy at the sample surface.

This was indeed observed experimentally, except for the thinner melted layers (~10 nm), where the temperature increase is slow, and the dopant desorption is favored to the detriment of incorporation, thus limiting the boron dose [19].

The incorporation of unwanted atoms, simultaneous to that of the dopant atoms, motivated the UHV conditions and the HF deoxidation before the laser anneal.

Once in the liquid phase, the B and Cl atoms diffuse rapidly down to the solid Si underneath, where the diffusion is negligible. During the fast crystallization, nearly all the B present in the liquid phase is incorporated in the solid phase of the forming Si:B crystal, the nonequilibrium segregation coefficient of B in Si being close to 1 (see also Chapter 6). The Cl, with a segregation coefficient close to 0, does not enter the solid phase and is entirely expelled when the solidification front reaches the surface.

### 9.2.4 Dopant content and distribution

Since all the B is incorporated from the liquid into the solid phase, the B dose in the Si:B crystal is determined by the (constant) density of chemisorption sites. The total dose grows thus linearly with the number of process repetitions with an incorporated dose of [1.2–2] $\times$ $10^{14}$ $cm^{-2}$/laser shot (the range being due to different measurement techniques). This corresponds to about ¼ of the atoms' density at the surface of a (100) Si sample, the difference observed being probably related to the size of the Cl atoms.

For concentrations below ~3 at.%, the total number of B atoms introduced is found to be active, i.e., is integrally incorporated, substitutionally, in sites of the crystal lattice. Thus, the *active* B concentration increases linearly with the number of laser shots (Figs. 9.5B and 9.7C). When approaching concentrations ~4 at.%, the active incorporation rate starts to slow down, and a saturation of

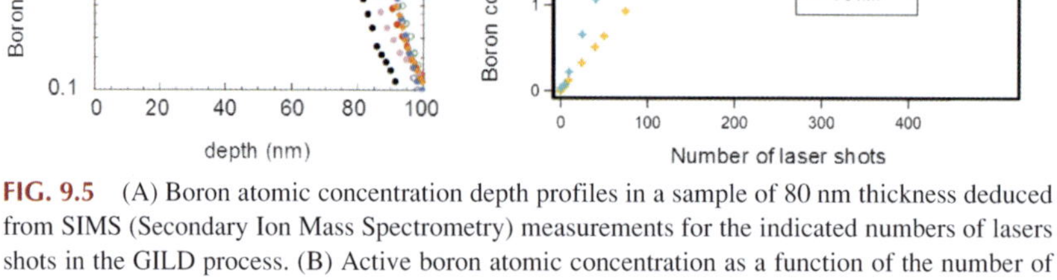


**FIG. 9.5** (A) Boron atomic concentration depth profiles in a sample of 80 nm thickness deduced from SIMS (Secondary Ion Mass Spectrometry) measurements for the indicated numbers of lasers shots in the GILD process. (B) Active boron atomic concentration as a function of the number of laser shots in two samples of 70 and 150 nm thickness measured in a Hall cross-geometry.

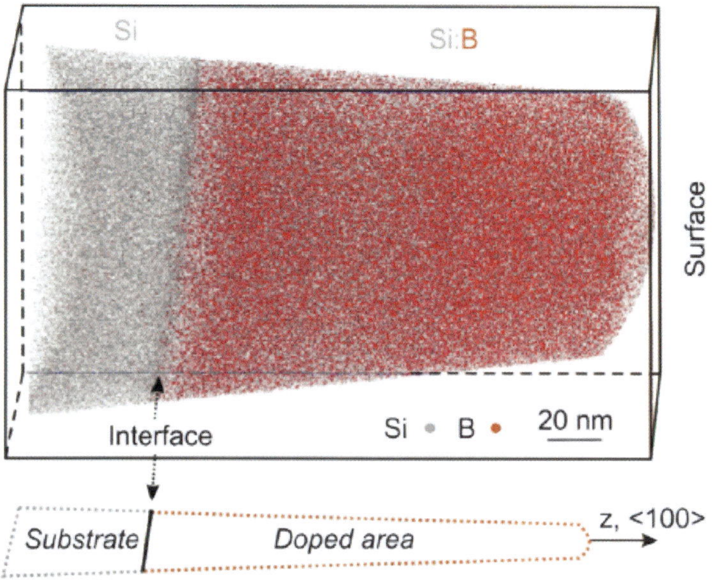


**FIG. 9.6** Reconstruction of a 60 × 60 × 200 $nm^3$ volume in heavily boron-doped silicon epilayers analyzed by atom probe tomography. Each dot corresponds to an atom. Representative of the whole distribution, only 1% (10%) of the captured silicon (boron) atoms is shown. A sharp interface is clearly visible between undoped and doped regions [15].

the concentration at 10 at.% ($5 \times 10^{21}$ $cm^{-3}$) was highlighted by [17], along with an increase in the interstitial B concentration.

Thanks to the fast B diffusion in the liquid Si, the liquid and consequently the solid phases are homogeneous in B. Indeed, the concentration profile obtained in the Si:B crystalline layer is highly uniform up to concentrations ~5 at.%, with a plateau followed by a sharp dopant decrease over a 2–9 nm at the Si:B/Si interface (Figs. 9.5 and 9.6) [20, 22]. However, in doping layers thicker than ~50 nm, a few laser shots are necessary to give the B time to attain the bottom of the melted region, and obtain fully homogeneous layers. For example, 5–10 shots are necessary for an 80-nm-thick layer [21]. Above 5 at.%, the dopants are not homogeneous any more through the entire doped layer as two homogeneous layers are formed: one, with smaller concentration, epitaxied over the Si substrate; and the second, with larger doping, on top (Figs. 9.5A and 9.6).

To investigate the possible presence of boron aggregates, APT (atom probe tomography) measurements were performed by [15]. The APT analysis consists in slowly etching a thin Si:B tip by the application of ps laser pulses inducing the surface atoms' evaporation. Similar to SIMS measurements, the nanometer-precise APT technique is furthermore improved by the measurement of the time of flight of the ejected atoms, which allows to track down the position of the Si and B atoms in the original material, and to check the presence of B aggregates in the 3D Si:B layer image.

Fig. 9.6 shows the result for an 80-nm-thick layer submitted to 500 GILD pulses, corresponding to the highest doping studied. The remarkably sharp and straight interface between the Si:B and Si can be noticed, with an average roughness of 0.7 nm. Within the doped layer, a modulation of the doping creates two regions of different concentrations (8 at.% and 10 at.%), comparable with those observed by SIMS for the same number of laser shots (Fig. 9.5). The boron atoms are randomly distributed, following a binomial distribution around the average value, and no aggregate could be observed despite the very high doping concentrations. Even though the temperature was high enough to form stable $SiB_3$ precipitates, such process was indeed hindered by the high recrystallization velocity and lack of implantation defects.

### 9.2.5 Dopant-induced strain

When increasing the boron concentration, a strong increase in the tensile biaxial strain of the doped Si:B epitaxied layer was observed (Fig. 9.7). Indeed, boron atoms are smaller than Si ones, and their substitutional incorporation induces a decrease in the lattice volume, with a resulting decrease of the out-of-plane lattice parameter, as the in-plane lattice is set by the Si one during the epitaxy. X-ray diffraction (XRD) was performed using the Kα1 Cu line ($\lambda = 1.5598$ Å) in the [004] direction; a cartography of the reciprocal space was also performed around the asymmetric reflexion [224]. These measurements showed a diffraction peak associated with the Si:B layer, of FWHM ~0.1 degrees, confirming its monocrystalline nature. The angular position of the Si:B peak moves away from the Si peak when the concentration increases, as a result of the reduction of the out-of-plane lattice parameter $a_\perp$ (Fig. 9.7B). The linear decrease of $a_\perp$ with the boron concentration is in agreement with the Vegard's law: $a_{SiB} = n_B\ a_B + (1 - n_B)\ a_{Si}$. The maximum strain is found for a shift of the XRD peak of $\Delta\theta = 1.3$ degrees, corresponding to an out-of-plane lattice parameter decrease of −3.5%. Above a concentration of ~5 at.%, the fully strained Si:B layer experiences a relaxation leading to a strain sharing between the out-of-plane and in-plane lattice. It can be noticed that above 6 at.% two layers result from the relaxation, one fully strained, with smaller concentration and epitaxied over the Si substrate; and one, with larger doping, on top, confirming what was already observed by SIMS and APT (Figs. 9.5–9.6).

## 9.3 Silicon: A BCS superconductor tunable with doping

Superconducting boron-doped silicon belongs to the family of covalent superconductors, and can be described in the frame of weak-coupling BCS (Bardeen-Cooper-Schrieffer) theory with hole Cooper pairs. Indeed, despite the low density of states at Fermi level $N(0)$ (due in Si to the low concentration of charge carriers as compared with a metal), the covalent bonds lead to a strong electron-phonon coupling potential $V$. The resulting electron-phonon coupling

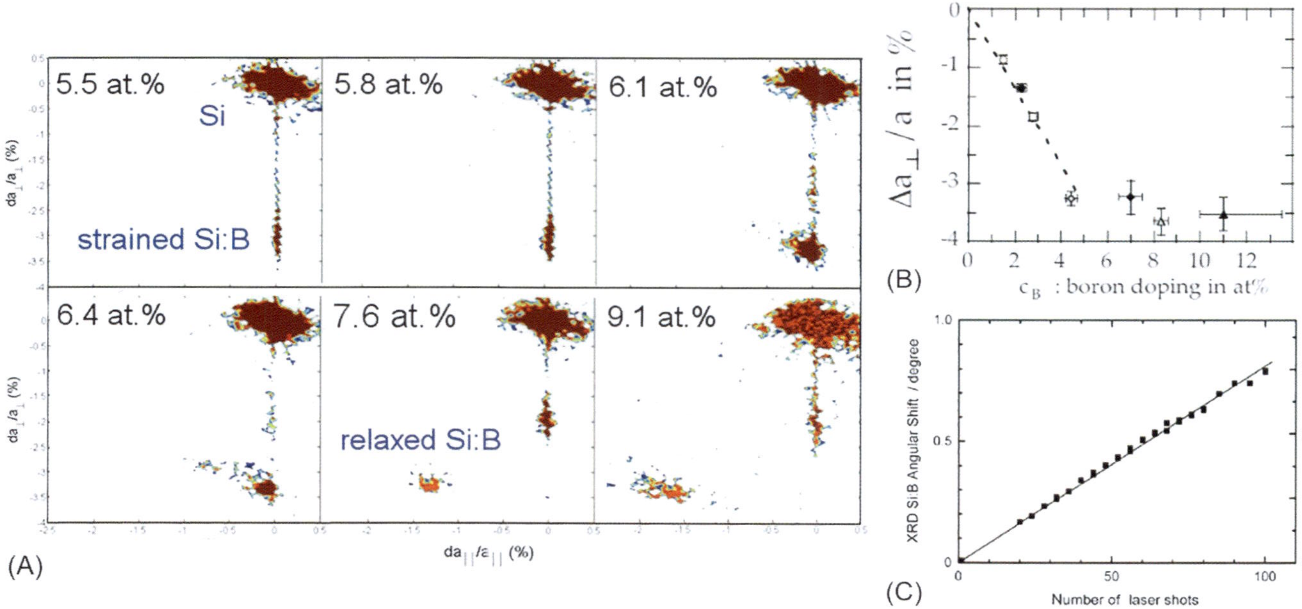


**FIG. 9.7** (A) Reciprocal space maps around the asymmetric [224] reflection for a 80-nm layer, showing that, up to 5.5 at.% (180 laser shots), GILD realizes strained and crystalline Si:B layers. Consecutive to a pseudomorphic epitaxy, the in-plane parameter $a_{||}$ for the Si substrate and the doped layer are equal ($\varepsilon_{||}$=0), while the modification of the out-of-plane lattice parameter $a_\perp$ is revealed by an XRD angular shift up to a relative variation of $\varepsilon_\perp = -3.5\%$. (B) High-resolution XRD measurements of Si:B films at [004] Bragg reflection. Relative variation in the lattice parameter perpendicular to the surface in % as a function of the boron concentration $c_B$. *Broken line*: linear dependence (Vegard's law) at low doping [20]. (C) XRD angular shift of the Si:B related peak and deduced boron concentration of boron-doped silicon layers of 80 nm thickness, doped with 20–100 laser pulses [21].

constant $\lambda = N(0)V$ is thus strong enough to induce superconductivity, as, for superconductors described by a $\delta$-like Eliashberg function

$$T_c = \theta_D \, e^{-(1+\lambda)/(\lambda-\mu(1+\lambda))} \tag{9.1}$$

where $T_c$ is the critical temperature for the superconducting transition, $\theta_D \sim 450$ K is the Si:B Debye temperature, and $\mu \sim 0.15$ is the screened and retarded Coulomb repulsion potential [3].

Scanning tunnel microscopy (STM) was performed on Si:B layers of different thicknesses (30–130 nm) and doping (200–500 laser shots) to measure the superconducting gap [23]. Indeed, the differential conductance of a tunnel contact created between the STM tip and the superconducting sample is the convolution of the superconductor density of states (DOS) and the Fermi-Dirac distribution in the tip. The good agreement with BCS theory of (i) the measured DOS as a function of energy, (ii) the temperature dependence of the deduced gap $\Delta$ (Fig. 9.8A), and (iii) the ratio $\Delta(T=0)/k_B T_c = 1.76 \pm 0.35$, confirms the validity of BCS model for superconducting silicon. STM measurements also allowed to estimate the spatial gap variations at the sample surface, finding $\delta\Delta/\Delta = \pm 10\%$, a variation compatible with a disordered metal.

The temperature dependence of the resistance featured a single, sharp ($\Delta T_c \sim 20$ mK) transition, as expected for a macroscopically homogeneous material (Fig. 9.8B). Field-cooling resistive transitions were measured to estimate the critical magnetic field $H_{c2}$ and extract the coherence length $\xi_{\mathrm{eff}}$, as $\mu_0 H_{c2}/T = \Phi_0/2\pi\xi_{\mathrm{eff}}^2 T_c$ (Fig. 9.8C). The coherence length is one of the most characteristic parameters of superconductors, roughly representing the size of Cooper pairs. In the case of a material rich in impurities (dirty limit), the coherence length is reduced while the magnetic field penetration depth is increased, their effective values being given by $\xi_{\mathrm{eff}} = \sqrt{\xi_0 l_e}$ and $\lambda_{L,\mathrm{eff}} = \lambda_L \sqrt{\xi_0 / l_e}$, with $l_e$ the electronic mean free path and $\xi_0$ the coherence length in the absence of any impurity scattering. This is indeed the case for ultra-doped silicon layers, which are disorder-induced type II superconductors. As an example, below are reported typical Si:B thin-layer parameters and the deduced characteristic lengths:

| | |
|---|---|
| Doping concentration $n_B = 4 \times 10^{21}$ cm$^{-3}$ | |
| Thickness $d = 40$ nm | Mean free path $l_e \sim 5$ nm |
| Critical temperature $T_c = 0.47$ K $\rightarrow\rightarrow$ | Effective coherence length $\xi_{\mathrm{eff}} \sim 60$ nm |
| Resistivity $\rho = 110\ \mu\Omega$ cm | London effective penetration length $\lambda_{L,\mathrm{eff}} \sim 1.6\ \mu m$ |
| Critical field $B_{c2} = 94$ mT | |

As $d \gg l_e$, the transport is diffusive. The strong disorder comes from the random position of the substitutional B atoms, inherent to the doping technique and observed by APT.

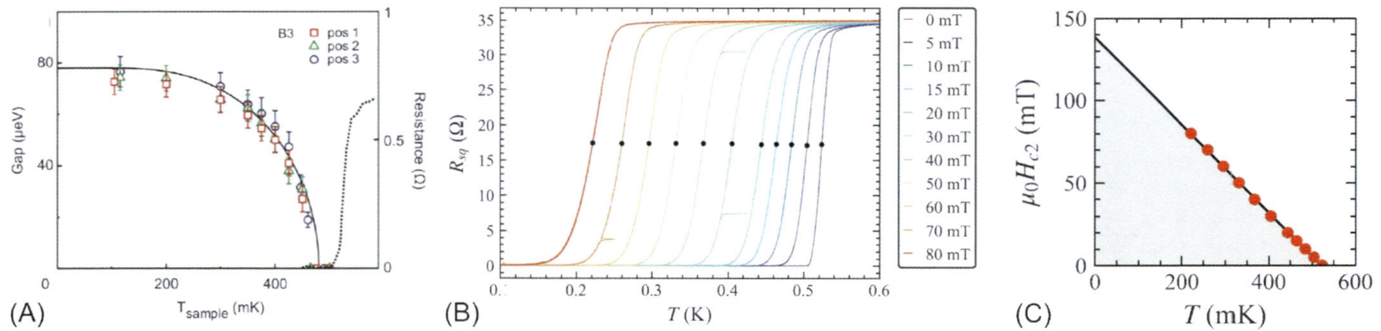


**FIG. 9.8** (A) *Left axis*: variation in the superconducting energy gap versus sample temperature, recorded for three different positions in a sample doped with 500 laser shots and 130 nm thickness. Position 1 and 2 are 15 nm apart, whereas position 3 is 100 nm away. The continuous line is to a fit to the BCS theory. *Right axis*: electrical resistance of the same sample versus temperature showing a macroscopic superconductive transition with a critical temperature $T_c = 510$ mK [23]. (B) Resistance vs temperature of a sample of 30 nm thickness and doped with 200 laser shots in the presence of a perpendicular magnetic field between 0 and 80 mT. The $T_c$, measured as the temperature corresponding to a resistance half the normal state resistance *(black dots)*, is then plotted against the magnetic field (C) to extract the superconducting coherence length as $\mu_0 H_{c2}/T = \Phi_0/2\pi\xi_{\text{eff}}^2\, T_c$ [17].

The layers can have either a 2D or 3D geometry, as $d \sim \xi_{eff}$. Finally, the magnetic field penetrates fully in the superconducting layers ($d \ll \lambda_{L,eff}$). Thus, because of disorder, the Ginzburg-Landau parameter κ is strongly increased: $\kappa_{dirty} = \lambda_{L,eff}/\xi_{eff} > 1/\sqrt{2} \gg \kappa_{clean}$, indicating that Si:B is a type II superconductor, characterized by vortex penetration.

The superconducting behavior in Si:B is controlled by doping. This was experimentally demonstrated by Ref. [24], where Si:B layers of thickness in the range 20–210 nm and concentrations $3 \times 10^{20}$–$6 \times 10^{21}$ cm$^{-3}$ were measured (Fig. 9.9). The $T_c$ was observed to increase linearly with the concentration, and then to progressively saturate as a result of the saturation of the active concentration (Fig. 9.5B). When plotting all the samples not as a function of concentration but as a function of the boron surface dose (Fig. 9.9B), all curves collapse into one, suggesting that the boron dose, and not the boron concentration, determines both the superconductivity threshold and the critical temperature. Indeed, the superconductivity threshold is constant for all thicknesses at $\sim 5 \times 10^{15}$ cm$^{-2}$, and the critical temperature $T_c$ increases monotonically with the dopant dose from 0 to 0.7 K, with a slope at low doping of $5 \times 10^{-14}$ mK/cm$^{-2}$.

The importance of the dose was also confirmed in samples realized with PLIE, where a constant $T_c$ was found for a fixed implanted dose, independently on the melted layer thickness and thus dopant concentration; when doubling the implanted dose,the $T_c$ was observed to increase [24].

To model the $T_c$ vs boron dose dependence, the electron-phonon coupling constant was predicted from ab-initio calculations using a supercell model (a periodic dopant distribution) combined with a density functional theory with a contraction of unit cell lattice parameter to account for B-induced strain, or by virtual crystal approximation [24, 25]. However, the experimental behavior could not be reproduced (thick line in Fig. 9.9A).

Alternatively, a bilayer structure was hypothesized within the doped layer, with a superconducting layer in parallel with a weaker superconducting (S′) or metallic (N) layer. McMillan-de Gennes model allows to calculate the electron-phonon coupling constant in such an inhomogeneous bilayer, when the two layers are thinner than the coherence length, with

$$\lambda_{eff} = \frac{N_1^2 V_1 d_1 + N_2^2 V_2 d_2}{N_1 d_1 + N_2 d_2} \sim \lambda_1 \left(1 - \frac{d_{eff}}{d_1 + d_2}\right) \tag{9.2}$$

where $N_i$, $V_i$, and $d_i$ are the DOS, electron-phonon potential, and thickness of the two layers 1, 2. Similar density of states were assumed in the two layers. As $d_{eff} = (V_2/V_1 - 1)d_2$, we find for a S/N bilayer $d_{eff} = d_2$, the thickness of the normal layer.

The experimental data could then be modeled by $\lambda_1 = 0.45$, a value compatible with the theoretical estimations [3, 25] and $d_{eff} = \alpha/n_B$, with $\alpha = 1.2 \times 10^{15}$ cm$^{-2}$, compatible with the expected maximal thickness of a pseudomorphic layer before the appearance of the relaxation. This gives a dose-dependent coupling parameter $\lambda_{eff} = \lambda_1$ $(1 - \alpha/\text{dose})$, identifying a nonsuperconducting layer a few nm thick.

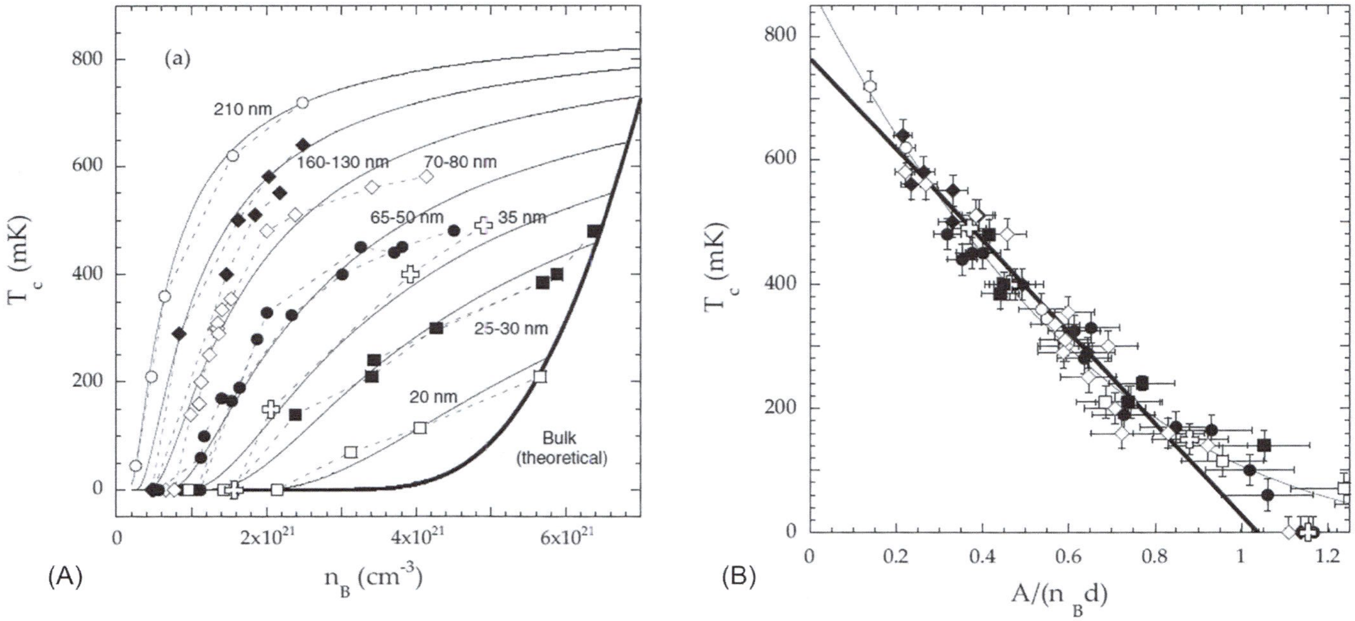


**FIG. 9.9** (A) Doping dependence of the critical temperature of Si:B epilayers at the indicated values of the layer thickness. The *thick solid line* is the critical temperature expected from ab initio calculations of the electron-phonon coupling constant $\lambda$. The *thin lines* correspond to the values expected from Eq. (9.2) [24].

Multiple evidences of bilayers can be found in the structural Si:B characterization: the doping concentration decreases below the superconductivity threshold at the bottom of the doped layer, creating a few nanometer-thick metallic layer; a lower concentration of *active* boron at the Si:B/Si interface may also be induced by the fusion front as it slows and stops at the bottom of the melted layer before starting in the opposite direction as a solidification front; finally, SIMS, APT, and XRD evidenced two regions of different doping concentrations within the melted layer, probably resulting from strain relaxation, but present only at large doping (>5 at.%). Despite a few solid hypotheses, however, no full understanding of the origin of the bilayer has been achieved at the moment. Ongoing measurements of the depth-resolved concentration of active boron atoms may give useful information in relating the superconducting properties to the structure of Si:B ultra-doped layers.

In conclusion, thin homogeneous ultra-doped Si:B epilayers were grown by GILD on Si substrates with a sharp interface of low roughness. The doped depth could be finely modulated in the 10- to 300-nm range, while the doping was tuned in the 1–10 at.% range. The boron is mostly introduced as an active atom in substitutional sites, and does not create aggregates thanks to the fast laser-induced solidification. Above a dose threshold of $5 \times 10^{15}$ $cm^{-2}$, Si:B displays BCS superconductivity with hole Cooper pairs. Si:B is a dirty type II superconductor, with a $T_c$ tuneable with the doped dose from 0 to 0.7 K.

## 9.4 All-silicon superconducting devices

Thanks to the extensive characterization and global understanding of the structural and low temperature transport properties of ultra-doped Si:B, proof-of-concept quantum devices could be demonstrated in the last 6 years. On one side, the realized devices allowed to investigate in more depth the physics of the superconducting boron-doped Si, especially at high frequency. On the other side, all-silicon quantum devices represented a first step toward an integrated Si quantum electronics and ultra-sensitive astronomic detection of certain interest owing to silicon potential.

Quantum information is actually one of the main challenges in the search of faster, more efficient, data treatment. In this frame, spin qubits made of isotopically purified silicon have recently emerged as one of the most promising technologies. Indeed, silicon is an extremely well-known material, where chemical purification, crystal growth, defect control, etc., have been developed to a high extent. Similarly, micro- and nanofabrication clean-room processes have been optimized for decades, allowing the realization of scalable devices. Finally, thanks to a low spin-orbit coupling and to the presence of stable isotopes without a nuclear magnetic moment, it is possible to obtain inimitable electron and nuclear spin coherence times [26]. Silicon thus appears as a choice material to realize coherent quantum circuits, even more so when coupled to superconductivity, whose robustness and absence of dissipation were proven a great asset in the realization of advanced solid-state structures for quantum information [27]. Superconducting silicon offers then a perfect solution to combine the advantages of Si and superconductivity.

In the following part, the fabrication of three main superconducting silicon devices is described: (i) the superconducting quantum interference device (SQUID), which allowed demonstrating the macroscopic nature of the quantum wave function of superconductivity in silicon, as opposite to a percolation-controlled transport; (ii) the all-silicon Josephson junction, a first step toward the Josephson transistor, which, thanks to the extremely transparent epitaxial interface between superconducting Si:B (S) and doped Si (N), allowed demonstrating long phase coherence times and a full proximity effect in all-silicon SNS junctions; (iii) the superconducting silicon resonator, which highlighted the large Si:B kinetic inductance, and its strong nonlinearity with power, proposing superconducting silicon as a choice material for large tunable inductors, sought after in quantum information.

### 9.4.1 Superconducting quantum interference device (SQUID)

The first demonstration of a superconducting device made entirely of silicon was published in 2015 [28]. DC SQUIDs were fabricated with standard e-beam lithography and RIE etching from single layers of superconducting Si, 80 nm thick and with a B concentration of 5 at.%. A SQUID is formed by two Josephson junctions connected in parallel in a superconducting loop. The Josephson junctions implemented were Dayem bridges, realized by designing a constriction 100 nm wide and 100 nm long in the superconducting wire (highlighted in Fig. 9.10A), which effectively decreased the critical temperature creating a weak link (Fig. 9.10A). A perpendicular DC magnetic flux applied in the loop dephases the current circulating in the two branches, thus inducing a modulation of the SQUID current with period $\Phi_0$ ($\Phi_0 = h/2e$, the superconducting quantum flux). The critical current (the maximum superconducting current) $I_c$ measured had an amplitude of a few μA, determined by the ratio of the weak link length and superconducting coherence length ($\xi \sim 50$ nm). $I_c$ was shown to be modulated by 15%–30% in the 100–200 mK range ($T/T_c \sim 0.4$–$0.77$), and its dependence with magnetic field was well modeled by the Ginzburg-Landau equations (Fig. 9.10B). The shape of the measured oscillations could be even more precisely understood by considering, in a symmetric SQUID (where the two junctions in parallel are identical), the inductance of the loop [29], as shown in Fig. 9.10C; the residual differences seen can probably be attributed to a slight junction asymmetry.

The latter modelization allowed a first estimation of the high kinetic inductance (the inductance associated with the inertia of Cooper pairs) of superconducting silicon: indeed, a kinetic inductance $L_{k,sq} = 140$ pH, much larger than the geometric inductance $L_{g,sq} = 0.5$ pH, was found, comparable with that of strongly disordered superconductors such as TiN [30], NbN [31], W [32], and NbSi [33].

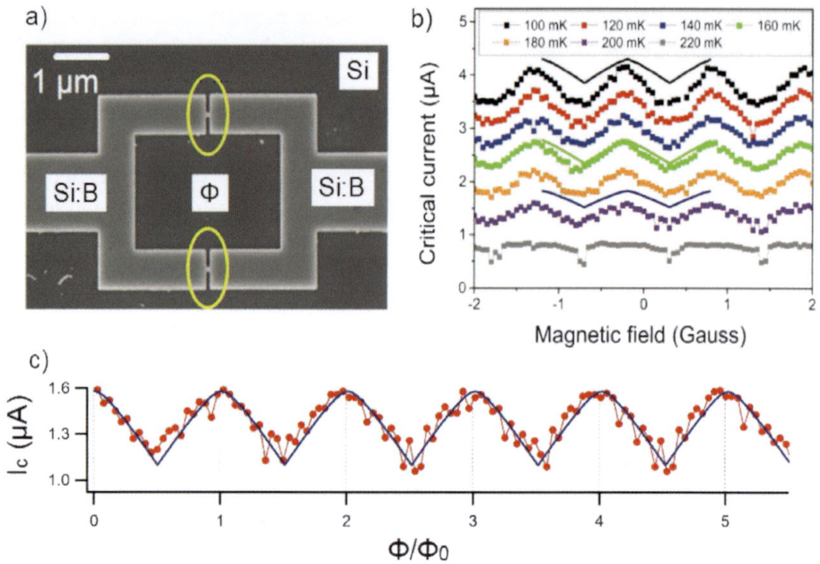


**FIG. 9.10** (A) SEM micrograph of a SQUID etched from a single Si:B layer: a superconducting Si loop (S) is interrupted by two constrictions of weaker $T_c$ (S′) creating SS′S Dayem bridges *(highlighted in yellow)*. (B) Critical current oscillations with magnetic field (creating a magnetic flux in the loop) for temperatures in the 100–220 mK range, and theoretical expectations from Ginzburg-Landau equations [28]. (C) Critical current modulation at 200 mK, fitted with the model from Ref. [29] for a symmetric SQUID, taking into account the inductance of the loop.

These results, showing a well-modeled flux modulation at low temperature and low field, demonstrated the macroscopic nature of the quantum wave function of superconductivity in silicon, and highlighted its high kinetic inductance, opening routes toward silicon-based superconducting quantum engineering.

### 9.4.2 Proximity effect and all-silicon superconductor/normal metal Josephson junctions

As a function of doping, silicon can assume a semiconducting (sc), metallic (N), or superconducting (S) state. This allows the conception of all-silicon complex devices, exploiting the advantages of each state. One of the most interesting devices is the Josephson transistor (JoFET), in which a semiconducting wire is coupled to superconducting contacts, creating a nondissipative, gate-tunable, Josephson junction. Indeed, thanks to the proximity effect, superconducting correlations can penetrate in the semiconducting wire over the phase coherence length, if the interface is transparent enough, and a supercurrent can thus circulate in the semiconductor and be controlled by a gate. JoFET has been realized using metallic superconductors (Nb, PbInAu) and carbon nanotubes,

2D electron gases (2DEG), or nanowires (NW) as semiconductors (InAs, GaAs.) [34]. Even though gate modulation of the critical current was demonstrated in a few papers, these devices faced two main problems: a complex geometry and structure, e.g., in order to fabricate a contact without affecting the 2DEG/NW mobility and ballistic transport; and a low transmission, due to the Schottky barrier at the S/sc interface avoidable only for particular combinations of superconductors and semiconductors. Moreover, even in those cases where an ohmic contact could be achieved, the interface transparency was limited by the difference of Fermi velocities in the metallic superconductor on one side and the semiconductor on the other, due to the large difference in carrier density.

On the contrary, all-silicon SNS Josephson junctions were shown to achieve extremely transparent interfaces, owing to the epitaxial growth of the superconductor over the semiconducting Si and to similar Fermi velocities, while at the same time benefitting from the well-known Si technology [35, 36]. While the Si JoFET has not yet been realized, the results discussed below show the demonstration of full proximity effect in doped Si wires 100–500 nm long, thanks to an excellent interface transparency and a long phase coherence length, constituting the first step toward the JoFET.

Fig. 9.11A (inset) shows a scheme of the superconducting Si/doped Si structure realized in Ref. [36]. The SN bilayers consisted in the superposition of a superconducting layer on the top of a normal (nonsuperconducting) one. First, a normal layer of thickness $d$ was realized by GILD. The laser energy density was then reduced to dope a thinner layer above the superconducting threshold. The bilayers fabricated had an upper superconducting layers of thickness $d_S = [30, 80]$ nm with a critical temperature $T_{c,S} = 470$ mK, on the top of normal layers of thickness $d_N = d - d_S = [10, 90]$ nm and concentration $3 \times 10^{20}$ cm$^{-3}$.

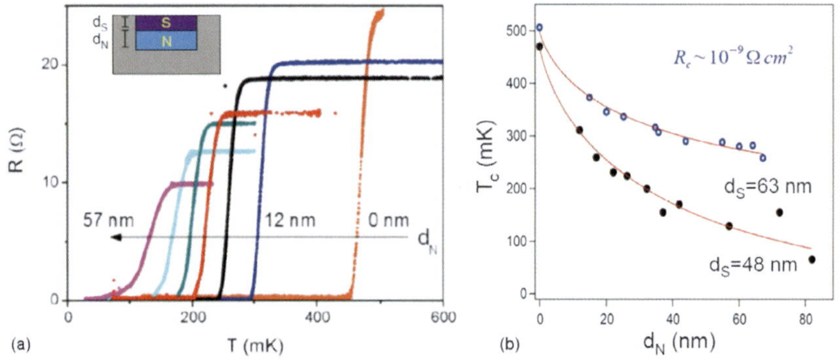


**FIG. 9.11** (A) Resistive transition of superconducting Si/metallic-doped Si bilayers for increasing thickness of the metallic layer. (B) Critical temperature vs metallic layer thickness for two series of bilayers of different superconductor thickness. The extracted contact resistance is $\sim R_c = 10^{-9}\ \Omega\ \mathrm{cm}^2$ [36].

Characteristic resistive transitions of the SN bilayers are shown in Fig. 9.11A. The transitions are quite sharp, with width varying between 25 and 58 mK, and show a single transition indicating a homogeneous superconductor. A drastic reduction of the bilayers' $T_c$ was observed with the increase of the normal layer thickness, leading to a saturation at large $d_N$. The $T_c$ reduction, as high as 86%, was more marked in the series of samples with the thinner superconducting layer (Fig. 9.11B). This is expected, if the transparency is good enough, for an S layer thinner than the superconducting coherence length $\xi$ (the "size" of a Cooper pair, $\xi \sim 80$ nm). Indeed, in this case, the Cooper pairs explore both S and the N layers, but experience an attractive potential only in the S layer. The effective attractive potential will therefore be on average reduced for thicker normal layers as Cooper pairs spend a longer time in the N region. The saturation can be qualitatively understood as the Cooper pairs explore the normal layer on an average depth $\xi$: when $d_N > \xi$, the superconductivity is no longer sensitive to an increased thickness of the normal layer.

The strong influence of the normal layer over the superconducting one supported the expectation of a transparent S/N interface, as the interface transparency is the determinant parameter to describe the $T_c$ suppression with $d_N$. The variation of $T_c$ with $d_N$ (at fixed $d_S$) and with $d_S$ (at fixed $d_N$) was modeled with the analytical solution of the linearized Usadel equations near $T_c$ in the limit of good transparency ($\tau \gg k_B T_c$) [37]. The estimated interface resistance was $R_c \sim 10^{-9}\ \Omega\,\text{cm}^2$, smaller by an order of magnitude than the experimental [38] or theoretical [39] determinations of silicon/metallic superconductor contact resistance.

SNS junctions were realized from the described SN bilayers with $d_S = 57$ nm, $d_N = 58$ nm, $T_c = 200$ mK, and a normal layer concentration of $5.2 \times 10^{20}\ \text{cm}^{-3}$. The N wire width $w = 3.8$–$8\ \mu$m was defined by reactive ion etching. Then, the superconducting upper layer was completely etched by RIE over the junction length of 150–350 nm (see inset of Fig. 9.12B).

The SNS junctions' $V(I)$ curves (Fig. 9.12A) showed a zero-resistance state, evidence of a proximity effect extended over the entire length of the weak link, a result of a long phase coherence length, estimated over 300 nm. The obtained IV characteristics were standard: a sharp jump when the junction switches to the resistive state at a critical current $I_c \approx 0.5$–$3\ \mu$A, and a hysteresis cycle characteristic of junction heating, as, reducing the bias current, the junction switches back to the superconducting state at the retrapping current $I_r < 1\ \mu$A.

The decrease in the critical current with temperature, shown in Fig. 9.12B, could be successfully modeled with the de Gennes expression for a dirty doped semiconductor weak link. A good agreement of the critical current temperature dependence could also be obtained with the models developed in Ref. [42], describing a SN-N-NS "sandwich" junction in the dirty limit with no potential barrier at the SN interface, or in Ref. [43], describing a SS'S/SNS junction in the dirty limit with small interface resistance and $L$ of the order of the coherence length. As before, this pointed toward the presence of a highly transparent

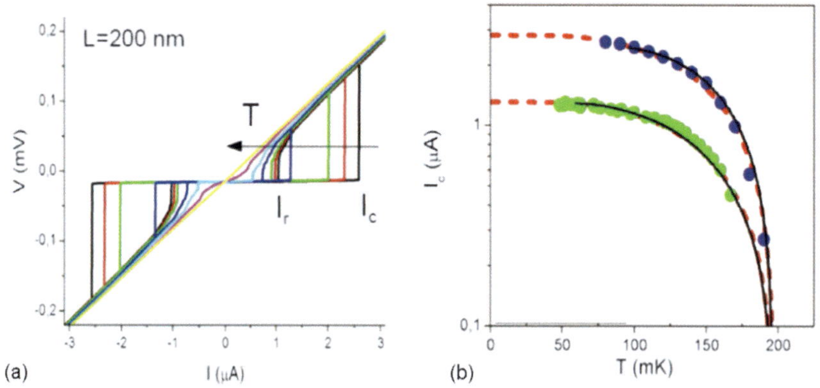


**FIG. 9.12** (A) $V(I)$ curves for a junction with weak-link length $L = 200$ nm in the temperature range 70–210 mK (from *black to yellow*) for increasing temperature and up and down current sweeps. (B) Critical current dependence on temperature for two junctions with junction length $L = 200$ nm *(blue)* and $L = 285$ nm *(green)*. Continuous lines: de Gennes expression for a dirty doped semiconductor weak link [40]. *Dotted lines* are Dubos formula for a short SNS junction [36, 41].

interface between the superconducting silicon contacts and the doped silicon wire, and demonstrated a good understanding of Josephson junctions made entirely of silicon.

### 9.4.3 Superconducting microwave resonators

Superconducting coplanar waveguide (CPW) resonators are present in many fields: quantum information, astronomic detection, and mesoscopic physics. For example, they are used in arrays of thousands of multiplexed photon detectors [44], appear as read-out, interconnections and memory elements in quantum processors [45], and are proposed in the design of a scalable 3D architecture for quantum computers, as they constitute lossless connections preserving the system coherence [27].

As the research on quantum information goes beyond the few-qubits operations and approaches larger circuit scales [46], the problems connected to scalability and costs arise. In this aspect, superconducting silicon CPW resonators are an interesting alternative to metallic superconductors, as the fabrication would profit from the low costs, already developed microelectronics industry. Admittedly the laser doping necessary for the fabrication of superconducting silicon is not standard; however, the use of laser processes is becoming well established in the microelectronic industry and may not constitute an insuperable obstacle. The possibility to integrate laser doping in the process flow of CMOS fabrication has been already demonstrated and constitutes the first step toward more complex integration [47]. As a further incentive, doped silicon benefits from a low carrier density, leading to the possible gate control of the superconducting properties, a precious tool for the conception of circuits and devices.

Superconducting silicon CPW resonators were realized with a central Si:B conductor flanked by two Al ground planes (Fig. 9.13E) and connected to the external circuit through coupling capacitors (transmission geometry, Fig. 9.13C) or by coupling to a feedline (shunt coupling, Fig. 9.13D) [48]. Half- or quarter-wavelength CPW resonators were fabricated from 20- to 80-nm-thick Si:B

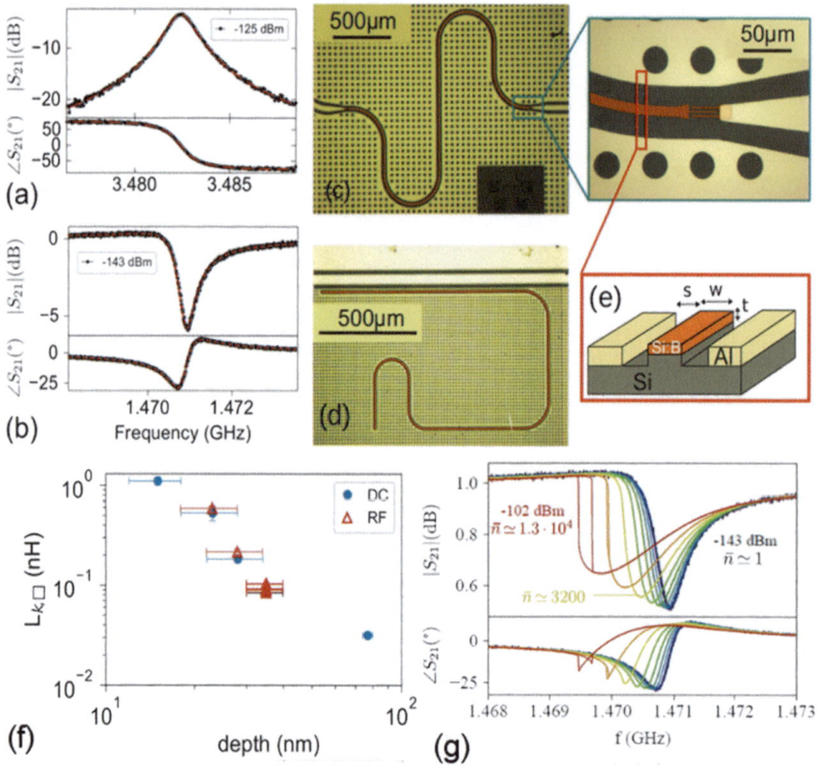


**FIG. 9.13** (A) Typical resonance curve measurement *(black dots)* in a transmission-coupled geometry at $T = 14$ mK and read-out power $P = -125$ dBm. The setup background transmission has been removed. (B) Typical resonance curve measurement for shunt-coupled $P = -143$ dBm. *Red dashed lines*: modeled transmission coefficient. (C) Optical micrograph of the transmission geometry. (D) Optical micrograph of the shunt geometry. (E) Scheme of the CPW section. (F) Sheet kinetic inductance $L_{k,\Upsilon}$ as a function of Si:B depth inferred from measurements at DC *(blue dots)* or at RF *(red triangles)*. The error bars attributed to the thickness are associated with the uncertainty of the thickness of the superconducting layer within the doped layer. (G) Amplitude and phase of the transmission coefficient representing the resonance curves in the nonlinear regime attained when increasing the input power, for a number of photons in the resonator varying from 1 *(black line)* to 13,000 *(red line)*. Both up- and down-frequency sweeps are plotted, showing hysteretic behavior at high power [48].

films ($d \sim \xi$) epitaxied on a high-resistivity (5 kΩ cm) Si substrate. The [5–50]-μm-wide Si:B line was etched into the doped layer with fluorine-based reactive ion etching (RIE), while the capacitances (~1 fF), feedline, and ground plane were realized with evaporated Al. The resonant frequencies ($f \sim$ 1–12 GHz) and the quality factors ($Q \sim$ 1000–4000) were deduced from the fit of the resonant transmitted RF power (Lorentizian) and phase (Fig. 9.13A and B). The sheet kinetic inductance was determined from the resonant frequency and confirmed with SONNET electromagnetic simulations. As expected, superconducting silicon resonators showed a high kinetic inductance $L_k$, attaining values as high as 500 pH/ϒ (Fig. 9.13F) and completely dominating the geometric inductance $L_g \sim 0.5$ pH/ϒ. Indeed, the kinetic inductance at zero temperature follows $L_{k,\Upsilon} = (hR_\Upsilon)/(2\pi^2 \Delta_0)$, where $\Delta_0$ is the superconducting gap at zero temperature, so that $L_{k,\Upsilon}$ is proportional to the square resistance $R_\Upsilon$ (large in doped silicon due to the low carrier concentration and disorder) and inversely proportional to the critical temperature $T_c$ (low in Si:B) through the superconducting gap $\Delta_0 = 1.76k_BT_c$. Moreover, $L_k$ is modulable with the dopant density, via the modulation of the superconducting critical temperature. Fig. 9.13F shows $L_k$ for different resonators of varying superconducting layer thickness in the [20–80]-nm range, compared with the BCS expectations estimated using the DC-measured $R_\Upsilon$ and $T_c$. The good agreement observed confirmed the validity of BCS description (as does the temperature dependence of the kinetic inductance) and suggested that values of the order of 1 nH/ϒ could be obtained in Si:B resonators patterned from thinner films.

This large, modulable kinetic inductance is a very interesting feature, as any perturbation to the superconducting condensate, caused by electrical currents, phonons, electromagnetic radiation, or magnetic vortices, modifies the kinetic inductance. The sensitivity to external perturbations enables a parametric tuning of the kinetic inductance, which can be exploited, e.g., for parametric amplification [49, 50] or tunable inductors [51]. Such applications exploit in particular the nonlinearity of the kinetic inductance with current. Disorder, by weakening the superconductivity, yields a higher susceptibility to pair breaking and thus enhances the kinetic inductance nonlinearity. It was found that, owing to its low carrier density and moderate level of disorder, superconducting silicon displays a high kinetic inductance with an unexpectedly strong nonlinearity with RF power.

Indeed, when increasing the read-out power, the resonances shift toward the low frequencies, then become progressively skewed, up to a bistable state where a hysteresis is observed between frequency sweeps up or down (Fig. 9.13G). Alongside with this change of resonance frequency and shape, an increase in the internal losses was noticeable, with a reduction of the resonance amplitude. In Si:B, the nonlinearity appeared at extremely low power: as an example, the first mode ($f_1 = 1.48$ GHz) of a 5-μm-wide, shunt-coupled, quarter wavelength resonator displayed frequency shifts of the order of 100 kHz at read-out powers of the order of fW, corresponding to about 200 photons in the resonator, while the hysteresis appeared for pW powers. The measured nonlinearity was unexpectedly large, up to three orders of magnitude higher than the nonlinearity induced by depairing or quasiparticle heating predicted in the framework of BCS superconductivity. Instead, the measurements pointed toward a mechanism involving vortices motion by the RF current, which may also be responsible for the observed internal losses. This huge nonlinearity, coupled to the large superconducting silicon kinetic inductance, makes superconducting silicon a promising material for detection applications and possibly quantum circuits provided that an increase of the quality factor can be achieved.

## 9.5 Toward large-scale applications: Laser-annealed superconducting layers on SOI

As shown in the previous sections, proof-of-concept quantum devices made of superconducting silicon have been successfully demonstrated, even though optimization is still needed. However, a sizeable amount of work is required to implement such devices in industrial facilities. To do so, it is necessary to integrate superconducting silicon on standard microelectronic circuits, and to fabricate scalable devices within the actual clean-room facilities without impinging on the already-implemented processes.

In this section, the realization of superconducting silicon on SOI (silicon on insulator) substrates by GILD and PLIE is discussed. SOI structures consist of a film of single crystalline Si separated from the bulk Si substrate by a layer of $SiO_2$, called buried oxide (BOX). SOI is used for a large-scale integration of electronic circuits, as the transistor bodies are isolated from the Si substrate and the surrounding transistors by the amorphous dielectric BOX layer. SOI structures are interesting for industrial applications as they minimize the impact of ionizing radiation on the device performance, providing isolation from the generated charges; insure fast transistor switching; reduce the power consumption by lowering the operating voltage; and contain the "short channel effects" improving the prospects of building 10-nm channel lengths [52–54].

Laser annealing allows activating dopants in the very few nanometers from the surface, by triggering localized and depth-confined fusion. As the SOI layers widely used in microelectronics are typically 20–30 nm thick, annealing these structures by pulsed laser is a real challenge. Indeed, to fabricate an epitaxial fully crystalline Si:B layer, the melted Si region needs to be smaller than the SOI thickness to begin on a crystalline seed; otherwise, the final layer will be amorphous or polycrystalline. Moreover, oxygen atoms may diffuse in the melted Si from the BOX, introducing electrical impurities in the doped layer. Thus, the melted region and consequently the laser energy density need to be controlled with great precision at the nanometer level. Such control, based on the measure of the time-resolved reflectivity (TRR), becomes more complicated on SOI, due to the interferences of the probing laser in the four-layer structure as compared with the two-layer system of a bulk Si substrate. In general, different wavelengths are needed to probe the TTR as the interference is strongly dependent on the SOI thickness. Finally, the homogeneity of the energy density needs to be very high to avoid punctually reaching the oxide.

As an added difficulty, to achieve the highest critical temperatures, it is necessary to maximize the doped depth and master laser melting down to a few nm from the oxide, without touching it. Indeed, the highest critical temperatures correspond to the highest boron doses, the maximum dose being found at the largest thickness as a consequence of the concentration saturation (Fig. 9.5B).

### 9.5.1 Superconductivity in SOI with GILD

In her PhD thesis, A. Francheteau describes the first results of gas immersion laser doping on SOI substrates, obtained with the same setup described in Section 9.2 [55]. To monitor at best the silicon fusion dynamics in SOI, time-resolved reflectivity measurements were performed with a blue laser in addition to the original red one. The 23-nm-thick active top Si layer had an initial implanted boron dose of $10^{15}$ $cm^{-2}$, well below the superconductivity requirements. The SOI was then doped with GILD over two different thickness and for a variable number of laser pulses determining the final active B dose. The resulting samples showed a superconducting behavior with $T_c$ varying from 120 to 230 mK, increasing with the number of laser pulses. These $T_c$ are comparable with those obtained on bulk silicon for similar thickness and at low doping dose (i.e., low number of pulses). However, above 100 laser pulses, in SOI the $T_c$ saturated around 230 mK, while in bulk Si it increased up to 400 mK. This difference may be explained by a larger impurity level in the layer doped on SOI, or by the presence of deep dislocations due to the implantation process.

Nevertheless, these first experiments showed that, with an adapted in situ time-resolved reflectometry and thanks to the sharp interface specific to laser annealing, it was possible to achieve an excellent control of the doped thickness down to 1- to 2-nm precision (Fig. 9.14).

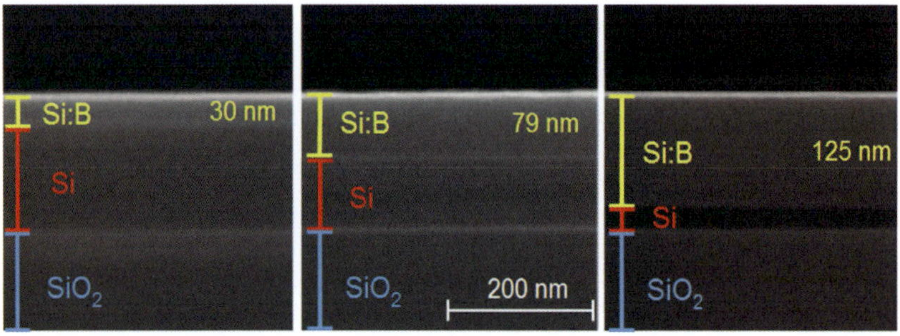


**FIG. 9.14** Selection of scanning electron micrographs showing a cut of a Si:B layer doped on a thick SOI substrate, for different laser energies allowing to melt and recrystallize 30 nm *(left)*, 79 nm *(center)*, or 125 nm *(right)*. Note the sharp, straight interfaces.

GILD doping generally requires a large number of laser pulses (200–300) to reach the concentrations of interest for superconductivity, and the energy density needs to be constantly tuned to keep a constant doped thickness. The difficulties of such control may be avoided, and a better reproducibility attained, by pulsed laser-induced epitaxy (PLIE), where a single laser shot can activate preimplanted boron atoms.

### 9.5.2 PLIE on SOI under ultra-high vacuum

PLIE was first performed on bulk Si by A. Grockowiack et al. [13]. $B_2$ was implanted with a dose of $2.5 \times 10^{16}\ cm^{-2}$ and an energy of 3 kV, to keep the implantation-induced defects close to the surface. Indeed, the maximum of the dose was located in the first ~15 nm from the surface, which was thus partially amorphized. The samples were then annealed over a thickness varying from 10 to 140 nm, and for a number of laser pulses from 1 to 50. SIMS, APT, and XRD analyses showed that a homogeneous crystalline layer with at most ~0.05% B dimers could be obtained for an optimum of 5–10 laser pulses. For a smaller number of pulses, the concentration profile was not homogeneous, the B not having yet reached the bottom of the melted layer, while for a larger number of pulses, an inhomogeneous redistribution of the strain was observed. All the layers, with thickness [20–100] nm and concentration [2.7–9] at.%, showed superconducting characteristics with an almost constant $T_c \sim 250$ mK, comparable with $T_c \sim 300$ mK obtained by GILD for a similar dose. This was a supplementary confirmation that the boron dose, in PLIE fixed by the implantation, controls the superconducting properties.

The study of pulsed laser-induced epitaxy on SOI samples was started by A. Francheteau and collaborators [55]. The chosen substrate was a 34-nm-thick SOI with a 20-nm-thick BOX, implanted with a $B_2$ dose of $2.5 \times 10^{16}\ cm^{-2}$ at an energy of 3 keV. The resulting SOI was formed by a partially amorphized layer of ~15 nm thickness and containing ~75% of the implanted dose (estimated with crystal TRIM simulations [56]) on the top of a crystalline Si layer. The B was activated by 1 or 5 laser pulses at various energy densities, from the fusion threshold to the entire melt of the SOI layer.

Fig. 9.15 shows the dependence of the time-resolved reflectivity (TRR), square resistance, and X-ray reflectivity for one laser shot as a function of the annealing energy density (ED). Every point corresponds to a different $2 \times 2\ mm^2$ laser-annealed region on the same SOI substrate. Three main regimes were identified:

(1) $0.35\ J/cm^2 < ED < 0.425\ J/cm^2$: *Explosive crystallization and melting of the a-Si.*

The TTR shown in Fig. 9.15A and C in this energy range presents a reflectivity peak in the first few nanoseconds of the energy rise of the laser pulse. This first peak corresponds to the explosive crystallization (EC) [57]

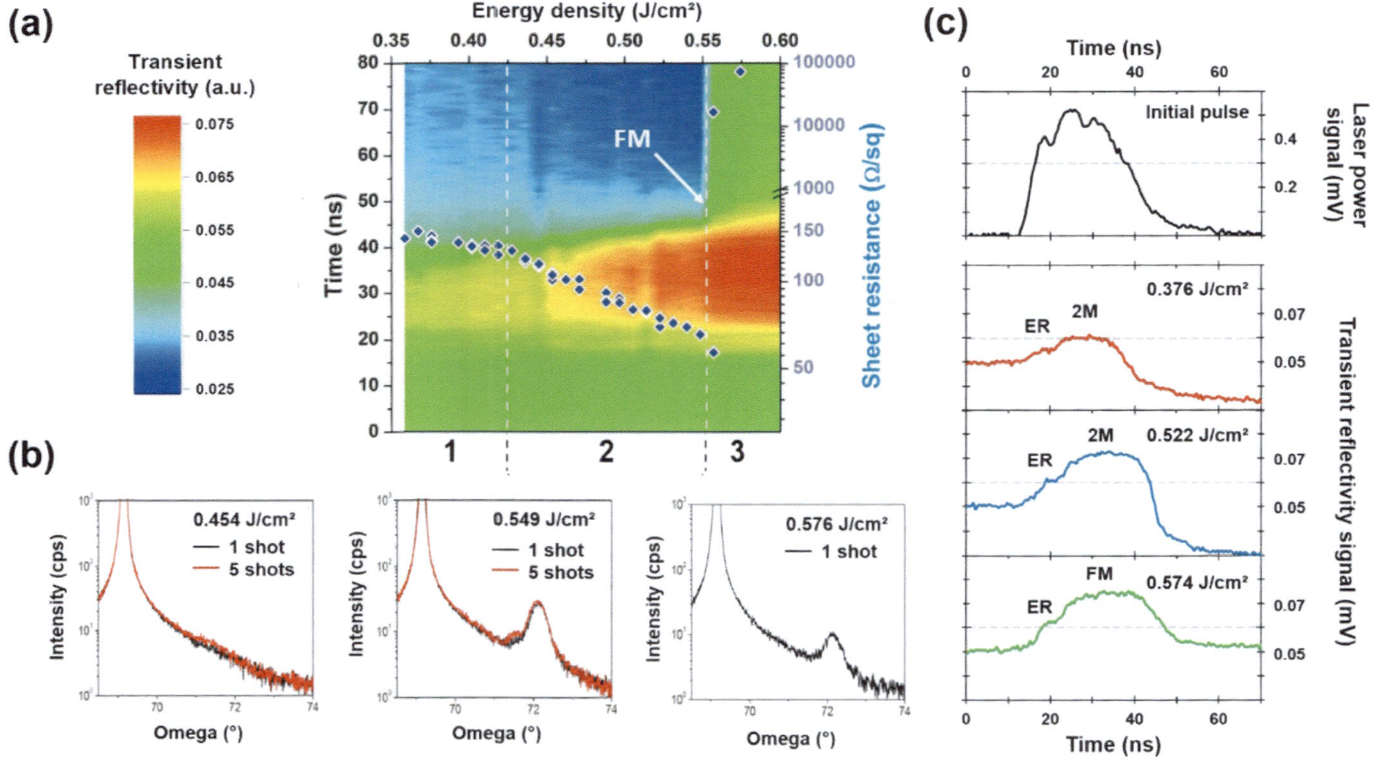


**FIG. 9.15** (A) Time-resolved reflectometry (TRR) map as a function of time and energy density for single-pulse anneal. The corresponding sheet resistance is superimposed. The different regimes identified are numbered from 1 to 3. ER, 2 M, and FM refer to explosive recrystallization, second melt, and full-melt, respectively. (B) X-ray diffraction $\omega - 2\theta$ spectra for 1 and 5 laser shots displaying a peak associated with the Si:B layer, tensile strained, on the right of the substrate Si peak. (C) Amplitude as a function of time (measured by TRR) for single-pulse anneal at various energy densities (first curve shows the pulse time profile).

of the implanted layer. During the EC, the release of the fusion latent heat in a-Si induces a fast fusion-solidification cycle, without an underneath crystal to insure its coherence. This results in an erratic crystallization over the implanted region. As the laser impulsion continues, a fusion of the polycrystalline layer takes place, and the recrystallization starting from random seeds leads to a polycrystalline layer with a weak dopant activation.

The final TTR is lower than the initial one, as a result of the passage from an amorphous to a polycrystalline structure. No XRD peak associated with the Si:B layer can be observed. When the energy density increases, the resulting polycrystalline layer thickens, and the square resistance slightly decreases at nearly constant active dose (Fig. 9.16), probably an effect of a slight reduction of the material defects at larger temperatures.

**(2)** 0.425 J/cm$^2$ < ED < 0.55 J/cm$^2$: *Ultra-doped crystalline Si:B layers*

Above 425 mJ/cm$^2$, the behavior changes drastically. The TRR signal presents two peaks: one, of short duration (~3 ns), followed by a wider (~30 ns) one with larger amplitude. The first corresponds as before to the explosive crystallization of the a-Si. The second peak corresponds to a fusion front which attains the c-Si beyond the polycrystalline region created by EC from the a-Si. The solidification starts then from a monocrystalline

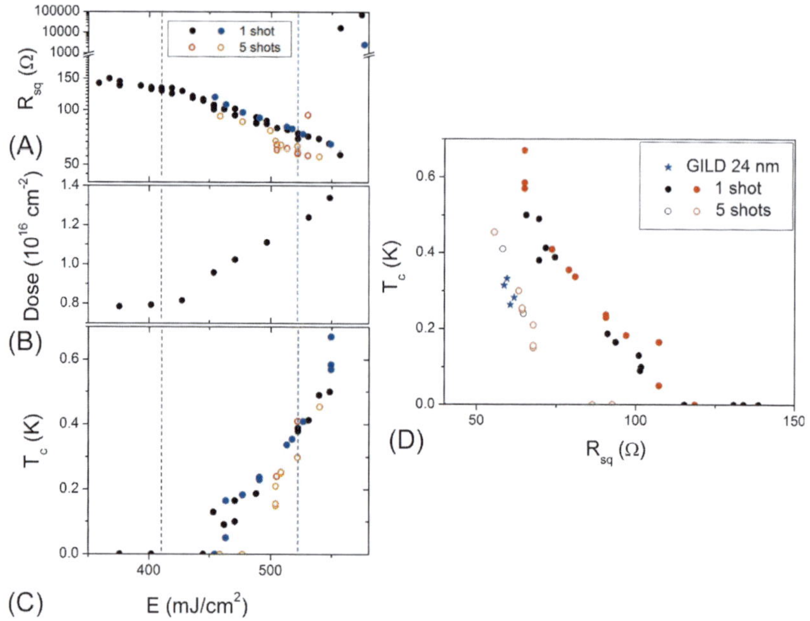


**FIG. 9.16** Sheet resistance (A) compared with active boron dose (B) and critical temperature (C) as a function of energy density for 1–5 pulses laser annealing. Regimes 1–3 are separated by *dotted lines*. (D) Critical temperature as a function of sheet resistance for 1 and 5 laser pulses, compared with Si:B layers of similar thickness made by GILD on bulk Si.

seed and is similar to bulk Si. The final reflectivity is about half the reflectivity of the untreated sample, and lower than the reflectivity for ED < 425 mJ/cm$^2$, indicating better crystalline properties of the annealed layer. This is confirmed by the presence of a peak associated with Si:B in the X-ray diffraction spectra (Fig. 9.15B), which can already be observed for ED = 450 mJ/cm$^2$ at $\theta = 35.7$ degrees for 1 or 5 laser pulses. The amplitude for five laser pulses is slightly larger as a result of better crystallization upon multiple anneals. The position of the Si:B peak, shifted from the substrate Si peak by $\Delta\theta \sim 1.5$ degrees, indicates strained layers with $\Delta a_\perp / a_\perp = -3.5\%$, corresponding to the maximum strain induced in bulk Si by GILD. Increasing the energy induces an increased amplitude in the XRD peak observed, as the thickness of the crystalline layer increases. A second peak can be observed on the left of the main Si:B one, especially for five shots. It may either correspond to a thickness fringe, resulting from the sharp Si:B/Si interface (such fringes were observed in bulk Si layers thick from 30 to 200 nm and employed to extract the layer thickness), or correspond to a doping inhomogeneity within the doped layer with the creation of two layers (as already observed at large doping in bulk Si).

This second regime corresponds thus to fully crystalline, doped Si:B layers, thicker than the a-Si region (~20 nm). The active B dose linearly increases with the laser energy density, with a maximum B dose activated by a single laser pulse $\sim 1.1 \times 10^{16}$ cm$^{-2}$, corresponding to the saturation concentration of $\sim 4 \times 10^{21}$ cm$^{-3}$ (the total implanted dose being extremely large, the saturation concentration is attained for each energy/melted thickness). The sheet resistance decreases with the energy density as a consequence of the increased active boron dose, reaching a minimum of 56 Ω for a single laser pulse. For five laser pulses, the measured sheet resistance is 25% smaller on average, as a result of a better crystallization following multiple fast annealing.

(3) $E > 0.55$ J/cm$^2$: *Beyond full-melt: reaching the BOX*

A threshold is reached when the melted region reaches the BOX. A sharp transition can be observed in both the sheet resistance and the reflectivity. $R_s$ diverges up to 10–100 kΩ as the doped layer loses its crystallinity (the Si:B XRD peak losing already a factor of 3 in amplitude at ED = 557 mJ/cm$^2$) and is enriched with oxygen. The reflectivity of the layer after solidification is twice that of the previous regime and is comparable with the reflectivity of the SOI before the laser treatment, characteristic of an amorphous material.

### 9.5.3 Superconductivity in SOI and role of disorder

Low-temperature measurements were performed on all the layers, and superconductivity was found only in the intermediate regime corresponding to crystalline layers with large boron active dose. The dependence of the critical temperature on energy density is shown in Fig. 9.16: above an energy threshold, the $T_c$ increases linearly with the energy up to 670 mK for a layer of 32 nm thickness and dose $\sim 1 \times 10^{16}\ cm^{-2}$. Since the dose increases linearly with the laser energy, these measurements confirm that the $T_c$ is globally controlled by the active B dose, in agreement with previous experiments on bulk Si [58]. Indeed, superconductivity appears when the dose overcomes a threshold of $\sim 9 \times 10^{15}\ cm^{-2}$ active dose. The increase in $T_c$, $1.1 \times 10^{-16}$ K/cm$^{-2}$, is also comparable with the one observed for GILD samples.

The coherence length was extracted from the $(T, H)$ phase diagram using the 2D layer expression $H_{c,2} = \Phi_0/[2\pi\xi^2\ (1 - T/T_c)]$. As a function of the energy density, the coherence length varied in the range $\xi \sim$ [38–56] nm, values comparable with the ones found for GILD on Si for similar thickness [24].

Interestingly, even though the crystallinity is larger for five laser pulses than for a single pulse, the $T_c$ seems comparable or lower at similar energy densities. Indeed, despite a roughly common global dependence of the $T_c$ with the energy density, neither the energy density alone, nor the dose or the crystallinity seems to fully determine the superconducting $T_c$. Instead, the sheet resistance, controlled by both the active boron dose and the disorder level, was found to be a better choice. When plotting the $T_c$ against the sheet resistance, the series annealed with five laser pulses clearly separates from the series annealed with one laser pulse. The change of the $T_c$ with the resistance for five laser pulses becomes faster than for one pulse, and is comparable with that of bulk Si samples of similar thickness doped with GILD. Indeed, as previously discussed, the effect of repeated anneals increases the vertical homogeneity of the dopants, reduces the defects and thus the sheet resistance, and improves the crystallinity. The influence of the crystallinity on superconductivity was already observed for disordered metallic superconductors, such as vanadium, and is currently under study.

In conclusion, superconducting Si:B layers were realized with PLIE on SOI substrates with critical temperatures and sheet resistances comparable with layers realized with GILD on Si. This demonstrated the fine control achieved for thin layers anneals in complex environments (the BOX modifying the heat dynamics and complicating the reflectivity monitoring). Moreover, the larger disorder in the Si:B layers, associated with the larger concentration of interstitial B and the implantation-induced defects, allowed to highlight the sensitivity of superconductivity on the crystalline degree of the doped structure. This behavior is currently under investigation.

## 9.6 Superconducting silicon by PLIE in industrial conditions

The very promising demonstrations discussed so far in this chapter were obtained using laboratory systems with very specific process conditions far from being compatible with an industrial environment and its cost constraints. To implement a new quantum electronics combining superconducting silicon with

CMOS technology, the historical GILD or PLIE process conditions will have to be strongly simplified. First, an atmospheric pressure process would be highly preferable to the current ultra-high-vacuum environment ($\sim10^{-9}$ mbar). Also, the multipulse treatment (>100 pulses) combined with a low repetition rate and a small beam ($2 \times 2$ mm$^2$) is not suitable for an industrial purpose in terms of productivity. For this reason, process conditions were recently adapted in the direction of a more compatible conditions in industrial environment, and preliminary studies were done with a 300-mm industrial pulsed laser annealing system with the same wavelength (XeCl, 308 nm), a large beam and at atmospheric pressure under $N_2$ flux [14].

Samples similar to those already submitted to PLIE in Section 9.5.2 were fabricated starting from 300-mm SOI wafers with a 34-nm-thin layer on top of a 20-nm BOX, itself on top of bulk (100) silicon (CZ, p-type). Then, boron implantation at 3 keV with a total dose of $2.5 \times 10^{16}$ cm$^{-2}$ was performed. At this stage, the implanted SOI layer was partially amorphized (15-nm upper region) with a 19-nm-thick crystal seed below. The last step consisted in the UV-nano laser annealing (NLA) ($\lambda = 308$ nm, pulse duration $= 160$ ns (Fig. 9.17C), 4 Hz repetition rate, $15 \times 15$ mm$^2$ laser beam) under $N_2$ at atmospheric pressure on a chuck stabilized at 200°C to facilitate contaminant expulsion from the surface. Taking into account the drastic change in the laser pulse duration from 25 to 160 ns, the laser energy density was increased compared with the setting used for GILD or PLIE on the laboratory system. Single-pulse and multipulse conditions were investigated: one pulse with ED ranging from 0.400 to 1.500 J/cm$^2$ and 1, 3, 10 or 100 cumulative pulses with *ED* ranging from 0.600 to 1.500 J/cm$^2$.

## 9.6.1 Room temperature characterization after single laser pulse annealing

Thanks to optical characterization (TRR), surface morphology observations (haze), electrical measurements (4-point resistance), structure (TEM), and composition (XRD, not shown) analysis, five laser annealing regimes as a function of the energy density were identified. Fig. 9.17A plots the evolution of the SOI layer sheet resistance and haze superimposed to TRR mapping as a function of the ED. Some of the corresponding TEM observations and TRR signals are reported in Fig. 9.17B and C, respectively.

**(1)** 0.4 J/cm$^2$ < ED < 0.775 J/cm$^2$: *Submelt regime*

At low energy, the low and constant TRR signal indicates a submelt regime (Fig. 9.17A and C). At these conditions, the stack remains unchanged, with a-Si on top of c-Si. The concomitant low haze level indicates that the surface roughness is unchanged, in agreement with no phase change upon annealing. The decreasing sheet resistance with the increase of the energy density demonstrates that even a low thermal budget is sufficient to activate dopants in c-Si.

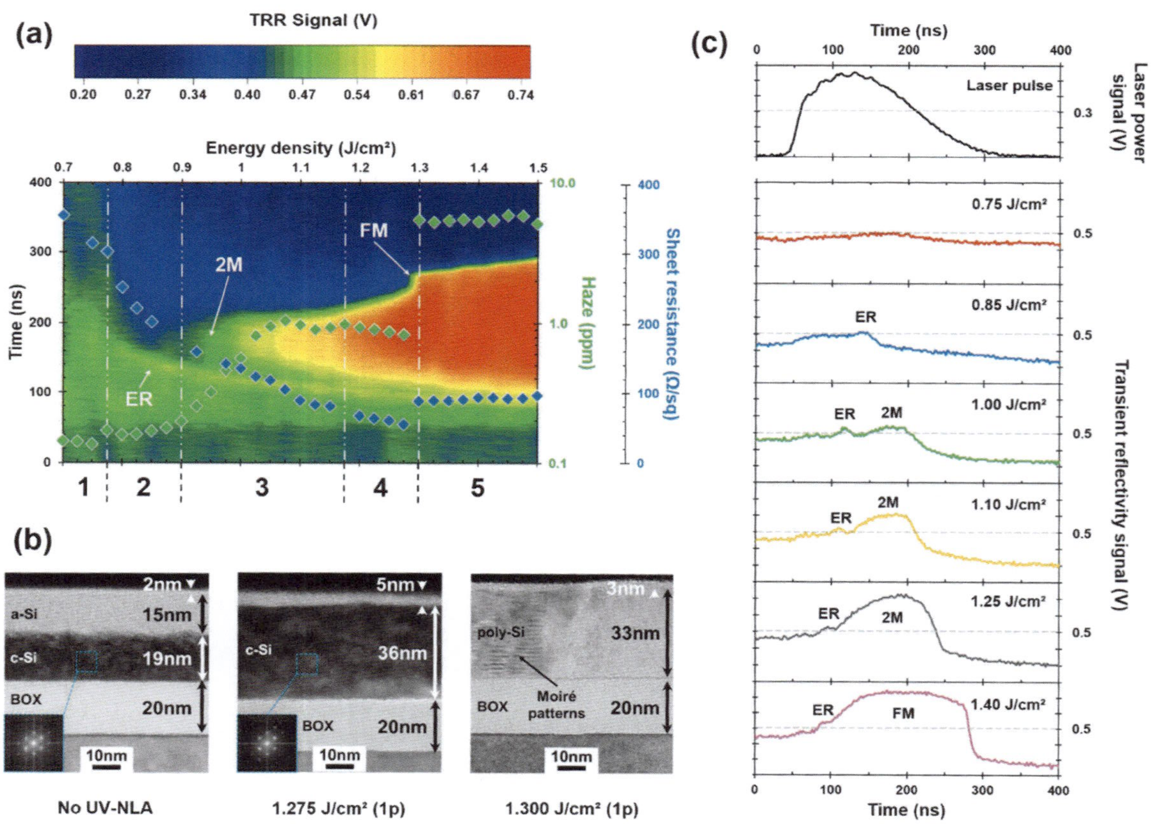


**FIG. 9.17** (A) Time-resolved reflectometry (TRR) map as a function of time and energy density for single-pulse NLA. The corresponding haze and sheet resistance plots are superimposed. The different regimes identified are numbered from 1 to 5. ER, 2 M, and FM refers to explosive recrystallization, second melt, and full-melt, respectively. (B) Cross-sectional TEM observations of the SOI layer after boron implantation, and subsequent UV-NLA (one pulse) at 1.275 J/cm$^2$ and at 1.300 J/cm$^2$. (C) Amplitude as a function of time (measured by TRR) for single-pulse NLA at various energy densities (first curve shows the initial pulse). *(Adapted from R. Daubriac, P. Acosta Alba, C. Marcenat, S. Lequien, T.D. Vethaak, F. Nemouchi, F. Lefloch, S. Kerdilès, ECS J. Solid State Sci. Technol. 10 (2021) 014004.)*

**(2)** 0.775 J/cm$^2$ < ED < 0.9 J/cm$^2$: *Explosive crystallization*

Above 0.775 J/cm$^2$, a short (~20 ns)- and high-amplitude (>0.5 V) TRR peak appears. Such a signal is representative of explosive recrystallization (ER) during which a fast and exclusive melt of the a-Si occurs before a fast solidification into poly-Si [59] (Fig. 9.17A and C). The sheet resistance decreasing might be the result of grain size enlargement and/or activation enhancement as the thermal budget becomes more important. Concerning haze measurements, the slight discontinuity at 0.775 J/cm$^2$ can be explained by the phase change from a-Si to poly-Si.

**(3)** 0.9 J/cm$^2$ < ED < 1.15 J/cm$^2$: *Second melt: polycrystalline Si:B*

Above 0.9 J/cm$^2$, the ER is directly followed by a high-amplitude (>0.5 V) and long-duration (50–200 ns) TRR peak corresponding to a second melt (2 M), a phase modification from solid to liquid (Fig. 9.17A and C). As the energy density increases, the second melt lasts longer and intensifies. Below 1.150 J/cm$^2$, the second melt front is confined inside the poly-Si:B layer, so that the solidification leads to a polycrystalline layer, with an improved electrical activation (Fig. 9.17A). The haze increases from 0.15 up to 1.00 ppm, which corresponds to the apparition of surface asperities, due to polycrystalline recrystallization. XRD measurements ($\omega - 2\theta$) only show a typical silicon peak around 69 degrees coming from bulk and SOI seed layer but no distinctive signature from the poly-Si layer.

**(4)** 1.15 J/cm$^2$ < ED < 1.275 J/cm$^2$: *Second melt: monocrystalline Si:B*

When at least part of the c-Si seed layer is melted, the final layer after recrystallization is monocrystalline (Fig. 9.17A and B). Boron concentrations above the solid solubility are reached thanks to the fast recrystallization (~30 ns) induced by UV-NLA. The deeper the melt-front goes, the thicker the active layer becomes and the more boron atoms are incorporated in substitutional sites, resulting in a decrease of the sheet resistance down to a minimum value of 55 Ω/sq. at 1.275 J/cm$^2$ (Fig. 9.17A). Diffraction measurements exhibit, in addition to the contribution from the Si substrate, a peak around 71 degrees that can be assigned to the heavily boron-doped silicon region.

**(5)** 1.3 J/cm$^2$ < ED < 1.5 J/cm$^2$: *Full-melt: polycrystalline Si:B*

Finally, the last regime corresponds to the melt of the entire SOI layer, leaving no seed for recrystallization, so that the final layer after solidification is entirely polycrystalline. TEM observations at 1.300 J/cm$^2$ show a polycrystalline layer with amorphous regions and some stacking faults (Fig. 9.17B). In this last regime, TRR measurements show a sharp expansion of the melt duration at 1.300 J/cm$^2$ (abrupt extension of the red region in the TRR map, indicated by a FM acronym) due to enhanced heat confinement by the buried oxide that acts as a thermal barrier (regime 5, Fig. 9.17A). Then, the melting time increases linearly with the ED. The sharp haze increase at 1.300 J/cm$^2$ also indicates the transition to a polycrystalline material. Simultaneously, sheet resistance jumps from 55 Ω/sq. at 1.275 J/cm$^2$ up to 87–90 Ω/sq. in the 1.300–1.500 J/cm$^2$ ED range. Thanks to XRD, the transition from partial to full-melt of the SOI layer was confirmed by the disappearance of the peak at 71 degrees.

Based on the previous studies with equivalent samples annealed in ultra-high vacuum using a XeCl laser (25 ns), superconducting silicon was expected in the regime 4, i.e., when the recrystallization leads to a monocrystalline SOI layer. The following section will show that superconducting Si samples can indeed be obtained using this industrial-compatible PLIE process, but not in the expected regime.

### 9.6.2 Low-temperature characterization after single laser pulse

The normalized resistance (derived from 4-point probe measurements) as a function of the temperature is plotted in Fig. 9.18A for samples irradiated with a single pulse at various ED. From these curves, three behaviors were identified: resistive, partially superconducting, and superconducting. For ED lower than 1.25 J/cm$^2$, the material showed no superconducting transition down to 100 mK (black curve). At 1.275 J/cm$^2$, a partial superconducting transition was observed, starting at 340 mK and reaching 50% of the initial resistance at 100 mK (pink curve). For an ED of 1.3 J/cm$^2$ and beyond (regime 5), complete superconducting transitions were detected. Thus, superconductivity was obtained not for the conditions leading to monocrystalline layers (regime 4) but beyond the full-melt threshold resulting in superconducting heavily boron-doped *poly*-SOI layers. For a single-laser pulse treatment, the average critical temperature was found around 170 mK with almost no dependence on the ED in the 1.3–1.5 J/cm$^2$ range (Fig. 9.18B).

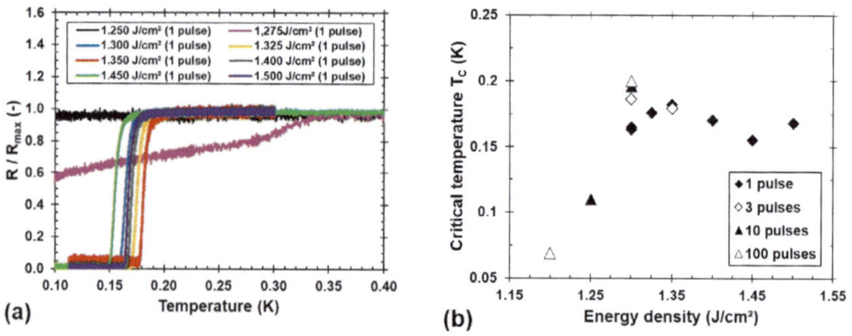


FIG. 9.18 (A) Normalized resistance of samples laser annealed with a single pulse at various laser ED. Three regimes are identifiable: resistive (ED ≤ 1.25 J/cm$^2$), partially superconducting (at 1.275 J/cm$^2$), and superconducting regime (ED ≥1.3 J/cm$^2$). (B) Variation of critical temperature $T_c$ for various laser ED and number of pulses. *Black diamond points* and *vertical bars* are temperature values corresponding to 50%, 10%, and 90% of resistance drop during superconducting transition, respectively. *(Adapted from R. Daubriac, P. Acosta Alba, C. Marcenat, S. Lequien, T.D. Vethaak, F. Nemouchi, F. Lefloch, S. Kerdilès, ECS J. Solid State Sci. Technol. 10 (2021) 014004.)*

### 9.6.3 Impact of multipulse laser annealing in the industrial tool

Cumulative effects are to be expected when irradiating the same location with multiple laser pulses at a given energy density. In the case of boron-implanted SOI layers, such multipulse laser treatment caused a slight reduction of the ED resulting in the full SOI layer melt. The melt duration derived from TRR measurements as a function of the number of laser pulses (up to 100) is plotted in Fig. 9.19A for energy densities in the 1.15–1.35 J/cm$^2$ range. Constant TRR signals signal below 120 ns and above 140 ns correspond to partial melt and full-melt (FM), respectively. Transition from the partial to the full-melt regime depending on the number of laser pulses is observable at 1.200 and 1.250 J/cm$^2$. For example, at 1.2 J/cm$^2$, between 40 and 60 pulses, the melting duration increases from 110-ns up to a 140-ns plateau, corresponding to a partial melt for 1 to 60 pulses and a complete melt above 60 pulses.

This slight shift of the full-melt threshold when increasing the number of laser pulses is also detectable in the electrical properties of the corresponding layers. Laser annealing at 1.25 J/cm$^2$ with 1 to 3 pulses resulted in resistive layers even at 100 mK, while the same treatment with 10 pulses or more enabled the formation of superconducting layers with a critical temperature around 170 mK (see Fig. 9.19B), equivalent to what can be obtained with 1 pulse at a slightly higher ED value. This specific result confirms that, in this experiment, superconducting Si was obtained only when the SOI layer was fully melted and

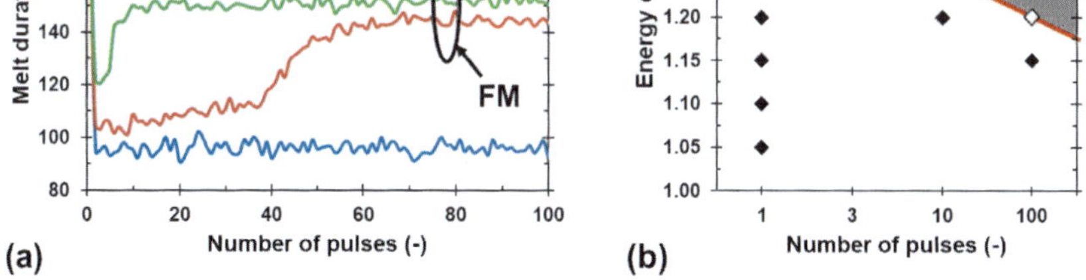


**FIG. 9.19** (A) Melting time as a function of the number of laser pulses for energy densities between 1.15 and 1.35 J/cm$^2$. (B) Plot showing superconducting *(open diamonds)* and not superconducting *(full diamonds)* behaviors (at 100 mK) as a function of UV-NLA energy density and number of laser pulses. The *shaded and white areas* indicate full-melt and partial melt conditions, respectively. The *red line* represents the full-melt threshold. *(From R. Daubriac, P. Acosta Alba, C. Marcenat, S. Lequien, T.D. Vethaak, F. Nemouchi, F. Lefloch, S. Kerdilès, ECS J. Solid State Sci. Technol. 10 (2021) 014004.)*

converted into a polycrystalline film. Particularly, at 1.300 J/cm$^2$, the average critical temperature between 1, 3, 10, and 100 pulses was 182 mK with a standard deviation of 17 mK, which validates that the number of pulses has no impact on the critical temperature.

Fig. 9.19B shows an overview of the normal (i.e., still resistive at 100 mK) and superconducting states measured at low temperature as a function of laser conditions (energy density and number of pulses). For any number of pulses, energy density conditions leading to a complete melt of the SOI layer match with the observation of superconducting behavior. Also, all laser conditions that give a partial melt correspond to a resistive behavior. For this particular fabrication process, exceeding the full SOI layer melt threshold with UV-NLA is mandatory to obtain superconductive heavily boron-doped poly-SOI layers (red line, Fig. 9.19B).

Single-pulse and multipulse results at laser conditions that give superconductivity, the sheet resistance, and haze values are unvarying (~90 Ω/sq. and ~5.0 ppm). For this reason and in the perspective of a large-scale integration of superconducting silicon, there is no particular interest to adopt a multipulse laser treatment since all measured parameters, including the critical temperature, are equivalent to those obtained after a single pulse with the appropriate ED setting. Though polycrystalline, the resulting superconducting layers obtained with this industrial laser annealing tool can be compared with the ones obtained using the laboratory system and starting with the same initial structure. In the $T_c = f(R_s)$ plot from Fig. 9.16D, the corresponding PLIE data (~90 Ω/sq.; 0.18 K) perfectly integrate in the trend derived from the PLIE samples fabricated with one laser shot. Despite the different pulse durations (160 ns vs 25 ns), the process transfer seems possible.

### 9.6.4 Critical field measurements

Low-temperature measurements coupled with magnetic fields were performed to study in depth the superconducting properties of the poly-SOI layers for various ED and number of pulses. The normalized electrical resistance evolution of a layer annealed at 1.3 J/cm$^2$ (10 pulses) is shown in Fig. 9.20A as a function of temperature for various magnetic fields. The critical temperature decreases when the magnetic field increases with a slight widening of the transition shape. The upper critical magnetic field $H_{c2}$ is plotted in Fig. 9.20B as a function of the normalized critical temperature. The very similar slope coefficients (between −0.47 and −0.35 T/K) confirm that the superconducting boron-doped poly-SOI layers have similar and reproducible superconducting properties. Calculations from a simple one-band free electron model and extractions from upper critical field curves ($H_{c2}(0) \sim 75 \pm 15$ mT) give: $\xi_0 \sim 4$ μm $\gg l_e \sim 2$ nm, consistent with the linear dependence between $H_{c2}$ and $T_c$ that follows the standard microscopic theory in the very dirty limit [60, 61]. This is in good agreement with previous investigations on monocrystalline superconducting silicon (Section 9.3 and [13, 20, 24]). These measurements confirmed that heavily

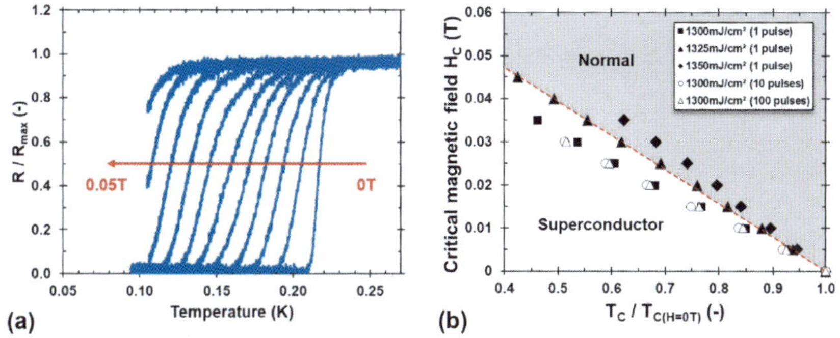


**FIG. 9.20** (A) Normalized resistance as a function of the temperature for magnetic fields varying from 0 to 0.05 T with 0.005 T increments (laser condition: 1.30 J/cm$^2$, 10 pulses). (B) Upper critical field as a function of the normalized critical temperature. *(From R. Daubriac, P. Acosta Alba, C. Marcenat, S. Lequien, T.D. Vethaak, F. Nemouchi, F. Lefloch, S. Kerdilès, ECS J. Solid State Sci. Technol. 10 (2021) 014004.)*

doped polysilicon is a type II superconductor due to strong impurity effect, as was the case of crystalline Si [58, 62, 63], with an average effective coherence length $\xi_{eff}$ around 80 nm.

### 9.6.5 Conclusion on the superconducting Si process transfer onto an industrial platform

Starting from PLIE process conditions defined in an academic environment, a first attempt to generate superconducting silicon layers by boron ion implantation followed by nanosecond laser annealing (using a commercially available system under $N_2$ flow with a 160-ns pulse duration) was successfully made. Simplified and cost-effective process conditions were identified, since superconducting Si was obtained with a single laser pulse at atmospheric pressure with a large and uniform laser beam (15 × 15 mm$^2$). Surprisingly, the functioning point was found to be the full-melt regime, leading to a highly boron-doped *polycrystalline* material. This is a major difference with the previous studies performed with shorter laser pulses and leading to monocrystalline superconducting Si layers with a uniform distribution of boron atoms despite a concentration well above the solubility limit. Boron distribution still has to be elucidated to confirm whether superconductivity originates from the highly boron-doped polysilicon grains or grain boundaries. Low-temperature measurements under magnetic field revealed a type II superconductor dominated by impurity diffusion, with an effective coherence length around 80 nm. Process conditions, mainly the implanted dose, still have to be adapted to further improve the critical temperature and the crystalline quality of the resulting superconducting Si layers. However, these first experiments demonstrated the possibility to transfer such process on an industrial platform compatible with state-of-the-art CMOS fabrication.

## 9.7 Conclusions

During the past decade, quantum devices have brought many technical challenges in the field of materials as well as in terms of processes. In 2006, as expected from theoretical studies on superconductivity in covalent semiconductors, superconducting silicon was discovered for high boron doping, bringing the promise of a scalable, integrable material for quantum information and ultra-sensitive detection. However, the extreme doping concentration required to trigger superconductivity is more than three times the solubility limit of boron in silicon, and impossible to reach using conventional microelectronic techniques. To achieve these levels of incorporation, nanosecond laser annealing was identified as the best candidate as it enables fast recrystallization of silicon, quenching its heavily boron-concentrated liquid phase. The boron is mostly introduced as an active atom in substitutional sites and does not create aggregates thanks to the rapidity of the laser-induced recrystallization. Gas immersion laser doping (GILD), relying on $BCl_3$ chemisorption and UV laser annealing (308 nm, 25 ns) in melt condition and under ultra-high vacuum, was the first method to provide defect-free and depth-controlled homogeneous superconducting heavily boron-doped silicon epilayers on bulk wafers. Above a threshold dose of $5 \times 10^{15}\ cm^{-2}$, Si:B displays BCS superconductivity carried by hole Cooper pairs. Si:B is a disorder-induced dirty type II superconductor, with a coherence length $\xi \sim 50$ nm. By tuning GILD parameters, superconducting silicon has been studied in depth, exploring particularly the dependence on the boron dose of the critical temperature, tunable up to 0.7 K for concentrations of ~10 at.%.

After exploring this new material properties, efforts have been made to integrate superconducting silicon in typical quantum devices. In 2015, the first superconducting quantum interference device (SQUID) entirely made with silicon has been fabricated with GILD and standard clean-room nanofabrication processes. The observed $\Phi_0$-periodic oscillations of the critical current confirmed the macroscopic nature of the superconducting quantum wavefunction, ruling out the percolation-based transport. An added benefit of silicon is its versatility: as a function of the doping level, the semiconductor (sc) turns metallic (N), and then superconducting (S). One can thus take advantage of semiconducting, metal and/or superconducting properties to design complex devices, such as all-Si Josephson junctions, formed by a metallic or semiconducting Si weak link between two superconducting Si contacts. If the superconductor/weak link interface is highly transparent, a full proximity effect can be induced in the weak link, resulting in a nondissipative supercurrent flowing between the contacts. GILD was used to fabricate, study, and optimize the S/sc interface, and it was demonstrated that high-quality ohmic contacts with $R_C \sim 10^{-9}\ \Omega\ cm^2$, an order of magnitude lower than all superconductor/silicon contacts, could be reached. All-silicon Josephson junctions were then realized, demonstrating the full proximity effect and long coherence lengths up to few hundreds of nm in silicon. The modulation of such supercurrent is the missing ingredient for the superconducting transistor, the Josephson FET.

Finally, microwave superconducting coplanar waveguide (CPW) resonators were investigated in order to better understand Si:B high-frequency properties and for their potential for applications in quantum information and sensitive astronomic detection. CPW resonators are indeed intensively used as read-outs, interconnections, or memories in quantum circuits, as well as multiplexed sensitive photon detectors for astronomic applications. In addition to its low cost, the process fabrication maturity acquired in the microelectronics industry makes superconducting silicon a good alternative to metallic superconductors, as quantum information goes beyond the few-qubits operations and starts worrying about scalability and costs. Silicon CPW resonators have been fabricated by combining GILD and standard microelectronic fabrication, with two different designs: in the transmission geometry, the resonator is coupled at each end to a transmission line via coupling capacitances; in the shunt geometry, multiple resonators are side-coupled to the transmission line. Resonators were demonstrated in the frequency range [1–12] GHz, and had internal quality factors $\sim$4000. These devices showed a high sheet kinetic inductance $\sim$500 pH, three orders of magnitude larger than the geometric inductance. This was expected in the frame of BCS theory, as inherently linked to the low carrier density, disordered Si:B material. In addition, an unexpectedly large nonlinearity of the kinetic inductance with current was observed, highlighting the extreme sensitivity of Si:B to any perturbation of the superconducting condensate, opening interesting possibilities for sensitive detection or as tunable large impedance elements for quantum information.

On the road toward the large-scale integration of superconducting silicon, a first step from GILD to clean-room compatible process was successfully made by coupling ion implantation and laser annealing in ultra-high vacuum. Superconducting boron-doped silicon epilayers with a homogeneous concentration were obtained, with typical critical temperatures around 0.25 K.

In order to fit the new standard architectures, this same process was then applied on thin silicon-on-insulator (SOI) layers ($\sim$30 nm). PLIE led to a partial melt of the SOI layer by the liquid-phase epitaxy regrowth, and as a consequence, to the formation of monocrystalline heavily boron-doped layers. These layers had superconductive properties comparable with the GILD layers on bulk Si, with the higher critical temperatures (0.67 K) corresponding to the laser treatment of nearly the whole SOI layer. All these previous results were obtained in an academic environment, with restrictive conditions: ultra-high-vacuum, small irradiated zone ($2 \times 2\ mm^2$) and a 100-mm wafer-compatible chamber.

To take a step forward, the same preimplanted wafers were annealed in an industrial laser tool at atmospheric pressure, with larger irradiated areas ($15 \times 15\ mm^2$), 300-mm compatible chamber, the same XeCl laser wavelength but a longer laser pulse (160 ns). Superconducting poly-Si layers were successfully fabricated with single- and multipulse laser conditions corresponding to the entire melt of the SOI layer. The resulting layers showed an average critical temperature of 0.17 K and a type II superconductor behavior due to impurity effects. Given the calculated incorporated boron concentration and thanks to structural characterizations, it appeared that a significant amount of boron might be in interstitial position, as clusters. Even if process conditions need to be optimized to reach monocrystalline quality (particularly through the implanted dose), and the necessary investigations of the origin of superconductivity in poly-Si, these preliminary results demonstrated that superconducting silicon is transferrable in industrial platforms.

## Acknowledgments

The authors would like to thank SCREEN Company and its French subsidiary LASSE for their assistance in the maintenance and operation of the LT-3100 laser annealing system, and D. Débarre, F. Lefloch, C. Marcenat, and P. Acosta Alba for their technical support for the low-temperature measurements and valuable discussions that greatly boosted the SILARUS Project. F. Chiodi acknowledges support from G. Hallais, S. Lequien, and X. Leroux for the structural characterisations of the Si:B layers, D. Débarre, P. Bonnet, L. Desvignes, and Y. Baron for fruitful discussions and insight on nanosecond laser doping and the electrical properties of the Si:B layers, F. Lefloch and C. Marcenat for their help in measuring and understanding the superconducting properties of Si:B. F. Chiodi is grateful for the support of the Réseau RENATECH and the French National Research Agency (ANR) under contract number ANR-16-CE24-0016-01, and as part of the "Investissements d'Avenir" program (Labex NanoSaclay, ANR-10-LABX-0035).

## References

[1] M. Cohen, Superconductivity in many-valley semiconductors and in semimetals, Phys. Rev. 134 (1964) A511.

[2] E. Ekimov, V. Sidorov, A. Rakhmanini, N. Mel'nik, M. Timofeev, R. Sadykov, Superconductivity in diamond, Nature 428 (2004) 542–545.

[3] L. Boeri, J. Kortus, O. Andersen, Three-dimensional $MgB_2$-type superconductivity in hole-doped diamond, Phys. Rev. Lett. 93 (23) (2004), 237002.

[4] V. Borisenko, S. Yudin, Steady-state solubility of substitutional impurities in silicon, Phys. Status Solidi A 101 (1987) 123.

[5] F. Trumbore, Solid solubilities of impurity elements in germanium and silicon, Bell Syst. Tech. J. 39 (1960) 205.

[6] E. Landi, A. Armigliato, S. Solmi, R. Kögler, E. Wieser, Electrical activation of boron implanted-silicon during rapid thermal annealing, Appl. Phys. A Solids Surf 47 (4) (1988) 359–366.

[7] S. Solmi, E. Landi, F. Baruffaldi, High-concentration boron diffusion in silicon: simulation of the precipitation phenomena, J. Appl. Phys. 68 (7) (1990) 3250–3258.

[8] H. Ryssel, K. Müller, K. Haberger, R. Henkelmann, F. Jahnel, High concentration effects of ion implanted boron in silicon, Appl. Phys. 22 (1) (1980) 35–38.

[9] P. Tsien, S. Tsou, M. Takai, D. Röschenthaler, M. Ramin, H. Ryssel, I. Ruge, K. Wittmaack, Annealing of boron-implanted silicon using a CW $CO_2$ laser, Phys. Status Solidi (a) 63 (2) (1981) 547–555.

[10] P. Lill, M. Dahlinger, J. Köhler, Boron partitioning coefficient above unity in laser crystallized silicon, Materials 10 (2) (2017) 189.

[11] J. Boulmer, B. Bourguignon, J.P. Budin, D. Débarre, Time of flight study of low pressure laser etching of silicon by chlorine, Appl. Surf. Sci. 43 (1–4) (1989) 424–431.

[12] E. Bustarret, C. Marcenat, P. Achatz, J. Kačmarčik, F. Lévy, A. Huxley, L. Ortéga, E. Bourgeois, X. Blase, D. Débarre, J. Boulmer, Superconductivity in doped cubic silicon, Nature 444 (2006) 465.

[13] A. Grockowiak, T. Klein, E. Bustarret, J. Kačmarčik, C. Dubois, G. Prudon, K. Hoummada, D. Mangelinck, T. Kociniewski, D. Débarre, J. Boulmer, C. Marcenat, Superconducting properties of laser annealed implanted Si:B layers, Supercond. Sci. Technol. 26 (2013), 045009.

[14] R. Daubriac, P. Acosta Alba, C. Marcenat, S. Lequien, T.D. Vethaak, F. Nemouchi, F. Lefloch, S. Kerdilès, Superconducting polycrystalline silicon layer obtained by boron implantation and nanosecond laser annealing, ECS J. Solid State Sci. Technol. 10 (2021), 014004.

[15] K. Hoummada, F. Dahlem, T. Kociniewski, J. Boulmer, C. Dubois, G. Prudon, E. Bustarret, H. Courtois, D. Débarre, D. Mangelinck, Absence of boron aggregates in superconducting silicon confirmed by atom probe tomography, Appl. Phys. Lett. 101 (2012) 182602, https://doi.org/10.1063/1.4760261.

[16] B. Dragnea, Modification de Surfaces Induites par Laser et Sondées par Reflectivité et Somme de Frequences (PhD thesis), Orsay, 1997.

[17] P. Bonnet, Mesures Résonantes des Propriétés Hautes Fréquences du Silicium Supraconducteur Ultra-Dopé au Bore par Laser (PhD thesis), Université Paris Saclay, 2019.

[18] L. Dagault, S. Kerdilès, A.P. Acosta, J.-M. Hartmann, J.-P. Barnes, P. Gergaud, E. Scheid, F. Cristiano, Investigation of recrystallization and stress relaxation in nanosecond laser annealed Si1-xGex/Si epilayers, Appl. Surf. Sci. 527 (2020) 146752.

[19] G. Kerrien, Etude du Traitement laser du Silicium Appliqué à la Réalisation de couches Dopées Nanométriques Pour la Microélectronique (PhD thesis), Orsay, 2004.

[20] C. Marcenat, J. Kačmarčik, R. Piquerel, P. Achatz, G. Prudon, C. Dubois, B. Gautier, J. Dupuy, E. Bustarret, L. Ortega, T. Klein, J. Boulmer, T. Kociniewski, D. Débarre, Low-temperature transition to a superconducting phase in boron-doped silicon films grown on (001)-oriented silicon wafers, Phys. Rev. B 81 (2010), https://doi.org/10.1103/PhysRevB.81.020501, 020501.

[21] A. Bhaduri, T. Kociniewski, F. Fossard, J. Boulmer, D. Débarre, Optical and electrical properties of laser doped Si:B in the alloy range, Appl. Surf. Sci. 258 (23) (2012) 9228–9232.

[22] G. Kerrien, J. Boulmer, D. Débarre, D. Bouchier, A. Grouillet, D. Lenoble, Ultrashallow, super-doped and box-like junctions realized by laser-induced doping, Appl. Surf. Sci. 186 (1–4) (2002) 45–51.

[23] F. Dahlem, T. Kociniewski, C. Marcenat, A. Grockowiak, L.M.A. Pascal, P. Achatz, J. Boulmer, D. Débarre, T. Klein, E. Bustarret, H. Courtois, Subkelvin tunneling spectroscopy showing Bardeen-Cooper-Schrieffer superconductivity in heavily boron-doped silicon epilayers, Phys. Rev. B 82 (14) (2010), 140505.

[24] A. Grockowiak, T. Klein, H. Cercellier, F. Lévy-Bertrand, X. Blasé, J. Kačmarčik, T. Kociniewski, F. Chiodi, D. Débarre, G. Prudon, C. Dubois, C. Marcenat, Thickness dependence of the superconducting critical temperature in heavily doped Si:B epilayers, Phys. Rev. B 88 (2013), https://doi.org/10.1103/PhysRevB.88.064508, 064508.

[25] E. Bourgeois, X. Blase, Superconductivity in doped cubic silicon: an ab initio study, Appl. Phys. Lett. 90 (14) (2007), 142511.

[26] J.T. Muhonen, et al., Storing quantum information for 30 seconds in a nanoelectronic device, Nat. Nanotechnol. 9 (2014) 986.

[27] T. Brecht, et al., Multilayer microwave integrated quantum circuits for scalable quantum computing, npj Quantum Inf. 2 (2016), 16002.
[28] J.E. Duvauchelle, A. Francheteau, C. Marcenat, F. Chiodi, D. Débarre, K. Hasselbach, J.R. Kirtley, F. Lefloch, Silicon superconducting quantum interference device, Appl. Phys. Lett. 107 (7) (2015), 072601.
[29] F. Balestro, Dynamique quantique d'un SQUID-DC (PhD thesis), Université Joseph Fourier, Grenoble, 2013.
[30] A. Shearrow, G. Koolstra, S.J. Whiteley, N. Earnest, P.S. Barry, F.J. Heremans, D.D. Awschalom, E. Shirokoff, D.I. Schuster, Atomic layer deposition of titanium nitride for quantum circuits, Appl. Phys. Lett. 113 (2018), 212601.
[31] J.R. Gao, M. Hajenius, F.D. Tichelaar, T.M. Klapwijk, B. Voronov, E. Grishin, G. Gol'tsman, C.A. Zorman, M. Mehregany, Monocrystalline NbN nanofilms on a 3C-SiCSi substrate, Appl. Phys. Lett. 91 (2007), 062504.
[32] J. Basset, D. Watfa, G. Aiello, M. Féchant, A. Morvan, J. Estève, J. Gabelli, M. Aprili, R. Weil, A. Kasumov, H. Bouchiat, R. Deblock, High kinetic inductance microwave resonators made by He-Beam assisted deposition of tungsten nanowires, Appl. Phys. Lett. 114 (2019), 102601.
[33] M. Calvo, A. D'Addabbo, A. Monfardini, A. Benoit, N. Boudou, O. Bourrion, A. Catalano, L. Dumoulin, J. Goupy, H. le Sueur, S. Marnieros, Niobium silicon alloys for kinetic inductance detectors, J. Low Temp. Phys. 176 (2014) 518.
[34] T. Akazaki, A Josephson field effect transistor using an InAs-inserted-channel In0.52Al0.48As/In0.53Ga0.47As inverted modulation-doped structure, Appl. Phys. Lett. 68 (1996) 418.
[35] F. Chiodi, A. Grockowiak, J.E. Duvauchelle, F. Fossard, F. Lefloch, T. Klein, C. Marcenat, D. Débarre, Gas immersion laser doping for superconducting nanodevices, Appl. Surf. Sci. 302 (2014) 209–212.
[36] F. Chiodi, J.E. Duvauchelle, C. Marcenat, D. Débarre, F. Lefloch, Proximity induced superconductivity in all-silicon superconductor/normal-metal junctions, Phys. Rev. B 96 (2017), 024503.
[37] Y.V. Fominov, M.V. Feigel'man, Superconductive properties of thin dirty superconductor–normal-metal bilayers, Phys. Rev. B 63 (2001), 094518.
[38] B. Buonomo, R. Leoni, M.G. Castellano, F. Mattioli, G. Torrioli, Electron thermometry and refrigeration with doped silicon and superconducting electrodes, J. Appl. Phys. 94 (2003) 7784.
[39] C.Y. Chang, Y.K. Fang, S.M. Sze, Specific contact resistance of metal-semiconductor barriers, Solid State Electron. 14 (1971) 541.
[40] J. Seto, T. Van Duzer, Theory and measurements on lead-tellurium-lead supercurrent junctions, Low Temp. Phys. LT-13 3 (1974) 328.
[41] P. Dubos, H. Courtois, B. Pannetier, F.K. Wilhelm, A.D. Zaikin, G. Schon, Josephson critical current in a long mesoscopic S-N-S junction, Phys. Rev. B 63 (2001), 064502. Y.-P. Dubos, (Ph.D. thesis), Centre de Recherches sur les Très Basses Températures, Grenoble, 2000.
[42] A.A. Golubov, M.Y. Kupriyanov, V.F. Lukichev, Influence of proximity effect and electrodes on stationary properties of SN-N-NS variable-thickness microbridges, Mikroelektronika 12 (1983) 342.
[43] M.Y. Kuprianov, V.F. Lukichev, Influence of boundary transparency on the critical current of "dirty" SS'S structures, Sov. Phys. JETP 67 (1988) 1163.
[44] A. Monfardini, et al., A dual-band millimeter-wave Kinetic Inductance Camera for the IRAM 30 m telescope, ApJs 194 (2011) 24.
[45] A. Blais, et al., Cavity quantum electrodynamics for superconducting electrical circuits: an architecture for quantum computation, Phys. Rev. A 69 (2004), 062320.

[46] M.H. Devoret, R.J. Schoelkopf, Superconducting circuits for quantum information: an outlook, Science 339 (2013) 1169.
[47] C. Fenouillet-Béranger, B. Mathieu, B. Prévitali, M.P. Samson, N. Rambal, V. Benevent, S. Kerdiles, J.P. Barnes, J.M. Hartmann, et al., New insights on bottom layer stability and laser annealing promises for high performances 3D VLSI, 7047120, Technical Digest IEDM Conference, 2015, pp. 27.5.1–27.5.4.
[48] P. Bonnet, F. Chiodi, D. Flanigan, R. Delagrange, N. Brochu, D. Débarre, H. le Sueur, Strongly non-linear superconducting silicon resonators, 2021. arXiv:2101.11125 [cond-mat.supr-con].
[49] B. Ho Eom, P.K. Day, H.G. LeDuc, J. Zmuidzinas, A wideband, low-noise superconducting amplifier with high dynamic range, Nat. Phys. 8 (2012) 623.
[50] B. Yurke, E. Buks, Performance of cavity-parametric amplifiers, employing Kerr nonlinearites, in the presence of two-photon loss, J. Lightwave Technol. 24 (2006) 5054.
[51] A.J. Annunziata, D.F. Santavicca, L. Frunzio, G. Catelani, M.J. Rooks, A. Frydman, D.E. Prober, Tunable superconducting nanoinductors, Nanotechnology 21 (44) (2010), 445202.
[52] G.K. Celler, S. Cristoloveanu, Frontiers of silicon-on-insulator, J. Appl. Phys. 93 (9) (2003) 4955–4978.
[53] O. Faynot, M. Vinet, C. Fenouillet, O. Weber, P. Perreau, L. Grenouillet, F. Andrieu, T. Poiroux, S. Deleonibus, FDSOI technology: a power efficient solution down to 10 nm, ECS Trans. 58 (9) (2013) 3–8.
[54] R. Ritzenthaler, Architectures Avancées des Transistors FinFETs: Réalisation, Caractérisation et Modélisation (PhD thesis), 2006.
[55] A. Francheteau, Superconducting Silicon on Insulator and Silicide-Based Superconducting MOSFET for Quantum Technologies, Université Grenoble Alpes, 2017.
[56] M. Posselt, J.P. Biersack, Computer simulation of ion implantation into crystalline targets, Nucl. Inst. Methods Phys. Res. B 64 (1992) 706–710.
[57] S.F. Lombardo, S. Boninelli, F. Cristiano, I. Deretzis, M.G. Grimaldi, K. Huet, E. Napolitani, A. La Magna, Phase field model of the nanoscale evolution during the explosive crystallization phenomenon, J. Appl. Phys. 123 (10) (2018) 105105.
[58] A. Grockowiak, Supraconductivité et Propriétés Physiques du Silicium très Fortement Dopé (PhD thesis), Université Grenoble-Alpes, 2012.
[59] L. Nikolova, M. Stern, J. MacLeod, B. Reed, H. Ibrahim, G. Campbell, F. Rosei, T. LaGrange, B. Siwick, In situ investigation of explosive crystallization in a-Ge: experimental determination of the interface response function using dynamic transmission electron microscopy, J. Appl. Phys. 116 (2014), https://doi.org/10.1063/1.4894397, 093512.
[60] P.-G. De Gennes, Behavior of dirty superconductors in high magnetic fields, J. Phys. Condens. Matter 3 (1964) 79–90, https://doi.org/10.1007/BF02422354.
[61] K. Maki, The magnetic properties of superconducting alloys I, Physics 1 (1964) 21–30.
[62] M. Kriener, Y. Maeno, T. Oguchi, Z.-A. Ren, J. Kato, T. Muranaka, J. Akimitsu, Specific heat and electronic states of superconducting boron-doped silicon carbide, Phys. Rev. B 78 (2008), https://doi.org/10.1103/PhysRevB.78.024517, 024517.
[63] M. Tinkham, Introduction to inhomogeneous superconductors, in: AIP Conference Proceedings, 1979, p. 58, https://doi.org/10.1063/1.32141.
[64] S.H. Jain, P.B. Griffin, J.D. Plummer, S. Mccoy, J. Gelpey, T. Selinger, D.F. Downey, Metastable boron active concentrations in Si using flash assisted solid phase epitaxy, J. Appl. Phys. 96 (2004) 7357.
[65] C. Fuller, J. Ditzenberger, Diffusion of boron and phosphorus into silicon, J. Appl. Phys. 25 (11) (1954) 1439–1440.

[66] F. Schwettmann, Characterization of incomplete activation of high dose boron implants in silicon, J. Appl. Phys. 45 (4) (1974) 1918–1920.
[67] M. Tsai, B. Streetman, Recrystallization of implanted amorphous silicon layers. I Electrical properties of silicon implanted with $BF_2^+$ or $Si^+ + B^+$, J. Appl. Phys. 50 (1) (1979) 183–187.
[68] G. Vick, K. Whittle, Solid solubility and diffusion coefficient of boron in silicon, J. Electrochem. Soc. 116 (8) (1969) 1142–1144.
[69] G. Masetti, M. Severi, S. Solmi, Modeling of carrier mobility against carrier concentration in arsenic-, phosphorus-, and boron-doped silicon, IEEE Trans. Electr. Dev. 30 (7) (1983) 764.
[70] R.T. Young, J. Narayan, Laser annealing of diffusion-induced imperfections in silicon, Appl. Phys. Lett. 33 (1978) 14.

# Index

Note: Page numbers followed by *f* indicate figures and *t* indicate tables.

## P



## R

Printed in Dunstable, United Kingdom

88265654R00239